T0178581

Quantum Theory, Groups and Representations

Peter Woit

Quantum Theory, Groups and Representations

An Introduction

 Springer

Peter Woit
Department of Mathematics
Columbia University
New York, NY
USA

ISBN 978-3-319-87835-5 ISBN 978-3-319-64612-1 (eBook)
DOI 10.1007/978-3-319-64612-1

Mathematics Subject Classification (2010): 81Rxx, 82SXX, 8101

Printed on acid-free paper

This Springer imprint is published by Springer Nature
The registered company is Springer International Publishing AG
The registered company address is: Gewerbestrasse 11, 6330 Cham, Switzerland

Preface

This book began as course notes prepared for a class taught at Columbia University during the 2012–13 academic year. The intent was to cover the basics of quantum mechanics, up to and including relativistic quantum field theory of free fields, from a point of view emphasizing the role of unitary representations of Lie groups in the foundations of the subject. It has been significantly rewritten and extended since that time, partially based upon experience teaching the same material during 2014–15.

The approach to this material is simultaneously rather advanced, using crucially some fundamental mathematical structures discussed, if at all, only in graduate mathematics courses, while at the same time trying to do this in as elementary terms as possible. The Lie groups needed are (with one crucial exception) ones that can be described simply in terms of matrices. Much of the representation theory will also just use standard manipulations of matrices. The only prerequisite for the course as taught was linear algebra and multivariable calculus (while a full appreciation of the topics covered would benefit from quite a bit more than this). My hope is that this level of presentation will simultaneously be useful to mathematics students trying to learn something about both quantum mechanics and Lie groups and their representations, as well as to physics students who already have seen some quantum mechanics, but would like to know more about the mathematics underlying the subject, especially that relevant to exploiting symmetry principles.

The topics covered emphasize the mathematical structure of the subject and often intentionally avoid overlap with the material of standard physics courses in quantum mechanics and quantum field theory, for which many excellent textbooks are available. This document is best read in conjunction with such a text. In particular, some experience with the details of the physics not covered here is needed to truly appreciate the subject. Some of the main differences with standard physics presentations include the following:

- The role of Lie groups, Lie algebras, and their unitary representations is systematically emphasized, including not just the standard use of these to derive consequences for the theory of a "symmetry" generated by operators commuting with the Hamiltonian.

- Symplectic geometry and the role of the Lie algebra of functions on phase space in the classical theory of Hamiltonian mechanics are emphasized. "Quantization" is then the passage to a unitary representation (unique by the Stone–von Neumann theorem) of a subalgebra of this Lie algebra.

- The role of the metaplectic representation and the subtleties of the projective factor involved are described in detail. This includes phenomena depending on the choice of a complex structure, a topic known to physicists as "Bogoliubov transformations."

- The closely parallel story of the Clifford algebra and spinor representation is extensively investigated. These are related to the Heisenberg Lie algebra and the metaplectic representation by interchanging commutative ("bosonic") and anticommutative ("fermionic") generators, introducing the notion of a "Lie superalgebra" generalizing that of a Lie algebra.

- Many topics usually first encountered in physics texts in the context of relativistic quantum field theory are instead first developed in simpler non-relativistic or finite dimensional contexts. Non-relativistic quantum field theory based on the Schrödinger equation is described in detail before moving on to the relativistic case. The topic of irreducible representations of space–time symmetry groups is first addressed with the case of the Euclidean group, where the implications for the non-relativistic theory are explained. The analogous problem for the relativistic case, that of the irreducible representations of the Poincaré group, is then worked out later on.

- The emphasis is on the Hamiltonian formalism and its representation-theoretical implications, with the Lagrangian formalism (the basis of most quantum field theory textbooks) de-emphasized. In particular, the operators generating symmetry transformations are derived using the moment map for the action of such transformations on phase space, not by invoking Noether's theorem for transformations that leave invariant a Lagrangian.

- Care is taken to keep track of the distinction between vector spaces and their duals. It is the dual of phase space (linear coordinates on phase space) that appears in the Heisenberg Lie algebra, with quantization a representation of this Lie algebra by linear operators.

- The distinction between real and complex vector spaces, along with the role of complexification and choice of a complex structure, is systematically emphasized. A choice of complex structure plays a crucial part in quantization using annihilation and creation operator methods, especially in relativistic quantum field theory, where a different sort of choice than in the non-relativistic case is responsible for the existence of antiparticles.

Some differences with other mathematics treatments of this material are as follows:

- A fully rigorous treatment of the subject is not attempted. At the same time, an effort is made to indicate where significant issues arise should one pursue such a treatment, and to provide references to rigorous discussions of these issues. An attempt is also made to make clear the difference between where a rigorous treatment could be pursued relatively straightforwardly and where there are serious problems of principle making a rigorous treatment hard to achieve.

- The discussion of Lie groups and their representations is focused on specific examples, not the general theory. For compact Lie groups, emphasis is on the groups $U(1), SO(3), SU(2)$ and their finite dimensional representations. Central to the basic structure of quantum mechanics are the Heisenberg group, the symplectic groups $Sp(2n, \mathbf{R})$ and the metaplectic representation, as well as the spinor groups and the spin representation. The geometry of space–time leads to the study of Euclidean groups in two and three dimensions, and the Lorentz (SO (3,1)) and Poincaré groups, together with their representations. These examples of non-compact Lie groups are a fundamental feature of quantum mechanics, but not a conventional topic in the mathematics curriculum.

- A central example studied thoroughly and in some generality is that of the metaplectic representation of the double cover of $Sp(2n, \mathbf{R})$ (in the commutative case), or spin representation of the double cover of $SO(2n, \mathbf{R})$ (anticommutative case). This specific example of a representation provides the foundation of quantum theory, with quantum field theory involving a generalization to the case of n infinite.

- No attempt is made to pursue a general notion of quantization, despite the great mathematical interest of such generalizations. In particular, attention is restricted to the case of linear symplectic manifolds. The linear structure plays a crucial role, with quantization given by a representation of a Heisenberg algebra in the commutative case and a Clifford algebra in the anticommutative case. The very explicit methods used (staying close to the physics formalism) mostly do not apply to more general conceptions of quantization (e.g., geometric quantization) of mathematical interest for their applications in representation theory.

The scope of material covered in later sections of the book is governed by a desire to give some explanation of what the central mathematical objects are that occur in the Standard Model of particle physics, while staying within the bounds of a one-year course. The Standard Model embodies our best current understanding of the fundamental nature of reality, making a better understanding of its mathematical nature a central problem for anyone who believes that mathematics and physics are intimately connected at their deepest levels. The author hopes that the treatment of this subject here will be helpful to anyone interested in pursuing a better understanding of this connection.

0.1 Acknowledgements

The students of Mathematics W4391-2 at Columbia during 2012–13 and 2014–15 deserve much of the credit for the existence of this book and for whatever virtues it might have. Their patience with and the interest they took in what I was trying to do were a great encouragement, and the many questions they asked were often very helpful. The reader should be aware that the book they have in their hands, whatever its faults, is a huge improvement over what these students had to put up with.

The quality of the manuscript was dramatically improved over that of early versions through the extreme diligence of Michel Talagrand, who early on took an interest in what I was doing, and over a long period of time carefully read over many versions. His combination of encouragement and extensive detailed criticism was invaluable. He will at some point be publishing his own take on many of the same topics covered here ([91]), which I can't recommend enough.

At some point I started keeping a list of those who provided specific suggestions, it includes Kimberly Clinch, Art Brown, Jason Ezra Williams, Mateusz Wasilewski, Gordon Watson, Cecilia Jarlskog, Alex Purins, James Van Meter, Thomas Tallant, Todor Popov, Stephane T'Jampens, Cris Moore, Noah Miller, Ben Israeli, Nigel Green, Charles Waldman, Peter Grieve, Kevin McCann, Chris Weed, Fernando Chamizo, and various anonymous commenters on my blog. My apologies to others who I'm sure that I've forgotten.

The illustrations were done in TikZ by Ben Dribus, who was a great pleasure to work with.

Much early enthusiasm and encouragement for this project was provided by Eugene Ha at Springer. Marc Strauss and Loretta Bartolini have been the ones there who have helped to finally bring this to a conclusion.

Thanks also to all my colleagues in the mathematics department at Columbia, who have over the years provided a very supportive environment for me to work in and learn more every day about mathematics.

Finally, I'm grateful for the daily encouragement and unfailing support over the years from my partner Pamela Cruz that has been invaluable for surviving getting to the end of this project and will make possible whatever the next one might be.

Contents

Chapter 1
Introduction and Overview

1.1 Introduction

A famous quote from Richard Feynman goes "I think it is safe to say that
no one understands quantum mechanics." [22]. In this book, we will pursue
one possible route to such an understanding, emphasizing the deep connections of quantum mechanics to fundamental ideas of modern mathematics.
The strangeness inherent in quantum theory that Feynman was referring
to has two rather different sources. One of them is the striking disjunction
and incommensurability between the conceptual framework of the classical
physics which governs our everyday experience of the physical world, and the
very different framework which governs physical reality at the atomic scale.
Familiarity with the powerful formalisms of classical mechanics and electromagnetism provides deep understanding of the world at the distance scales
familiar to us. Supplementing these with the more modern (but still "classical" in the sense of "not quantum") subjects of special and general relativity
extends our understanding into other much less familiar regimes, while still
leaving atomic physics a mystery.

   Read in context though, Feynman was pointing to a second source of difficulty, contrasting the mathematical formalism of quantum mechanics with
that of the theory of general relativity, a supposedly equally hard to understand subject. General relativity can be a difficult subject to master, but its
mathematical and conceptual structure involves a fairly straightforward extension of structures that characterize nineteenth-century physics. The fundamental physical laws (Einstein's equations for general relativity) are expressed as partial differential equations, a familiar if difficult mathematical
subject. The state of a system is determined by a set of fields satisfying these
equations, and observable quantities are functionals of these fields. The mathematics is largely that of the usual calculus: differential equations and their
real-valued solutions.

© Peter Woit 2017
P. Woit, *Quantum Theory, Groups and Representations*,
DOI 10.1007/978-3-319-64612-1_1

In quantum mechanics, the state of a system is best thought of as a different sort of mathematical object: a vector in a complex vector space with a Hermitian inner product, the so-called "state space". Such a state space will sometimes be a space of functions known as wavefunctions. While these may, like classical fields, satisfy a differential equation, one non-classical feature is that wavefunctions are complex-valued. What's completely different about quantum mechanics is the treatment of observable quantities, which correspond to self-adjoint linear operators on the state space. When such operators do not commute, our intuitions about how physics should work are violated, as we can no longer simultaneously assign numerical values to the observables.

During the earliest days of quantum mechanics, the mathematician Hermann Weyl quickly recognized that the mathematical structures being used were ones he was quite familiar with from his work in the field of representation theory. From the point of view that takes representation theory as a central theme in mathematics, the framework of quantum mechanics looks perfectly natural. Weyl soon wrote a book expounding such ideas [100], but this got a mixed reaction from physicists unhappy with the penetration of unfamiliar mathematical structures into their subject (with some of them characterizing the situation as the "Gruppenpest," the group theory plague). One goal of this book will be to try and make some of this mathematics as accessible as possible, boiling down part of Weyl's exposition to its essentials while updating it in light of many decades of progress toward better understanding of the subject.

Weyl's insight that quantization of a classical system crucially involves understanding the Lie groups that act on the classical phase space and the unitary representations of these groups has been vindicated by later developments which dramatically expanded the scope of these ideas. The use of representation theory to exploit the symmetries of a problem has become a powerful tool that has found uses in many areas of science, not just quantum mechanics. I hope that readers whose main interest is physics will learn to appreciate some of such mathematical structures that lie behind the calculations of standard textbooks, helping them understand how to effectively exploit them in other contexts. Those whose main interest is mathematics will hopefully gain some understanding of fundamental physics, at the same time as seeing some crucial examples of groups and representations. These should provide a good grounding for appreciating more abstract presentations of the subject that are part of the standard mathematical curriculum. Anyone curious about the relation of fundamental physics to mathematics and what Eugene Wigner described as "The Unreasonable Effectiveness of Mathematics in the Natural Sciences" [101] should benefit from an exposure to this remarkable story at the intersection of the two subjects.

The following sections give an overview of the fundamental ideas behind much of the material to follow. In this sketchy and abstract form, they will likely seem rather mystifying to those meeting them for the first time. As we work through basic examples in the coming chapters, a better understanding of the overall picture described here should start to emerge.

1.2 Basic principles of quantum mechanics

We will divide the conventional list of basic principles of quantum mechanics into two parts, with the first covering the fundamental mathematics structures.

1.2.1 Fundamental axioms of quantum mechanics

In classical physics, the state of a system is given by a point in a "phase space," which can be thought of equivalently as the space of solutions of an equation of motion, or as (parametrizing solutions by initial value data) the space of coordinates and momenta. Observable quantities are just functions on this space (e.g., functions of the coordinates and momenta). There is one distinguished observable, the energy or Hamiltonian, and it determines how states evolve in time through Hamilton's equations.

The basic structure of quantum mechanics is quite different, with the formalism built on the following simple axioms:

Axiom (States). *The state of a quantum mechanical system is given by a nonzero vector in a complex vector space $\mathcal{H}$ with Hermitian inner product $\langle \cdot, \cdot \rangle$.*

We will review in chapter 4 some linear algebra, including the properties of inner products on complex vector spaces. $\mathcal{H}$ may be finite or infinite dimensional, with further restrictions required in the infinite dimensional case (e.g., we may want to require $\mathcal{H}$ to be a Hilbert space). Note two very important differences with classical mechanical states:

- The state space is always linear: A linear combination of states is also a state.

- The state space is a *complex* vector space: these linear combinations can and do crucially involve complex numbers, in an inescapable way. In the classical case, only real numbers appear, with complex numbers used only as an inessential calculational tool.

We will sometimes use the notation introduced by Dirac for vectors in the state space $\mathcal{H}$: Such a vector with a label ψ is denoted

$$|\psi\rangle$$

Axiom (Observables). *The observables of a quantum mechanical system are given by self-adjoint linear operators on* $\mathcal{H}$.

We will review the definition of self-adjointness for $\mathcal{H}$ finite dimensional in chapter 4. For $\mathcal{H}$ infinite dimensional, the definition becomes much more subtle, and we will not enter into the analysis needed.

Axiom (Dynamics). *There is a distinguished observable, the Hamiltonian* H. *Time evolution of states* $|\psi(t)\rangle \in \mathcal{H}$ *is given by the Schrödinger equation*

$$i\hbar\frac{d}{dt}|\psi(t)\rangle = H|\psi(t)\rangle \qquad (1.1)$$

The operator H has eigenvalues that are bounded below.

The Hamiltonian observable H will have a physical interpretation in terms of energy, with the boundedness condition necessary in order to assure the existence of a stable lowest energy state.

$\hbar$ is a dimensional constant, called Planck's constant, the value of which depends on what units one uses for time and for energy. It has the dimensions $[energy] \cdot [time]$, and its experimental values are

$$1.054571726(47) \times 10^{-34}\,\text{Joule} \cdot \text{seconds} = 6.58211928(15) \times 10^{-16}\text{eV} \cdot \text{seconds}$$

(eV is the unit of "electron-Volt," the energy acquired by an electron moving through a one-Volt electric potential). The most natural units to use for quantum mechanical problems would be energy and time units chosen so that $\hbar = 1$. For instance, one could use seconds for time and measure energies in the very small units of 6.6×10^{-16} eV, or use eV for energies, and then the very small units of 6.6×10^{-16} seconds for time. Schrödinger's equation implies that if one is looking at a system where the typical energy scale is an eV, one's state vector will be changing on the very short time scale of 6.6×10^{-16} seconds. When we do computations, usually we will set $\hbar = 1$, implicitly going to a unit system natural for quantum mechanics. After calculating a final result, appropriate factors of $\hbar$ can be inserted to get answers in more conventional unit systems.

It is sometimes convenient, however, to carry along factors of $\hbar$, since this can help make clear which terms correspond to classical physics behavior, and which ones are purely quantum mechanical in nature. Typically, classical physics comes about in the limit where

$$\frac{(energy\ scale)(time\ scale)}{\hbar}$$

is large. This is true for the energy and time scales encountered in everyday life, but it can also always be achieved by taking $\hbar \to 0$, and this is what will often be referred to as the "classical limit." One should keep in mind though that the manner in which classical behavior emerges out of quantum theory in such a limit can be a very complicated phenomenon.

1.2.2   Principles of measurement theory

The above axioms characterize the mathematical structure of a quantum theory, but they do not address the "measurement problem." This is the question of how to apply this structure to a physical system interacting with some sort of macroscopic, human-scale experimental apparatus that "measures" what is going on. This is a highly thorny issue, requiring in principle the study of two interacting quantum systems (the one being measured and the measurement apparatus) in an overall state that is not just the product of the two states, but is highly "entangled" (for the meaning of this term, see chapter 9). Since a macroscopic apparatus will involve something like 10^{23} degrees of freedom, this question is extremely hard to analyze purely within the quantum mechanical framework (requiring for instance the solution of a Schrödinger equation in 10^{23} variables).

Instead of trying to resolve in general this problem of how macroscopic classical physics behavior emerges in a measurement process, one can adopt the following two principles as providing a phenomenological description of what will happen, and these allow one to make precise statistical predictions using quantum theory:

Principle (Observables). *States for which the value of an observable can be characterized by a well-defined number are the states that are eigenvectors for the corresponding self-adjoint operator. The value of the observable in such a state will be a real number, the eigenvalue of the operator.*

This principle identifies the states we have some hope of sensibly associating a label to (the eigenvalue), a label which in some contexts corresponds to an observable quantity characterizing states in classical mechanics. The observables with important physical significance (for instance, the energy, momentum, angular momentum, or charge) will turn out to correspond to some group action on the physical system.

Principle (The Born rule). *Given an observable O and two unit-norm states $|\psi_1\rangle$ and $|\psi_2\rangle$ that are eigenvectors of O with distinct eigenvalues λ_1 and λ_2*

$$O|\psi_1\rangle = \lambda_1|\psi_1\rangle, \quad O|\psi_2\rangle = \lambda_2|\psi_2\rangle$$

the complex linear combination state

$$c_1|\psi_1\rangle + c_2|\psi_2\rangle$$

will not have a well-defined value for the observable O. If one attempts to measure this observable, one will get either λ_1 or λ_2, with probabilities

$$\frac{|c_1^2|}{|c_1^2| + |c_2^2|}$$

and

$$\frac{|c_2^2|}{|c_1^2| + |c_2^2|}$$

respectively.

The Born rule is sometimes raised to the level of an axiom of the theory, but it is plausible to expect that, given a full understanding of how measurements work, it can be derived from the more fundamental axioms of the previous section. Such an understanding though of how classical behavior emerges in experiments is a very challenging topic, with the notion of "decoherence" playing an important role. See the end of this chapter for some references that discuss these issues in detail.

Note that the state $c|\psi\rangle$ will have the same eigenvalues and probabilities as the state $|\psi\rangle$, for any complex number c. It is conventional to work with states of norm fixed to the value 1, which fixes the amplitude of c, leaving a remaining ambiguity which is a phase $e^{i\theta}$. By the above principles, this phase will not contribute to the calculated probabilities of measurements. We will, however, not take the point of view that this phase information can just be ignored. It plays an important role in the mathematical structure, and the relative phase of two different states certainly does affect measurement probabilities.

1.3 Unitary group representations

The mathematical framework of quantum mechanics is closely related to what mathematicians describe as the theory of "unitary group representations." We will be examining this notion in great detail and working through many examples in the coming chapters, but here is a quick summary of the general theory.

1.3.1 Lie groups

A fundamental notion that appears throughout different fields of mathematics is that of a group:

Definition (Group). *A group G is a set with an associative multiplication, such that the set contains an identity element, as well as the multiplicative inverse of each element.*

If the set has a finite number of elements, this is called a "finite group." The theory of these and their use in quantum mechanics is a well-developed subject, but one we mostly will bypass in favor of the study of "Lie groups," which have an infinite number of elements. The elements of a Lie group make up a geometrical space of some dimension, and choosing local coordinates on the space, the group operations are given by differentiable functions. Most of the Lie groups we will consider are "matrix groups," meaning subgroups of the group of n by n invertible matrices (with real or complex matrix entries). The group multiplication in this case is matrix multiplication. An example we will consider in great detail is the group of all rotations about a point in three-dimensional space, in which case such rotations can be identified with 3 by 3 matrices, with composition of rotations corresponding to the multiplication of matrices.

Digression. *A standard definition of a Lie group is as a smooth manifold, with group laws given by smooth (infinitely differentiable) maps. More generally, one might consider topological manifolds and continuous maps, but this gives nothing new (by the solution to Hilbert's Fifth problem). Most of the finite dimensional Lie groups of interest are matrix Lie groups, which can be defined as closed subgroups of the group of invertible matrices of some fixed dimension. One particular group of importance in quantum mechanics (the metaplectic group, see chapter 20) is not a matrix group, so the more general definition is needed to include this case.*

1.3.2 Group representations

Groups often occur as "transformation groups," meaning groups of elements acting as transformations of some particular geometric object. In the example mentioned above of the group of three-dimensional rotations, such rotations are linear transformations of $\mathbf{R}^3$. In general:

Definition (Group action on a set). *An action of a group G on a set M is given by a map*

$$(g, x) \in G \times M \to g \cdot x \in M$$

that takes a pair (g, x) of a group element $g \in G$ and an element $x \in M$ to another element $g \cdot x \in M$ such that

$$g_1 \cdot (g_2 \cdot x) = (g_1 g_2) \cdot x \qquad (1.2)$$

and

$$e \cdot x = x$$

where e is the identity element of G

A good example to keep in mind is that of three-dimensional space $M = \mathbf{R}^3$ with the standard inner product. This comes with two different group actions preserving the inner product

- An action of the group $G_1 = \mathbf{R}^3$ on $\mathbf{R}^3$ by translations.

- An action of the group $G_2 = O(3)$ of three-dimensional orthogonal transformations of $\mathbf{R}^3$. These are the rotations about the origin (possibly combined with a reflection). Note that in this case order matters: For non-commutative groups like $O(3)$, one has $g_1 g_2 \neq g_2 g_1$ for some group elements g_1, g_2.

A fundamental principle of modern mathematics is that the way to understand a space M, given as some set of points, is to look at $F(M)$, the set of functions on this space. This "linearizes" the problem, since the function space is a vector space, no matter what the geometrical structure of the original set is. If the set has a finite number of elements, the function space will be a finite dimensional vector space. In general, though, it will be infinite dimensional and one will need to further specify the space of functions (e.g., continuous functions, differentiable functions, functions with finite norm) under consideration.

Given a group action of G on M, functions on M come with an action of G by linear transformations, given by

$$(g \cdot f)(x) = f(g^{-1} \cdot x) \qquad (1.3)$$

where f is some function on M.

The order in which elements of the group act may matter, so the inverse is needed to get the group action property 1.2, since

$$
\begin{aligned}
g_1 \cdot (g_2 \cdot f)(x) &= g_2 \cdot f(g_1^{-1} \cdot x) \\
&= f(g_2^{-1} \cdot (g_1^{-1} \cdot x)) \\
&= f((g_2^{-1} g_1^{-1}) \cdot x) \\
&= f((g_1 g_2)^{-1} \cdot x) \\
&= (g_1 g_2) \cdot f(x)
\end{aligned}
$$

This calculation would not work out properly for non-commutative G if one defined $(g \cdot f)(x) = f(g \cdot x)$.

One can abstract from this situation and define as follows a representation as an action of a group by linear transformations on a vector space:

Definition (Representation). *A representation (π, V) of a group G is a homomorphism*

$$\pi : g \in G \to \pi(g) \in GL(V)$$

where $GL(V)$ is the group of invertible linear maps $V \to V$, with V a vector space.

Saying the map π is a homomorphism means

$$\pi(g_1)\pi(g_2) = \pi(g_1 g_2)$$

for all $g_1, g_2 \in G$, i.e., it satisfies the property needed to get a group action. We will mostly be interested in the case of complex representations, where V is a complex vector space, so one should assume from now on that a representation is complex unless otherwise specified (there will be cases where the representations are real).

When V is finite dimensional and a basis of V has been chosen, then linear maps and matrices can be identified (see the review of linear algebra in chapter 4). Such an identification provides an isomorphism

$$GL(V) \simeq GL(n, \mathbf{C})$$

of the group of invertible linear maps of V with $GL(n, \mathbf{C})$, the group of invertible n by n complex matrices. We will begin by studying representations that are finite dimensional and will try to make rigorous statements. Later on, we will get to representations on function spaces, which are infinite dimensional, and will then often neglect rigor and analytical difficulties. Note that only in the case of M a finite set of points will we get an action by finite dimensional matrices this way, since then $F(M)$ will be a finite dimensional vector space ($\mathbf{C}^{\#\text{ of points in M}}$).

A good example to consider to understand this construction in the finite dimensional case is the following:

- Take M to be a set of 3 elements x_1, x_2, x_3. So $F(M) = \mathbf{C}^3$. For $f \in F(M)$, f is a vector in $\mathbf{C}^3$, with components $(f(x_1), f(x_2), f(x_3))$.

- Take $G = S_3$, the group of permutations of 3 elements. This group has $3! = 6$ elements.

- Take G to act on M by permuting the 3 elements

$$(g, x_j) \to g \cdot x_j$$

• This group action provides a representation of G on $F(M)$ by the linear maps
$$(\pi(g)f)(x_j) = f(g^{-1} \cdot x_j)$$

Taking the standard basis of $F(M) = \mathbf{C}^3$, the j'th basis element will correspond to the function f that takes value 1 on x_j and 0 on the other two elements. With respect to this basis, the $\pi(g)$ give six 3 by 3 complex matrices, which under multiplication of matrices satisfy the same relations as the elements of the group under group multiplication. In this particular case, all the entries of the matrix will be 0 or 1, but that is special to the permutation representation.

A common source of confusion is that representations (π, V) are sometimes referred to by the map π, leaving implicit the vector space V that the matrices $\pi(g)$ act on, but at other times referred to by specifying the vector space V, leaving implicit the map π. One reason for this is that the map π may be the identity map: Often, G is a matrix group, so a subgroup of $GL(n, \mathbf{C})$, acting on $V \simeq \mathbf{C}^n$ by the standard action of matrices on vectors. One should keep in mind though that just specifying V is generally not enough to specify the representation, since it may not be the standard one. For example, it could very well be the trivial representation on V, where

$$\pi(g) = \mathbf{1}_n$$

i.e., each element of G acts on V as the identity.

1.3.3 Unitary group representations

The most interesting classes of complex representations are often those for which the linear transformations $\pi(g)$ are "unitary," preserving the notion of length given by the standard Hermitian inner product and thus taking unit vectors to unit vectors. We have the definition:

Definition (Unitary representation). *A representation (π, V) on a complex vector space V with Hermitian inner product $\langle \cdot, \cdot \rangle$ is a unitary representation if it preserves the inner product, i.e.,*

$$\langle \pi(g)v_1, \pi(g)v_2 \rangle = \langle v_1, v_2 \rangle$$

for all $g \in G$ and $v_1, v_2 \in V$.

For a unitary representation, the matrices $\pi(g)$ take values in a subgroup $U(n) \subset GL(n, \mathbf{C})$. In our review of linear algebra (chapter 4), we will see that $U(n)$ can be characterized as the group of n by n complex matrices U such that

$$U^{-1} = U^\dagger$$

where $U^\dagger$ is the conjugate-transpose of U. Note that we will be using the notation "$\dagger$" to mean the "adjoint" or conjugate-transpose matrix. This notation is pretty universal in physics, whereas mathematicians prefer to use "$*$" instead of "$\dagger$."

1.4   Representations and quantum mechanics

The fundamental relationship between quantum mechanics and representation theory is that whenever we have a physical quantum system with a group G acting on it, the space of states $\mathcal{H}$ will carry a unitary representation of G (at least up to a phase factor ambiguity). For physicists working with quantum mechanics, this implies that representation theory provides information about quantum mechanical state spaces when G acts on the system. For mathematicians studying representation theory, this means that physics is a very fruitful source of unitary representations to study: any physical system with a group G acting on it will provide one.

For a representation π and group elements g that are close to the identity, exponentiation can be used to write $\pi(g) \in GL(n, \mathbf{C})$ as

$$\pi(g) = e^A$$

where A is also a matrix, close to the zero matrix. We will study this situation in much more detail and work extensively with examples, showing in particular that if $\pi(g)$ is unitary (i.e., in the subgroup $U(n) \subset GL(n, \mathbf{C})$), then A will be skew-adjoint:

$$A^\dagger = -A$$

where $A^\dagger$ is the conjugate-transpose matrix. Defining $B = iA$, we find that B is self-adjoint

$$B^\dagger = B$$

We thus see that, at least in the case of finite dimensional $\mathcal{H}$, the unitary representation π of G on $\mathcal{H}$ coming from an action of G of our physical system gives us not just unitary matrices $\pi(g)$, but also corresponding self-adjoint operators B on $\mathcal{H}$. Lie group actions thus provide us with a class of quantum mechanical observables, with the self-adjointness property of these operators corresponding to the unitarity of the representation on state space. It is a remarkable fact that for many physical systems, the class of observables that arise in this way include the ones of most physical interest.

In the following chapters, we will see many examples of this phenomenon. A fundamental example that we will study in detail is that of action by translation in time. Here, the group is $G = \mathbf{R}$ (with the additive group law) and we get a unitary representation of $\mathbf{R}$ on the space of states $\mathcal{H}$. The corresponding self-adjoint operator is the Hamiltonian operator H (divided by $\hbar$), and the representation is given by

$$t \in \mathbf{R} \to \pi(t) = e^{-\frac{i}{\hbar}Ht}$$

which one can check is a group homomorphism from the additive group $\mathbf{R}$ to a group of unitary operators. This unitary representation gives the dynamics of the theory, with the Schrödinger equation 1.1 just the statement that $-\frac{i}{\hbar}H\Delta t$ is the skew-adjoint operator that gets exponentiated to give the unitary transformation that moves states $\psi(t)$ ahead in time by an amount Δt.

One way to construct quantum mechanical state spaces $\mathcal{H}$ is as "wavefunctions," meaning complex-valued functions on space–time. Given any group action on space–time, we get a representation π on the state space $\mathcal{H}$ of such wavefunctions by the construction of equation 1.3. Many of the representations of interest will, however, not come from this construction, and we will begin our study of the subject in the next few chapters with such examples, which are simpler because they are finite dimensional. In later chapters, we will turn to representations induced from group actions on space–time, which will be infinite dimensional.

1.5 Groups and symmetries

The subject we are considering is often described as the study of "symmetry groups," since the groups may occur as groups of elements acting by transformations of a space M preserving some particular structure (thus, a "symmetry transformation"). We would like to emphasize though that it is not necessary that the transformations under consideration preserve any particular structure. In the applications to physics, the term "symmetry" is best restricted to the case of groups acting on a physical system in a way that preserves the equations of motion (e.g., by leaving the Hamiltonian function unchanged in the case of a classical mechanical system). For the case of groups of such symmetry transformations, the use of the representation theory of the group to derive implications for the behavior of a quantum mechanical system is an important application of the theory. We will see, however, that the role of representation theory in quantum mechanics is quite a bit deeper than this, with the overall structure of the theory determined by group actions that are not symmetries (in the sense of not preserving the Hamiltonian).

1.6 For further reading

We will be approaching the subject of quantum theory from a different direction than the conventional one, starting with the role of symmetry and with the simplest possible finite dimensional quantum systems, systems which are purely quantum mechanical, with no classical analog. This means that the early discussion found in most physics textbooks is rather different from the one here. They will generally include the same fundamental principles described here, but often begin with the theory of motion of a quantized particle, trying to motivate it from classical mechanics. The state space is then a space of wavefunctions, which is infinite dimensional and necessarily brings some analytical difficulties.

Quantum mechanics is inherently a quite different conceptual structure than classical mechanics. The relationship between the two subjects is rather complicated, but it is clear that quantum mechanics cannot be derived from classical mechanics, so attempts to motivate it that way are unconvincing, although they correspond to the very interesting historical story of how the subject evolved. We will come to the topic of the quantized motion of a particle only in chapter 10, at which point it should become much easier to follow the standard books.

There are many good physics quantum mechanics textbooks available, aimed at a wide variety of backgrounds, and a reader of this book should look for one at an appropriate level to supplement the discussions here. One example would be [81], which is not really an introductory text, but it includes the physicist's version of many of the standard calculations we will also be considering. Some useful textbooks on the subject aimed at mathematicians are [20], [41], [43], [57], and [90]. The first few chapters of [28] provide an excellent while very concise summary of both basic physics and quantum mechanics. One important topic we will not discuss is that of the application of the representation theory of finite groups in quantum mechanics. For this as well as a discussion that overlaps quite a bit with the point of view of this book while emphasizing different topics, see [84]. For another textbook at the level of this one emphasizing the physicist's point of view, see [106].

For the difficult issue of how measurements work and how classical physics emerges from quantum theory, an important part of the story is the notion of "decoherence." Good places to read about this are Wojciech Zurek's updated version of his 1991 Physics Today article [110], as well as his more recent work on "quantum Darwinism" [111]. There is an excellent book on the subject by Schlosshauer [75], and for the details of what happens in real experimental setups, see the book by Haroche and Raimond [44]. For a review of how classical physics emerges from quantum physics written from the mathematical point of view, see Landsman [54]. Finally, to get an idea of the wide variety of points of view available on the topic of the "interpretation" of quantum mechanics, there is a volume of interviews [76] with experts on the topic.

The topic of Lie groups and their representation theory is a standard part of the mathematical curriculum at a more advanced level. As we work through examples in later chapters, we will give references to textbooks covering this material.

Chapter 2
The Group $U(1)$ and its Representations

The simplest example of a Lie group is the group of rotations of the plane, with elements parametrized by a single number, the angle of rotation θ. It is useful to identify such group elements with unit vectors $e^{i\theta}$ in the complex plane. The group is then denoted $U(1)$, since such complex numbers can be thought of as 1 by 1 unitary matrices. We will see in this chapter how the general picture described in chapter 1 works out in this simple case. State spaces will be unitary representations of the group $U(1)$, and we will see that any such representation decomposes into a sum of one-dimensional representations. These one-dimensional representations will be characterized by an integer q, and such integers are the eigenvalues of a self-adjoint operator we will call Q, which is an observable of the quantum theory.

One motivation for the notation Q is that this is the conventional physics notation for electric charge, and this is one of the places where a $U(1)$ group occurs in physics. Examples of $U(1)$ groups acting on physical systems include:

- Quantum particles can be described by a complex-valued "wavefunction" (see chapter 10), and $U(1)$ acts on such wavefunctions by pointwise phase transformations of the value of the function. This phenomenon can be used to understand how particles interact with electromagnetic fields, and in this case, the physical interpretation of the eigenvalue of the Q operator will be the electric charge of the state. We will discuss this in detail in chapter 45.

- If one chooses a particular direction in three-dimensional space, then the group of rotations about that axis can be identified with the group $U(1)$. The eigenvalues of Q will have a physical interpretation as the quantum version of angular momentum in the chosen direction. The fact that such eigenvalues are not continuous, but integral, shows that quantum angular momentum has quite different behavior than classical angular momentum.

© Peter Woit 2017
P. Woit, *Quantum Theory, Groups and Representations*,
DOI 10.1007/978-3-319-64612-1_2

- When we study the harmonic oscillator (chapter 22), we will find that it
  has a $U(1)$ symmetry (rotations in the position-momentum plane) and
  that the Hamiltonian operator is a multiple of the operator Q for this
  case. This implies that the eigenvalues of the Hamiltonian (which give
  the energy of the system) will be integers times some fixed value. When
  one describes multiparticle systems in terms of quantum fields, one finds
  a harmonic oscillator for each momentum mode, and then the Q for that
  mode counts the number of particles with that momentum.

We will sometimes refer to the operator Q as a "charge" operator, assigning
a much more general meaning to the term than that of the specific example
of electric charge. $U(1)$ representations are also ubiquitous in mathematics,
where often the integral eigenvalues of the Q operator will be called "weights."

In a very real sense, the reason for the "quantum" in "quantum mechanics"
is precisely because of the role of $U(1)$ groups acting on the state space. Such
an action implies observables that characterize states by an integer eigenvalue
of an operator Q, and it is this "quantization" of observables that motivates
the name of the subject.

2.1 Some representation theory

Recall the definition of a group representation:

Definition (Representation). *A (complex) representation (π, V) of a group
G on a complex vector space V (with a chosen basis identifying $V \simeq \mathbf{C}^n$) is
a homomorphism*

$$\pi : G \to GL(n, \mathbf{C})$$

This is just a set of n by n matrices, one for each group element, satisfying
the multiplication rules of the group elements. n is called the dimension of
the representation.

We are mainly interested in the case of G a Lie group, where G is a differentiable manifold of some dimension. In such a case, we will restrict attention
to representations given by differentiable maps π. As a space, $GL(n, \mathbf{C})$ is the
space $\mathbf{C}^{n^2}$ of all n by n complex matrices, with the locus of non-invertible
(zero determinant) elements removed. Choosing local coordinates on G, π
will be given by $2n^2$ real functions on G, and the condition that G is a differentiable manifold means that the derivative of π is consistently defined.
Our focus will be not on the general case, but on the study of certain specific
Lie groups and representations π which are of central interest in quantum
mechanics. For these representations, one will be able to readily see that the
maps π are differentiable.

To understand the representations of a group G, one proceeds by first identifying the irreducible ones:

Definition (Irreducible representation). *A representation π is called irreducible if it is has no subrepresentations, meaning nonzero proper subspaces $W \subset V$ such that $(\pi_{|W}, W)$ is a representation. A representation that does have such a subrepresentation is called reducible.*

Given two representations, their direct sum is defined as:

Definition (Direct sum representation). *Given representations π_1 and π_2 of dimensions n_1 and n_2, there is a representation of dimension $n_1 + n_2$ called the direct sum of the two representations, denoted by $\pi_1 \oplus \pi_2$. This representation is given by the homomorphism*

$$(\pi_1 \oplus \pi_2) : g \in G \rightarrow \begin{pmatrix} \pi_1(g) & \mathbf{0} \\ \mathbf{0} & \pi_2(g) \end{pmatrix}$$

In other words, representation matrices for the direct sum are block-diagonal matrices with π_1 and π_2 giving the blocks. For unitary representations

Theorem 2.1. *Any unitary representation π can be written as a direct sum*

$$\pi = \pi_1 \oplus \pi_2 \oplus \cdots \oplus \pi_m$$

where the π_j are irreducible.

Proof. If (π, V) is not irreducible, there exists a $W \subset V$ such that $(\pi_{|W}, W)$ is a representation, and

$$(\pi, V) = (\pi_{|W}, W) \oplus (\pi_{|W^\perp}, W^\perp)$$

Here, $W^\perp$ is the orthogonal complement of W in V (with respect to the Hermitian inner product on V). $(\pi_{|W^\perp}, W^\perp)$ is a subrepresentation since, by unitarity, the representation matrices preserve the Hermitian inner product. The same argument can be applied to W and $W^\perp$, and continue until (π, V) is decomposed into a direct sum of irreducibles. $\square$

Note that non-unitary representations may not be decomposable in this way. For a simple example, consider the group of upper triangular 2 by 2 matrices, acting on $V = \mathbf{C}^2$. The subspace $W \subset V$ of vectors proportional to $\begin{pmatrix} 1 \\ 0 \end{pmatrix}$ is a subrepresentation, but there is no complement to W in V that is also a subrepresentation (the representation is not unitary, so there is no orthogonal complement subrepresentation).

Finding the decomposition of an arbitrary unitary representation into irreducible components can be a very non-trivial problem. Recall that one gets explicit matrices for the $\pi(g)$ of a representation (π, V) only when a basis for V is chosen. To see if the representation is reducible, one cannot just look to see if the $\pi(g)$ are all in block-diagonal form. One needs to find out whether there is some basis for V for which they are all in such form, something very non-obvious from just looking at the matrices themselves.

The following theorem provides a criterion that must be satisfied for a representation to be irreducible:

Theorem (Schur's lemma). *If a complex representation (π, V) is irreducible, then the only linear maps $M : V \to V$ commuting with all the $\pi(g)$ are $\lambda\mathbf{1}$, multiplication by a scalar $\lambda \in \mathbf{C}$.*

Proof. Assume that M commutes with all the $\pi(g)$. We want to show that (π, V) irreducible implies $M = \lambda\mathbf{1}$. Since we are working over the field $\mathbf{C}$ (this does not work for $\mathbf{R}$), we can always solve the eigenvalue equation

$$\det(M - \lambda\mathbf{1}) = 0$$

to find the eigenvalues λ of M. The eigenspaces

$$V_\lambda = \{v \in V : Mv = \lambda v\}$$

are *nonzero* vector subspaces of V and can also be described as $ker(M - \lambda\mathbf{1})$, the kernel of the operator $M - \lambda\mathbf{1}$. Since this operator and all the $\pi(g)$ commute, we have

$$v \in ker(M - \lambda\mathbf{1}) \implies \pi(g)v \in ker(M - \lambda\mathbf{1})$$

so $ker(M - \lambda\mathbf{1}) \subset V$ is a representation of G. If V is irreducible, we must have either $ker(M - \lambda\mathbf{1}) = V$ or $ker(M - \lambda\mathbf{1}) = 0$. Since λ is an eigenvalue, $ker(M - \lambda\mathbf{1}) \neq 0$, so $ker(M - \lambda\mathbf{1}) = V$, and thus, $M = \lambda\mathbf{1}$ as a linear operator on V.   □

More concretely Schur's lemma says that for an irreducible representation, if a matrix M commutes with all the representation matrices $\pi(g)$, then M must be a scalar multiple of the unit matrix. Note that the proof crucially uses the fact that eigenvalues exist. This will only be true in general if one works with $\mathbf{C}$ and thus with complex representations. For the theory of representations on real vector spaces, Schur's lemma is no longer true.

An important corollary of Schur's lemma is the following characterization of irreducible representations of G when G is commutative.

Theorem 2.2. *If G is commutative, all of its irreducible representations are one dimensional.*

Proof. For G commutative, $g \in G$, any representation will satisfy

$$\pi(g)\pi(h) = \pi(h)\pi(g)$$

for all $h \in G$. If π is irreducible, Schur's lemma implies that, since they commute with all the $\pi(g)$, the matrices $\pi(h)$ are all scalar matrices, i.e., $\pi(h) = \lambda_h \mathbf{1}$ for some $\lambda_h \in \mathbf{C}$. π is then irreducible when it is the one-dimensional representation given by $\pi(h) = \lambda_h$. $\qquad\square$

2.2 The group $U(1)$ and its representations

One might think that the simplest Lie group is the one-dimensional additive group $\mathbf{R}$, a group that we will study together with its representations beginning in chapter 10. It turns out that one gets a much easier to analyze Lie group by adding a periodicity condition (which removes the problem of what happens as you go to $\pm\infty$), getting the "circle group" of points on a unit circle. Each such point is characterized by an angle, and the group law is addition of angles.

The circle group can be identified with the group of rotations of the plane $\mathbf{R}^2$, in which case it is called $SO(2)$, for reasons discussed in chapter 4. It is quite convenient, however, to identify $\mathbf{R}^2$ with the complex plane $\mathbf{C}$ and work with the following group (which is isomorphic to $SO(2)$):

Definition (The group $U(1)$). *The elements of the group $U(1)$ are points on the unit circle, which can be labeled by a unit complex number $e^{i\theta}$, or an angle $\theta \in \mathbf{R}$ with θ and $\theta + N2\pi$ labeling the same group element for $N \in \mathbf{Z}$. Multiplication of group elements is complex multiplication, which by the properties of the exponential satisfies*

$$e^{i\theta_1} e^{i\theta_2} = e^{i(\theta_1 + \theta_2)}$$

so in terms of angles the group law is addition (mod 2π).

The name "$U(1)$" is used since complex numbers $e^{i\theta}$ are 1 by 1 unitary matrices.

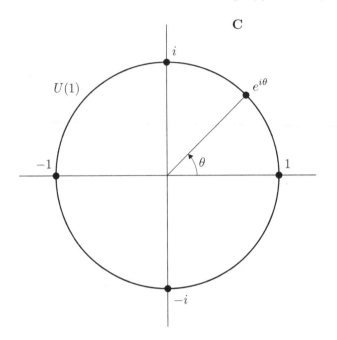

Figure 2.1: $U(1)$ viewed as the unit circle in the complex plane $\mathbf{C}$

By theorem 2.2, since $U(1)$ is a commutative group, all irreducible representations will be one dimensional. Such an irreducible representation will be given by a differentiable map

$$\pi : U(1) \to GL(1, \mathbf{C})$$

$GL(1, \mathbf{C})$ is the group of invertible complex numbers, also called $\mathbf{C}^*$. A differentiable map π that is a representation of $U(1)$ must satisfy homomorphism and periodicity properties which can be used to show:

Theorem 2.3. *All irreducible representations of the group $U(1)$ are unitary and given by*

$$\pi_k : e^{i\theta} \in U(1) \to \pi_k(\theta) = e^{ik\theta} \in U(1) \subset GL(1, \mathbf{C}) \simeq \mathbf{C}^*$$

for $k \in \mathbf{Z}$.

Proof. We will write the π_k as a function of an angle $\theta \in \mathbf{R}$, so satisfying the periodicity property

$$\pi_k(2\pi) = \pi_k(0) = 1$$

Since it is a representation, π will satisfy the homomorphism property

$$\pi_k(\theta_1 + \theta_2) = \pi_k(\theta_1)\pi_k(\theta_2)$$

We need to show that any differentiable map

$$f : U(1) \to \mathbf{C}^*$$

satisfying the homomorphism and periodicity properties is of this form. Computing the derivative $f'(\theta) = \frac{df}{d\theta}$, we find

$$
\begin{aligned}
f'(\theta) &= \lim_{\Delta\theta \to 0} \frac{f(\theta + \Delta\theta) - f(\theta)}{\Delta\theta} \\
&= f(\theta) \lim_{\Delta\theta \to 0} \frac{(f(\Delta\theta) - 1)}{\Delta\theta} \quad \text{(using the homomorphism property)} \\
&= f(\theta)f'(0)
\end{aligned}
$$

Denoting the constant $f'(0)$ by c, the only solutions to this differential equation satisfying $f(0) = 1$ are

$$f(\theta) = e^{c\theta}$$

Requiring periodicity, we find

$$f(2\pi) = e^{c2\pi} = f(0) = 1$$

which implies $c = ik$ for $k \in \mathbf{Z}$, and $f = \pi_k$ for some integer k. $\square$

The representations we have found are all unitary, with π_k taking values in $U(1) \subset \mathbf{C}^*$. The complex numbers $e^{ik\theta}$ satisfy the condition to be a unitary 1 by 1 matrix, since

$$(e^{ik\theta})^{-1} = e^{-ik\theta} = \overline{e^{ik\theta}}$$

These representations are restrictions to the unit circle $U(1)$ of irreducible representations of the group $\mathbf{C}^*$, which are given by

$$\pi_k : z \in \mathbf{C}^* \to \pi_k(z) = z^k \in \mathbf{C}^*$$

Such representations are not unitary, but they have an extremely simple form, so it sometimes is convenient to work with them, later restricting to the unit circle, where the representation is unitary.

2.3 The charge operator

Recall from chapter 1, the claim of a general principle that, when the state space $\mathcal{H}$ is a unitary representation of a Lie group, we get an associated self-adjoint operator on $\mathcal{H}$. We will now illustrate this for the simple case of $G = U(1)$, where the self-adjoint operator we construct will be called the charge operator and denoted Q.

If the representation of $U(1)$ on $\mathcal{H}$ is irreducible, by theorem 2.2 it must be one dimensional with $\mathcal{H} = \mathbf{C}$. By theorem 2.3, it must be of the form $(\pi_q, \mathbf{C})$ for some $q \in \mathbf{Z}$. In this case, the self-adjoint operator Q is multiplication of elements of $\mathcal{H}$ by the integer q. Note that the integrality condition on q is needed because of the periodicity condition on θ, corresponding to the fact that we are working with the group $U(1)$, not the group $\mathbf{R}$.

For a general $U(1)$ representation, by theorems 2.1 and 2.3 we have

$$\mathcal{H} = \mathcal{H}_{q_1} \oplus \mathcal{H}_{q_2} \oplus \cdots \oplus \mathcal{H}_{q_n}$$

for some set of integers $q_1, q_2, \ldots, q_n$ (n is the dimension of $\mathcal{H}$, the q_j may not be distinct), where $\mathcal{H}_{q_j}$ is a copy of $\mathbf{C}$, with $U(1)$ acting by the π_{q_j} representation. One can then define

Definition. *The charge operator Q for the $U(1)$ representation $(\pi, \mathcal{H})$ is the self-adjoint linear operator on $\mathcal{H}$ that acts by multiplication by q_j on the irreducible subrepresentation $\mathcal{H}_{q_j}$. Taking basis elements in $\mathcal{H}_{q_j}$ it acts on $\mathcal{H}$ as the matrix*

$$Q = \begin{pmatrix} q_1 & 0 & \cdots & 0 \\ 0 & q_2 & \cdots & 0 \\ \cdots & & & \cdots \\ 0 & 0 & \cdots & q_n \end{pmatrix}$$

Thinking of $\mathcal{H}$ as a quantum mechanical state space, Q is our first example of a quantum mechanical observable, a self-adjoint operator on $\mathcal{H}$. States in the subspaces $\mathcal{H}_{q_j}$ will be eigenvectors for Q and will have a well-defined numerical value for this observable, the integer q_j. A general state will be a linear superposition of state vectors from different $\mathcal{H}_{q_j}$, and there will not be a well-defined numerical value for the observable Q on such a state.

From the action of Q on $\mathcal{H}$, the representation can be recovered. The action of the group $U(1)$ on $\mathcal{H}$ is given by multiplying by i and exponentiating, to get

$$\pi(e^{i\theta}) = e^{iQ\theta} = \begin{pmatrix} e^{iq_1\theta} & 0 & \cdots & 0 \\ 0 & e^{iq_2\theta} & \cdots & 0 \\ \cdots & & & \cdots \\ 0 & 0 & \cdots & e^{iq_n\theta} \end{pmatrix} \in U(n) \subset GL(n, \mathbf{C})$$

The standard physics terminology is that "Q is the generator of the $U(1)$ action by unitary transformations on the state space $\mathcal{H}$."

The general abstract mathematical point of view (which we will discuss in much more detail in chapter 5) is that a representation π is a map between manifolds, from the Lie group $U(1)$ to the Lie group $GL(n, \mathbf{C})$, that takes the identity of $U(1)$ to the identity of $GL(n, \mathbf{C})$. As such it has a differential π',

which is a linear map from the tangent space at the identity of $U(1)$ (which
here is $i\mathbf{R}$) to the tangent space at the identity of $GL(n, \mathbf{C})$ (which is the
space $M(n, \mathbf{C})$ of n by n complex matrices). The tangent space at the identity
of a Lie group is called a "Lie algebra." In later chapters, we will study many
different examples of such Lie algebras and such maps π', with the linear map
π' often determining the representation π.

In the $U(1)$ case, the relation between the differential of π and the operator
Q is

$$\pi' : i\theta \in i\mathbf{R} \to \pi'(i\theta) = iQ\theta$$

The following drawing illustrates the situation:

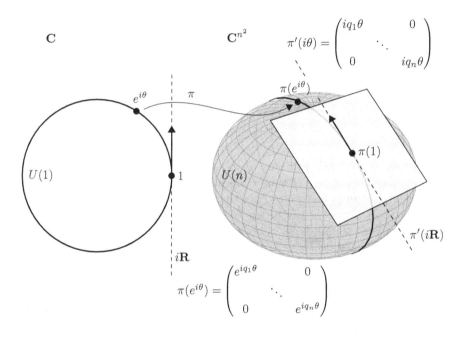

Figure 2.2: Visualizing a representation $\pi : U(1) \to U(n)$, along with its differential

The spherical figure in the right-hand side of the picture is supposed to
indicate the space $U(n) \subset GL(n, \mathbf{C})$ ($GL(n, \mathbf{C})$ is the n by n complex matrices, $\mathbf{C}^{n^2}$, minus the locus of matrices with zero determinant, which are those
that cannot be inverted). It has a distinguished point, the identity. The representation π takes the circle $U(1)$ to a circle inside $U(n)$. Its derivative π'
is a linear map taking the tangent space $i\mathbf{R}$ to the circle at the identity to a
line in the tangent space to $U(n)$ at the identity.

In the very simple example $G = U(1)$, this abstract picture is over-kill and likely confusing. We will see the same picture though occurring in many other much more complicated examples in later chapters. Just like in this $U(1)$ case, for finite dimensional representations the linear maps π' will be matrices, and the representation matrices π can be found by exponentiating the π'.

2.4 Conservation of charge and $U(1)$ symmetry

The way we have defined observable operators in terms of a group representation on $\mathcal{H}$, the action of these operators has nothing to do with the dynamics. If we start at time $t = 0$ in a state in $\mathcal{H}_{q_j}$, with definite numerical value q_j for the observable, there is no reason that time evolution should preserve this. Recall from one of our basic axioms that time evolution of states is given by the Schrödinger equation

$$\frac{d}{dt}|\psi(t)\rangle = -iH|\psi(t)\rangle$$

(we have set $\hbar = 1$). We will later more carefully study the relation of this equation to the symmetry of time translation (the Hamiltonian operator H generates an action of the group $\mathbf{R}$ of time translations, just as the operator Q generates an action of the group $U(1)$). For now though, note that for time-independent Hamiltonian operators H, the solution to this equation is given by exponentiating H, with

$$|\psi(t)\rangle = U(t)|\psi(0)\rangle$$

where

$$U(t) = e^{-itH} = 1 - itH + \frac{(-it)^2}{2!}H^2 + \cdots$$

The commutator of two operators O_1, O_2 is defined by

$$[O_1, O_2] := O_1 O_2 - O_2 O_1$$

and such operators are said to commute if $[O_1, O_2] = 0$. If the Hamiltonian operator H and the charge operator Q commute, then Q will also commute with all powers of H

$$[H^k, Q] = 0$$

and thus with the exponential of H, so

$$[U(t), Q] = 0$$

This condition

$$U(t)Q = QU(t) \qquad (2.1)$$

implies that if a state has a well-defined value q_j for the observable Q at time $t = 0$, it will continue to have the same value at any other time t, since

$$Q|\psi(t)\rangle = QU(t)|\psi(0)\rangle = U(t)Q|\psi(0)\rangle = U(t)q_j|\psi(0)\rangle = q_j|\psi(t)\rangle$$

This will be a general phenomenon: if an observable commutes with the Hamiltonian observable, one gets a conservation law. This conservation law says that if one starts in a state with a well-defined numerical value for the observable (an eigenvector for the observable operator), one will remain in such a state, with the value not changing, i.e., "conserved."

When $[Q, H] = 0$, the group $U(1)$ is said to act as a "symmetry group" of the system, with $\pi(e^{i\theta})$ the "symmetry transformations." Equation 2.1 implies that

$$U(t)e^{iQ\theta} = e^{iQ\theta}U(t)$$

so the action of the $U(1)$ group on the state space of the system commutes with the time evolution law determined by the choice of Hamiltonian. It is only when a representation determined by Q has this particular property that the action of the representation is properly called an action by symmetry transformations and that one gets conservation laws. In general $[Q, H] \neq 0$, with Q then generating a unitary action on $\mathcal{H}$ that does not commute with time evolution and does not imply a conservation law.

2.5   Summary

To summarize the situation for $G = U(1)$, we have found

- Irreducible representations π are one dimensional and characterized by their derivative π' at the identity. If $G = \mathbf{R}$, π' could be any complex number. If $G = U(1)$, periodicity requires that π' must be $iq, q \in \mathbf{Z}$, so irreducible representations are labeled by an integer.

- An arbitrary representation π of $U(1)$ is of the form

$$\pi(e^{i\theta}) = e^{i\theta Q}$$

where Q is a matrix with eigenvalues a set of integers q_j. For a quantum system, Q is the self-adjoint observable corresponding to the $U(1)$ group action on the system and is said to be a "generator" of the group action.

- If $[Q, H] = 0$, the $U(1)$ group acts on the state space as "symmetries." In this case, the q_j will be "conserved quantities," numbers that characterize the quantum states, and do not change as the states evolve in time.

Note that we have so far restricted attention to finite dimensional representations. In section 11.1, we will consider an important infinite dimensional case, a representation on functions on the circle which is essentially the theory of Fourier series. This comes from the action of $U(1)$ on the circle by rotations, giving an induced representation on functions by equation 1.3.

2.6 For further reading

I've had trouble finding another source that covers the material here. Most quantum mechanics books consider it somehow too trivial to mention, starting their discussion of group actions and symmetries with more complicated examples.

Chapter 3
Two-state Systems and $SU(2)$

The simplest truly non-trivial quantum systems have state spaces that are inherently two-complex dimensional. This provides a great deal more structure than that seen in chapter 2, which could be analyzed by breaking up the space of states into one-dimensional subspaces of given charge. We will study these two-state systems in this section, encountering for the first time the implications of working with representations of a non-commutative group. Since they give the simplest non-trivial realization of many quantum phenomena, such systems are the fundamental objects of quantum information theory (the "qubit") and the focus of attempts to build a quantum computer (which would be built out of multiple copies of this sort of fundamental object). Many different possible two-state quantum systems could potentially be used as the physical implementation of a qubit.

One of the simplest possibilities to take would be the idealized situation of a single electron, somehow fixed so that its spatial motion could be ignored, leaving its quantum state described solely by its so-called "spin degree of freedom," which takes values in $\mathcal{H} = \mathbf{C}^2$. The term "spin" is supposed to call to mind the angular momentum of an object spinning about some axis, but such classical physics has nothing to do with the qubit, which is a purely quantum system.

In this chapter, we will analyze what happens for general quantum systems with $\mathcal{H} = \mathbf{C}^2$ by first finding the possible observables. Exponentiating these will give the group $U(2)$ of unitary 2 by 2 matrices acting on $\mathcal{H} = \mathbf{C}^2$. This is a specific representation of $U(2)$, the "defining" representation. By restricting to the subgroup $SU(2) \subset U(2)$ of elements of determinant one, one gets a representation of $SU(2)$ on $\mathbf{C}^2$ often called the "spin $\frac{1}{2}$" representation.

Later on, in chapter 8, we will find all the irreducible representations of $SU(2)$. These are labeled by a natural number

$$N = 0, 1, 2, 3, \ldots$$

© Peter Woit 2017
P. Woit, *Quantum Theory, Groups and Representations*,
DOI 10.1007/978-3-319-64612-1_3

and have dimension $N + 1$. The corresponding quantum systems are said to have "spin $N/2$." The case $N = 0$ is the trivial representation on $\mathbf{C}$, and the case $N = 1$ is the case of this chapter. In the limit $N \to \infty$, one can make contact with classical notions of spinning objects and angular momentum, but the spin $\frac{1}{2}$ case is at the other limit, where the behavior is purely quantum-mechanical.

3.1 The two-state quantum system

3.1.1 The Pauli matrices: observables of the two-state quantum system

For a quantum system with two-dimensional state space $\mathcal{H} = \mathbf{C}^2$, observables are self-adjoint linear operators on $\mathbf{C}^2$. With respect to a chosen basis of $\mathbf{C}^2$, these are 2 by 2 complex matrices M satisfying the condition $M = M^\dagger$ ($M^\dagger$ is the conjugate-transpose of M). Any such matrix will be a (real) linear combination of four matrices:

$$M = c_0 \mathbf{1} + c_1 \sigma_1 + c_2 \sigma_2 + c_3 \sigma_3$$

with $c_j \in \mathbf{R}$ and the standard choice of basis elements given by

$$\mathbf{1} = \begin{pmatrix} 1 & 0 \\ 0 & 1 \end{pmatrix}, \quad \sigma_1 = \begin{pmatrix} 0 & 1 \\ 1 & 0 \end{pmatrix}, \quad \sigma_2 = \begin{pmatrix} 0 & -i \\ i & 0 \end{pmatrix}, \quad \sigma_3 = \begin{pmatrix} 1 & 0 \\ 0 & -1 \end{pmatrix}$$

where the σ_j are called the "Pauli matrices." This choice of basis is a convention, with one aspect of this convention that of taking the basis element in the 3-direction to be diagonal. In common physical situations and conventions, the third direction is the distinguished "up-down" direction in space, so often chosen when a distinguished direction in $\mathbf{R}^3$ is needed.

Recall that the basic principle of how measurements are supposed to work in quantum theory says that the only states that have well-defined values for these four observables are the eigenvectors for these matrices, where the value is the eigenvalue, real since the operator is self-adjoint. The first matrix gives a trivial observable (the identity on every state), whereas the last one, σ_3, has the two eigenvectors

$$\sigma_3 \begin{pmatrix} 1 \\ 0 \end{pmatrix} = \begin{pmatrix} 1 \\ 0 \end{pmatrix}$$

and

$$\sigma_3 \begin{pmatrix} 0 \\ 1 \end{pmatrix} = - \begin{pmatrix} 0 \\ 1 \end{pmatrix}$$

with eigenvalues $+1$ and -1. In quantum information theory, where this is the qubit system, these two eigenstates are labeled $|0\rangle$ and $|1\rangle$ because of the analogy with a classical bit of information. When we get to the theory of spin in chapter 7, we will see that the observable $\frac{1}{2}\sigma_3$ corresponds (in a non-trivial way) to the action of the group $SO(2) = U(1)$ of rotations about the third spatial axis, and the eigenvalues $-\frac{1}{2}, +\frac{1}{2}$ of this operator will be used to label the two eigenstates, so

$$|+\frac{1}{2}\rangle = \begin{pmatrix} 1 \\ 0 \end{pmatrix} \quad \text{and} \quad |-\frac{1}{2}\rangle = \begin{pmatrix} 0 \\ 1 \end{pmatrix}$$

Such eigenstates $|+\frac{1}{2}\rangle$ and $|-\frac{1}{2}\rangle$ provide a basis for $\mathbf{C}^2$, so an arbitrary vector in $\mathcal{H}$ can be written as

$$|\psi\rangle = \alpha |+\frac{1}{2}\rangle + \beta |-\frac{1}{2}\rangle$$

for $\alpha, \beta \in \mathbf{C}$. Only if α or β is 0 does the observable σ_3 correspond to a well-defined number that characterizes the state and can be measured. This will be either $\frac{1}{2}$ (if $\beta = 0$ so the state is an eigenvector $|+\frac{1}{2}\rangle$), or $-\frac{1}{2}$ (if $\alpha = 0$ so the state is an eigenvector $|-\frac{1}{2}\rangle$).

An easy to check fact is that $|+\frac{1}{2}\rangle$ and $|-\frac{1}{2}\rangle$ are NOT eigenvectors for the operators σ_1 and σ_2. One can also check that no pair of the three σ_j commutes, which implies that there are no vectors that are simultaneous eigenvectors for more than one σ_j. This non-commutativity of the operators is responsible for the characteristic paradoxical property of quantum observables: There exist states with a well-defined number for the measured value of one observable σ_j, but such states will not have a well-defined number for the measured value of the other two non-commuting observables.

The physical description of this phenomenon in the realization of this system as a spin$\frac{1}{2}$ particle is that if one prepares states with a well-defined spin component in the j-direction, the two other components of the spin cannot be assigned a numerical value in such a state. Any attempt to prepare states that simultaneously have specific chosen numerical values for the 3 observables corresponding to the σ_j is doomed to failure. So is any attempt to simultaneously measure such values: If one measures the value for a particular observable σ_j, then going on to measure one of the other two will ensure that the first measurement is no longer valid (repeating it will not necessarily give the same thing). There are many subtleties in the theory of measurement for quantum systems, but this simple two-state example already shows some of

the main features of how the behavior of observables is quite different from
that of classical physics.

While the basis vectors $\begin{pmatrix} 1 \\ 0 \end{pmatrix}$ and $\begin{pmatrix} 0 \\ 1 \end{pmatrix}$ are eigenvectors of σ_3, σ_1, and σ_2,
take these basis vectors to non-trivial linear combinations of basis vectors. It
turns out that there are two specific linear combinations of σ_1 and σ_2 that
do something very simple to the basis vectors. Since

$$(\sigma_1 + i\sigma_2) = \begin{pmatrix} 0 & 2 \\ 0 & 0 \end{pmatrix} \quad \text{and} \quad (\sigma_1 - i\sigma_2) = \begin{pmatrix} 0 & 0 \\ 2 & 0 \end{pmatrix}$$

we have

$$(\sigma_1 + i\sigma_2)\begin{pmatrix} 0 \\ 1 \end{pmatrix} = 2\begin{pmatrix} 1 \\ 0 \end{pmatrix} \quad (\sigma_1 + i\sigma_2)\begin{pmatrix} 1 \\ 0 \end{pmatrix} = \begin{pmatrix} 0 \\ 0 \end{pmatrix}$$

and

$$(\sigma_1 - i\sigma_2)\begin{pmatrix} 1 \\ 0 \end{pmatrix} = 2\begin{pmatrix} 0 \\ 1 \end{pmatrix} \quad (\sigma_1 - i\sigma_2)\begin{pmatrix} 0 \\ 1 \end{pmatrix} = \begin{pmatrix} 0 \\ 0 \end{pmatrix}$$

$(\sigma_1 + i\sigma_2)$ is called a "raising operator": On eigenvectors of σ_3, it either
increases the eigenvalue by 2, or annihilates the vector. $(\sigma_1 - i\sigma_2)$ is called a
"lowering operator": On eigenvectors of σ_3, it either decreases the eigenvalue
by 2, or annihilates the vector. Note that these linear combinations are not
self-adjoint and are not observables, $(\sigma_1 + i\sigma_2)$ is the adjoint of $(\sigma_1 - i\sigma_2)$
and vice versa.

3.1.2 Exponentials of Pauli matrices: unitary transformations of the two-state system

We saw in chapter 2 that in the $U(1)$ case, knowing the observable operator Q on $\mathcal{H}$ determined the representation of $U(1)$, with the representation
matrices found by exponentiating $i\theta Q$. Here, we will find the representation corresponding to the two-state system observables by exponentiating
the observables in a similar way.

Taking the identity matrix first, multiplication by $i\theta$ and exponentiation
gives the diagonal unitary matrix

$$e^{i\theta \mathbf{1}} = \begin{pmatrix} e^{i\theta} & 0 \\ 0 & e^{i\theta} \end{pmatrix}$$

This is exactly the case studied in chapter 2, for a $U(1)$ group acting on
$\mathcal{H} = \mathbf{C}^2$, with

$$Q = \begin{pmatrix} 1 & 0 \\ 0 & 1 \end{pmatrix}$$

This matrix commutes with any other 2 by 2 matrix, so we can treat its action on $\mathcal{H}$ independently of the action of the σ_j.

Turning to the other three basis elements of the space of observables, the Pauli matrices, it turns out that since all the σ_j satisfy $\sigma_j^2 = \mathbf{1}$, their exponentials also take a simple form.

$$\begin{aligned}
e^{i\theta\sigma_j} &= 1 + i\theta\sigma_j + \frac{1}{2}(i\theta)^2\sigma_j^2 + \frac{1}{3!}(i\theta)^3\sigma_j^3 + \cdots \\
&= 1 + i\theta\sigma_j - \frac{1}{2}\theta^2\mathbf{1} - i\frac{1}{3!}\theta^3\sigma_j + \cdots \\
&= (1 - \frac{1}{2!}\theta^2 + \cdots)\mathbf{1} + i(\theta - \frac{1}{3!}\theta^3 + \cdots)\sigma_j \\
&= (\cos\theta)\mathbf{1} + i\sigma_j(\sin\theta)
\end{aligned} \tag{3.1}$$

As θ goes from $\theta = 0$ to $\theta = 2\pi$, this exponential traces out a circle in the space of unitary 2 by 2 matrices, starting and ending at the unit matrix. This circle is a group, isomorphic to $U(1)$. So, we have found three different $U(1)$ subgroups inside the unitary 2 by 2 matrices, but only one of them (the case $j = 3$) will act diagonally on $\mathcal{H}$, with the $U(1)$ representation determined by

$$Q = \begin{pmatrix} 1 & 0 \\ 0 & -1 \end{pmatrix}$$

For the other two cases $j = 1$ and $j = 2$, by a change of basis either one could be put in the same diagonal form, but doing this for one value of j makes the other two no longer diagonal. To understand the $SU(2)$ action on $\mathcal{H}$, one needs to consider not just the $U(1)$ subgroups, but the full three-dimensional $SU(2)$ group one gets by exponentiating general linear combinations of Pauli matrices.

To compute such exponentials, one can check that these matrices satisfy the following relations, useful in general for doing calculations with them instead of multiplying out explicitly the 2 by 2 matrices:

$$[\sigma_j, \sigma_k]_+ \equiv \sigma_j\sigma_k + \sigma_k\sigma_j = 2\delta_{jk}\mathbf{1} \tag{3.2}$$

Here, $[\cdot, \cdot]_+$ is called the anticommutator. This relation says that all σ_j satisfy $\sigma_j^2 = \mathbf{1}$ and distinct σ_j anticommute (e.g., $\sigma_j\sigma_k = -\sigma_k\sigma_j$ for $j \neq k$).

Notice that the anticommutation relations imply that, if we take a vector $\mathbf{v} = (v_1, v_2, v_3) \in \mathbf{R}^3$ and define a 2 by 2 matrix by

$$\mathbf{v}\cdot\sigma = v_1\sigma_1 + v_2\sigma_2 + v_3\sigma_3 = \begin{pmatrix} v_3 & v_1 - iv_2 \\ v_1 + iv_2 & -v_3 \end{pmatrix}$$

then taking powers of this matrix we find

$$(\mathbf{v} \cdot \boldsymbol{\sigma})^2 = (v_1^2 + v_2^2 + v_3^2)\mathbf{1} = |\mathbf{v}|^2 \mathbf{1}$$

If $\mathbf{v}$ is a unit vector, we have

$$(\mathbf{v} \cdot \boldsymbol{\sigma})^n = \begin{cases} \mathbf{1} & n \text{ even} \\ \mathbf{v} \cdot \boldsymbol{\sigma} & n \text{ odd} \end{cases}$$

Replacing σ_j by $\mathbf{v} \cdot \boldsymbol{\sigma}$, the same calculation as for equation 3.1 gives (for $\mathbf{v}$ a unit vector)

$$e^{i\theta \mathbf{v} \cdot \boldsymbol{\sigma}} = (\cos\theta)\mathbf{1} + i(\sin\theta)\mathbf{v} \cdot \boldsymbol{\sigma} \qquad (3.3)$$

Notice that the inverse of this matrix can easily be computed by taking θ to $-\theta$

$$(e^{i\theta \mathbf{v} \cdot \boldsymbol{\sigma}})^{-1} = (\cos\theta)\mathbf{1} - i(\sin\theta)\mathbf{v} \cdot \boldsymbol{\sigma}$$

We will review linear algebra and the notion of a unitary matrix in chapter 4, but one form of the condition for a matrix M to be unitary is

$$M^\dagger = M^{-1}$$

so the self-adjointness of the σ_j implies unitarity of $e^{i\theta \mathbf{v} \cdot \boldsymbol{\sigma}}$ since

$$\begin{aligned}
(e^{i\theta \mathbf{v} \cdot \boldsymbol{\sigma}})^\dagger &= ((\cos\theta)\mathbf{1} + i(\sin\theta)\mathbf{v} \cdot \boldsymbol{\sigma})^\dagger \\
&= ((\cos\theta)\mathbf{1} - i(\sin\theta)\mathbf{v} \cdot \boldsymbol{\sigma}^\dagger) \\
&= ((\cos\theta)\mathbf{1} - i(\sin\theta)\mathbf{v} \cdot \boldsymbol{\sigma}) \\
&= (e^{i\theta \mathbf{v} \cdot \boldsymbol{\sigma}})^{-1}
\end{aligned}$$

The determinant of $e^{i\theta \mathbf{v} \cdot \boldsymbol{\sigma}}$ can also easily be computed

$$\begin{aligned}
\det(e^{i\theta \mathbf{v} \cdot \boldsymbol{\sigma}}) &= \det((\cos\theta)\mathbf{1} + i(\sin\theta)\mathbf{v} \cdot \boldsymbol{\sigma}) \\
&= \det \begin{pmatrix} \cos\theta + i(\sin\theta)v_3 & i(\sin\theta)(v_1 - iv_2) \\ i(\sin\theta)(v_1 + iv_2) & \cos\theta - i(\sin\theta)v_3 \end{pmatrix} \\
&= \cos^2\theta + (\sin^2\theta)(v_1^2 + v_2^2 + v_3^2) \\
&= 1
\end{aligned}$$

So, we see that by exponentiating i times linear combinations of the self-adjoint Pauli matrices (which all have trace zero), we get unitary matrices of determinant one. These are invertible and form the group named $SU(2)$, the group of unitary 2 by 2 matrices of determinant one. If we exponentiated not just $i\theta \mathbf{v} \cdot \boldsymbol{\sigma}$, but $i(\phi\mathbf{1} + \theta\mathbf{v} \cdot \boldsymbol{\sigma})$ for some real constant ϕ (such matrices will not have trace zero unless $\phi = 0$), we would get a unitary matrix with determinant $e^{i2\phi}$. The group of all unitary 2 by 2 matrices is called $U(2)$. It contains as subgroups $SU(2)$ as well as the $U(1)$ described at the beginning of this section. $U(2)$ is slightly different from the product of these two subgroups, since the group element

$$\begin{pmatrix} -1 & 0 \\ 0 & -1 \end{pmatrix}$$

is in both subgroups. In chapter 4, we will encounter the generalization to $SU(n)$ and $U(n)$, groups of unitary n by n complex matrices.

To get some more insight into the structure of the group $SU(2)$, consider an arbitrary 2 by 2 complex matrix

$$\begin{pmatrix} \alpha & \beta \\ \gamma & \delta \end{pmatrix}$$

Unitarity implies that the rows are orthonormal. This results from the condition that the matrix times its conjugate-transpose is the identity

$$\begin{pmatrix} \alpha & \beta \\ \gamma & \delta \end{pmatrix} \begin{pmatrix} \overline{\alpha} & \overline{\gamma} \\ \overline{\beta} & \overline{\delta} \end{pmatrix} = \begin{pmatrix} 1 & 0 \\ 0 & 1 \end{pmatrix}$$

Orthogonality of the two rows gives the relation

$$\gamma\overline{\alpha} + \delta\overline{\beta} = 0 \implies \delta = -\frac{\gamma\overline{\alpha}}{\overline{\beta}}$$

The condition that the first row has length one gives

$$\alpha\overline{\alpha} + \beta\overline{\beta} = |\alpha|^2 + |\beta|^2 = 1$$

Using these two relations and computing the determinant (which has to be 1) gives

$$\alpha\delta - \beta\gamma = -\frac{\alpha\overline{\alpha}\gamma}{\overline{\beta}} - \beta\gamma = -\frac{\gamma}{\overline{\beta}}(\alpha\overline{\alpha} + \beta\overline{\beta}) = -\frac{\gamma}{\overline{\beta}} = 1$$

so one must have

$$\gamma = -\overline{\beta}, \ \delta = \overline{\alpha}$$

and an $SU(2)$ matrix will have the form

$$\begin{pmatrix} \alpha & \beta \\ -\overline{\beta} & \overline{\alpha} \end{pmatrix}$$

where $(\alpha, \beta) \in \mathbf{C}^2$ and

$$|\alpha|^2 + |\beta|^2 = 1$$

The elements of $SU(2)$ are thus parametrized by two complex numbers, with the sum of their length-squared equal to one. Identifying $\mathbf{C}^2 = \mathbf{R}^4$, these are vectors of length one in $\mathbf{R}^4$. Just as $U(1)$ could be identified as a space with the unit circle S^1 in $\mathbf{C} = \mathbf{R}^2$, $SU(2)$ can be identified with the unit three-sphere S^3 in $\mathbf{R}^4$.

3.2 Commutation relations for Pauli matrices

An important set of relations satisfied by Pauli matrices are their commutation relations:

$$[\sigma_j, \sigma_k] \equiv \sigma_j \sigma_k - \sigma_k \sigma_j = 2i \sum_{l=1}^{3} \epsilon_{jkl} \sigma_l \qquad (3.4)$$

where ϵ_{jkl} satisfies $\epsilon_{123} = 1$, is antisymmetric under permutation of two of its subscripts, and vanishes if two of the subscripts take the same value. More explicitly, this says:

$$[\sigma_1, \sigma_2] = 2i\sigma_3, \ [\sigma_2, \sigma_3] = 2i\sigma_1, \ [\sigma_3, \sigma_1] = 2i\sigma_2$$

These relations can easily be checked by explicitly computing with the matrices. Putting together equations 3.2 and 3.4 gives a formula for the product of two Pauli matrices:

$$\sigma_j \sigma_k = \delta_{jk} \mathbf{1} + i \sum_{l=1}^{3} \epsilon_{jkl} \sigma_l$$

While physicists prefer to work with the self-adjoint Pauli matrices and their real eigenvalues, the skew-adjoint matrices

$$X_j = -i \frac{\sigma_j}{2}$$

can instead be used. These satisfy the slightly simpler commutation relations

$$[X_j, X_k] = \sum_{l=1}^{3} \epsilon_{jkl} X_l$$

or more explicitly

$$[X_1, X_2] = X_3, \ [X_2, X_3] = X_1, \ [X_3, X_1] = X_2 \qquad (3.5)$$

The non-triviality of the commutators reflects the non-commutativity of the group. Group elements $U \in SU(2)$ near the identity satisfy

$$U \simeq 1 + \epsilon_1 X_1 + \epsilon_2 X_2 + \epsilon_3 X_3$$

for ϵ_j small and real, just as group elements $z \in U(1)$ near the identity satisfy

$$z \simeq 1 + i\epsilon$$

The X_j and their commutation relations can be thought of as an infinitesimal version of the full group and its group multiplication law, valid near the identity. In terms of the geometry of manifolds, recall that $SU(2)$ is the space S^3. The X_j give a basis of the tangent space $\mathbf{R}^3$ to the identity of $SU(2)$, just as i gives a basis of the tangent space to the identity of $U(1)$.

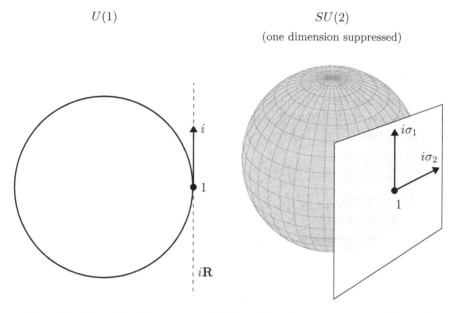

Figure 3.1: Comparing the geometry of $U(1)$ as S^1 to the geometry of $SU(2)$ as S^3.

3.3 Dynamics of a two-state system

Recall that the time-dependence of states in quantum mechanics is given by
the Schrödinger equation

$$\frac{d}{dt}|\psi(t)\rangle = -iH|\psi(t)\rangle$$

where H is a particular self-adjoint linear operator on $\mathcal{H}$, the Hamiltonian
operator. Considering the case of H time-independent, the most general such
operator H on $\mathbf{C}^2$ will be given by

$$H = h_0\mathbf{1} + h_1\sigma_1 + h_2\sigma_2 + h_3\sigma_3$$

for four real parameters h_0, h_1, h_2, h_3. The solution to the Schrödinger equation is then given by exponentiation:

$$|\psi(t)\rangle = U(t)|\psi(0)\rangle$$

where

$$U(t) = e^{-itH}$$

The $h_0\mathbf{1}$ term in H contributes an overall phase factor e^{-ih_0t}, with the
remaining factor of $U(t)$ an element of the group $SU(2)$ rather than the
larger group $U(2)$ of all 2 by 2 unitaries.

Using our equation 3.3, valid for a unit vector $\mathbf{v}$, our $U(t)$ is given by
taking $\mathbf{h} = (h_1, h_2, h_3)$, $\mathbf{v} = \frac{\mathbf{h}}{|\mathbf{h}|}$, and $\theta = -t|\mathbf{h}|$, so we find

$$
\begin{aligned}
U(t) =& e^{-ih_0t}\left(\cos(-t|\mathbf{h}|)\mathbf{1} + i\sin(-t|\mathbf{h}|)\frac{h_1\sigma_1 + h_2\sigma_2 + h_3\sigma_3}{|\mathbf{h}|}\right) \\
=& e^{-ih_0t}\left(\cos(t|\mathbf{h}|)\mathbf{1} - i\sin(t|\mathbf{h}|)\frac{h_1\sigma_1 + h_2\sigma_2 + h_3\sigma_3}{|\mathbf{h}|}\right) \\
=& e^{-ih_0t}\begin{pmatrix} \cos(t|\mathbf{h}|) - i\frac{h_3}{|\mathbf{h}|}\sin(t|\mathbf{h}|) & -i\sin(t|\mathbf{h}|)\frac{h_1-ih_2}{|\mathbf{h}|} \\ -i\sin(t|\mathbf{h}|)\frac{h_1+ih_2}{|\mathbf{h}|} & \cos(t|\mathbf{h}|) + i\frac{h_3}{|\mathbf{h}|}\sin(t|\mathbf{h}|) \end{pmatrix}
\end{aligned}
$$

In the special case $\mathbf{h} = (0, 0, h_3)$, we have

$$U(t) = \begin{pmatrix} e^{-it(h_0+h_3)} & 0 \\ 0 & e^{-it(h_0-h_3)} \end{pmatrix}$$

so if our initial state is

$$|\psi(0)\rangle = \alpha|+\tfrac{1}{2}\rangle + \beta|-\tfrac{1}{2}\rangle$$

for $\alpha, \beta \in \mathbf{C}$, at later times the state will be

$$|\psi(t)\rangle = \alpha e^{-it(h_0+h_3)}|+\tfrac{1}{2}\rangle + \beta e^{-it(h_0-h_3)}|-\tfrac{1}{2}\rangle$$

In this special case, the eigenvalues of the Hamiltonian are $h_0 \pm h_3$.

In the physical realization of this system by a spin $\frac{1}{2}$ particle (ignoring its spatial motion), the Hamiltonian is given by

$$H = \frac{ge}{4mc}(B_1\sigma_1 + B_2\sigma_2 + B_3\sigma_3) \qquad (3.6)$$

where the B_j are the components of the magnetic field, and the physical constants are the gyromagnetic ratio (g), the electric charge (e), the mass (m), and the speed of light (c). By computing $U(t)$ above, we have solved the problem of finding the time evolution of such a system, setting $h_j = \frac{ge}{4mc}B_j$. For the special case of a magnetic field in the 3-direction $(B_1 = B_2 = 0)$, we see that the two different states with well-defined energy $(|+\tfrac{1}{2}\rangle$ and $|-\tfrac{1}{2}\rangle$, recall that the energy is the eigenvalue of the Hamiltonian) will have an energy difference between them of

$$2h_3 = \frac{ge}{2mc}B_3$$

This is known as the Zeeman effect and is readily visible in the spectra of atoms subjected to a magnetic field. We will consider this example in more detail in chapter 7, seeing how the group of rotations of $\mathbf{R}^3$ enters into the story. Much later, in chapter 45, we will derive the Hamiltonian 3.6 from general principles of how electromagnetic fields couple to spin $\frac{1}{2}$ particles.

3.4 For further reading

Many quantum mechanics textbooks now begin with the two-state system, giving a much more detailed treatment than the one given here, including much more about the physical interpretation of such systems (see, e.g., [96]). Volume III of Feynman's *Lectures on Physics* [25] is a quantum mechanics text with much of the first half devoted to two-state systems. The field of "Quantum Information Theory" gives a perspective on quantum theory that puts such systems (in this context called the "qubit") front and center. One possible reference for this material is John Preskill's notes on quantum computation [69].

Chapter 4
Linear Algebra Review, Unitary and Orthogonal Groups

A significant background in linear algebra will be assumed in later chapters, and we will need a range of specific facts from that subject. These will include some aspects of linear algebra not emphasized in a typical linear algebra course, such as the role of the dual space and the consideration of various classes of invertible matrices as defining a group. For now, our vector spaces will be finite dimensional. Later on, we will come to state spaces that are infinite dimensional and will address the various issues that this raises at that time.

4.1 Vector spaces and linear maps

A vector space V over a field k is a set with a consistent way to take linear combinations of elements with coefficients in k. We will only be using the cases $k = \mathbf{R}$ and $k = \mathbf{C}$, so such finite dimensional V will just be $\mathbf{R}^n$ or $\mathbf{C}^n$. Choosing a basis (set of n linearly independent vectors) $\{\mathbf{e}_j\}$, an arbitrary vector $v \in V$ can be written as follows:

$$v = v_1\mathbf{e}_1 + v_2\mathbf{e}_2 + \cdots + v_n\mathbf{e}_n$$

giving an explicit identification of V with n-tuples v_j of real or complex numbers which we will usually write as column vectors

$$v = \begin{pmatrix} v_1 \\ v_2 \\ \vdots \\ v_n \end{pmatrix}$$

© Peter Woit 2017
P. Woit, *Quantum Theory, Groups and Representations*,
DOI 10.1007/978-3-319-64612-1_4

The choice of a basis $\{\mathbf{e}_j\}$ also allows us to express the action of a linear operator L on V

$$L : v \in V \to Lv \in V$$

as multiplication by an n by n matrix:

$$\begin{pmatrix} v_1 \\ v_2 \\ \vdots \\ v_n \end{pmatrix} \to \begin{pmatrix} L_{11} & L_{12} & \cdots & L_{1n} \\ L_{21} & L_{22} & \cdots & L_{2n} \\ \vdots & \vdots & \vdots & \vdots \\ L_{n1} & L_{n2} & \cdots & L_{nn} \end{pmatrix} \begin{pmatrix} v_1 \\ v_2 \\ \vdots \\ v_n \end{pmatrix}$$

The reader should be warned that we will often not notationally distinguish between a linear operator L and its matrix with matrix entries L_{jk} with respect to some unspecified basis, since we are often interested in properties of operators L that, for the corresponding matrix, are basis-independent (e.g., is the operator or matrix invertible?). The invertible linear operators on V form a group under composition, a group we will sometimes denote $GL(V)$, with "GL" indicating "General Linear." Choosing a basis identifies this group with the group of invertible matrices, with group law matrix multiplication. For V n-dimensional, we will denote this group by $GL(n, \mathbf{R})$ in the real case, $GL(n, \mathbf{C})$ in the complex case.

Note that when working with vectors as linear combinations of basis vectors, we can use matrix notation to write a linear transformation as follows:

$$v \to Lv = \begin{pmatrix} \mathbf{e}_1 & \cdots & \mathbf{e}_n \end{pmatrix} \begin{pmatrix} L_{11} & L_{12} & \cdots & L_{1n} \\ L_{21} & L_{22} & \cdots & L_{2n} \\ \vdots & \vdots & \vdots & \vdots \\ L_{n1} & L_{n2} & \cdots & L_{nn} \end{pmatrix} \begin{pmatrix} v_1 \\ v_2 \\ \vdots \\ v_n \end{pmatrix}$$

We see from this that we can think of the transformed vector as we did above in terms of transformed coefficients v_j with respect to fixed basis vectors but also could leave the v_j unchanged and transform the basis vectors. At times, we will want to use matrix notation to write formulas for how the basis vectors transform in this way and then will write

$$\begin{pmatrix} \mathbf{e}_1 \\ \mathbf{e}_2 \\ \vdots \\ \mathbf{e}_n \end{pmatrix} \to \begin{pmatrix} L_{11} & L_{21} & \cdots & L_{n1} \\ L_{12} & L_{22} & \cdots & L_{n2} \\ \vdots & \vdots & \vdots & \vdots \\ L_{1n} & L_{2n} & \cdots & L_{nn} \end{pmatrix} \begin{pmatrix} \mathbf{e}_1 \\ \mathbf{e}_2 \\ \vdots \\ \mathbf{e}_n \end{pmatrix}$$

Note that putting the basis vectors $\mathbf{e}_j$ in a column vector like this causes the matrix for L to act on them by the transposed matrix.

4.2   Dual vector spaces

To any vector space V, we can associate a new vector space, its dual:

Definition (Dual vector space). *For V a vector space over a field k, the dual vector space V^* is the vector space of all linear maps $V \to k$, i.e.,*

$$V^* = \{l : V \to k \text{ such that } l(\alpha v + \beta w) = \alpha l(v) + \beta l(w)\}$$

for $\alpha, \beta \in k, \ v, w \in V$.

Given a linear transformation L acting on V, we can define:

Definition (Transpose transformation). *The transpose of L is the linear transformation*

$$L^t : V^* \to V^*$$

given by

$$(L^t l)(v) = l(Lv) \tag{4.1}$$

for $l \in V^*, v \in V$.

For any choice of basis $\{\mathbf{e}_j\}$ of V, there is a dual basis $\{\mathbf{e}_j^*\}$ of V^* that satisfies

$$\mathbf{e}_j^*(\mathbf{e}_k) = \delta_{jk}$$

Coordinates on V with respect to a basis are linear functions, and thus elements of V^*. The coordinate function v_j can be identified with the dual basis vector $\mathbf{e}_j^*$ since

$$\mathbf{e}_j^*(v) = \mathbf{e}_j^*(v_1 \mathbf{e}_1 + v_2 \mathbf{e}_2 + \cdots + v_n \mathbf{e}_n) = v_j$$

It can easily be shown that the elements of the matrix for L in the basis $\mathbf{e}_j$ are given by

$$L_{jk} = \mathbf{e}_j^*(L\mathbf{e}_k)$$

and that the matrix for the transpose map (with respect to the dual basis) is the matrix transpose

$$(L^T)_{jk} = L_{kj}$$

Matrix notation can be used to write elements

$$l = l_1 \mathbf{e}_1^* + l_2 \mathbf{e}_2^* + \cdots + l_n \mathbf{e}_n^* \in V^*$$

of V^* as row vectors

$$\begin{pmatrix} l_1 & l_2 & \cdots & l_n \end{pmatrix}$$

of coordinates on V^*. Evaluation of l on a vector v is then given by matrix multiplication

$$l(v) = \begin{pmatrix} l_1 & l_2 & \cdots & l_n \end{pmatrix} \begin{pmatrix} v_1 \\ v_2 \\ \vdots \\ v_n \end{pmatrix} = l_1 v_1 + l_2 v_2 + \cdots + l_n v_n$$

For any representation (π, V) of a group G on V, we can define a corresponding representation on V^*:

Definition (Dual or contragredient representation). *The dual or contragredient representation on V^* is given by taking as linear operators*

$$(\pi^{-1})^t(g) : V^* \to V^* \tag{4.2}$$

These satisfy the homomorphism property since

$$(\pi^{-1}(g_1))^t(\pi^{-1}(g_2))^t = (\pi^{-1}(g_2)\pi^{-1}(g_1))^t = ((\pi(g_1)\pi(g_2))^{-1})^t$$

One way to characterize this representation is as the action on V^* such that pairings between elements of V^* and V are invariant, since

$$l(v) \to ((\pi^{-1}(g))^t l)(\pi(g)v) = l(\pi(g)^{-1}\pi(g)v) = l(v)$$

For a given representation operator $\pi(g)$, acting (with respect to a chosen basis) on V by

$$\begin{pmatrix} v_1 \\ v_2 \\ \vdots \\ v_n \end{pmatrix} \to P \begin{pmatrix} v_1 \\ v_2 \\ \vdots \\ v_n \end{pmatrix}$$

for a matrix P, we will have actions on the dual space V^* by

$$\begin{pmatrix} l_1 & l_2 & \cdots & l_n \end{pmatrix} \to \begin{pmatrix} l_1 & l_2 & \cdots & l_n \end{pmatrix} (P^{-1})^T$$

or in terms of column vectors by

$$\begin{pmatrix} l_1 \\ l_2 \\ \vdots \\ l_n \end{pmatrix} \to P^{-1} \begin{pmatrix} l_1 \\ l_2 \\ \vdots \\ l_n \end{pmatrix}$$

4.3   Change of basis

Any invertible transformation A on V can be used to change the basis $\mathbf{e}_j$ of V to a new basis $\mathbf{e}'_j$ by taking

$$\mathbf{e}_j \to \mathbf{e}'_j = A\mathbf{e}_j$$

The matrix for a linear transformation L transforms under this change of basis as follows:

$$\begin{aligned}
L_{jk} = \mathbf{e}^*_j(L\mathbf{e}_k) \to (\mathbf{e}'_j)^*(L\mathbf{e}'_k) &= (A\mathbf{e}_j)^*(LA\mathbf{e}_k) \\
&= (A^T)^{-1}(\mathbf{e}^*_j)(LA\mathbf{e}_k) \\
&= \mathbf{e}^*_j(A^{-1}LA\mathbf{e}_k) \\
&= (A^{-1}LA)_{jk}
\end{aligned}$$

In the second step, we are using the fact that elements of the dual basis transform as the dual representation. This is what is needed to ensure the relation

$$(\mathbf{e}'_j)^*(\mathbf{e}'_k) = \delta_{jk}$$

The change of basis formula shows that if two matrices L_1 and L_2 are related by conjugation by a third matrix A

$$L_2 = A^{-1}L_1 A$$

then they represent the same linear transformation, with respect to two different choices of basis. Recall that a finite dimensional representation is given by a set of matrices $\pi(g)$, one for each group element. If two representations are related by

$$\pi_2(g) = A^{-1}\pi_1(g)A$$

(for all g, A does not depend on g), then we can think of them as being the same representation, with different choices of basis. In such a case, the representations π_1 and π_2 are called "equivalent," and we will often implicitly identify representations that are equivalent.

4.4   Inner products

An inner product on a vector space V is an additional structure that provides a notion of length for vectors, of angle between vectors, and identifies $V^* \simeq V$. In the real case:

Definition (Inner product, real case). *An inner product on a real vector space V is a symmetric $((v, w) = (w, v))$ map*

$$(\cdot, \cdot) : V \times V \to \mathbf{R}$$

that is non-degenerate and linear in both variables.

Our real inner products will usually be positive-definite $((v, v) \geq 0$ and $(v, v) = 0 \implies v = 0)$, with indefinite inner products only appearing in the context of special relativity, where an indefinite inner product on four-dimensional space–time is used.

In the complex case:

Definition (Inner product, complex case). *A Hermitian inner product on a complex vector space V is a map*

$$\langle \cdot, \cdot \rangle : V \times V \to \mathbf{C}$$

that is conjugate symmetric

$$\langle v, w \rangle = \overline{\langle w, v \rangle}$$

non-degenerate in both variables, linear in the second variable, and antilinear in the first variable: for $\alpha \in \mathbf{C}$ and $u, v, w \in V$

$$\langle u + v, w \rangle = \langle u, w \rangle + \langle v, w \rangle, \quad \langle \alpha u, v \rangle = \overline{\alpha} \langle u, v \rangle$$

An inner product gives a notion of length-squared $|| \cdot ||^2$ for vectors, with

$$||v||^2 = \langle v, v \rangle$$

Note that whether to specify antilinearity in the first or second variable is a matter of convention. The choice we are making is universal among physicists, with the opposite choice common among mathematicians. Our Hermitian inner products will be positive-definite $(||v||^2 > 0$ for $v \neq 0)$ unless specifically noted otherwise (i.e., characterized explicitly as an indefinite Hermitian inner product).

An inner product also provides an isomorphism $V \simeq V^*$ by the map

$$v \in V \to l_v \in V^* \tag{4.3}$$

where l_v is defined by

$$l_v(w) = (v, w)$$

in the real case, and
$$l_v(w) = \langle v, w \rangle$$

in the complex case (where this is a complex antilinear rather than linear isomorphism).

Physicists have a useful notation due to Dirac for elements of a vector space and its dual, for the case when V is a complex vector space with a Hermitian inner product (such as the state space $\mathcal{H}$ for a quantum theory). An element of such a vector space V is written as a "ket vector"

$$|\alpha\rangle$$

where α is a label for a vector in V. Sometimes the vectors in question will be eigenvectors for some observable operator, with the label α the eigenvalue.

An element of the dual vector space V^* is written as a "bra vector"

$$\langle\alpha|$$

with the labeling in terms of α determined by the isomorphism 4.3, i.e.,

$$\langle\alpha| = l_{|\alpha\rangle}$$

Evaluating $\langle\alpha| \in V^*$ on $|\beta\rangle \in V$ gives an element of $\mathbf{C}$, written

$$\langle\alpha|(|\beta\rangle) = \langle\alpha|\beta\rangle$$

Note that in the inner product, the angle bracket notation means something different than in the bra-ket notation. The similarity is intentional though since $\langle\alpha|\beta\rangle$ is the inner product of a vector labeled by α and a vector labeled by β (with "bra-ket" a play on words based on this relation to the inner product bracket notation). Recalling what happens when one interchanges vectors in a Hermitian inner product, one has

$$\langle\beta|\alpha\rangle = \overline{\langle\alpha|\beta\rangle}$$

For a choice of orthonormal basis $\{\mathbf{e}_j\}$, i.e., satisfying

$$\langle\mathbf{e}_j, \mathbf{e}_k\rangle = \delta_{jk}$$

a useful choice of label is the index j, so

$$|j\rangle = \mathbf{e}_j$$

Because of orthonormality, coefficients of vectors $|\alpha\rangle$ with respect to the basis $\{\mathbf{e}_j\}$ are

$$\langle j|\alpha\rangle$$

and the expansion of a vector in terms of the basis is written

$$|\alpha\rangle = \sum_{j=1}^{n} |j\rangle\langle j|\alpha\rangle \tag{4.4}$$

Similarly, for elements $\langle\alpha| \in V^*$,

$$\langle\alpha| = \sum_{j=1}^{n} \langle\alpha|j\rangle\langle j|$$

The column vector expression for $|\alpha\rangle$ is thus

$$\begin{pmatrix} \langle 1|\alpha\rangle \\ \langle 2|\alpha\rangle \\ \vdots \\ \langle n|\alpha\rangle \end{pmatrix}$$

and the row vector form of $\langle\alpha|$ is

$$\left(\langle\alpha|1\rangle \quad \langle\alpha|2\rangle \quad \cdots \quad \langle\alpha|n\rangle \right) = \left(\overline{\langle 1|\alpha\rangle} \quad \overline{\langle 2|\alpha\rangle} \quad \cdots \quad \overline{\langle n|\alpha\rangle} \right)$$

The inner product is the usual matrix product

$$\langle\alpha|\beta\rangle = \left(\langle\alpha|1\rangle \quad \langle\alpha|2\rangle \quad \cdots \quad \langle\alpha|n\rangle \right) \begin{pmatrix} \langle 1|\beta\rangle \\ \langle 2|\beta\rangle \\ \vdots \\ \langle n|\beta\rangle \end{pmatrix}$$

If L is a linear operator $L : V \rightarrow V$, then with respect to the basis $\{\mathbf{e}_j\}$ it becomes a matrix with matrix elements

$$L_{kj} = \langle k|L|j\rangle$$

The expansion 4.4 of a vector $|\alpha\rangle$ in terms of the basis can be interpreted as multiplication by the identity operator

$$\mathbf{1} = \sum_{j=1}^{n} |j\rangle\langle j|$$

and this kind of expression is referred to by physicists as a "completeness relation," since it requires that the set of $|j\rangle$ be a basis with no missing elements. The operator

$$P_j = |j\rangle\langle j|$$

is the projection operator onto the jth basis vector.

Digression. *In this book, all of our indices will be lower indices. One way to keep straight the difference between vectors and dual vectors is to use upper indices for components of vectors and lower indices for components of dual vectors. This is quite useful in Riemannian geometry and general relativity, where the inner product is given by a metric that can vary from point to point, causing the isomorphism between vectors and dual vectors to also vary. For quantum mechanical state spaces, we will be using a single, standard, fixed inner product, so there will be a single isomorphism between vectors and dual vectors. In this case, the bra-ket notation can be used to provide a notational distinction between vectors and dual vectors.*

4.5 Adjoint operators

When V is a vector space with inner product, the adjoint of L can be defined by:

Definition (Adjoint operator). *The adjoint of a linear operator $L : V \to V$ is the operator $L^\dagger$ satisfying*

$$\langle Lv, w \rangle = \langle v, L^\dagger w \rangle$$

for all $v, w \in V$.

Note that mathematicians tend to favor L^* as notation for the adjoint of L, as opposed to the physicist's notation $L^\dagger$ that we are using.

In terms of explicit matrices, since l_{Lv} is the conjugate-transpose of Lv, the matrix for $L^\dagger$ will be given by the conjugate-transpose $\overline{L^T}$ of the matrix for L:

$$L^\dagger_{jk} = \overline{L_{kj}}$$

In the real case, the matrix for the adjoint is just the transpose matrix. We will say that a linear transformation is self-adjoint if $L^\dagger = L$, skew-adjoint if $L^\dagger = -L$.

4.6 Orthogonal and unitary transformations

A special class of linear transformations will be invertible transformations that preserve the inner product, i.e., satisfying

$$\langle Lv, Lw \rangle = \langle v, w \rangle$$

for all $v, w \in V$. Such transformations take orthonormal bases to orthonormal bases, so one role in which they appear is as a change of basis between two orthonormal bases.

In terms of adjoints, this condition becomes

$$\langle Lv, Lw \rangle = \langle v, L^{\dagger}Lw \rangle = \langle v, w \rangle$$

so

$$L^{\dagger}L = \mathbf{1}$$

or equivalently

$$L^{\dagger} = L^{-1}$$

In matrix notation, this first condition becomes

$$\sum_{k=1}^{n} (L^{\dagger})_{jk} L_{kl} = \sum_{k=1}^{n} \overline{L_{kj}} L_{kl} = \delta_{jl}$$

which says that the column vectors of the matrix for L are orthonormal vectors. Using instead the equivalent condition

$$LL^{\dagger} = \mathbf{1}$$

we find that the row vectors of the matrix for L are also orthonormal. Since such linear transformations preserving the inner product can be composed and are invertible, they form a group, and some of the basic examples of Lie groups are given by these groups for the cases of real and complex vector spaces.

4.6.1 Orthogonal groups

We will begin with the real case, where these groups are called orthogonal groups:

Definition (Orthogonal group). *The orthogonal group $O(n)$ in n dimensions is the group of invertible transformations preserving an inner product on a real n-dimensional vector space V. This is isomorphic to the group of n by n real invertible matrices L satisfying*

$$L^{-1} = L^T$$

The subgroup of $O(n)$ of matrices with determinant 1 (equivalently, the subgroup preserving orientation of orthonormal bases) is called $SO(n)$.

Recall that for a representation π of a group G on V, there is a dual representation on V^* given by taking the transpose-inverse of π. If G is an orthogonal group, then π and its dual are the same matrices, with V identified by V^* by the inner product.

Since the determinant of the transpose of a matrix is the same as the determinant of the matrix, we have

$$L^{-1}L = 1 \implies \det(L^{-1})\det(L) = \det(L^T)\det(L) = (\det(L))^2 = 1$$

so

$$\det(L) = \pm 1$$

$O(n)$ is a continuous Lie group, with two components distinguished by the sign of the determinant: $SO(n)$, the subgroup of orientation-preserving transformations, which include the identity, and a component of orientation-changing transformations.

The simplest non-trivial example is for $n = 2$, where all elements of $SO(2)$ are given by matrices of the form

$$\begin{pmatrix} \cos\theta & -\sin\theta \\ \sin\theta & \cos\theta \end{pmatrix}$$

These matrices give counterclockwise rotations in $\mathbf{R}^2$ by an angle θ. The other component of $O(2)$ will be given by matrices of the form

$$\begin{pmatrix} \cos\theta & \sin\theta \\ \sin\theta & -\cos\theta \end{pmatrix}$$

which describe a reflection followed by a rotation. Note that the group $SO(2)$ is isomorphic to the group $U(1)$ by

$$\begin{pmatrix} \cos\theta & -\sin\theta \\ \sin\theta & \cos\theta \end{pmatrix} \leftrightarrow e^{i\theta}$$

so the representation theory of $SO(2)$ is just as for $U(1)$, with irreducible complex representations one-dimensional and classified by an integer.

In chapter 6, we will consider in detail the case of $SO(3)$, which is crucial for physical applications because it is the group of rotations in the physical three-dimensional space.

4.6.2 Unitary groups

In the complex case, groups of invertible transformations preserving the Hermitian inner product are called unitary groups:

Definition (Unitary group). *The unitary group $U(n)$ in n dimensions is the group of invertible transformations preserving a Hermitian inner product on a complex n-dimensional vector space V. This is isomorphic to the group of n by n complex invertible matrices satisfying*

$$L^{-1} = \overline{L}^T = L^\dagger$$

The subgroup of $U(n)$ of matrices with determinant 1 is called $SU(n)$.

In the unitary case, the dual of a representation π has representation matrices that are transpose-inverses of those for π, but

$$(\pi(g)^T)^{-1} = \overline{\pi(g)}$$

so the dual representation is given by conjugating all elements of the matrix.

The same calculation as in the real case here gives

$$\det(L^{-1})\det(L) = \det(L^\dagger)\det(L) = \overline{\det(L)}\det(L) = |\det(L)|^2 = 1$$

so $\det(L)$ is a complex number of modulus one. The map

$$L \in U(n) \to \det(L) \in U(1)$$

is a group homomorphism.

We have already seen the examples $U(1)$, $U(2)$, and $SU(2)$. For general values of n, the study of $U(n)$ can be split into that of its determinant, which lies in $U(1)$ so is easy to deal with, followed by the subgroup $SU(n)$, which is a much more complicated story.

Digression. *Note that it is not quite true that the group $U(n)$ is the product group $SU(n) \times U(1)$. If one tries to identify the $U(1)$ as the subgroup of $U(n)$ of elements of the form $e^{i\theta}\mathbf{1}$, then matrices of the form*

$$e^{i\frac{m}{n}2\pi}\mathbf{1}$$

for m an integer will lie in both $SU(n)$ and $U(1)$, so $U(n)$ is not a product of those two groups (it is an example of a semi-direct product, these will be discussed in chapter 18).

We saw at the end of section 3.1.2 that $SU(2)$ can be identified with the three-sphere S^3, since an arbitrary group element can be constructed by specifying one row (or one column), which must be a vector of length one in $\mathbf{C}^2$. For the case $n = 3$, the same sort of construction starts by picking a row of length one in $\mathbf{C}^3$, which will be a point in S^5. The second row must be orthonormal, and it can be shown that the possibilities lie in a three-sphere S^3. Once the first two rows are specified, the third row is uniquely determined. So as a manifold, $SU(3)$ is eight-dimensional, and one might think it could be identified with $S^5 \times S^3$. It turns out that this is not the case, since the S^3 varies in a topologically non-trivial way as one varies the point in S^5. As spaces, the $SU(n)$ are topologically "twisted" products of odd-dimensional spheres, providing some of the basic examples of quite non-trivial topological manifolds.

4.7   Eigenvalues and eigenvectors

We have seen that the matrix for a linear transformation L of a vector space V changes by conjugation, when we change our choice of basis of V. To get basis-independent information about L, one considers the eigenvalues of the matrix. Complex matrices behave in a much simpler fashion than real matrices, since in the complex case the eigenvalue equation

$$\det(\lambda\mathbf{1} - L) = 0 \tag{4.5}$$

can always be factored into linear factors. For an arbitrary n by n complex matrix, there will be n solutions (counting repeated eigenvalues with multiplicity). A basis will exist for which the matrix will be in upper triangular form.

The case of self-adjoint matrices L is much more constrained, since transposition relates matrix elements. One has:

Theorem 4.1 (Spectral theorem for self-adjoint matrices). *Given a self-adjoint complex n by n matrix L, there exists a unitary matrix U such that*

$$ULU^{-1} = D$$

where D is a diagonal matrix with entries $D_{jj} = \lambda_j, \lambda_j \in \mathbf{R}$.

Given L, its eigenvalues λ_j are the solutions to the eigenvalue equation 4.5 and U is determined by the eigenvectors. For distinct eigenvalues, the corresponding eigenvectors are orthogonal.

This spectral theorem here is a theorem about finite dimensional vector spaces and matrices, but there are analogous theorems for self-adjoint operators on infinite dimensional state spaces. Such a theorem is of crucial importance in quantum mechanics, where for L an observable, the eigenvectors are the states in the state space with well-defined numerical values characterizing the state, and these numerical values are the eigenvalues. The theorem tells us that, given an observable, we can use it to choose distinguished orthonormal bases for the state space by picking a basis of eigenvectors, normalized to length one.

Using the bra-ket notation in this case, we can label elements of such a basis by their eigenvalues, so

$$|j\rangle = |\lambda_j\rangle$$

(the λ_j may include repeated eigenvalues). A general state is written as a linear combination of basis states

$$|\psi\rangle = \sum_j |j\rangle\langle j|\psi\rangle$$

which is sometimes written as a "resolution of the identity operator"

$$\sum_j |j\rangle\langle j| = \mathbf{1} \tag{4.6}$$

Turning from self-adjoint to unitary matrices, unitary matrices can also be diagonalized by conjugation by another unitary. The diagonal entries will all be complex numbers of unit length, so of the form $e^{i\lambda_j}, \lambda_j \in \mathbf{R}$. For the simplest examples, consider the cases of the groups $SU(2)$ and $U(2)$. Any matrix in $U(2)$ can be conjugated by a unitary matrix to the diagonal matrix

$$\begin{pmatrix} e^{i\lambda_1} & 0 \\ 0 & e^{i\lambda_2} \end{pmatrix}$$

which is the exponential of a corresponding diagonalized skew-adjoint matrix

$$\begin{pmatrix} i\lambda_1 & 0 \\ 0 & i\lambda_2 \end{pmatrix}$$

For matrices in the subgroup $SU(2)$, $\lambda_1 = -\lambda_2 = \lambda$, so in diagonal form an $SU(2)$ matrix will be

$$\begin{pmatrix} e^{i\lambda} & 0 \\ 0 & e^{-i\lambda} \end{pmatrix}$$

which is the exponential of a corresponding diagonalized skew-adjoint matrix that has trace zero

$$\begin{pmatrix} i\lambda & 0 \\ 0 & -i\lambda \end{pmatrix}$$

4.8   For further reading

Almost any of the more advanced linear algebra textbooks should cover the material of this chapter.

Chapter 5
Lie Algebras and Lie Algebra Representations

In this chapter, we will introduce Lie algebras and Lie algebra representations, which provide a tractable linear construction that captures much of the behavior of Lie groups and Lie group representations. We have so far seen the case of $U(1)$, for which the Lie algebra is trivial, and a little bit about the $SU(2)$ case, where the first non-trivial Lie algebra appears. Chapters 6 and 8 will provide details showing how the general theory works out for the basic examples of $SU(2)$, $SO(3)$, and their representations. The very general nature of the material in this chapter may make it hard to understand until one has some experience with examples that only appear in later chapters. The reader is thus advised that it may be a good idea to first skim the material of this chapter, returning for a deeper understanding and better insight into these structures after first seeing them in action later on in more concrete contexts.

For a group G, we have defined unitary representations (π, V) for finite dimensional vector spaces V of complex dimension n as homomorphisms

$$\pi : G \to U(n)$$

Recall that in the case of $G = U(1)$ (see the proof of theorem 2.3) we could use the homomorphism property of π to determine π in terms of its derivative at the identity. This turns out to be a general phenomenon for Lie groups G: We can study their representations by considering the derivative of π at the identity, which we will call π'. Because of the homomorphism property, knowing π' is often sufficient to characterize the representation π it comes from. π' is a linear map from the tangent space to G at the identity to the tangent space of $U(n)$ at the identity. The tangent space to G at the identity will carry some extra structure coming from the group multiplication, and this vector space with this structure will be called the Lie algebra of G. The linear map π' will be an example of a Lie algebra representation.

The subject of differential geometry gives many equivalent ways of defining the tangent space at a point of manifolds like G, but we do not want to enter

© Peter Woit 2017
P. Woit, *Quantum Theory, Groups and Representations*,
DOI 10.1007/978-3-319-64612-1_5

here into the subject of differential geometry in general. One of the standard definitions of the tangent space is as the space of tangent vectors, with tangent vectors defined as the possible velocity vectors of parametrized curves $g(t)$ in the group G.

More advanced treatments of Lie group theory develop this point of view (see, e.g., [98]) which applies to arbitrary Lie groups, whether or not they are groups of matrices. In our case though, since we are interested in specific groups that are usually explicitly given as groups of matrices, in such cases we can give a more concrete definition, using the exponential map on matrices. For a more detailed exposition of this subject, using the same concrete definition of the Lie algebra in terms of matrices, see, for instance, [42] or the abbreviated online version [40].

5.1 Lie algebras

If a Lie group G is defined as a differentiable manifold with a group law, one can consider the tangent space at the identity, and that will be the Lie algebra of G. We are, however, interested mainly in cases where G is a matrix group, and in such cases, the Lie algebra can be defined more concretely:

Definition (Lie algebra). *For G a Lie group of n by n invertible matrices, the Lie algebra of G (written $Lie(G)$ or $\mathfrak{g}$) is the space of n by n matrices X such that $e^{tX} \in G$ for $t \in \mathbf{R}$.*

Here, the exponential of a matrix is given by usual power series formula for the exponential

$$e^A = 1 + A + \frac{1}{2}A^2 + \cdots + \frac{1}{n!}A^n + \cdots$$

which can be shown to converge (like the usual exponential), for any matrix A. While this definition is more concrete than defining a Lie algebra as a tangent space, it does not make obvious some general properties of a Lie algebra, in particular that a Lie algebra is a real vector space (see theorem 3.20 of [42]). Our main interest will be in using it to recognize certain specific Lie algebras corresponding to specific Lie groups.

Notice that while the group G determines the Lie algebra $\mathfrak{g}$, the Lie algebra does not determine the group. For example, $O(n)$ and $SO(n)$ have the same tangent space at the identity and thus the same Lie algebra, but elements in $O(n)$ not in the component of the identity (i.e., with determinant -1) cannot be written in the form e^{tX} (since then you could make a path of matrices connecting such an element to the identity by shrinking t to zero).

Note also that, for a given X, different values of t may give the same group element, and this may happen in different ways for different groups sharing the same Lie algebra. For example, consider $G = U(1)$ and $G = \mathbf{R}$, which both have the same Lie algebra $\mathfrak{g} = \mathbf{R}$. In the first case, an infinity of values of t gives the same group element, and in the second, only one does. In chapter 6, we will see a more subtle example of this: $SU(2)$ and $SO(3)$ are different groups with the same Lie algebra.

We have $G \subset GL(n, \mathbf{C})$ and $X \in M(n, \mathbf{C})$, the space of n by n complex matrices. For all $t \in \mathbf{R}$, the exponential e^{tX} is an invertible matrix (with inverse e^{-tX}), so in $GL(n, \mathbf{C})$. For each X, we thus have a path of elements of $GL(n, \mathbf{C})$ going through the identity matrix at $t = 0$, with velocity vector

$$\frac{d}{dt} e^{tX} = X e^{tX}$$

which takes the value X at $t = 0$:

$$\frac{d}{dt} (e^{tX})_{|t=0} = X$$

To calculate this derivative, use the power series expansion for the exponential and differentiate term-by-term.

For the case $G = GL(n, \mathbf{C})$, we have $\mathfrak{gl}(n, \mathbf{C}) = M(n, \mathbf{C})$, which is a linear space of the right dimension to be the tangent space to G at the identity, so this definition is consistent with our general motivation. For subgroups $G \subset GL(n, \mathbf{C})$ given by some condition (e.g., that of preserving an inner product), we will need to identify the corresponding condition on $X \in M(n, \mathbf{C})$ and check that this defines a linear space.

The existence of such a linear space $\mathfrak{g} \subset M(n, \mathbf{C})$ will provide us with a distinguished representation on a real vector space, called the "adjoint representation":

Definition (Adjoint representation). *The adjoint representation* $(Ad, \mathfrak{g})$ *is given by the homomorphism*

$$Ad : g \in G \to Ad(g) \in GL(\mathfrak{g})$$

where $Ad(g)$ *acts on* $X \in \mathfrak{g}$ *by*

$$(Ad(g))(X) = gXg^{-1}$$

To show that this is well defined, one needs to check that $gXg^{-1} \in \mathfrak{g}$ when $X \in \mathfrak{g}$, but this can be shown using the identity

$$e^{tgXg^{-1}} = ge^{tX}g^{-1}$$

which implies that $e^{tgXg^{-1}} \in G$ if $e^{tX} \in G$. To check this identity, expand the exponential and use

$$(gXg^{-1})^k = (gXg^{-1})(gXg^{-1}) \cdots (gXg^{-1}) = gX^k g^{-1}$$

It is also easy to check that this is a homomorphism, with

$$Ad(g_1)Ad(g_2) = Ad(g_1 g_2)$$

A Lie algebra $\mathfrak{g}$ is not just a real vector space, but comes with an extra structure on the vector space:

Definition (Lie bracket). *The Lie bracket operation on $\mathfrak{g}$ is the bilinear antisymmetric map given by the commutator of matrices*

$$[\cdot,\cdot] : (X,Y) \in \mathfrak{g} \times \mathfrak{g} \to [X,Y] = XY - YX \in \mathfrak{g}$$

We need to check that this is well defined, i.e., that it takes values in $\mathfrak{g}$.

Theorem. *If $X, Y \in \mathfrak{g}$, $[X,Y] = XY - YX \in \mathfrak{g}$.*

Proof. Since $X \in \mathfrak{g}$, we have $e^{tX} \in G$ and we can act on $Y \in \mathfrak{g}$ by the adjoint representation

$$Ad(e^{tX})Y = e^{tX}Ye^{-tX} \in \mathfrak{g}$$

As t varies, this gives us a parametrized curve in $\mathfrak{g}$. Its velocity vector will also be in $\mathfrak{g}$, so

$$\frac{d}{dt}(e^{tX}Ye^{-tX}) \in \mathfrak{g}$$

One has (by the product rule, which can easily be shown to apply in this case)

$$\frac{d}{dt}(e^{tX}Ye^{-tX}) = \left(\frac{d}{dt}(e^{tX}Y)\right)e^{-tX} + e^{tX}Y\left(\frac{d}{dt}e^{-tX}\right)$$
$$= Xe^{tX}Ye^{-tX} - e^{tX}YXe^{-tX}$$

Evaluating this at $t = 0$ gives

$$XY - YX$$

which is thus, from the definition, shown to be in $\mathfrak{g}$.   □

The relation

$$\frac{d}{dt}(e^{tX}Ye^{-tX})_{|t=0} = [X,Y] \tag{5.1}$$

used in this proof will be continually useful in relating Lie groups and Lie algebras.

To do calculations with a Lie algebra, one can choose a basis $X_1, X_2, \ldots, X_n$ for the vector space $\mathfrak{g}$ and use the fact that the Lie bracket can be written in terms of this basis as

$$[X_j, X_k] = \sum_{l=1}^{n} c_{jkl} X_l \qquad (5.2)$$

where c_{jkl} is a set of constants known as the "structure constants" of the Lie algebra. For example, in the case of $\mathfrak{su}(2)$, the Lie algebra of $SU(2)$ has a basis X_1, X_2, X_3 satisfying

$$[X_j, X_k] = \sum_{l=1}^{3} \epsilon_{jkl} X_l$$

(see equation 3.5) so the structure constants of $\mathfrak{su}(2)$ are the totally antisymmetric ϵ_{jkl}.

5.2   Lie algebras of the orthogonal and unitary groups

The groups we are most interested in are the groups of linear transformations preserving an inner product: the orthogonal and unitary groups. We have seen that these are subgroups of $GL(n, \mathbf{R})$ or $GL(n, \mathbf{C})$, consisting of those elements Ω satisfying the condition

$$\Omega \Omega^\dagger = 1$$

In order to see what this condition becomes on the Lie algebra, write $\Omega = e^{tX}$, for some parameter t, and X a matrix in the Lie algebra. Since the transpose of a product of matrices is the product (order-reversed) of the transposed matrices, i.e.,

$$(XY)^T = Y^T X^T$$

and the complex conjugate of a product of matrices is the product of the complex conjugates of the matrices, one has

$$(e^{tX})^\dagger = e^{tX^\dagger}$$

The condition

$$\Omega \Omega^\dagger = 1$$

thus becomes

$$e^{tX}(e^{tX})^\dagger = e^{tX}e^{tX^\dagger} = 1$$

Taking the derivative of this equation gives

$$e^{tX}X^\dagger e^{tX^\dagger} + Xe^{tX}e^{tX^\dagger} = 0$$

Evaluating this at $t = 0$ gives

$$X + X^\dagger = 0$$

so the matrices we want to exponentiate must be skew-adjoint (it can be shown that this is also a sufficient condition), satisfying

$$X^\dagger = -X$$

Note that physicists often choose to define the Lie algebra in these cases as self-adjoint matrices, then multiplying by i before exponentiating to get a group element. We will not use this definition, with one reason that we want to think of the Lie algebra as a real vector space, so want to avoid an unnecessary introduction of complex numbers at this point.

5.2.1 Lie algebra of the orthogonal group

Recall that the orthogonal group $O(n)$ is the subgroup of $GL(n, \mathbf{R})$ of matrices Ω satisfying $\Omega^T = \Omega^{-1}$. We will restrict attention to the subgroup $SO(n)$ of matrices with determinant 1, which is the component of the group containing the identity, with elements that can be written as

$$\Omega = e^{tX}$$

These give a path connecting Ω to the identity (taking $e^{sX}, s \in [0, t]$). We saw above that the condition $\Omega^T = \Omega^{-1}$ corresponds to skew-symmetry of the matrix X

$$X^T = -X$$

So in the case of $G = SO(n)$, we see that the Lie algebra $\mathfrak{so}(n)$ is the space of skew-symmetric $(X^T = -X)$ n by n real matrices, together with the bilinear, antisymmetric product given by the commutator:

$$(X, Y) \in \mathfrak{so}(n) \times \mathfrak{so}(n) \to [X, Y] \in \mathfrak{so}(n)$$

The dimension of the space of such matrices will be

$$1 + 2 + \cdots + (n-1) = \frac{n^2 - n}{2}$$

and a basis will be given by the matrices ϵ_{jk}, with $j, k = 1, \ldots, n, j < k$ defined as

$$(\epsilon_{jk})_{lm} = \begin{cases} -1 & \text{if } j = l, k = m \\ +1 & \text{if } j = m, k = l \\ 0 & \text{otherwise} \end{cases} \tag{5.3}$$

In chapter 6, we will examine in detail the $n = 3$ case, where the Lie algebra $\mathfrak{so}(3)$ is $\mathbf{R}^3$, realized as the space of antisymmetric real 3 by 3 matrices, with a basis the three matrices $\epsilon_{12}, \epsilon_{13}, \epsilon_{23}$.

5.2.2 Lie algebra of the unitary group

For the case of the group $U(n)$, the unitarity condition implies that X is skew-adjoint (also called skew-Hermitian), satisfying

$$X^\dagger = -X$$

So the Lie algebra $\mathfrak{u}(n)$ is the space of skew-adjoint n by n complex matrices, together with the bilinear, antisymmetric product given by the commutator:

$$(X, Y) \in \mathfrak{u}(n) \times \mathfrak{u}(n) \to [X, Y] \in \mathfrak{u}(n)$$

Note that these matrices form a subspace of $\mathbf{C}^{n^2}$ of half the dimension, so of real dimension n^2. $\mathfrak{u}(n)$ is a real vector space of dimension n^2, but it is NOT a space of real n by n matrices. It is the space of skew-Hermitian matrices, which in general are complex. While the matrices are complex, only real linear combinations of skew-Hermitian matrices are skew-Hermitian (recall that multiplication by i changes a skew-Hermitian matrix into a Hermitian matrix). Within this space of skew-Hermitian complex matrices, if one looks at the subspace of real matrices one gets the sub-Lie algebra $\mathfrak{so}(n)$ of anti-symmetric matrices (the Lie algebra of $SO(n) \subset U(n)$).

Any complex matrix $Z \in M(n, \mathbf{C})$ can be written as a sum of

$$Z = \frac{1}{2}(Z + Z^\dagger) + \frac{1}{2}(Z - Z^\dagger)$$

where the first term is self-adjoint and the second skew-Hermitian. This second term can also be written as i times a self-adjoint matrix

$$\frac{1}{2}(Z - Z^\dagger) = i\left(\frac{1}{2i}(Z - Z^\dagger)\right)$$

so we see that we can get all of $M(n, \mathbf{C})$ by taking all complex linear combinations of self-adjoint matrices.

There is an identity relating the determinant and the trace of a matrix

$$\det(e^X) = e^{trace(X)}$$

which can be proved by conjugating the matrix to upper triangular form and using the fact that the trace and the determinant of a matrix are conjugation invariant. Since the determinant of an $SU(n)$ matrix is 1, this shows that the Lie algebra $\mathfrak{su}(n)$ of $SU(n)$ will consist of matrices that are not only skew-Hermitian, but also of trace zero. So in this case, $\mathfrak{su}(n)$ is again a real vector space, with the trace zero condition a single linear condition giving a vector space of real dimension $n^2 - 1$.

One can show that $U(n)$ and $\mathfrak{u}(n)$ matrices can be diagonalized by conjugation by a unitary matrix and thus show that any $U(n)$ matrix can be written as an exponential of something in the Lie algebra. The corresponding theorem is also true for $SO(n)$ but requires looking at diagonalization into 2 by 2 blocks. It is not true for $O(n)$ (you cannot reach the disconnected component of the identity by exponentiation). It also turns out to not be true for the groups $SL(n, \mathbf{R})$ and $SL(n, \mathbf{C})$ for $n \geq 2$ (while the groups are connected, they have elements that are not exponentials of any matrix in $\mathfrak{sl}(n, \mathbf{R})$ or $\mathfrak{sl}(2, \mathbf{C})$, respectively).

5.3 A summary

Before turning to Lie algebra representations, we will summarize here the classes of Lie groups and Lie algebras that we have discussed and that we will be studying specific examples of in later chapters:

- The general linear groups $GL(n, \mathbf{R})$ and $GL(n, \mathbf{C})$ are the groups of all invertible matrices, with real or complex entries, respectively. Their Lie algebras are $\mathfrak{gl}(n, \mathbf{R}) = M(n, \mathbf{R})$ and $\mathfrak{gl}(n, \mathbf{C}) = M(n, \mathbf{C})$. These are the vector spaces of all n by n matrices, with Lie bracket the matrix commutator.

  Other Lie groups will be subgroups of these, with Lie algebras sub-Lie algebras of these Lie algebras.

- The special linear groups $SL(n, \mathbf{R})$ and $SL(n, \mathbf{C})$ are the groups of invertible matrices with determinant one. Their Lie algebras $\mathfrak{sl}(n, \mathbf{R})$ and $\mathfrak{sl}(n, \mathbf{C})$ are the Lie algebras of all n by n matrices with zero trace.

- The orthogonal group $O(n) \subset GL(n, \mathbf{R})$ is the group of n by n real matrices Ω satisfying $\Omega^T = \Omega^{-1}$. Its Lie algebra $\mathfrak{o}(n)$ is the Lie algebra of n by n real matrices X satisfying $X^T = -X$.

- The special orthogonal group $SO(n) \subset SL(n, \mathbf{R})$ is the subgroup of $O(n)$ with determinant one. It has the same Lie algebra as $O(n)$: $\mathfrak{so}(n) = \mathfrak{o}(n)$.

- The unitary group $U(n) \subset GL(n, \mathbf{C})$ is the group of n by n complex matrices Ω satisfying $\Omega^\dagger = \Omega^{-1}$. Its Lie algebra $\mathfrak{u}(n)$ is the Lie algebra of n by n skew-Hermitian matrices X, those satisfying $X^\dagger = -X$.

- The special unitary group $SU(n) \subset SL(n, \mathbf{C})$ is the subgroup of $U(n)$ of matrices of determinant one. Its Lie algebra $\mathfrak{su}(n)$ is the Lie algebra of n by n skew-Hermitian matrices X with trace zero.

In later chapters, we will encounter some other examples of matrix Lie groups, including the symplectic group $Sp(2d, \mathbf{R})$ (see chapter 16) and the pseudo-orthogonal groups $O(r, s)$ (see chapter 29).

5.4 Lie algebra representations

We have defined a group representation as a homomorphism (a map of groups preserving group multiplication)

$$\pi : G \to GL(n, \mathbf{C})$$

We can similarly define a Lie algebra representation as a map of Lie algebras preserving the Lie bracket:

Definition (Lie algebra representation). *A (complex) Lie algebra representation (ϕ, V) of a Lie algebra $\mathfrak{g}$ on an n-dimensional complex vector space V is given by a real linear map*

$$\phi : X \in \mathfrak{g} \to \phi(X) \in \mathfrak{gl}(n, \mathbf{C}) = M(n, \mathbf{C})$$

satisfying
$$\phi([X, Y]) = [\phi(X), \phi(Y)]$$

Such a representation is called unitary if its image is in $\mathfrak{u}(n)$, i.e., if it satisfies

$$\phi(X)^\dagger = -\phi(X)$$

More concretely, given a basis $X_1, X_2, \ldots, X_d$ of a Lie algebra $\mathfrak{g}$ of dimension d with structure constants c_{jkl}, a representation is given by a choice of d complex n-dimensional matrices $\phi(X_j)$ satisfying the commutation relations

$$[\phi(X_j), \phi(X_k)] = \sum_{l=1}^{d} c_{jkl}\phi(X_l)$$

The representation is unitary when the matrices are skew-adjoint.

The notion of a Lie algebra representation is motivated by the fact that the homomorphism property causes the map π to be largely determined by its behavior infinitesimally near the identity and thus by the derivative π'. One way to define the derivative of such a map is in terms of velocity vectors of paths, and this sort of definition in this case associates to a representation $\pi : G \to GL(n, \mathbf{C})$ a linear map

$$\pi' : \mathfrak{g} \to M(n, \mathbf{C})$$

where

$$\pi'(X) = \frac{d}{dt}(\pi(e^{tX}))_{|t=0}$$

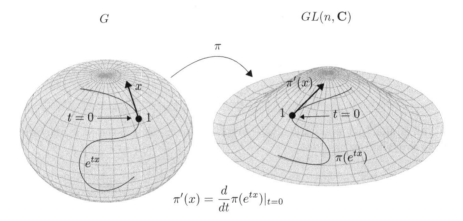

$$\pi'(x) = \frac{d}{dt}\pi(e^{tx})|_{t=0}$$

Figure 5.1: Derivative of a representation $\pi : G \to GL(n, \mathbf{C})$, illustrated in terms of "velocity" vectors along paths.

For the case of $U(1)$, we classified in theorem 2.3 all irreducible representations (homomorphisms $U(1) \to GL(1, \mathbf{C}) = \mathbf{C}^*$) by looking at the derivative of the map at the identity. For general Lie groups G, something similar can be done, showing that a representation π of G gives a representation of the Lie algebra (by taking the derivative at the identity) and then trying to classify Lie algebra representations.

Theorem. *If* $\pi : G \to GL(n, \mathbf{C})$ *is a group homomorphism, then*

$$\pi' : X \in \mathfrak{g} \to \pi'(X) = \frac{d}{dt}(\pi(e^{tX}))_{|t=0} \in \mathfrak{gl}(n, \mathbf{C}) = M(n, \mathbf{C})$$

satisfies

1.
$$\pi(e^{tX}) = e^{t\pi'(X)}$$

2. *For* $g \in G$
$$\pi'(gXg^{-1}) = \pi(g)\pi'(X)(\pi(g))^{-1}$$

3. π' *is a Lie algebra homomorphism:*
$$\pi'([X,Y]) = [\pi'(X), \pi'(Y)]$$

Proof. 1. We have

$$\begin{aligned}
\frac{d}{dt}\pi(e^{tX}) &= \frac{d}{ds}\pi(e^{(t+s)X})_{|s=0}\\
&= \frac{d}{ds}\pi(e^{tX}e^{sX})_{|s=0}\\
&= \pi(e^{tX})\frac{d}{ds}\pi(e^{sX})_{|s=0}\\
&= \pi(e^{tX})\pi'(X)
\end{aligned}$$

So $f(t) = \pi(e^{tX})$ satisfies the differential equation $\frac{d}{dt}f = f\pi'(X)$ with initial condition $f(0) = 1$. This has the unique solution $f(t) = e^{t\pi'(X)}$

2. We have

$$\begin{aligned}
e^{t\pi'(gXg^{-1})} &= \pi(e^{tgXg^{-1}})\\
&= \pi(ge^{tX}g^{-1})\\
&= \pi(g)\pi(e^{tX})\pi(g)^{-1}\\
&= \pi(g)e^{t\pi'(X)}\pi(g)^{-1}
\end{aligned}$$

Differentiating with respect to t at $t = 0$ gives

$$\pi'(gXg^{-1}) = \pi(g)\pi'(X)(\pi(g))^{-1}$$

3. Recall 5.1:
$$[X,Y] = \frac{d}{dt}(e^{tX}Ye^{-tX})_{|t=0}$$

so

$$\pi'([X,Y]) = \pi'\left(\frac{d}{dt}(e^{tX}Ye^{-tX})_{|t=0}\right)$$

$$= \frac{d}{dt}\pi'(e^{tX}Ye^{-tX})_{|t=0} \quad \text{(by linearity)}$$

$$= \frac{d}{dt}(\pi(e^{tX})\pi'(Y)\pi(e^{-tX}))_{|t=0} \quad \text{(by 2.)}$$

$$= \frac{d}{dt}(e^{t\pi'(X)}\pi'(Y)e^{-t\pi'(X)})_{|t=0} \quad \text{(by 1.)}$$

$$= [\pi'(X), \pi'(Y)]$$

□

This theorem shows that we can study Lie group representations (π, V) by studying the corresponding Lie algebra representation (π', V). This will generally be much easier since the π' are linear maps. Unlike the nonlinear maps π, the map π' is determined by its value on basis elements X_j of $\mathfrak{g}$. The $\pi'(X_j)$ will satisfy the same bracket relations as the X_j (see equation 5.2). We will proceed in this manner in chapter 8 when we construct and classify all $SU(2)$ and $SO(3)$ representations, finding that the corresponding Lie algebra representations are much simpler to analyze. Note though that representations of the Lie algebra $\mathfrak{g}$ do not necessarily correspond to representations of the group G (when they do they are called "integrable"). For a simple example, looking at the proof of theorem 2.3, one gets unitary representations of the Lie algebra of $U(1)$ for any value of the constant k, but these are only representations of the group $U(1)$ when k is integral.

For any Lie group G, we have seen that there is a distinguished representation, the adjoint representation $(Ad, \mathfrak{g})$. The corresponding Lie algebra representation is also called the adjoint representation, but written as $(Ad', \mathfrak{g}) = (ad, \mathfrak{g})$. From the fact that

$$Ad(e^{tX})(Y) = e^{tX}Ye^{-tX}$$

we can differentiate with respect to t and use equation 5.1 to get the Lie algebra representation

$$ad(X)(Y) = \frac{d}{dt}(e^{tX}Ye^{-tX})_{|t=0} = [X,Y] \tag{5.4}$$

This leads to the definition:

Definition (Adjoint Lie algebra representation). $(ad, \mathfrak{g})$ *is the Lie algebra representation given by*

$$X \in \mathfrak{g} \to ad(X)$$

where ad(X) is defined as the linear map from $\mathfrak{g}$ to itself given by

$$Y \to [X, Y]$$

Note that this linear map $ad(X)$, which can be written as $[X, \cdot]$, can be thought of as the infinitesimal version of the conjugation action

$$(\cdot) \to e^{tX}(\cdot)e^{-tX}$$

The Lie algebra homomorphism property of ad says that

$$ad([X, Y]) = ad(X) \circ ad(Y) - ad(Y) \circ ad(X)$$

where these are linear maps on $\mathfrak{g}$, with $\circ$ composition of linear maps, so operating on $Z \in \mathfrak{g}$ we have

$$ad([X, Y])(Z) = (ad(X) \circ ad(Y))(Z) - (ad(Y) \circ ad(X))(Z)$$

Using our expression for ad as a commutator, we find

$$[[X, Y], Z] = [X, [Y, Z]] - [Y, [X, Z]]$$

This is called the Jacobi identity. It could have been more simply derived as an identity about matrix multiplication, but here we see that it is true for a more abstract reason, reflecting the existence of the adjoint representation. It can be written in other forms, rearranging terms using antisymmetry of the commutator, with one example the sum of cyclic permutations

$$[[X, Y], Z] + [[Z, X], Y] + [[Y, Z], X] = 0$$

Lie algebras can be defined much more abstractly as follows:

Definition (Abstract Lie algebra). *An abstract Lie algebra over a field k is a vector space A over k, with a bilinear operation*

$$[\cdot, \cdot] : (X, Y) \in A \times A \to [X, Y] \in A$$

satisfying

1. *Antisymmetry:*
$$[X, Y] = -[Y, X]$$

2. *Jacobi identity:*

$$[[X, Y], Z] + [[Z, X], Y] + [[Y, Z], X] = 0$$

Such Lie algebras do not need to be defined as matrices, and their Lie bracket operation does not need to be defined in terms of a matrix commutator (although the same notation continues to be used). Later on, we will encounter important examples of Lie algebras that are defined in this more abstract way.

5.5 Complexification

Conventional physics discussion of Lie algebra representations proceed by assuming complex coefficients are allowed in all calculations, since we are interested in complex representations. An important subtlety is that the Lie algebra is a real vector space, often in a confusing way, as a subspace of complex matrices. To properly keep track of what is going on, one needs to understand the notion of "complexification" of a vector space or Lie algebra. In some cases, this is easily understood as just going from real to complex coefficients, but in other cases, a more complicated construction is necessary. The reader is advised that it might be a good idea to just skim this section at first reading, coming back to it later only as needed to make sense of exactly how things work when these subtleties make an appearance in a concrete problem.

The way we have defined a Lie algebra $\mathfrak{g}$, it is a real vector space, not a complex vector space. Even if G is a group of complex matrices, its tangent space at the identity will not necessarily be a complex vector space. Consider, for example, the cases $G = U(1)$ and $G = SU(2)$, where $\mathfrak{u}(1) = \mathbf{R}$ and $\mathfrak{su}(2) = \mathbf{R}^3$. While the tangent space to the group $GL(n, \mathbf{C})$ of all invertible complex matrices is a complex vector space ($M(n, \mathbf{C})$, all n by n matrices), imposing some condition such as unitarity picks out a subspace of $M(n, \mathbf{C})$ which generally is just a real vector space, not a complex one. So the adjoint representation $(Ad, \mathfrak{g})$ is in general not a complex representation, but a real representation, with

$$Ad(g) \in GL(\mathfrak{g}) = GL(dim \ \mathfrak{g}, \mathbf{R})$$

The derivative of this is the Lie algebra representation

$$ad : X \in \mathfrak{g} \to ad(X) \in \mathfrak{gl}(dim \ \mathfrak{g}, \mathbf{R})$$

and once we pick a basis of $\mathfrak{g}$, we can identify $\mathfrak{gl}(dim \ \mathfrak{g}, \mathbf{R}) = M(dim \ \mathfrak{g}, \mathbf{R})$. So, for each $X \in \mathfrak{g}$, we get a real linear operator on a real vector space.

We most often would like to work with not real representations, but complex representations, since it is for these that Schur's lemma applies (the proof of 2.1 also applies to the Lie algebra case), and representation operators can be diagonalized. To get from a real Lie algebra representation to a complex one, we can "complexify," extending the action of real scalars to complex scalars. If we are working with real matrices, complexification is nothing but allowing complex entries and using the same rules for multiplying matrices as before.

More generally, for any real vector space, we can define:

Definition. *The complexification $V_{\mathbf{C}}$ of a real vector space V is the space of pairs (v_1, v_2) of elements of V with multiplication by $a + bi \in \mathbf{C}$ given by*

$$(a + ib)(v_1, v_2) = (av_1 - bv_2, av_2 + bv_1)$$

One should think of the complexification of V as

$$V_{\mathbf{C}} = V + iV$$

with v_1 in the first copy of V and v_2 in the second copy. Then, the rule for multiplication by a complex number comes from the standard rules for complex multiplication.

Given a real Lie algebra $\mathfrak{g}$, the complexification $\mathfrak{g}_{\mathbf{C}}$ is pairs of elements (X, Y) of $\mathfrak{g}$, with the above rule for multiplication by complex scalars, which can be thought of as

$$\mathfrak{g}_{\mathbf{C}} = \mathfrak{g} + i\mathfrak{g}$$

The Lie bracket on $\mathfrak{g}$ extends to a Lie bracket on $\mathfrak{g}_{\mathbf{C}}$ by the rule

$$[(X_1, Y_1), (X_2, Y_2)] = ([X_1, X_2] - [Y_1, Y_2], [X_1, Y_2] + [Y_1, X_2])$$

which can be understood by the calculation

$$[X_1 + iY_1, X_2 + iY_2] = [X_1, X_2] - [Y_1, Y_2] + i([X_1, Y_2] + [Y_1, X_2])$$

With this Lie bracket, $\mathfrak{g}_{\mathbf{C}}$ is a Lie algebra over the complex numbers.

For many of the cases we will be interested in, this level of abstraction is not really needed, since they have the property that V will be given as a subspace of a complex vector space, with the property that $V \cap iV = 0$, in which case $V_{\mathbf{C}}$ will just be the larger subspace you get by taking complex linear combinations of elements of V. For example, $\mathfrak{gl}(n, \mathbf{R})$, the Lie algebra of real n by n matrices, is a subspace of $\mathfrak{gl}(n, \mathbf{C})$, the complex matrices, and one can see that

$$\mathfrak{gl}(n, \mathbf{R})_{\mathbf{C}} = \mathfrak{gl}(n, \mathbf{C})$$

Recalling our discussion from section 5.2.2 of $\mathfrak{u}(n)$, a real Lie algebra, with elements certain complex matrices (the skew-Hermitian ones), multiplication by i gives the Hermitian ones, and complexifying will give all complex matrices so

$$\mathfrak{u}(n)_{\mathbf{C}} = \mathfrak{gl}(n, \mathbf{C})$$

This example shows that two different real Lie algebras ($\mathfrak{u}(n)$ and $\mathfrak{gl}(n, \mathbf{R})$) may have the same complexification. For yet another example, $\mathfrak{so}(n)$ is the Lie algebra of all real antisymmetric matrices, $\mathfrak{so}(n)_{\mathbf{C}}$ is the Lie algebra of all complex antisymmetric matrices.

For an example where the general definition is needed and the situation becomes easily confusing, consider the case of $\mathfrak{gl}(n, \mathbf{C})$, thinking of it as a Lie algebra and thus a real vector space. The complexification of this real vector space will have twice the (real) dimension, so

$$\mathfrak{gl}(n, \mathbf{C})_{\mathbf{C}} = \mathfrak{gl}(n, \mathbf{C}) + i\mathfrak{gl}(n, \mathbf{C})$$

will not be what you get by just allowing complex coefficients ($\mathfrak{gl}(n, \mathbf{C})$), but something built out of two copies of this.

Given a representation π' of a real Lie algebra $\mathfrak{g}$, it can be extended to a representation of $\mathfrak{g}_{\mathbf{C}}$ by complex linearity, defining

$$\pi'(X + iY) = \pi'(X) + i\pi'(Y)$$

If the original representation was on a complex vector space V, the extended one will act on the same space. If the original representation was on a real vector space V, the extended one will act on the complexification $V_{\mathbf{C}}$. Some of the examples of these phenomena that we will encounter are the following:

- The adjoint representation

$$ad : \mathfrak{g} \to \mathfrak{gl}(dim\ \mathfrak{g}, \mathbf{R}) = M(dim\ \mathfrak{g}, \mathbf{R})$$

  extends to a complex representation

$$ad : \mathfrak{g}_{\mathbf{C}} \to \mathfrak{gl}(dim\ \mathfrak{g}, \mathbf{C}) = M(dim\ \mathfrak{g}, \mathbf{C})$$

- Complex n-dimensional representations

$$\pi' : \mathfrak{su}(2) \to M(n, \mathbf{C})$$

  of $\mathfrak{su}(2)$ extend to representations

$$\pi' : \mathfrak{su}(2)_{\mathbf{C}} = \mathfrak{sl}(2, \mathbf{C}) \to M(n, \mathbf{C})$$

Doing this allows one to classify the finite dimensional irreducible representations of $\mathfrak{su}(2)$ by studying $\mathfrak{sl}(2, \mathbf{C})$ representations (see section 8.1.2).

- We will see that complex representations of a real Lie algebra called the Heisenberg Lie algebra play a central role in quantum theory and in quantum field theory. An important technique for constructing such representations (using so-called "annihilation" and "creation" operators) does so by extending the representation to the complexification of the Heisenberg Lie algebra (see section 22.4).

- Quantum field theories based on complex fields start with a Heisenberg Lie algebra that is already complex (see chapter 37 for the case of non-relativistic fields and section 44.1.2 for relativistic fields). The use of annihilation and creation operators for such theories thus involves complexifying a Lie algebra that is already complex, requiring the use of the general notion of complexification discussed in this section.

5.6 For further reading

The material of this section is quite conventional mathematics, with many good expositions, although most aimed at a higher level than ours. Examples at a similar level to this one are [85] and [93], which cover basics of Lie groups and Lie algebras, but without representations. The notes [40] and book [42] of Brian Hall are a good source for the subject at a somewhat more sophisticated level than adopted here. Some parts of the proofs given in this chapter are drawn from those two sources.

Chapter 6
The Rotation and Spin Groups in Three and Four Dimensions

Among the basic symmetry groups of the physical world is the orthogonal group $SO(3)$ of rotations about a point in three-dimensional space. The observables one gets from this group are the components of angular momentum, and understanding how the state space of a quantum system behaves as a representation of this group is a crucial part of the analysis of atomic physics examples and many others. This is a topic one will find in some version or other in every quantum mechanics textbook, and in chapter 8, we will discuss it in detail.

Remarkably, it is an experimental fact that the quantum systems in nature are often representations not of $SO(3)$, but of a larger group called $Spin(3)$, one that has two elements corresponding to every element of $SO(3)$. Such a group exists in any dimension n, always as a "doubled" version of the orthogonal group $SO(n)$, one that is needed to understand some of the more subtle aspects of geometry in n dimensions. In the $n = 3$ case, it turns out that $Spin(3) \simeq SU(2)$, and in this chapter, we will study in detail the relationship between $SO(3)$ and $SU(2)$. This appearance of the unitary group $SU(2)$ is special to geometry in 3 and 4 dimensions, and the theory of quaternions will be used to provide an explanation for this.

6.1 The rotation group in three dimensions

Rotations in $\mathbf{R}^2$ about the origin are given by elements of $SO(2)$, with a counterclockwise rotation by an angle θ given by the matrix

$$R(\theta) = \begin{pmatrix} \cos\theta & -\sin\theta \\ \sin\theta & \cos\theta \end{pmatrix}$$

© Peter Woit 2017
P. Woit, *Quantum Theory, Groups and Representations*,
DOI 10.1007/978-3-319-64612-1_6

This can be written as an exponential, $R(\theta) = e^{\theta L} = \cos\theta\, \mathbf{1} + L\sin\theta$ for

$$L = \begin{pmatrix} 0 & -1 \\ 1 & 0 \end{pmatrix}$$

Here, $SO(2)$ is a commutative Lie group with Lie algebra $\mathfrak{so}(2) = \mathbf{R}$. Note that we have a representation on $V = \mathbf{R}^2$ here, but it is a real representation, not one of the complex ones we have when we have a representation on a quantum mechanical state space.

In three dimensions, the group $SO(3)$ is three dimensional and non-commutative. Choosing a unit vector $\mathbf{w}$ and angle θ, one gets an element $R(\theta, \mathbf{w})$ of $SO(3)$, rotation by an angle θ about the $\mathbf{w}$-axis. Using standard basis vectors $\mathbf{e}_j$, rotations about the coordinate axes are given by

$$R(\theta, \mathbf{e}_1) = \begin{pmatrix} 1 & 0 & 0 \\ 0 & \cos\theta & -\sin\theta \\ 0 & \sin\theta & \cos\theta \end{pmatrix}, \quad R(\theta, \mathbf{e}_2) = \begin{pmatrix} \cos\theta & 0 & \sin\theta \\ 0 & 1 & 0 \\ -\sin\theta & 0 & \cos\theta \end{pmatrix}$$

$$R(\theta, \mathbf{e}_3) = \begin{pmatrix} \cos\theta & -\sin\theta & 0 \\ \sin\theta & \cos\theta & 0 \\ 0 & 0 & 1 \end{pmatrix}$$

A standard parametrization for elements of $SO(3)$ is in terms of 3 "Euler angles" ϕ, θ, ψ with a general rotation given by

$$R(\phi, \theta, \psi) = R(\psi, \mathbf{e}_3) R(\theta, \mathbf{e}_1) R(\phi, \mathbf{e}_3) \tag{6.1}$$

i.e., first a rotation about the z-axis by an angle ϕ, then a rotation by an angle θ about the new x-axis, followed by a rotation by ψ about the new z-axis. Multiplying out the matrices gives a rather complicated expression for a rotation in terms of the three angles, and one needs to figure out what range to choose for the angles to avoid multiple counting.

The infinitesimal picture near the identity of the group, given by the Lie algebra structure on $\mathfrak{so}(3)$, is much easier to understand. Recall that for orthogonal groups, the Lie algebra can be identified with the space of anti-symmetric matrices, so in this case there is a basis

$$l_1 = \begin{pmatrix} 0 & 0 & 0 \\ 0 & 0 & -1 \\ 0 & 1 & 0 \end{pmatrix} \quad l_2 = \begin{pmatrix} 0 & 0 & 1 \\ 0 & 0 & 0 \\ -1 & 0 & 0 \end{pmatrix} \quad l_3 = \begin{pmatrix} 0 & -1 & 0 \\ 1 & 0 & 0 \\ 0 & 0 & 0 \end{pmatrix}$$

which satisfy the commutation relations

$$[l_1, l_2] = l_3, \quad [l_2, l_3] = l_1, \quad [l_3, l_1] = l_2$$

Note that these are exactly the same commutation relations (equation 3.5) satisfied by the basis vectors X_1, X_2, X_3 of the Lie algebra $\mathfrak{su}(2)$, so $\mathfrak{so}(3)$ and $\mathfrak{su}(2)$ are isomorphic Lie algebras. They both are the vector space $\mathbf{R}^3$ with the same Lie bracket operation on pairs of vectors. This operation is familiar in yet another context, that of the cross product of standard basis vectors $\mathbf{e}_j$ in $\mathbf{R}^3$:

$$\mathbf{e}_1 \times \mathbf{e}_2 = \mathbf{e}_3, \ \mathbf{e}_2 \times \mathbf{e}_3 = \mathbf{e}_1, \ \mathbf{e}_3 \times \mathbf{e}_1 = \mathbf{e}_2$$

We see that the Lie bracket operation

$$(X, Y) \in \mathbf{R}^3 \times \mathbf{R}^3 \to [X, Y] \in \mathbf{R}^3$$

that makes $\mathbf{R}^3$ a Lie algebra $\mathfrak{so}(3)$ is the cross product on vectors in $\mathbf{R}^3$.

So far, we have three different isomorphic ways of putting a Lie bracket on $\mathbf{R}^3$, making it into a Lie algebra:

1. Identify $\mathbf{R}^3$ with antisymmetric real 3 by 3 matrices and take the matrix commutator as Lie bracket.

2. Identify $\mathbf{R}^3$ with skew-adjoint, traceless, complex 2 by 2 matrices and take the matrix commutator as Lie bracket.

3. Use the vector cross product on $\mathbf{R}^3$ to get a Lie bracket, i.e., define

$$[\mathbf{v}, \mathbf{w}] = \mathbf{v} \times \mathbf{w}$$

Something very special that happens for orthogonal groups only in dimension $n = 3$ is that the vector representation (the defining representation of $SO(n)$ matrices on $\mathbf{R}^n$) is isomorphic to the adjoint representation. Recall that any Lie group G has a representation $(Ad, \mathfrak{g})$ on its Lie algebra $\mathfrak{g}$. $\mathfrak{so}(n)$ can be identified with the antisymmetric n by n matrices, so is of (real) dimension $\frac{n^2-n}{2}$. Only for $n = 3$ is this equal to n, the dimension of the representation on vectors in $\mathbf{R}^n$. This corresponds to the geometrical fact that only in 3 dimensions is a plane (in all dimensions rotations are built out of rotations in various planes) determined uniquely by a vector (the vector perpendicular to the plane). Equivalently, only in 3 dimensions is there a cross product $\mathbf{v} \times \mathbf{w}$ which takes two vectors determining a plane to a unique vector perpendicular to the plane.

The isomorphism between the vector representation $(\pi_{vector}, \mathbf{R}^3)$ on column vectors and the adjoint representation $(Ad, \mathfrak{so}(3))$ on antisymmetric matrices is given by

$$\begin{pmatrix} v_1 \\ v_2 \\ v_3 \end{pmatrix} \leftrightarrow v_1 l_1 + v_2 l_2 + v_3 l_3 = \begin{pmatrix} 0 & -v_3 & v_2 \\ v_3 & 0 & -v_1 \\ -v_2 & v_1 & 0 \end{pmatrix}$$

or in terms of bases by

$$\mathbf{e}_j \leftrightarrow l_j$$

For the vector representation on column vectors, $\pi_{vector}(g) = g$ and $\pi'_{vector}(X) = X$, where X is an antisymmetric 3 by 3 matrix, and $g = e^X$ is an orthogonal 3 by 3 matrix. Both act on column vectors by the usual multiplication.

For the adjoint representation on antisymmetric matrices

$$Ad(g) \begin{pmatrix} 0 & -v_3 & v_2 \\ v_3 & 0 & -v_1 \\ -v_2 & v_1 & 0 \end{pmatrix} = g \begin{pmatrix} 0 & -v_3 & v_2 \\ v_3 & 0 & -v_1 \\ -v_2 & v_1 & 0 \end{pmatrix} g^{-1}$$

The corresponding Lie algebra representation is given by

$$ad(X) \begin{pmatrix} 0 & -v_3 & v_2 \\ v_3 & 0 & -v_1 \\ -v_2 & v_1 & 0 \end{pmatrix} = [X, \begin{pmatrix} 0 & -v_3 & v_2 \\ v_3 & 0 & -v_1 \\ -v_2 & v_1 & 0 \end{pmatrix}]$$

where X is a 3 by 3 antisymmetric matrix.

One can explicitly check that these representations are isomorphic, for instance by calculating how basis elements $l_j \in \mathfrak{so}(3)$ act. On vectors, these l_j act by matrix multiplication, giving for instance, for $j = 1$

$$l_1 \mathbf{e}_1 = 0, \ l_1 \mathbf{e}_2 = \mathbf{e}_3, \ l_1 \mathbf{e}_3 = -\mathbf{e}_2$$

On antisymmetric matrices, one has instead the isomorphic relations

$$(ad(l_1))(l_1) = 0, \ (ad(l_1))(l_2) = l_3, \ (ad(l_1))(l_3) = -l_2$$

6.2   Spin groups in three and four dimensions

A subtle and remarkable property of the orthogonal groups $SO(n)$ is that they come with an associated group, called $Spin(n)$, with every element of $SO(n)$ corresponding to two distinct elements of $Spin(n)$. There is a surjective group homomorphism

$$\Phi : Spin(n) \to SO(n)$$

with the inverse image of each element of $SO(n)$ given by two distinct elements of $Spin(n)$.

Digression. *The topological reason for this is that (for $n > 2$), the funda-mental group of $SO(n)$ is non-trivial, with $\pi_1(SO(n)) = \mathbf{Z}_2$ (in particular, there is a non-contractible loop in $SO(n)$, contractible if you go around it twice). $Spin(n)$ is topologically the simply connected double cover of $SO(n)$, and the covering map $\Phi : Spin(n) \to SO(n)$ can be chosen to be a group homomorphism.*

$Spin(n)$ is a Lie group of the same dimension as $SO(n)$, with an isomor-phic tangent space at the identity, so the Lie algebras of the two groups are isomorphic: $\mathfrak{so}(n) \simeq \mathfrak{spin}(n)$.

In chapter 29, we will explicitly construct the groups $Spin(n)$ for any n, but here we will only do this for $n = 3$ and $n = 4$, using methods specific to these two cases. In the cases $n = 5$ (where $Spin(5) = Sp(2)$, the 2 by 2 norm-preserving quaternionic matrices) and $n = 6$ (where $Spin(6) = SU(4)$), special methods can be used to identify $Spin(n)$ with other matrix groups. For $n > 6$, the group $Spin(n)$ will be a matrix group, but distinct from other classes of such groups.

Given such a construction of $Spin(n)$, we also need to explicitly construct the homomorphism Φ, and show that its derivative Φ' is an isomorphism of Lie algebras. We will see that the simplest construction of the spin groups here uses the group $Sp(1)$ of unit-length quaternions, with $Spin(3) = Sp(1)$ and $Spin(4) = Sp(1) \times Sp(1)$. By identifying quaternions and pairs of complex numbers, we can show that $Sp(1) = SU(2)$ and thus work with these spin groups as either 2 by 2 complex matrices (for $Spin(3)$), or pairs of such matrices (for $Spin(4)$).

6.2.1 Quaternions

The quaternions are a number system (denoted by $\mathbf{H}$) generalizing the com-plex number system, with elements $q \in \mathbf{H}$ that can be written as

$$q = q_0 + q_1\mathbf{i} + q_2\mathbf{j} + q_3\mathbf{k}, \quad q_j \in \mathbf{R}$$

with $\mathbf{i}, \mathbf{j}, \mathbf{k} \in \mathbf{H}$ satisfying

$$\mathbf{i}^2 = \mathbf{j}^2 = \mathbf{k}^2 = -1, \mathbf{ij} = -\mathbf{ji} = \mathbf{k}, \mathbf{ki} = -\mathbf{ik} = \mathbf{j}, \mathbf{jk} = -\mathbf{kj} = \mathbf{i}$$

and a conjugation operation that takes

$$q \to \bar{q} = q_0 - q_1\mathbf{i} - q_2\mathbf{j} - q_3\mathbf{k}$$

This operation satisfies (for $u, v \in \mathbf{H}$)

$$\overline{uv} = \bar{v}\bar{u}$$

As a vector space over $\mathbf{R}$, $\mathbf{H}$ is isomorphic with $\mathbf{R}^4$. The length-squared function on this $\mathbf{R}^4$ can be written in terms of quaternions as

$$|q|^2 = q\bar{q} = q_0^2 + q_1^2 + q_2^2 + q_3^2$$

and is multiplicative since

$$|uv|^2 = uv\overline{uv} = uv\bar{v}\bar{u} = |u|^2|v|^2$$

Using

$$\frac{q\bar{q}}{|q|^2} = 1$$

one has a formula for the inverse of a quaternion

$$q^{-1} = \frac{\bar{q}}{|q|^2}$$

The length one quaternions thus form a group under multiplication, called $Sp(1)$. There are also Lie groups called $Sp(n)$ for larger values of n, consisting of invertible matrices with quaternionic entries that act on quaternionic vectors preserving the quaternionic length-squared, but these play no significant role in quantum mechanics so we will not study them further. $Sp(1)$ can be identified with the three-dimensional sphere since the length one condition on q is

$$q_0^2 + q_1^2 + q_2^2 + q_3^2 = 1$$

the equation of the unit sphere $S^3 \subset \mathbf{R}^4$.

6.2.2 Rotations and spin groups in four dimensions

Pairs (u, v) of unit quaternions give the product group $Sp(1) \times Sp(1)$. An element (u, v) of this group acts on $q \in \mathbf{H} = \mathbf{R}^4$ by left and right quaternionic multiplication

$$q \to uqv^{-1}$$

This action preserves lengths of vectors and is linear in q, so it must correspond to an element of the group $SO(4)$. One can easily see that pairs (u, v) and $(-u, -v)$ give the same linear transformation of $\mathbf{R}^4$, so the same element of $SO(4)$ and show that $SO(4)$ is the group $Sp(1) \times Sp(1)$, with the two elements (u, v) and $(-u, -v)$ identified. The name $Spin(4)$ is given to the Lie group $Sp(1) \times Sp(1)$ that "double covers" $SO(4)$ in this manner, with the covering map

$$\Phi : (u, v) \in Sp(1) \times Sp(1) = Spin(4) \to \{q \to uqv^{-1}\} \in SO(4)$$

6.2.3   Rotations and spin groups in three dimensions

Later on we will encounter $Spin(4)$ and $SO(4)$ again, but for now we are interested in the subgroup $Spin(3)$ that only acts non-trivially on 3 of the dimensions, and double covers not $SO(4)$ but $SO(3)$. To find this, consider the subgroup of $Spin(4)$ consisting of pairs (u, v) of the form (u, u) (a subgroup isomorphic to $Sp(1)$, since elements correspond to a single unit-length quaternion u). This subgroup acts on quaternions by conjugation

$$q \to uqu^{-1}$$

an action which is trivial on the real quaternions (since $u(q_0 \mathbf{1})u^{-1} = q_0 \mathbf{1}$). It preserves and acts non-trivially on the space of "pure imaginary" quaternions of the form

$$q = \vec{v} = v_1 \mathbf{i} + v_2 \mathbf{j} + v_3 \mathbf{k}$$

which can be identified with the vector space $\mathbf{R}^3$. An element $u \in Sp(1)$ acts on $\vec{v} \in \mathbf{R}^3 \subset \mathbf{H}$ as

$$\vec{v} \to u\vec{v}u^{-1}$$

This is a linear action, preserving the length $|\vec{v}|$, so it corresponds to an element of $SO(3)$. We thus have a map (which can easily be checked to be a homomorphism)

$$\Phi : u \in Sp(1) \to \{\vec{v} \to u\vec{v}u^{-1}\} \in SO(3)$$

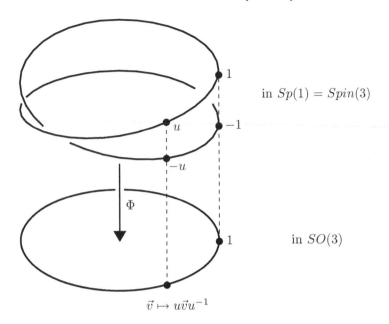

$$\vec{v} \mapsto u\vec{v}u^{-1}$$

Figure 6.1: Double cover $Sp(1) \rightarrow SO(3)$.

Both u and $-u$ act in the same way on $\vec{v}$, so we have two elements in $Sp(1)$ corresponding to the same element in $SO(3)$. One can show that Φ is a surjective map (any element of $SO(3)$ is Φ of something), so it is what is called a "covering" map, specifically a twofold cover. It makes $Sp(1)$ a double cover of $SO(3)$, and we give this group the name "$Spin(3)$." This also allows us to characterize more simply $SO(3)$ as a geometrical space. It is $S^3 = Sp(1) = Spin(3)$ with opposite points on the three-sphere identified. This space is known as $\mathbf{RP}^3$, real projective 3-space, which can also be thought of as the space of lines through the origin in $\mathbf{R}^4$ (each such line intersects S^3 in two opposite points).

Digression. *The covering map Φ is an example of a topologically non-trivial cover. Topologically, it is not true that $S^3 \simeq \mathbf{RP}^3 \times (+1, -1)$. S^3 is a connected space, not two disconnected pieces. This topological non-triviality implies that globally there is no possible homomorphism going in the opposite direction from Φ (i.e., $SO(3) \rightarrow Spin(3)$). This can be done locally, picking a local patch in $SO(3)$ and taking the inverse of Φ to a local patch in $Spin(3)$, but this will not work if we try and extend it globally to all of $SO(3)$.*

The identification $\mathbf{R}^2 = \mathbf{C}$ allowed us to represent elements of the unit circle group $U(1)$ as exponentials $e^{i\theta}$, where $i\theta$ was in the Lie algebra $\mathfrak{u}(1) = i\mathbf{R}$ of $U(1)$. $Sp(1)$ behaves in much the same way, with the Lie algebra $\mathfrak{sp}(1)$

now the space of all pure imaginary quaternions, which can be identified with $\mathbf{R}^3$ by

$$\mathbf{w} = \begin{pmatrix} w_1 \\ w_2 \\ w_3 \end{pmatrix} \in \mathbf{R}^3 \leftrightarrow \vec{w} = w_1\mathbf{i} + w_2\mathbf{j} + w_3\mathbf{k} \in \mathbf{H}$$

Unlike the $U(1)$ case, there is a non-trivial Lie bracket, the commutator of quaternions.

Elements of the group $Sp(1)$ are given by exponentiating such Lie algebra elements, which we will write in the form

$$u(\theta, \mathbf{w}) = e^{\theta\vec{w}}$$

where $\theta \in \mathbf{R}$ and $\vec{w}$ is a purely imaginary quaternion of unit length. Since

$$\vec{w}^2 = (w_1\mathbf{i} + w_2\mathbf{j} + w_3\mathbf{k})^2 = -(w_1^2 + w_2^2 + w_3^2) = -1$$

the exponential can be expanded to show that

$$e^{\theta\vec{w}} = \cos\theta + \vec{w}\sin\theta$$

Taking θ as a parameter, the $u(\theta, \mathbf{w})$ give paths in $Sp(1)$ going through the identity at $\theta = 0$, with velocity vector $\vec{w}$ since

$$\frac{d}{d\theta}u(\theta, \mathbf{w})_{|\theta=0} = (-\sin\theta + \vec{w}\cos\theta)_{|\theta=0} = \vec{w}$$

We can explicitly evaluate the homomorphism Φ on such elements $u(\theta, \mathbf{w}) \in Sp(1)$, with the result that Φ takes $u(\theta, \mathbf{w})$ to a rotation by an angle 2θ around the axis $\mathbf{w}$:

Theorem 6.1.
$$\Phi(u(\theta, \mathbf{w})) = R(2\theta, \mathbf{w})$$

Proof. First, consider the special case $\mathbf{w} = \mathbf{e}_3$ of rotations about the 3-axis.

$$u(\theta, \mathbf{e}_3) = e^{\theta\mathbf{k}} = \cos\theta + \mathbf{k}\sin\theta$$

and
$$u(\theta, \mathbf{e}_3)^{-1} = e^{-\theta\mathbf{k}} = \cos\theta - \mathbf{k}\sin\theta$$

so $\Phi(u(\theta, \mathbf{e}_3))$ is the rotation that takes $\mathbf{v}$ (identified with the quaternion $\vec{v} = v_1\mathbf{i} + v_2\mathbf{j} + v_3\mathbf{k}$) to

$$
\begin{aligned}
u(\theta, \mathbf{e}_3)\vec{v}u(\theta, \mathbf{e}_3)^{-1} &= (\cos\theta + \mathbf{k}\sin\theta)(v_1\mathbf{i} + v_2\mathbf{j} + v_3\mathbf{k})(\cos\theta - \mathbf{k}\sin\theta) \\
&= (v_1(\cos^2\theta - \sin^2\theta) - v_2(2\sin\theta\cos\theta))\mathbf{i} \\
&\quad + (2v_1\sin\theta\cos\theta + v_2(\cos^2\theta - \sin^2\theta))\mathbf{j} + v_3\mathbf{k} \\
&= (v_1\cos 2\theta - v_2\sin 2\theta)\mathbf{i} + (v_1\sin 2\theta + v_2\cos 2\theta)\mathbf{j} + v_3\mathbf{k}
\end{aligned}
$$

This is the orthogonal transformation of $\mathbf{R}^3$ given by

$$
\mathbf{v} = \begin{pmatrix} v_1 \\ v_2 \\ v_3 \end{pmatrix} \rightarrow \begin{pmatrix} \cos 2\theta & -\sin 2\theta & 0 \\ \sin 2\theta & \cos 2\theta & 0 \\ 0 & 0 & 1 \end{pmatrix} \begin{pmatrix} v_1 \\ v_2 \\ v_3 \end{pmatrix} \tag{6.2}
$$

The same calculation can readily be done for the case of $\mathbf{e}_1$, then use the Euler angle parametrization of equation 6.1 to show that a general $u(\theta, \mathbf{w})$ can be written as a product of the cases already worked out. $\qquad\square$

Notice that as θ goes from 0 to 2π, $u(\theta, \mathbf{w})$ traces out a circle in $Sp(1)$. The homomorphism Φ takes this to a circle in $SO(3)$, one that gets traced out twice as θ goes from 0 to 2π, explicitly showing the nature of the double covering above that particular circle in $SO(3)$.

The derivative of the map Φ will be a Lie algebra homomorphism, a linear map

$$
\Phi' : \mathfrak{sp}(1) \rightarrow \mathfrak{so}(3)
$$

It takes the Lie algebra $\mathfrak{sp}(1)$ of pure imaginary quaternions to the Lie algebra $\mathfrak{so}(3)$ of 3 by 3 antisymmetric real matrices. One can compute it easily on basis vectors, using, for instance, equation 6.2 above to find for the case $\vec{w} = \mathbf{k}$

$$
\begin{aligned}
\Phi'(\mathbf{k}) &= \frac{d}{d\theta}\Phi(\cos\theta + \mathbf{k}\sin\theta)_{|\theta=0} \\
&= \begin{pmatrix} -2\sin 2\theta & -2\cos 2\theta & 0 \\ 2\cos 2\theta & -2\sin 2\theta & 0 \\ 0 & 0 & 0 \end{pmatrix}_{|\theta=0} \\
&= \begin{pmatrix} 0 & -2 & 0 \\ 2 & 0 & 0 \\ 0 & 0 & 0 \end{pmatrix} = 2l_3
\end{aligned}
$$

Repeating this on other basis vectors, one finds that

$$
\Phi'(\mathbf{i}) = 2l_1, \ \Phi'(\mathbf{j}) = 2l_2, \ \Phi'(\mathbf{k}) = 2l_3
$$

Thus, Φ' is an isomorphism of $\mathfrak{sp}(1)$ and $\mathfrak{so}(3)$ identifying the bases

$$
\frac{\mathbf{i}}{2}, \frac{\mathbf{j}}{2}, \frac{\mathbf{k}}{2} \quad \text{and} \quad l_1, l_2, l_3
$$

Note that it is the $\frac{\mathbf{i}}{2}, \frac{\mathbf{j}}{2}, \frac{\mathbf{k}}{2}$ that satisfy simple commutation relations

$$\left[\frac{\mathbf{i}}{2}, \frac{\mathbf{j}}{2}\right] = \frac{\mathbf{k}}{2}, \quad \left[\frac{\mathbf{j}}{2}, \frac{\mathbf{k}}{2}\right] = \frac{\mathbf{i}}{2}, \quad \left[\frac{\mathbf{k}}{2}, \frac{\mathbf{i}}{2}\right] = \frac{\mathbf{j}}{2}$$

6.2.4 The spin group and $SU(2)$

Instead of doing calculations using quaternions with their non-commutativity and special multiplication laws, it is more conventional to choose an isomorphism between quaternions $\mathbf{H}$ and a space of 2 by 2 complex matrices, and work with matrix multiplication and complex numbers. The Pauli matrices can be used to give such an isomorphism, taking

$$\mathbf{1} \rightarrow 1 = \begin{pmatrix} 1 & 0 \\ 0 & 1 \end{pmatrix}, \quad \mathbf{i} \rightarrow -i\sigma_1 = \begin{pmatrix} 0 & -i \\ -i & 0 \end{pmatrix}, \quad \mathbf{j} \rightarrow -i\sigma_2 = \begin{pmatrix} 0 & -1 \\ 1 & 0 \end{pmatrix}$$

$$\mathbf{k} \rightarrow -i\sigma_3 = \begin{pmatrix} -i & 0 \\ 0 & i \end{pmatrix}$$

The correspondence between $\mathbf{H}$ and 2 by 2 complex matrices is then given by

$$q = q_0 + q_1\mathbf{i} + q_2\mathbf{j} + q_3\mathbf{k} \leftrightarrow \begin{pmatrix} q_0 - iq_3 & -q_2 - iq_1 \\ q_2 - iq_1 & q_0 + iq_3 \end{pmatrix}$$

Since

$$\det \begin{pmatrix} q_0 - iq_3 & -q_2 - iq_1 \\ q_2 - iq_1 & q_0 + iq_3 \end{pmatrix} = q_0^2 + q_1^2 + q_2^2 + q_3^2$$

we see that the length-squared function on quaternions corresponds to the determinant function on 2 by 2 complex matrices. Taking $q \in Sp(1)$, so of length one, the corresponding complex matrix is in $SU(2)$.

Under this identification of $\mathbf{H}$ with 2 by 2 complex matrices, we have an identification of Lie algebras $\mathfrak{sp}(1) = \mathfrak{su}(2)$ between pure imaginary quaternions and skew-Hermitian trace-zero 2 by 2 complex matrices

$$\vec{w} = w_1\mathbf{i} + w_2\mathbf{j} + w_3\mathbf{k} \leftrightarrow \begin{pmatrix} -iw_3 & -w_2 - iw_1 \\ w_2 - iw_1 & iw_3 \end{pmatrix} = -i\mathbf{w} \cdot \boldsymbol{\sigma}$$

The basis $\frac{\mathbf{i}}{2}, \frac{\mathbf{j}}{2}, \frac{\mathbf{k}}{2}$ of $\mathfrak{sp}(1)$ gets identified with a basis for the Lie algebra $\mathfrak{su}(2)$ which written in terms of the Pauli matrices is

$$X_j = -i\frac{\sigma_j}{2}$$

with the X_j satisfying the commutation relations

$$[X_1, X_2] = X_3, \ [X_2, X_3] = X_1, \ [X_3, X_1] = X_2$$

which are precisely the same commutation relations as for $\mathfrak{so}(3)$

$$[l_1, l_2] = l_3, \ [l_2, l_3] = l_1, \ [l_3, l_1] = l_2$$

We now have three isomorphic Lie algebras $\mathfrak{sp}(1) = \mathfrak{su}(2) = \mathfrak{so}(3)$, with elements that get identified as follows

$$\left(w_1 \frac{\mathbf{i}}{2} + w_2 \frac{\mathbf{j}}{2} + w_3 \frac{\mathbf{k}}{2}\right) \leftrightarrow -\frac{i}{2}\begin{pmatrix} w_3 & w_1 - iw_2 \\ w_1 + iw_2 & -w_3 \end{pmatrix} \leftrightarrow \begin{pmatrix} 0 & -w_3 & w_2 \\ w_3 & 0 & -w_1 \\ -w_2 & w_1 & 0 \end{pmatrix}$$

This isomorphism identifies basis vectors by

$$\frac{\mathbf{i}}{2} \leftrightarrow -i\frac{\sigma_1}{2} \leftrightarrow l_1$$

etc. The first of these identifications comes from the way we chose to identify $\mathbf{H}$ with 2 by 2 complex matrices. The second identification is Φ', the derivative at the identity of the covering map Φ.

On each of these isomorphic Lie algebras, we have adjoint Lie group (Ad) and Lie algebra (ad) representations. Ad is given by conjugation with the corresponding group elements in $Sp(1), SU(2)$, and $SO(3)$. ad is given by taking commutators in the respective Lie algebras of pure imaginary quaternions, skew-Hermitian trace-zero 2 by 2 complex matrices, and 3 by 3 real antisymmetric matrices.

Note that these three Lie algebras are all three-dimensional real vector spaces, so these are real representations. To get a complex representation, take complex linear combinations of elements. This is less confusing in the case of $\mathfrak{su}(2)$ than for $\mathfrak{sp}(1)$ since taking complex linear combinations of skew-Hermitian trace-zero 2 by 2 complex matrices gives all trace-zero 2 by 2 matrices (the Lie algebra $\mathfrak{sl}(2, \mathbf{C})$).

In addition, recall that there is a fourth isomorphic version of this representation, the representation of $SO(3)$ on column vectors. This is also a real representation, but can straightforwardly be complexified. Since $\mathfrak{so}(3)$ and $\mathfrak{su}(2)$ are isomorphic Lie algebras, their complexifications $\mathfrak{so}(3)_{\mathbf{C}}$ and $\mathfrak{sl}(2, \mathbf{C})$ will also be isomorphic.

In terms of 2 by 2 complex matrices, Lie algebra elements can be exponentiated to get group elements in $SU(2)$ and define

$$\Omega(\theta, \mathbf{w}) = e^{\theta(w_1 X_1 + w_2 X_2 + w_3 X_3)} = e^{-i\frac{\theta}{2}\mathbf{w} \cdot \boldsymbol{\sigma}} \tag{6.3}$$

$$= 1 \cos \frac{\theta}{2} - i(\mathbf{w} \cdot \boldsymbol{\sigma}) \sin \frac{\theta}{2} \tag{6.4}$$

Transposing the argument of Theorem 6.1 from $\mathbf{H}$ to complex matrices, one finds that, identifying

$$\mathbf{v} \leftrightarrow \mathbf{v} \cdot \boldsymbol{\sigma} = \begin{pmatrix} v_3 & v_1 - iv_2 \\ v_1 + iv_2 & -v_3 \end{pmatrix}$$

one has

$$\Phi(\Omega(\theta, \mathbf{w})) = R(\theta, \mathbf{w})$$

with $\Omega(\theta, \mathbf{w})$ acting by conjugation, taking

$$\mathbf{v} \cdot \boldsymbol{\sigma} \to \Omega(\theta, \mathbf{w})(\mathbf{v} \cdot \boldsymbol{\sigma})\Omega(\theta, \mathbf{w})^{-1} = (R(\theta, \mathbf{w})\mathbf{v}) \cdot \boldsymbol{\sigma} \tag{6.5}$$

Note that in changing from the quaternionic to complex case, we are treating the factor of 2 differently, since in the future we will want to use $\Omega(\theta, \mathbf{w})$ to perform rotations by an angle θ. In terms of the identification $SU(2) = Sp(1)$, we have $\Omega(\theta, \mathbf{w}) = u(\frac{\theta}{2}, \mathbf{w})$.

Recall that any $SU(2)$ matrix can be written in the form

$$\begin{pmatrix} \alpha & \beta \\ -\overline{\beta} & \overline{\alpha} \end{pmatrix}$$

$$\alpha = q_0 - iq_3, \quad \beta = -q_2 - iq_1$$

with $\alpha, \beta \in \mathbf{C}$ arbitrary complex numbers satisfying $|\alpha|^2 + |\beta|^2 = 1$. A somewhat unenlightening formula for the map $\Phi : SU(2) \to SO(3)$ in terms of such explicit $SU(2)$ matrices is given by

$$\Phi \begin{pmatrix} \alpha & \beta \\ -\overline{\beta} & \overline{\alpha} \end{pmatrix} = \begin{pmatrix} \operatorname{Im}(\beta^2 - \alpha^2) & \operatorname{Re}(\alpha^2 + \beta^2) & 2\operatorname{Im}(\alpha\beta) \\ \operatorname{Re}(\beta^2 - \alpha^2) & \operatorname{Im}(\alpha^2 + \beta^2) & 2\operatorname{Re}(\alpha\beta) \\ 2\operatorname{Re}(\alpha\overline{\beta}) & -2\operatorname{Im}(\alpha\overline{\beta}) & |\alpha|^2 - |\beta|^2 \end{pmatrix}$$

See [82], page 123-4, for a derivation.

6.3 A summary

To summarize, we have shown that in the three-dimensional case, we have two distinct Lie groups:

- $Spin(3)$, which geometrically is the space S^3. Its Lie algebra is $\mathbf{R}^3$ with Lie bracket the cross product. We have seen two different explicit constructions of $Spin(3)$, in terms of unit quaternions $(Sp(1))$, and in terms of 2 by 2 unitary matrices of determinant 1 $(SU(2))$.

- $SO(3)$, which has a Lie algebra isomorphic to that of $Spin(3)$.

There is a group homomorphism Φ that takes the first group to the second, which is a twofold covering map. Its derivative Φ' is an isomorphism of the Lie algebras of the two groups.

We can see from these constructions two interesting irreducible representations of these groups:

- A representation on $\mathbf{R}^3$ which can be constructed in two different ways: as the adjoint representation of either of the two groups, or as the defining representation of $SO(3)$. This is known to physicists as the "spin 1" representation.

- A representation of the first group on $\mathbf{C}^2$, which is most easily seen as the defining representation of $SU(2)$. It is not a representation of $SO(3)$, since going once around a non-contractible loop starting at the identity takes one to minus the identity, not back to the identity as required. This is called the "spin $\frac{1}{2}$" or "spinor" representation and will be studied in more detail in chapter 7.

6.4 For further reading

For another discussion of the relationship between $SO(3)$ and $SU(2)$ as well as a construction of the map Φ, see [82], sections 4.2 and 4.3, as well as [3], chapter 8, and [85], chapters 2 and 4.

Chapter 7
Rotations and the Spin $\frac{1}{2}$ Particle in a Magnetic Field

The existence of a non-trivial double cover $Spin(3)$ of the three-dimensional rotation group may seem to be a somewhat obscure mathematical fact. Remarkably though, the existence of fundamental spin $\frac{1}{2}$ particles shows that it is $Spin(3)$ rather than $SO(3)$ that is the symmetry group corresponding to rotations of fundamental quantum systems. Ignoring the degrees of freedom describing their motion in space, which we will examine in later chapters, states of elementary particles such as the electron are described by a state space $\mathcal{H} = \mathbf{C}^2$, with rotations acting on this space by the two-dimensional irreducible representation of $SU(2) = Spin(3)$.

This is the same two-state system studied in Chapter 3, with the $SU(2)$ action found there now acquiring an interpretation as corresponding to the double cover of the group of rotations of physical space. In this chapter, we will revisit that example, emphasizing the relation to rotations.

7.1 The spinor representation

In chapter 6, we examined in great detail various ways of looking at a particular three-dimensional irreducible real representation of the groups $SO(3), SU(2)$, and $Sp(1)$. This was the adjoint representation for those three groups, and isomorphic to the vector representation for $SO(3)$. In the $SU(2)$ and $Sp(1)$ cases, there is an even simpler non-trivial irreducible representation than the adjoint: the representation of 2 by 2 complex matrices in $SU(2)$ on column vectors $\mathbf{C}^2$ by matrix multiplication or the representation of unit quaternions in $Sp(1)$ on $\mathbf{H}$ by scalar multiplication. Choosing an identification $\mathbf{C}^2 = \mathbf{H}$, these are isomorphic representations on $\mathbf{C}^2$ of isomorphic groups, and for calculational convenience, we will use $SU(2)$ and its complex matrices rather than dealing with quaternions. We thus have:

© Peter Woit 2017
P. Woit, *Quantum Theory, Groups and Representations*,
DOI 10.1007/978-3-319-64612-1_7

Definition (Spinor representation). *The spinor representation of* $Spin(3) = SU(2)$ *is the representation on* $\mathbf{C}^2$ *given by*

$$g \in SU(2) \to \pi_{spinor}(g) = g$$

Elements of the representation space $\mathbf{C}^2$ *are called "spinors."*

The spin representation of $SU(2)$ is not a representation of $SO(3)$. The double cover map $\Phi : SU(2) \to SO(3)$ is a homomorphism, so given a representation (π, V) of $SO(3)$ one gets a representation $(\pi \circ \Phi, V)$ of $SU(2)$ by composition. One cannot go in the other direction: there is no homomorphism $SO(3) \to SU(2)$ that would allow one to make the spin representation of $SU(2)$ on $\mathbf{C}^2$ into an $SO(3)$ representation.

One could try and define a representation of $SO(3)$ by

$$g \in SO(3) \to \pi(g) = \pi_{spinor}(\tilde{g}) \in SU(2)$$

where $\tilde{g}$ is some choice of one of the two elements $\tilde{g} \in SU(2)$ satisfying $\Phi(\tilde{g}) = g$. The problem with this is that it will not quite give a homomorphism. Changing the choice of $\tilde{g}$ will introduce a minus sign, so π will only be a homomorphism up to sign

$$\pi(g_1)\pi(g_2) = \pm\pi(g_1 g_2)$$

The non-trivial nature of the double covering map Φ implies that there is no way to completely eliminate all minus signs, no matter how one chooses $\tilde{g}$ (since a continuous choice of $\tilde{g}$ is not possible for all g in a non-contractible loop of elements of $SO(3)$). Examples like this, which satisfy the representation property only one up to a sign ambiguity, are known as "projective representations." So, the spinor representation of $SU(2) = Spin(3)$ can be used to construct a projective representation of $SO(3)$, but not a true representation of $SO(3)$.

Quantum mechanics texts sometimes deal with this phenomenon by noting that there is an ambiguity in how one specifies physical states in $\mathcal{H}$, since multiplying a vector in $\mathcal{H}$ by a scalar does not change the eigenvalues of operators or the relative probabilities of observing these eigenvalues. As a result, the sign ambiguity noted above has no physical effect since arguably one should be working with states modulo the scalar ambiguity. It seems more straightforward though to not try and work with projective representations, but just use the larger group $Spin(3)$, accepting that this is the correct group reflecting the action of rotations on three-dimensional quantum systems.

The spin representation is more fundamental than the vector representation, in the sense that the spin representation cannot be found only knowing the vector representation, but the vector representation of $SO(3)$ can be constructed knowing the spin representation of $SU(2)$. We have seen this using the identification of $\mathbf{R}^3$ with 2 by 2 complex matrices, with equation 6.5 showing that rotations of $\mathbf{R}^3$ correspond to conjugation by spin representation matrices. Another way of seeing this uses the tensor product and is explained in section 9.4.3. Note that taking spinors as fundamental entails abandoning the description of three-dimensional geometry purely in terms of real numbers. While the vector representation is a real representation of $SO(3)$ or $Spin(3)$, the spinor representation is a complex representation.

7.2   The spin $\frac{1}{2}$ particle in a magnetic field

In chapter 3, we saw that a general quantum system with $\mathcal{H} = \mathbf{C}^2$ could be understood in terms of the action of $U(2)$ on $\mathbf{C}^2$. The self-adjoint observables correspond (up to a factor of i) to the corresponding Lie algebra representation. The $U(1) \subset U(2)$ subgroup commutes with everything else and can be analyzed separately; here, we will consider only the $SU(2)$ subgroup. For an arbitrary such system, the group $SU(2)$ has no particular geometric significance. When it occurs in its role as double cover of the rotational group, the quantum system is said to carry "spin," in particular "spin $\frac{1}{2}$" for the two-dimensional irreducible representation (in chapter 8, we will discuss state spaces of higher spin values).

As before, we take as a standard basis for the Lie algebra $\mathfrak{su}(2)$ the operators $X_j, j = 1, 2, 3$, where

$$X_j = -i\frac{\sigma_j}{2}$$

which satisfy the commutation relations

$$[X_1, X_2] = X_3, \ [X_2, X_3] = X_1, \ [X_3, X_1] = X_2$$

To make contact with the physics formalism, we will define self-adjoint operators

$$S_j = iX_j = \frac{\sigma_j}{2} \tag{7.1}$$

In general, to a skew-adjoint operator (which is what one gets from a unitary Lie algebra representation and what exponentiates to unitary operators) we will associate a self-adjoint operator by multiplying by i. These self-adjoint operators have real eigenvalues (in this case $\pm\frac{1}{2}$), so are favored by physicists as observables since experimental results are given by real numbers. In the other direction, given a physicist's observable self-adjoint operator, we will multiply by $-i$ to get a skew-adjoint operator (which may be an operator for a unitary Lie algebra representation).

Note that the conventional definition of these operators in physics texts includes a factor of $\hbar$:

$$S_j^{phys} = i\hbar X_j = \frac{\hbar \sigma_j}{2}$$

A compensating factor of $1/\hbar$ is then introduced when exponentiating to get group elements

$$\Omega(\theta, \mathbf{w}) = e^{-i\frac{\theta}{\hbar}\mathbf{w}\cdot\mathbf{S}^{phys}} \in SU(2)$$

which do not depend on $\hbar$. The reason for this convention has to do with the action of rotations on functions on $\mathbf{R}^3$ (see chapter 19) and the appearance of $\hbar$ in the definition of the momentum operator. Our definitions of S_j and of rotations using (see equation 6.3)

$$\Omega(\theta, \mathbf{w}) = e^{-i\theta\mathbf{w}\cdot\mathbf{S}} = e^{\theta\mathbf{w}\cdot\mathbf{X}}$$

will not include these factors of $\hbar$, but in any case they will be equivalent to the usual physics definitions when we make our standard choice of working with units such that $\hbar = 1$.

States in $\mathcal{H} = \mathbf{C}^2$ that have a well-defined value of the observable S_j will be the eigenvectors of S_j, with value for the observable the corresponding eigenvalue, which will be $\pm\frac{1}{2}$. Measurement theory postulates that if we perform the measurement corresponding to S_j on an arbitrary state $|\psi\rangle$, then we will

- with probability c_+ get a value of $+\frac{1}{2}$ and leave the state in an eigenvector $|j, +\frac{1}{2}\rangle$ of S_j with eigenvalue $+\frac{1}{2}$

- with probability c_- get a value of $-\frac{1}{2}$ and leave the state in an eigenvector $|j, -\frac{1}{2}\rangle$ of S_j with eigenvalue $-\frac{1}{2}$

where if

$$|\psi\rangle = \alpha|j, +\frac{1}{2}\rangle + \beta|j, -\frac{1}{2}\rangle$$

we have

$$c_+ = \frac{|\alpha|^2}{|\alpha|^2 + |\beta|^2}, \quad c_- = \frac{|\beta|^2}{|\alpha|^2 + |\beta|^2}$$

After such a measurement, any attempt to measure another $S_k, k \neq j$ will give $\pm\frac{1}{2}$ with equal probability (since the inner products of $|j, \pm\frac{1}{2}\rangle$ and $|k, \pm\frac{1}{2}\rangle$ are equal up to a phase) and put the system in a corresponding eigenvector of S_k.

If a quantum system is in an arbitrary state $|\psi\rangle$, it may not have a well-defined value for some observable A, but the "expected value" of A can be calculated. This is the sum over a basis of $\mathcal{H}$ consisting of eigenvectors (which will all be orthogonal) of the corresponding eigenvalues, weighted by the probability of their occurrence. The calculation of this sum in this case $(A = S_j)$ using expansion in eigenvectors of S_j gives

$$
\begin{aligned}
\frac{\langle\psi|A|\psi\rangle}{\langle\psi|\psi\rangle} &= \frac{(\overline{\alpha}\langle j, +\tfrac{1}{2}| + \overline{\beta}\langle j, -\tfrac{1}{2}|)A(\alpha|j, +\tfrac{1}{2}\rangle + \beta|j, -\tfrac{1}{2}\rangle)}{(\overline{\alpha}\langle j, +\tfrac{1}{2}| + \overline{\beta}\langle j, -\tfrac{1}{2}|)(\alpha|j, +\tfrac{1}{2}\rangle + \beta|j, -\tfrac{1}{2}\rangle)} \\
&= \frac{|\alpha|^2(+\tfrac{1}{2}) + |\beta|^2(-\tfrac{1}{2})}{|\alpha|^2 + |\beta|^2} \\
&= c_+(+\frac{1}{2}) + c_-(-\frac{1}{2})
\end{aligned}
$$

One often chooses to simplify such calculations by normalizing states so that the denominator $\langle\psi|\psi\rangle$ is 1. Note that the same calculation works in general for the probability of measuring the various eigenvalues of an observable A, as long as one has orthogonality and completeness of eigenvectors.

In the case of a spin $\frac{1}{2}$ particle, the group $Spin(3) = SU(2)$ acts on states by the spinor representation with the element $\Omega(\theta, \mathbf{w}) \in SU(2)$ acting as

$$|\psi\rangle \rightarrow \Omega(\theta, \mathbf{w})|\psi\rangle$$

As we saw in chapter 6, the $\Omega(\theta, \mathbf{w})$ also acts on self-adjoint matrices by conjugation, an action that corresponds to rotation of vectors when one makes the identification

$$\mathbf{v} \leftrightarrow \mathbf{v} \cdot \boldsymbol{\sigma}$$

(see equation 6.5). Under this identification, the S_j correspond (up to a factor of 2) to the basis vectors $\mathbf{e}_j$. Their transformation rule can be written as

$$S_j \rightarrow S_j' = \Omega(\theta, \mathbf{w})S_j\Omega(\theta, \mathbf{w})^{-1}$$

and

$$
\begin{pmatrix} S_1' \\ S_2' \\ S_3' \end{pmatrix} = R(\theta, \mathbf{w})^T \begin{pmatrix} S_1 \\ S_2 \\ S_3 \end{pmatrix}
$$

Note that, recalling the discussion in section 4.1, rotations on sets of basis vectors like this involve the transpose $R(\theta, \mathbf{w})^T$ of the matrix $R(\theta, \mathbf{w})$ that acts on coordinates.

Recalling the discussion in section 3.3, the spin degree of freedom that we are describing by $\mathcal{H} = \mathbf{C}^2$ has a dynamics described by the Hamiltonian

$$H = -\boldsymbol{\mu} \cdot \mathbf{B} \tag{7.2}$$

Here, $\mathbf{B}$ is the vector describing the magnetic field, and

$$\boldsymbol{\mu} = g\frac{-e}{2mc}\mathbf{S}$$

is an operator called the magnetic moment operator. The constants that appear are: $-e$ the electric charge, c the speed of light, m the mass of the particle, and g, a dimensionless number called the "gyromagnetic ratio," which is approximately 2 for an electron, about 5.6 for a proton.

The Schrödinger equation is

$$\frac{d}{dt}|\psi(t)\rangle = -i(-\boldsymbol{\mu}\cdot\mathbf{B})|\psi(t)\rangle$$

with solution

$$|\psi(t)\rangle = U(t)|\psi(0)\rangle$$

where

$$U(t) = e^{it\boldsymbol{\mu}\cdot\mathbf{B}} = e^{it\frac{-ge}{2mc}\mathbf{S}\cdot\mathbf{B}} = e^{t\frac{ge}{2mc}\mathbf{X}\cdot\mathbf{B}} = e^{t\frac{ge|\mathbf{B}|}{2mc}\mathbf{X}\cdot\frac{\mathbf{B}}{|\mathbf{B}|}}$$

The time evolution of a state is thus given at time t by the same $SU(2)$ element that, acting on vectors, gives a rotation about the axis $\mathbf{w} = \frac{\mathbf{B}}{|\mathbf{B}|}$ by an angle

$$\frac{ge|\mathbf{B}|t}{2mc}$$

so is a rotation about $\mathbf{w}$ taking place with angular velocity $\frac{ge|\mathbf{B}|}{2mc}$.

The amount of non-trivial physics that is described by this simple system is impressive, including:

- The Zeeman effect: this is the splitting of atomic energy levels that occurs when an atom is put in a constant magnetic field. With respect to the energy levels for no magnetic field, where both states in $\mathcal{H} = \mathbf{C}^2$ have the same energy, the term in the Hamiltonian given above adds

$$\pm\frac{ge|\mathbf{B}|}{4mc}$$

  to the two energy levels, giving a splitting between them proportional to the size of the magnetic field.

- The Stern–Gerlach experiment: here, one passes a beam of spin $\frac{1}{2}$ quantum systems through an inhomogeneous magnetic field. We have not yet discussed particle motion, so more is involved here than the simple two-state system. However, it turns out that one can arrange this in such a way as to pick out a specific direction $\mathbf{w}$, and split the beam into two components, of eigenvalue $+\frac{1}{2}$ and $-\frac{1}{2}$ for the operator $\mathbf{w}\cdot\mathbf{S}$.

- Nuclear magnetic resonance spectroscopy: a spin $\frac{1}{2}$ can be subjected to a time-varying magnetic field $\mathbf{B}(t)$, and such a system will be described by the same Schrödinger equation (although now the solution cannot be found just by exponentiating a matrix). Nuclei of atoms provide spin $\frac{1}{2}$ systems that can be probed with time and space-varying magnetic fields, allowing imaging of the material that they make up.

- Quantum computing: attempts to build a quantum computer involve trying to put together multiple systems of this kind (qubits), keeping them isolated from perturbations by the environment, but still allowing interaction with the system in a way that preserves its quantum behavior.

7.3   The Heisenberg picture

The treatment of time-dependence so far has used what physicists call the "Schrödinger picture" of quantum mechanics. States in $\mathcal{H}$ are functions of time, obeying the Schrödinger equation determined by a Hamiltonian observable H, while observable self-adjoint operators $\mathcal{O}$ are time-independent. Time evolution is given by a unitary transformation

$$U(t) = e^{-itH}, \quad |\psi(t)\rangle = U(t)|\psi(0)\rangle$$

$U(t)$ can instead be used to make a unitary transformation that puts the time-dependence in the observables, removing it from the states, giving something called the "Heisenberg picture." This is done as follows:

$$|\psi(t)\rangle \rightarrow |\psi(t)\rangle_H = U^{-1}(t)|\psi(t)\rangle = |\psi(0)\rangle, \quad \mathcal{O} \rightarrow \mathcal{O}_H(t) = U^{-1}(t)\mathcal{O}U(t)$$

where the "H" subscripts indicate the Heisenberg picture choice for the treatment of time-dependence. It can easily be seen that the physically observable quantities given by eigenvalues and expectations values are identical in the two pictures:

$$_H\langle\psi(t)|\mathcal{O}_H|\psi(t)\rangle_H = \langle\psi(t)|U(t)(U^{-1}(t)\mathcal{O}U(t))U^{-1}(t)|\psi(t)\rangle = \langle\psi(t)|\mathcal{O}|\psi(t)\rangle$$

In the Heisenberg picture, the dynamics is given by a differential equation not for the states but for the operators. Recall from our discussion of the adjoint representation (see equation 5.1) the formula

$$\frac{d}{dt}(e^{tX}Ye^{-tX}) = \left(\frac{d}{dt}(e^{tX}Y)\right)e^{-tX} + e^{tX}Y\left(\frac{d}{dt}e^{-tX}\right)$$
$$= Xe^{tX}Ye^{-tX} - e^{tX}Ye^{-tX}X$$

Using this with

$$Y = \mathcal{O}, \quad X = iH$$

we find

$$\frac{d}{dt}\mathcal{O}_H(t) = [iH, \mathcal{O}_H(t)] = i[H, \mathcal{O}_H(t)]$$

and this equation determines the time evolution of the observables in the Heisenberg picture.

Applying this to the case of the spin $\frac{1}{2}$ system in a magnetic field, and taking for our observable $\mathbf{S}$ (the S_j, taken together as a column vector), we find

$$\frac{d}{dt}\mathbf{S}_H(t) = i[H, \mathbf{S}_H(t)] = i\frac{eg}{2mc}[\mathbf{S}_H(t)\cdot\mathbf{B}, \mathbf{S}_H(t)] \tag{7.3}$$

We know from the discussion above that the solution will be

$$\mathbf{S}_H(t) = U(t)\mathbf{S}_H(0)U(t)^{-1}$$

for

$$U(t) = e^{-it\frac{ge|\mathbf{B}|}{2mc}\mathbf{S}\cdot\frac{\mathbf{B}}{|\mathbf{B}|}}$$

By equation 6.5 and the identification there of vectors and 2 by 2 matrices, the spin vector observable evolves in the Heisenberg picture by rotating about the magnetic field vector $\mathbf{B}$ with angular velocity $\frac{ge|\mathbf{B}|}{2mc}$.

7.4 Complex projective space

There is a different possible approach to characterizing states of a quantum system with $\mathcal{H} = \mathbf{C}^2$. Multiplication of vectors in $\mathcal{H}$ by a nonzero complex number does not change eigenvectors, eigenvalues, or expectation values, so arguably has no physical effect. Thus what is physically relevant is the quotient space $(\mathbf{C}^2 - \mathbf{0})/\mathbf{C}^*$, which is constructed by taking all nonzero elements of $\mathbf{C}^2$ and identifying those related by multiplication by a nonzero complex number.

For some insight into this construction, consider first the analog for real numbers, where $(\mathbf{R}^2 - \mathbf{0})/\mathbf{R}^*$ can be thought of as the space of all lines in the plane going through the origin.

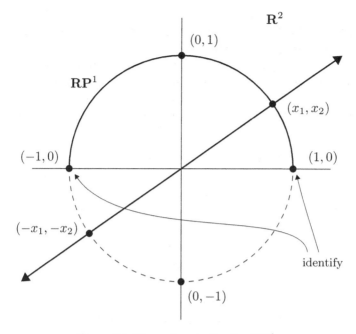

Figure 7.1: The real projective line $\mathbf{RP}^1$.

One sees that each such line hits the unit circle in two opposite points, so this set could be parametrized by a semicircle, identifying the points at the two ends. This space is given the name $\mathbf{RP}^1$ and called the "real projective line." In higher dimensions, the space of lines through the origin in $\mathbf{R}^n$ is called $\mathbf{RP}^{n-1}$ and can be thought of as the unit sphere in $\mathbf{R}^n$, with opposite points identified (recall from section 6.2.3 that $SO(3)$ can be identified with $\mathbf{RP}^3$).

What we are interested in is the complex analog $\mathbf{CP}^1$, which is quite a bit harder to visualize since in real terms it is a space of two-dimensional planes through the origin of a four-dimensional space. A standard way to choose coordinates on $\mathbf{CP}^1$ is to associate to the vector

$$\begin{pmatrix} z_1 \\ z_2 \end{pmatrix} \in \mathbf{C}^2$$

the complex number z_1/z_2. Overall multiplication by a complex number will drop out in this ratio, so one gets different values for the coordinate z_1/z_2 for each different coset element, and elements of $\mathbf{CP}^1$ correspond to points on the complex plane. There is, however, one problem with this coordinate: The point on the plane corresponding to

$$\begin{pmatrix} 1 \\ 0 \end{pmatrix}$$

does not have a well-defined value: as one approaches this point one moves off to infinity in the complex plane. In some sense, the space $\mathbf{CP}^1$ is the complex plane, but with a "point at infinity" added.

$\mathbf{CP}^1$ is better thought of not as a plane together with a point, but as a sphere (often called the "Riemann sphere"), with the relation to the plane and the point at infinity given by stereographic projection. Here, one creates a one-to-one mapping by considering the lines that go from a point on the sphere to the north pole of the sphere. Such lines will intersect the plane in a point, and give a one-to-one mapping between points on the plane and points on the sphere, except for the north pole. Now, the north pole can be identified with the "point at infinity," and thus, the space $\mathbf{CP}^1$ can be identified with the space S^2. The picture looks like this

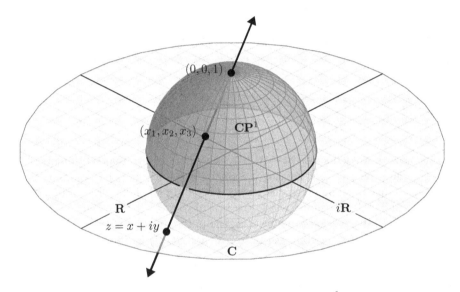

Figure 7.2: The complex projective line $\mathbf{CP}^1$.

and the equations relating coordinates (x_1, x_2, x_3) on the sphere and the complex coordinate $z_1/z_2 = z = x + iy$ on the plane are given by

$$x = \frac{x_1}{1 - x_3}, \; y = \frac{x_2}{1 - x_3}$$

and

$$x_1 = \frac{2x}{x^2 + y^2 + 1}, \; x_2 = \frac{2y}{x^2 + y^2 + 1}, \; x_3 = \frac{x^2 + y^2 - 1}{x^2 + y^2 + 1}$$

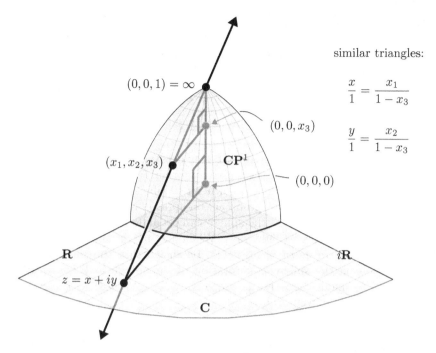

Figure 7.3: Complex-valued coordinates on $\mathbf{CP}^1 - \{\infty\}$ via stereographic projection.

Digression. *For another point of view on* $\mathbf{CP}^1$, *one constructs the quotient of* $\mathbf{C}^2$ *by complex scalars in two steps. Multiplication by a real scalar corresponds to a change in normalization of the state, and we will often use this freedom to work with normalized states, those satisfying*

$$\langle\psi|\psi\rangle = z_1\bar{z}_1 + z_2\bar{z}_2 = 1$$

Such normalized states are unit-length vectors in $\mathbf{C}^2$, *which are given by points on the unit sphere* $S^3 \subset \mathbf{R}^4 = \mathbf{C}^2$.

With such normalized states, one still must quotient out the action of multiplication by a phase, identifying elements of S^3 *that differ by multiplication by* $e^{i\theta}$. *The set of these elements forms a new geometrical space, often written* $S^3/U(1)$. *This structure is called a "fibering" of* S^3 *by circles (the action by phase multiplication traces out non-intersecting circles) and is known as the "Hopf fibration." Try an Internet search for various visualizations of the geometrical structure involved, a surprising decomposition of three-dimensional space (identifying points at infinity to get* S^3*) into non-intersecting curves.*

Acting on $\mathcal{H} = \mathbf{C}^2$ by linear maps

$$\begin{pmatrix} z_1 \\ z_2 \end{pmatrix} \rightarrow \begin{pmatrix} \alpha & \beta \\ \gamma & \delta \end{pmatrix} \begin{pmatrix} z_1 \\ z_2 \end{pmatrix}$$

takes

$$z = \frac{z_1}{z_2} \rightarrow \frac{\alpha z + \beta}{\gamma z + \delta}$$

Such transformations are invertible if the determinant of the matrix is nonzero, and one can show that these give conformal (angle-preserving) transformations of the complex plane known as "Möbius transformations." In chapter 40, we will see that this group action appears in the theory of special relativity, where the action on the sphere can be understood as transformations acting on the space of light rays. When the matrix above is in $SU(2)$ ($\gamma = -\bar{\beta}$, $\delta = \bar{\alpha}$, $\alpha\bar{\alpha} + \beta\bar{\beta} = 1$), it can be shown that the corresponding transformation on the sphere is a rotation of the sphere in $\mathbf{R}^3$, providing another way to understand the nature of $SU(2) = Spin(3)$ as the double cover of the rotation group $SO(3)$.

7.5 The Bloch sphere

For another point of view on the relation between the two-state system with $\mathcal{H} = \mathbf{C}^2$ and the geometry of the sphere (known to physicists as the "Bloch sphere" description of states), the unit sphere $S^2 \subset \mathbf{R}^3$ can be mapped to operators by

$$\mathbf{x} \rightarrow \boldsymbol{\sigma} \cdot \mathbf{x}$$

For each point $\mathbf{x} \in S^2$, $\boldsymbol{\sigma} \cdot \mathbf{x}$ has eigenvalues ± 1. Eigenvectors with eigenvalue $+1$ are the solutions to the equation

$$\boldsymbol{\sigma} \cdot \mathbf{x} |\psi\rangle = |\psi\rangle \tag{7.4}$$

and give a subspace $\mathbf{C} \subset \mathcal{H}$, giving another parametrization of the points in $\mathbf{CP}(1)$. Note that one could equivalently consider the operators

$$P_{\mathbf{x}} = \frac{1}{2}(1 - \boldsymbol{\sigma} \cdot \mathbf{x})$$

and look at the space of solutions to

$$P_{\mathbf{x}} |\psi\rangle = 0$$

It can easily be checked that $P_{\mathbf{x}}$ satisfies $P_{\mathbf{x}}^2 = P_{\mathbf{x}}$ and is a projection operator.

For a more physical interpretation of this in terms of the spin operators, one can multiply 7.4 by $\frac{1}{2}$ and characterize the $\mathbf{C} \subset \mathcal{H}$ corresponding to $\mathbf{x} \in S^2$ as the solutions to

$$\mathbf{S} \cdot \mathbf{x}|\psi\rangle = \frac{1}{2}|\psi\rangle$$

Then, the north pole of the sphere is a "spin-up" state, and the south pole is a "spin-down" state. Along the equator, one finds two points corresponding to states with definite values for S_1, as well as two for states that have definite values for S_2.

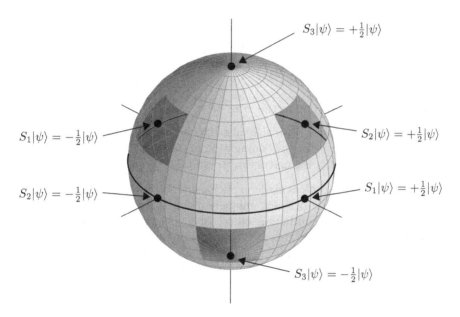

Figure 7.4: The Bloch sphere.

For later applications of the spin representation, we would like to make for each $\mathbf{x}$ a choice of solution to equation 7.4, getting a map

$$u_+ : \mathbf{x} \in S^2 \to |\psi\rangle = u_+(\mathbf{x}) \in \mathcal{H} = \mathbf{C}^2$$

such that

$$(\boldsymbol{\sigma} \cdot \mathbf{x})u_+(\mathbf{x}) = u_+(\mathbf{x}) \tag{7.5}$$

This equation determines u_+ only up to multiplication by an $\mathbf{x}$-dependent scalar.

A standard choice is

$$u_+(\mathbf{x}) = \frac{1}{\sqrt{2(1+x_3)}} \begin{pmatrix} 1+x_3 \\ x_1 + ix_2 \end{pmatrix} = \begin{pmatrix} \cos\frac{\theta}{2} \\ e^{i\phi}\sin\frac{\theta}{2} \end{pmatrix} \tag{7.6}$$

where θ, ϕ are standard spherical coordinates (which will be discussed in section 8.3). This particular choice has two noteworthy characteristics:

- One can check that it is invariant under the action of the rotation group, in the sense that

$$u_+(R\mathbf{x}) = \Omega u_+(\mathbf{x})$$

where $R = \Phi(\Omega)$ is the rotation corresponding to an $SU(2)$ element

$$\Omega = \begin{pmatrix} \cos\frac{\theta}{2} & -e^{i\phi}\sin\frac{\theta}{2} \\ e^{i\phi}\sin\frac{\theta}{2} & \cos\frac{\theta}{2} \end{pmatrix}$$

$u_+(\mathbf{x})$ is determined by setting it to be $\begin{pmatrix} 1 \\ 0 \end{pmatrix}$ at the north pole, and defining it at other points $\mathbf{x}$ on the sphere by acting on it by the element Ω which, acting on vectors by conjugation (as usual using the identification of vectors and complex matrices), would take an north pole to $\mathbf{x}$.

- With the specific choices made, $u_+(\mathbf{x})$ is discontinuous at the south pole, where $x_3 = -1$, and ϕ is not uniquely defined. For topological reasons, there cannot be a continuous choice of $u_+(\mathbf{x})$ with unit length for all $\mathbf{x}$. In applications one generally will be computing quantities that are independent of the specific choice of $u_+(\mathbf{x})$, so the discontinuity (which is choice-dependent) should not cause problems.

One can similarly pick a solution $u_-(\mathbf{x})$ to the equation

$$(\boldsymbol{\sigma} \cdot \mathbf{x})u_-(\mathbf{x}) = -u_-(\mathbf{x})$$

for eigenvectors with eigenvalue -1, with a standard choice

$$u_-(\mathbf{x}) = \frac{1}{\sqrt{2(1+x_3)}} \begin{pmatrix} -(x_1 + ix_2) \\ 1+x_3 \end{pmatrix} = \begin{pmatrix} -e^{-i\phi}\sin\frac{\theta}{2} \\ \cos\frac{\theta}{2} \end{pmatrix}$$

For each $\mathbf{x}$, $u_+(\mathbf{x})$ and $u_-(\mathbf{x})$ satisfy

$$u_+(\mathbf{x})^\dagger u_-(\mathbf{x}) = 0$$

so provide an orthonormal (for the Hermitian inner product) complex basis for $\mathbf{C}^2$.

Digression. *The association of a different vector space $\mathbf{C} \subset \mathcal{H} = \mathbf{C}^2$ to each point $\mathbf{x}$ by taking the solutions to equation 7.5 is an example of something called a "vector bundle" over the sphere of $\mathbf{x} \in S^2$. A specific choice for each $\mathbf{x}$ of a solution $u_+(\mathbf{x})$ is called a "section" of the vector bundle. It can be thought of as a sort of "twisted" complex-valued function on the sphere, taking values not in the same $\mathbf{C}$ for each $\mathbf{x}$ as would a usual function, but in copies of $\mathbf{C}$ that vary with $\mathbf{x}$.*

These copies of $\mathbf{C}$ move around in $\mathbf{C}^2$ in a topologically non-trivial way: they cannot all be identified with each other in a continuous manner. The vector bundle that appears here is perhaps the most fundamental example of a topologically non-trivial vector bundle. A discontinuity such as that found in the section u_+ of equation 7.6 is required because of this topological non-triviality. For a non-trivial bundle like this one, there cannot be continuous nonzero sections.

While the Bloch sphere provides a simple geometrical interpretation of the states of the two-state system, it should be noted that this association of points on the sphere with states does not at all preserve the notion of inner product. For example, the north and south poles of the sphere correspond to orthogonal vectors in $\mathcal{H}$, but of course $(0,0,1)$ and $(0,0,-1)$ are not at all orthogonal as vectors in $\mathbf{R}^3$.

7.6   For further reading

Just about every quantum mechanics textbook works out this example of a spin $\frac{1}{2}$ particle in a magnetic field. For one example, see chapter 14 of [81]. For an inspirational discussion of spin and quantum mechanics, together with more about the Bloch sphere, see chapter 22 of [65].

Chapter 8
Representations of $SU(2)$ and $SO(3)$

For the case of $G = U(1)$, in chapter 2 we were able to classify all complex irreducible representations by an element of $\mathbf{Z}$ and explicitly construct each irreducible representation. We would like to do the same thing here for representations of $SU(2)$ and $SO(3)$. The end result will be that irreducible representations of $SU(2)$ are classified by a nonnegative integer $n = 0, 1, 2, 3, \cdots$, and have dimension $n + 1$, so we will (hoping for no confusion with the irreducible representations $(\pi_n, \mathbf{C})$ of $U(1)$) denote them $(\pi_n, \mathbf{C}^{n+1})$. For even n, these will correspond to an irreducible representation ρ_n of $SO(3)$ in the sense that

$$\pi_n = \rho_n \circ \Phi$$

but this will not be true for odd n. It is common in physics to label these representations by $s = \frac{n}{2} = 0, \frac{1}{2}, 1, \cdots$ and call the representation labeled by s the "spin s representation." We already know the first three examples:

- Spin 0: π_0 or ρ_0 is the trivial representation for $SU(2)$ or $SO(3)$. In physics, this is sometimes called the "scalar representation." Saying that states transform under rotations as the scalar representation just means that they are invariant under rotations.

- Spin $\frac{1}{2}$: Taking

$$\pi_1(g) = g \in SU(2) \subset U(2)$$

  gives the defining representation on $\mathbf{C}^2$. This is the spinor representation discussed in chapter 7. It does not correspond to a representation of $SO(3)$.

- Spin 1: Since $SO(3)$ is a group of 3 by 3 matrices, it acts on vectors in $\mathbf{R}^3$. This is just the standard action on vectors by rotation. In other words, the representation is $(\rho_2, \mathbf{R}^3)$, with ρ_2 the identity homomorphism

$$g \in SO(3) \to \rho_2(g) = g \in SO(3)$$

© Peter Woit 2017
P. Woit, *Quantum Theory, Groups and Representations*,
DOI 10.1007/978-3-319-64612-1_8

This is sometimes called the "vector representation," and we saw in chapter 6 that it is isomorphic to the adjoint representation. Composing the homomorphisms Φ and ρ:

$$\pi_2 = \rho \circ \Phi : SU(2) \to SO(3) \subset GL(3, \mathbf{R})$$

gives a representation $(\pi_2, \mathbf{R}^3)$ of $SU(2)$, the adjoint representation. Complexifying gives a representation on $\mathbf{C}^3$, which in this case is just the action with $SO(3)$ matrices on complex column vectors, replacing the real coordinates of vectors by complex coordinates.

8.1   Representations of $SU(2)$: classification

8.1.1   Weight decomposition

If we make a choice of a $U(1) \subset SU(2)$, then given any representation (π, V) of $SU(2)$ of dimension m, we get a representation $(\pi_{|U(1)}, V)$ of $U(1)$ by restriction to the $U(1)$ subgroup. Since we know the classification of irreducibles of $U(1)$, we know that

$$(\pi_{|U(1)}, V) = \mathbf{C}_{q_1} \oplus \mathbf{C}_{q_2} \oplus \cdots \oplus \mathbf{C}_{q_m}$$

for some $q_1, q_2, \cdots, q_m \in \mathbf{Z}$, where $\mathbf{C}_q$ denotes the one-dimensional representation of $U(1)$ corresponding to the integer q (theorem 2.3). These q_j are called the "weights" of the representation V. They are exactly the same thing discussed in chapter 2 as "charges," but here we will favor the mathematician's terminology since the $U(1)$ here occurs in a context far removed from that of electromagnetism and its electric charges.

Since our standard choice of coordinates (the Pauli matrices) picks out the z-direction and diagonalizes the action of the $U(1)$ subgroup corresponding to rotation about this axis, this is the $U(1)$ subgroup we will choose to define the weights of the $SU(2)$ representation V. This is the subgroup of elements of $SU(2)$ of the form

$$\begin{pmatrix} e^{i\theta} & 0 \\ 0 & e^{-i\theta} \end{pmatrix}$$

Our decomposition of an $SU(2)$ representation (π, V) into irreducible representations of this $U(1)$ subgroup equivalently means that we can choose a basis of V so that

$$\pi\begin{pmatrix} e^{i\theta} & 0 \\ 0 & e^{-i\theta} \end{pmatrix} = \begin{pmatrix} e^{i\theta q_1} & 0 & \cdots & 0 \\ 0 & e^{i\theta q_2} & \cdots & 0 \\ \cdots & & & \\ 0 & 0 & \cdots & e^{i\theta q_m} \end{pmatrix}$$

An important property of the set of integers q_j is the following:

Theorem *If q is in the set $\{q_j\}$, so is $-q$.*

Proof. Recall that if we diagonalize a unitary matrix, the diagonal entries are the eigenvalues, but their order is undetermined: Acting by permutations on these eigenvalues, we get different diagonalizations of the same matrix. In the case of $SU(2)$, the matrix

$$P = \begin{pmatrix} 0 & 1 \\ -1 & 0 \end{pmatrix}$$

has the property that conjugation by it permutes the diagonal elements, in particular

$$P\begin{pmatrix} e^{i\theta} & 0 \\ 0 & e^{-i\theta} \end{pmatrix} P^{-1} = \begin{pmatrix} e^{-i\theta} & 0 \\ 0 & e^{i\theta} \end{pmatrix}$$

So

$$\pi(P)\pi(\begin{pmatrix} e^{i\theta} & 0 \\ 0 & e^{-i\theta} \end{pmatrix})\pi(P)^{-1} = \pi(\begin{pmatrix} e^{-i\theta} & 0 \\ 0 & e^{i\theta} \end{pmatrix})$$

and we see that $\pi(P)$ gives a change of basis of V such that the representation matrices on the $U(1)$ subgroup are as before, with $\theta \rightarrow -\theta$. Changing $\theta \rightarrow -\theta$ in the representation matrices is equivalent to changing the sign of the weights q_j. The elements of the set $\{q_j\}$ are independent of the basis, so the additional symmetry under sign change implies that for each nonzero element in the set, there is another one with the opposite sign.   □

Looking at our three examples so far, we see that, restricted to $U(1)$, the scalar or spin 0 representation of course is one-dimensional and of weight 0

$$(\pi_0, \mathbf{C}) = \mathbf{C}_0$$

and the spin $\frac{1}{2}$ representation decomposes into $U(1)$ irreducibles of weights $-1, +1$:

$$(\pi_1, \mathbf{C}^2) = \mathbf{C}_{-1} \oplus \mathbf{C}_{+1}$$

For the spin 1 representation, recall (theorem 6.1) that the double cover homomorphism Φ takes

$$\begin{pmatrix} e^{i\theta} & 0 \\ 0 & e^{-i\theta} \end{pmatrix} \in SU(2) \rightarrow \begin{pmatrix} \cos 2\theta & -\sin 2\theta & 0 \\ \sin 2\theta & \cos 2\theta & 0 \\ 0 & 0 & 1 \end{pmatrix} \in SO(3)$$

Acting with the $SO(3)$ matrix above on $\mathbf{C}^3$ will give a unitary transformation of $\mathbf{C}^3$, which therefore is in the group $U(3)$. One can show that the upper left diagonal 2 by 2 block acts on $\mathbf{C}^2$ with weights $-2, +2$, whereas the bottom right element acts trivially on the remaining part of $\mathbf{C}^3$, which is a one-dimensional representation of weight 0. So, restricted to $U(1)$, the spin 1 representation decomposes as

$$(\pi_2, \mathbf{C}^3) = \mathbf{C}_{-2} \oplus \mathbf{C}_0 \oplus \mathbf{C}_{+2}$$

Recall that the spin 1 representation of $SU(2)$ is often called the "vector" representation, since it factors in this way through the representation of $SO(3)$ by rotations on three-dimensional vectors.

8.1.2 Lie algebra representations: raising and lowering operators

To proceed further in characterizing a representation (π, V) of $SU(2)$, we need to use not just the action of the chosen $U(1)$ subgroup, but the action of group elements in the other two directions away from the identity. The non-commutativity of the group keeps us from simultaneously diagonalizing those actions and assigning weights to them. We can, however, work instead with the corresponding Lie algebra representation (π', V) of $\mathfrak{su}(2)$. As in the $U(1)$ case, the group representation is determined by the Lie algebra representation. We will see that for the Lie algebra representation, we can exploit the complexification (recall section 5.5) $\mathfrak{sl}(2, \mathbf{C})$ of $\mathfrak{su}(2)$ to further analyze the possible patterns of weights.

Recall that the Lie algebra $\mathfrak{su}(2)$ can be thought of as the tangent space $\mathbf{R}^3$ to $SU(2)$ at the identity element, with a basis given by the three skew-adjoint 2 by 2 matrices

$$X_j = -i\frac{1}{2}\sigma_j$$

which satisfy the commutation relations

$$[X_1, X_2] = X_3, \ [X_2, X_3] = X_1, \ [X_3, X_1] = X_2$$

We will often use the self-adjoint versions $S_j = iX_j$ that satisfy

$$[S_1, S_2] = iS_3, \ [S_2, S_3] = iS_1, \ [S_3, S_1] = iS_2$$

A unitary representation (π, V) of $SU(2)$ of dimension m is given by a homomorphism

$$\pi : SU(2) \rightarrow U(m)$$

We can take the derivative of this to get a map between the tangent spaces of $SU(2)$ and of $U(m)$, at the identity of both groups, and thus a Lie algebra representation

$$\pi' : \mathfrak{su}(2) \rightarrow \mathfrak{u}(m)$$

which takes skew-adjoint 2 by 2 matrices to skew-adjoint m by m matrices, preserving the commutation relations.

We have seen in section 8.1.1 that restricting the representation (π, V) to the diagonal $U(1)$ subgroup of $SU(2)$ and decomposing into irreducibles tells us that we can choose a basis of V so that

$$(\pi, V) = (\pi_{q_1}, \mathbf{C}) \oplus (\pi_{q_2}, \mathbf{C}) \oplus \cdots \oplus (\pi_{q_m}, \mathbf{C})$$

For our choice of $U(1)$ as matrices of the form

$$e^{i2\theta S_3} = \begin{pmatrix} e^{i\theta} & 0 \\ 0 & e^{-i\theta} \end{pmatrix}$$

with $e^{i\theta}$ going around $U(1)$ once as θ goes from 0 to 2π, this means we can choose a basis of V so that

$$\pi(e^{i2\theta S_3}) = \begin{pmatrix} e^{i\theta q_1} & 0 & \cdots & 0 \\ 0 & e^{i\theta q_2} & \cdots & 0 \\ \cdots & & & \cdots \\ 0 & 0 & \cdots & e^{i\theta q_m} \end{pmatrix}$$

Taking the derivative of this representation to get a Lie algebra representation, using

$$\pi'(X) = \frac{d}{d\theta}\pi(e^{\theta X})|_{\theta=0}$$

we find for $X = i2S_3$

$$\pi'(i2S_3) = \frac{d}{d\theta}\begin{pmatrix} e^{i\theta q_1} & 0 & \cdots & 0 \\ 0 & e^{i\theta q_2} & \cdots & 0 \\ \cdots & & & \cdots \\ 0 & 0 & \cdots & e^{i\theta q_m} \end{pmatrix}_{|\theta=0} = \begin{pmatrix} iq_1 & 0 & \cdots & 0 \\ 0 & iq_2 & \cdots & 0 \\ \cdots & & & \cdots \\ 0 & 0 & \cdots & iq_m \end{pmatrix}$$

Recall that π' is a real linear map from a real vector space $(\mathfrak{su}(2) = \mathbf{R}^3)$ to another real vector space $(\mathfrak{u}(n)$, the skew-Hermitian m by m complex matrices). As discussed in section 5.5, we can use complex linearity to extend any such map to a complex linear map from $\mathfrak{su}(2)_{\mathbf{C}}$ (the complexification of $\mathfrak{su}(2)$) to $\mathfrak{u}(m)_{\mathbf{C}}$ (the complexification of $\mathfrak{u}(m)$). Since $\mathfrak{su}(2) \cap i\mathfrak{su}(2) = 0$ and any element of $\mathfrak{sl}(2,\mathbf{C})$ be written as a complex number times an element of $\mathfrak{su}(2)$, we have

$$\mathfrak{su}(2)_{\mathbf{C}} = \mathfrak{su}(2) + i\mathfrak{su}(2) = \mathfrak{sl}(2,\mathbf{C})$$

Similarly,

$$\mathfrak{u}(m)_{\mathbf{C}} = \mathfrak{u}(m) + i\mathfrak{u}(m) = M(m,\mathbf{C}) = \mathfrak{gl}(m,\mathbf{C})$$

As an example, multiplying $X = i2S_3 \in \mathfrak{su}(2)$ by $-\frac{i}{2}$, we have $S_3 \in \mathfrak{sl}(2,\mathbf{C})$ and the diagonal elements in the matrix $\pi'(i2S_3)$ get also multiplied by $-\frac{i}{2}$ (since π' is now a complex linear map), giving

$$\pi'(S_3) = \begin{pmatrix} \frac{q_1}{2} & 0 & \cdots & 0 \\ 0 & \frac{q_2}{2} & \cdots & 0 \\ \cdots & & & \cdots \\ 0 & 0 & \cdots & \frac{q_m}{2} \end{pmatrix}$$

We see that $\pi'(S_3)$ will have half-integral eigenvalues, and make the following definitions:

Definition (Weights and Weight Spaces). *If $\pi'(S_3)$ has an eigenvalue $\frac{k}{2}$, we say that k is a weight of the representation (π, V).*
The subspace $V_k \subset V$ of the representation V satisfying

$$v \in V_k \implies \pi'(S_3)v = \frac{k}{2}v$$

is called the kth weight space of the representation. All vectors in it are eigenvectors of $\pi'(S_3)$ with eigenvalue $\frac{k}{2}$.
The dimension dim V_k is called the multiplicity of the weight k in the representation (π, V).

S_1 and S_2 do not commute with S_3, so they will not preserve the subspaces V_k and we cannot diagonalize them simultaneously with S_3. We can, however, exploit the fact that we are in the complexification $\mathfrak{sl}(2,\mathbf{C})$ to construct two complex linear combinations of S_1 and S_2 that do something interesting:

Definition (Raising and lowering operators). *Let*

$$S_+ = S_1 + iS_2 = \begin{pmatrix} 0 & 1 \\ 0 & 0 \end{pmatrix}, \ S_- = S_1 - iS_2 = \begin{pmatrix} 0 & 0 \\ 1 & 0 \end{pmatrix}$$

We have $S_+, S_- \in \mathfrak{sl}(2, \mathbf{C})$. These are neither self-adjoint nor skew-adjoint, but satisfy

$$(S_\pm)^\dagger = S_\mp$$

and similarly, we have

$$\pi'(S_\pm)^\dagger = \pi'(S_\mp)$$

We call $\pi'(S_+)$ a "raising operator" for the representation (π, V), and $\pi'(S_-)$ a "lowering operator."

The reason for this terminology is the following calculation:

$$[S_3, S_+] = [S_3, S_1 + iS_2] = iS_2 + i(-iS_1) = S_1 + iS_2 = S_+$$

which implies (since π' is a Lie algebra homomorphism)

$$\pi'(S_3)\pi'(S_+) - \pi'(S_+)\pi'(S_3) = \pi'([S_3, S_+]) = \pi'(S_+)$$

For any $v \in V_k$, we have

$$\pi'(S_3)\pi'(S_+)v = \pi'(S_+)\pi'(S_3)v + \pi'(S_+)v = (\frac{k}{2} + 1)\pi'(S_+)v$$

so

$$v \in V_k \implies \pi'(S_+)v \in V_{k+2}$$

The linear operator $\pi'(S_+)$ takes vectors with a well-defined weight to vectors with the same weight, plus 2 (thus the terminology "raising operator"). A similar calculation shows that $\pi'(S_-)$ takes V_k to V_{k-2}, lowering the weight by 2.

We are now ready to classify all finite dimensional irreducible unitary representations (π, V) of $SU(2)$. We define:

Definition (Highest weights and highest weight vectors). *A nonzero vector $v \in V_n \subset V$ such that*

$$\pi'(S_+)v = 0$$

is called a highest weight vector, with highest weight n.

Irreducible representations will be characterized by a highest weight vector, as follows

Theorem (Highest weight theorem). *Finite dimensional irreducible representations of $SU(2)$ have weights of the form*

$$-n, -n+2, \cdots, n-2, n$$

for n a nonnegative integer, each with multiplicity 1, with n a highest weight.

Proof. Finite dimensionality implies there is a highest weight n, and we can choose any highest weight vector $v_n \in V_n$. Repeatedly applying $\pi'(S_-)$ to v_n will give new vectors

$$v_{n-2j} = \pi'(S_-)^j v_n \in V_{n-2j}$$

with weights $n - 2j$.

Consider the span of the $v_{n-2j}, j \geq 0$. To show that this is a representation one needs to show that the $\pi'(S_3)$ and $\pi'(S_+)$ leave it invariant. For $\pi'(S_3)$, this is obvious; for $\pi'(S_+)$, one can show that

$$\pi'(S_+)v_{n-2j} = j(n - j + 1)v_{n-2(j-1)} \tag{8.1}$$

by an induction argument. For $j = 0$, this is the highest weight condition on v_n. Assuming validity for j, validity for $j + 1$ can be checked by

$$
\begin{aligned}
\pi'(S_+)v_{n-2(j+1)} &= \pi'(S_+)\pi'(S_-)v_{n-2j} \\
&= (\pi'([S_+, S_-]) + \pi'(S_-)\pi'(S_+))v_{n-2j} \\
&= (\pi'(2S_3) + \pi'(S_-)\pi'(S_+))v_{n-2j} \\
&= ((n - 2j)v_{n-2j} + \pi'(S_-)j(n - j + 1)v_{n-2(j-1)} \\
&= ((n - 2j) + j(n - j + 1))v_{n-2j} \\
&= (j + 1)(n - (j + 1) + 1)v_{n-2((j+1)-1)}
\end{aligned}
$$

where we have used the commutation relation

$$[S_+, S_-] = 2S_3$$

By finite dimensionality, there must be some integer k such that $v_{n-2j} \neq 0$ for $j \leq k$, and $v_{n-2j} = 0$ for $j = k + 1$. But then, for $j = k + 1$, we must have

$$\pi'(S_+)v_{n-2j} = 0$$

By equation 8.1, this will happen only for $k + 1 = n + 1$ (and we need to take n positive). We thus see that v_{-n} will be a "lowest weight vector," annihilated by $\pi'(S_-)$. As expected, the pattern of weights is invariant under change of sign, with nonzero weight spaces for

$$-n, -n + 2, \cdots, n - 2, n$$

$\square$

Digression *Dropping the requirement of finite dimensionality, the same construction starting with a highest weight vector and repeatedly applying the lowering operator can be used to produce infinite dimensional irreducible* *representations of the Lie algebras* $\mathfrak{su}(2)$ *or* $\mathfrak{sl}(2,\mathbf{C})$. *These occur when the highest weight is not a nonnegative integer, and they will be non-integrable representations (representations of the Lie algebra, but not of the Lie group).*

Since we saw in section 8.1.1 that representations can be studied by looking at the set of their weights under the action of our chosen $U(1) \subset SU(2)$, we can label irreducible representations of $SU(2)$ by a nonnegative integer n, the highest weight. Such a representation will be of dimension $n+1$, with weights

$$-n, -n+2, \cdots, n-2, n$$

Each weight occurs with multiplicity one, and we have

$$(\pi_{|U(1)}, V) = \mathbf{C}_{-n} \oplus \mathbf{C}_{-n+2} \oplus \cdots \mathbf{C}_{n-2} \oplus \mathbf{C}_n$$

Starting with a highest weight or lowest weight vector, a basis for the representation can be generated by repeatedly applying lowering or raising operators. The picture to keep in mind is this

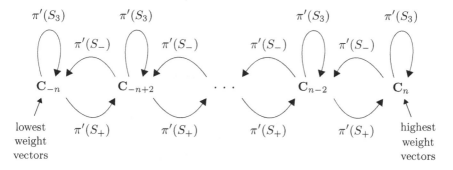

Figure 8.1: Basis for a representation of $SU(2)$ in terms of raising and lowering operators.

where all the vector spaces are copies of $\mathbf{C}$, and all the maps are isomorphisms (multiplications by various numbers).

In summary, we see that all irreducible finite dimensional unitary $SU(2)$ representations can be labeled by a nonnegative integer, the highest weight n. These representations have dimension $n+1$, and we will denote them $(\pi_n, V^n = \mathbf{C}^{n+1})$. Note that V_n is the nth weight space, V^n is the representation with highest weight n. The physicist's terminology for this uses not n, but $\frac{n}{2}$ and calls this number the "spin" of the representation. We have so far seen the lowest three examples $n = 0, 1, 2$, or spin $s = \frac{n}{2} = 0, \frac{1}{2}, 1$, but there is an infinite class of larger irreducibles, with dim $V_n = n+1 = 2s+1$.

8.2   Representations of $SU(2)$: construction

The argument of the previous section only tells us what properties possible finite dimensional irreducible representations of $SU(2)$ must have. It shows how to construct such representations given a highest weight vector, but does not provide any way to construct such highest weight vectors. We would like to find some method to explicitly construct an irreducible (π_n, V^n) for each highest weight n. There are several possible constructions, but perhaps the simplest one is the following, which gives a representation of highest weight n by looking at polynomials in two complex variables, homogeneous of degree n. This construction will produce representations not just of $SU(2)$, but of the larger group $GL(2, \mathbf{C})$.

Recall from equation 1.3 that if one has an action of a group on a space M, one can get a representation on functions f on M by taking

$$(\pi(g)f)(x) = f(g^{-1} \cdot x)$$

For the group $GL(2, \mathbf{C})$, we have an obvious action on $M = \mathbf{C}^2$ (by matrices acting on column vectors), and we look at a specific class of functions on this space, the polynomials. We can break up the infinite dimensional space of polynomials on $\mathbf{C}^2$ into finite dimensional subspaces as follows:

Definition (Homogeneous polynomials). *The complex vector space of homogeneous polynomials of degree n in two complex variables z_1, z_2 is the space of functions on $\mathbf{C}^2$ of the form*

$$f(z_1, z_2) = a_0 z_1^n + a_1 z_1^{n-1} z_2 + \cdots + a_{n-1} z_1 z_2^{n-1} + a_n z_2^n$$

The space of such functions is a complex vector space of dimension $n + 1$.

It turns out that this space of functions is exactly the representation space V^n that we need to get the irreducible representations of $SU(2) \subset GL(2, \mathbf{C})$. If we take

$$g = \begin{pmatrix} \alpha & \beta \\ \gamma & \delta \end{pmatrix} \in GL(2, \mathbf{C})$$

to act by $g^{-1} \cdot x = g^{-1}x$ on column vectors $x \in \mathbf{C}^2$, then its action on coordinates z_1, z_2 will be given by

$$g \cdot \begin{pmatrix} z_1 \\ z_2 \end{pmatrix} = g^T \begin{pmatrix} z_1 \\ z_2 \end{pmatrix} = \begin{pmatrix} \alpha & \beta \\ \gamma & \delta \end{pmatrix}^T \begin{pmatrix} z_1 \\ z_2 \end{pmatrix} = \begin{pmatrix} \alpha z_1 + \gamma z_2 \\ \beta z_1 + \delta z_2 \end{pmatrix}$$

This is because these coordinates are basis elements of the dual $\mathbf{C}^2$, see the discussion at the end of sections 4.1 and 4.2.

The representation $\pi_n(g)$ on homogeneous polynomial functions will be given by replacing

$$z_1 \rightarrow \alpha z_1 + \gamma z_2, \quad z_2 \rightarrow \beta z_1 + \delta z_2$$

in the expression for the polynomial in terms of z_1 and z_2.

Taking the derivative, the Lie algebra representation is given by

$$\pi'_n(X)f = \frac{d}{dt}\pi_n(e^{tX})f_{|t=0} = \frac{d}{dt}f(e^{tX} \cdot z_1, e^{tX} \cdot z_2)_{|t=0}$$

where $X \in \mathfrak{gl}(2, \mathbf{C})$ is any 2 by 2 complex matrix. By the chain rule, for

$$\begin{pmatrix} z_1(t) \\ z_2(t) \end{pmatrix} = (e^{tX})^T \begin{pmatrix} z_1 \\ z_2 \end{pmatrix} = e^{tX^T} \begin{pmatrix} z_1 \\ z_2 \end{pmatrix}$$

this is

$$\pi'_n(X)f = \left(\frac{\partial f}{\partial z_1}, \frac{\partial f}{\partial z_2} \right) \left(\frac{d}{dt} e^{tX^T} \begin{pmatrix} z_1 \\ z_2 \end{pmatrix} \right)_{|t=0}$$
$$= \frac{\partial f}{\partial z_1}(X_{11}z_1 + X_{21}z_2) + \frac{\partial f}{\partial z_2}(X_{12}z_1 + X_{22}z_2)$$

where the X_{jk} are the components of the matrix X.

Computing what happens for $X_j = -i\frac{\sigma_j}{2}$ (a basis of $\mathfrak{su}(2)$), we get

$$(\pi'_n(X_3)f)(z_1, z_2) = -\frac{i}{2} \left(\frac{\partial f}{\partial z_1} z_1 - \frac{\partial f}{\partial z_2} z_2 \right)$$

so

$$\pi'_n(X_3) = -\frac{i}{2} \left(z_1 \frac{\partial}{\partial z_1} - z_2 \frac{\partial}{\partial z_2} \right)$$

and similarly,

$$\pi'_n(X_1) = -\frac{i}{2} \left(z_1 \frac{\partial}{\partial z_2} + z_2 \frac{\partial}{\partial z_1} \right), \quad \pi'_n(X_2) = \frac{1}{2} \left(z_2 \frac{\partial}{\partial z_1} - z_1 \frac{\partial}{\partial z_2} \right)$$

The $z_1^{n-k} z_2^k$ are eigenvectors for $S_3 = iX_3$ with eigenvalue $\frac{1}{2}(n - 2k)$ since

$$\pi'_n(S_3) z_1^{n-k} z_2^k = \frac{1}{2}((n-k)z_1^{n-k} z_2^k - kz_1^{n-k} z_2^k) = \frac{1}{2}(n - 2k) z_1^{n-k} z_2^k$$

z_1^n will be an explicit highest weight vector for the representation (π_n, V^n).

An important thing to note here is that the formulas we have found for π'_n are not in terms of matrices. Instead, we have seen that when we construct our representations using functions on $\mathbf{C}^2$, for any $X \in \mathfrak{gl}(2, \mathbf{C})$, $\pi'_n(X)$ is given by a differential operator. These differential operators are independent of n, with the same operator $\pi'(X)$ on all the V^n. This is because the original definition of the representation

$$(\pi(g)f)(x) = f(g^{-1} \cdot x)$$

is on the full infinite dimensional space of polynomials on $\mathbf{C}^2$. While this space is infinite dimensional, issues of analysis do not really come into play here, since polynomial functions are essentially an algebraic construction.

Restricting the differential operators $\pi'(X)$ to V^n, the homogeneous polynomials of degree n, they become linear operators on a finite dimensional space. We now have an explicit highest weight vector, and an explicit construction of the corresponding irreducible representation. If one chooses a basis of V^n, then the linear operator $\pi'(X)$ will be given by a $n+1$ by $n+1$ matrix. Clearly though, the expression as a simple first-order differential operator is much easier to work with. In the examples we will be studying in later chapters, the representations under consideration will often be on function spaces, with Lie algebra representations appearing as differential operators. Instead of using linear algebra techniques to find eigenvalues and eigenvectors, the eigenvector equation will be a partial differential equation, with our focus on using Lie groups and their representation theory to solve such equations.

One issue we have not addressed yet is that of unitarity of the representation. We need Hermitian inner products on the spaces V^n, inner products that will be preserved by the action of $SU(2)$ that we have defined on these spaces. A standard way to define a Hermitian inner product on functions on a space M is to define them using an integral: for f, g complex-valued functions on M, take their inner product to be

$$\langle f, g \rangle = \int_M \overline{f} g$$

While for $M = \mathbf{C}^2$ this gives an $SU(2)$ invariant inner product on functions (one that is not invariant for the full group $GL(2, \mathbf{C})$), it is useless for f, g polynomial, since such integrals diverge. In this case, an inner product on polynomial functions on $\mathbf{C}^2$ can be defined by

$$\langle f, g \rangle = \frac{1}{\pi^2} \int_{\mathbf{C}^2} \overline{f(z_1, z_2)} g(z_1, z_2) e^{-(|z_1|^2 + |z_2|^2)} dx_1 dy_1 dx_2 dy_2 \qquad (8.2)$$

Here, $z_1 = x_1 + iy_1, z_2 = x_2 + iy_2$. Integrals of this kind can be done fairly easily since they factorize into separate integrals over z_1 and z_2, each of which can be treated using polar coordinates and standard calculus methods. One can check by explicit computation that the polynomials

$$\frac{z_1^j z_2^k}{\sqrt{j!k!}}$$

will be an orthonormal basis of the space of polynomial functions with respect to this inner product, and the operators $\pi'(X), X \in \mathfrak{su}(2)$ will be skew-adjoint.

Working out what happens for the first few examples of irreducible $SU(2)$ representations, one finds orthonormal bases for the representation spaces V^n of homogeneous polynomials as follows

• For $n = s = 0$

$$1$$

• For $n = 1, s = \frac{1}{2}$

$$z_1, \; z_2$$

• For $n = 2, s = 1$

$$\frac{1}{\sqrt{2}}z_1^2, \; z_1 z_2, \; \frac{1}{\sqrt{2}}z_2^2$$

• For $n = 3, s = \frac{3}{2}$

$$\frac{1}{\sqrt{6}}z_1^3, \; \frac{1}{\sqrt{2}}z_1^2 z_2, \; \frac{1}{\sqrt{2}}z_1 z_2^2, \; \frac{1}{\sqrt{6}}z_2^3$$

8.3 Representations of $SO(3)$ and spherical harmonics

We would like to now use the classification and construction of representations of $SU(2)$ to study the representations of the closely related group $SO(3)$. For any representation (ρ, V) of $SO(3)$, we can use the double covering homomorphism $\Phi : SU(2) \to SO(3)$ to get a representation

$$\pi = \rho \circ \Phi$$

of $SU(2)$. It can be shown that if ρ is irreducible, π will be too, so we must have $\pi = \rho \circ \Phi = \pi_n$, one of the irreducible representations of $SU(2)$ found in the last section. Using the fact that $\Phi(-1) = 1$, we see that

$$\pi_n(-1) = \rho \circ \Phi(-1) = 1$$

From knowing that the weights of π_n are $-n, -n+2, \cdots, n-2, n$, we know that

$$\pi_n(-1) = \pi_n \begin{pmatrix} e^{i\pi} & 0 \\ 0 & e^{-i\pi} \end{pmatrix} = \begin{pmatrix} e^{in\pi} & 0 & \cdots & 0 \\ 0 & e^{i(n-2)\pi} & \cdots & 0 \\ \cdots & & & \cdots \\ 0 & 0 & \cdots & e^{-in\pi} \end{pmatrix} = 1$$

which will only be true for n even, not for n odd. Since the Lie algebra of $SO(3)$ is isomorphic to the Lie algebra of $SU(2)$, the same Lie algebra argument using raising and lowering operators as in the last section also applies. The irreducible representations of $SO(3)$ will be $(\rho_l, V = \mathbf{C}^{2l+1})$ for $l = 0, 1, 2, \cdots$, of dimension $2l+1$ and satisfying

$$\rho_l \circ \Phi = \pi_{2l}$$

Just like in the case of $SU(2)$, we can explicitly construct these representations using functions on a space with an $SO(3)$ action. The obvious space to choose is $\mathbf{R}^3$, with $SO(3)$ matrices acting on $\mathbf{x} \in \mathbf{R}^3$ as column vectors. The induced representation is as usual

$$(\rho(g)f)(\mathbf{x}) = f(g^{-1}\mathbf{x})$$

and by the same argument as in the $SU(2)$ case, $g \in SO(3)$ acts on the coordinates (a basis of the dual $\mathbf{R}^3$) by

$$g \cdot \begin{pmatrix} x_1 \\ x_2 \\ x_3 \end{pmatrix} = g^T \begin{pmatrix} x_1 \\ x_2 \\ x_3 \end{pmatrix}$$

Taking the derivative, the Lie algebra representation on functions is given by

$$\rho'(X)f = \frac{d}{dt}\rho(e^{tX})f_{|t=0} = \frac{d}{dt}f(e^{tX} \cdot x_1, e^{tX} \cdot x_2, e^{tX} \cdot x_3)_{|t=0}$$

where $X \in \mathfrak{so}(3)$. Recall that a basis for $\mathfrak{so}(3)$ is given by

$$l_1 = \begin{pmatrix} 0 & 0 & 0 \\ 0 & 0 & -1 \\ 0 & 1 & 0 \end{pmatrix} \quad l_2 = \begin{pmatrix} 0 & 0 & 1 \\ 0 & 0 & 0 \\ -1 & 0 & 0 \end{pmatrix} \quad l_3 = \begin{pmatrix} 0 & -1 & 0 \\ 1 & 0 & 0 \\ 0 & 0 & 0 \end{pmatrix}$$

which satisfy the commutation relations

$$[l_1, l_2] = l_3, \ [l_2, l_3] = l_1, \ [l_3, l_1] = l_2$$

Digression. *A Note on Conventions*

We are using the notation l_j for the real basis of the Lie algebra $\mathfrak{so}(3) = \mathfrak{su}(2)$. For a unitary representation ρ, the $\rho'(l_j)$ will be skew-adjoint linear operators. For consistency with the physics literature, we will use the notation $L_j = i\rho'(l_j)$ for the self-adjoint version of the linear operator corresponding to l_j in this representation on functions. The L_j satisfy the commutation relations

$$[L_1, L_2] = iL_3, \quad [L_2, L_3] = iL_1, \quad [L_3, L_1] = iL_2$$

We will also use elements $l_{\pm} = l_1 \pm il_2$ of the complexified Lie algebra to create raising and lowering operators $L_{\pm} = i\rho'(l_{\pm})$.

As with the $SU(2)$ case, we will not include a factor of $\hbar$ as is usual in physics (the usual convention is $L_j = i\hbar\rho'(l_j)$), since for considerations of the action of the rotation group it would cancel out (physicists define rotations using $e^{\frac{i}{\hbar}\theta L_j}$). The factor of $\hbar$ is only of significance when L_j is expressed in terms of the momentum operator, a topic discussed in chapter 19.

In the $SU(2)$ case, the $\pi'(S_j)$ had half-integral eigenvalues, with the eigenvalues of $\pi'(2S_3)$ the integral weights of the representation. Here, the L_j will have integer eigenvalues, the weights will be the eigenvalues of $2L_3$, which will be even integers.

Computing $\rho'(l_1)$, we find

$$\rho'(l_1)f = \frac{d}{dt}f\left(e^{t\begin{pmatrix} 0 & 0 & 0 \\ 0 & 0 & 1 \\ 0 & -1 & 0 \end{pmatrix}}\begin{pmatrix} x_1 \\ x_2 \\ x_3 \end{pmatrix}\right)\Bigg|_{t=0}$$

$$= \frac{d}{dt}f\left(\begin{pmatrix} 1 & 0 & 0 \\ 0 & \cos t & \sin t \\ 0 & -\sin t & \cos t \end{pmatrix}\begin{pmatrix} x_1 \\ x_2 \\ x_3 \end{pmatrix}\right)\Bigg|_{t=0}$$

$$= \frac{d}{dt}f\left(\begin{pmatrix} x_1 \\ x_2 \cos t + x_3 \sin t \\ -x_2 \sin t + x_3 \cos t \end{pmatrix}\right)\Bigg|_{t=0}$$

$$= \left(\frac{\partial f}{\partial x_1}, \frac{\partial f}{\partial x_2}, \frac{\partial f}{\partial x_3}\right) \cdot \begin{pmatrix} 0 \\ x_3 \\ -x_2 \end{pmatrix}$$

$$= x_3\frac{\partial f}{\partial x_2} - x_2\frac{\partial f}{\partial x_3}$$

so

$$\rho'(l_1) = x_3\frac{\partial}{\partial x_2} - x_2\frac{\partial}{\partial x_3}$$

and similar calculations give

$$\rho'(l_2) = x_1\frac{\partial}{\partial x_3} - x_3\frac{\partial}{\partial x_1}, \ \rho'(l_3) = x_2\frac{\partial}{\partial x_1} - x_1\frac{\partial}{\partial x_2}$$

The space of all functions on $\mathbf{R}^3$ is much too big: It will give us an infinity of copies of each finite dimensional representation that we want. Notice that when $SO(3)$ acts on $\mathbf{R}^3$, it leaves the distance to the origin invariant. If we work in spherical coordinates (r, θ, ϕ) (see picture),

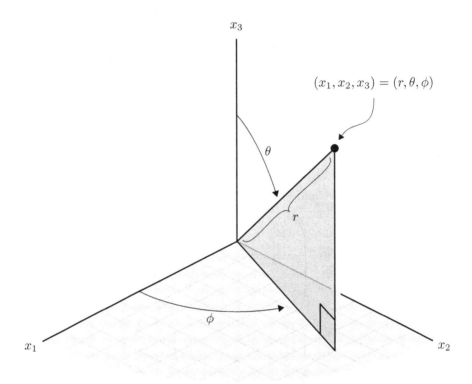

Figure 8.2: Spherical coordinates.

we will have

$$x_1 = r\sin\theta\cos\phi$$
$$x_2 = r\sin\theta\sin\phi$$
$$x_3 = r\cos\theta$$

Acting on $f(r, \phi, \theta)$, $SO(3)$ will leave r invariant, only acting non-trivially on θ, ϕ. It turns out that we can cut down the space of functions to something that will only contain one copy of the representation we want in various ways. One way to do this is to restrict our functions to the unit sphere, i.e., look at functions $f(\theta, \phi)$. We will see that the representations we are looking for can be found in simple trigonometric functions of these two angular variables.

We can construct our irreducible representations ρ_l' by explicitly constructing a function we will call $Y_l^l(\theta, \phi)$ that will be a highest weight vector of weight l. The weight l condition and the highest weight condition give two differential equations for $Y_l^l(\theta, \phi)$:

$$L_3 Y_l^l = l Y_l^l, \quad L_+ Y_l^l = 0$$

These will turn out to have a unique solution (up to scalars).

We first need to change coordinates from rectangular to spherical in our expressions for $L_3, L_\pm$. Using the chain rule to compute expressions like

$$\frac{\partial}{\partial r} f(x_1(r, \theta, \phi), x_2(r, \theta, \phi), x_3(r, \theta, \phi))$$

we find

$$
\begin{pmatrix} \frac{\partial}{\partial r} \\ \frac{\partial}{\partial \theta} \\ \frac{\partial}{\partial \phi} \end{pmatrix} = \begin{pmatrix} \sin\theta\cos\phi & \sin\theta\sin\phi & \cos\theta \\ r\cos\theta\cos\phi & r\cos\theta\sin\phi & -r\sin\theta \\ -r\sin\theta\sin\phi & r\sin\theta\cos\phi & 0 \end{pmatrix} \begin{pmatrix} \frac{\partial}{\partial x_1} \\ \frac{\partial}{\partial x_2} \\ \frac{\partial}{\partial x_3} \end{pmatrix}
$$

so

$$
\begin{pmatrix} \frac{\partial}{\partial r} \\ \frac{1}{r}\frac{\partial}{\partial \theta} \\ \frac{1}{r\sin\theta}\frac{\partial}{\partial \phi} \end{pmatrix} = \begin{pmatrix} \sin\theta\cos\phi & \sin\theta\sin\phi & \cos\theta \\ \cos\theta\cos\phi & \cos\theta\sin\phi & -\sin\theta \\ -\sin\phi & \cos\phi & 0 \end{pmatrix} \begin{pmatrix} \frac{\partial}{\partial x_1} \\ \frac{\partial}{\partial x_2} \\ \frac{\partial}{\partial x_3} \end{pmatrix}
$$

This is an orthogonal matrix, so can be inverted by taking its transpose, to get

$$
\begin{pmatrix} \frac{\partial}{\partial x_1} \\ \frac{\partial}{\partial x_2} \\ \frac{\partial}{\partial x_3} \end{pmatrix} = \begin{pmatrix} \sin\theta\cos\phi & \cos\theta\cos\phi & -\sin\phi \\ \sin\theta\sin\phi & \cos\theta\sin\phi & \cos\phi \\ \cos\theta & -\sin\theta & 0 \end{pmatrix} \begin{pmatrix} \frac{\partial}{\partial r} \\ \frac{1}{r}\frac{\partial}{\partial \theta} \\ \frac{1}{r\sin\theta}\frac{\partial}{\partial \phi} \end{pmatrix}
$$

So we finally have

$$L_1 = i\rho'(l_1) = i\left(x_3\frac{\partial}{\partial x_2} - x_2\frac{\partial}{\partial x_3}\right) = i\left(\sin\phi\frac{\partial}{\partial\theta} + \cot\theta\cos\phi\frac{\partial}{\partial\phi}\right)$$

$$L_2 = i\rho'(l_2) = i\left(x_1\frac{\partial}{\partial x_3} - x_3\frac{\partial}{\partial x_1}\right) = i\left(-\cos\phi\frac{\partial}{\partial\theta} + \cot\theta\sin\phi\frac{\partial}{\partial\phi}\right)$$

$$L_3 = i\rho'(l_3) = i\left(x_1\frac{\partial}{\partial x_3} - x_3\frac{\partial}{\partial x_1}\right) = -i\frac{\partial}{\partial\phi}$$

and

$$L_+ = i\rho'(l_+) = e^{i\phi}\left(\frac{\partial}{\partial\theta} + i\cot\theta\frac{\partial}{\partial\phi}\right), \; L_- = i\rho'(l_-) = e^{-i\phi}\left(-\frac{\partial}{\partial\theta} + i\cot\theta\frac{\partial}{\partial\phi}\right)$$

Now that we have expressions for the action of the Lie algebra on functions in spherical coordinates, our two differential equations saying our function $Y_l^l(\theta,\phi)$ is of weight l and in the highest weight space are

$$L_3 Y_l^l(\theta,\phi) = -i\frac{\partial}{\partial\phi}Y_l^l(\theta,\phi) = l Y_l^l(\theta,\phi)$$

and

$$L_+ Y_l^l(\theta,\phi) = e^{i\phi}\left(\frac{\partial}{\partial\theta} + i\cot\theta\frac{\partial}{\partial\phi}\right)Y_l^l(\theta,\phi) = 0$$

The first of these tells us that

$$Y_l^l(\theta,\phi) = e^{il\phi}F_l(\theta)$$

for some function $F_l(\theta)$, and using the second, we get

$$(\frac{\partial}{\partial\theta} - l\cot\theta)F_l(\theta) = 0$$

with solution

$$F_l(\theta) = C_{ll}\sin^l\theta$$

for an arbitrary constant C_{ll}. Finally,

$$Y_l^l(\theta,\phi) = C_{ll}e^{il\phi}\sin^l\theta$$

This is a function on the sphere, which is also a highest weight vector in a $2l+1$-dimensional irreducible representation of $SO(3)$. Repeatedly applying the lowering operator L_- gives vectors spanning the rest of the weight spaces, the functions

$$Y_l^m(\theta,\phi) = C_{lm}(L_-)^{l-m}Y_l^l(\theta,\phi)$$

$$= C_{lm}\left(e^{-i\phi}\left(-\frac{\partial}{\partial\theta} + i\cot\theta\frac{\partial}{\partial\phi}\right)\right)^{l-m}e^{il\phi}\sin^l\theta$$

for $m = l, l-1, l-2\cdots, -l+1, -l$

The functions $Y_l^m(\theta,\phi)$ are called "spherical harmonics," and they span the space of complex functions on the sphere in much the same way that the $e^{in\theta}$ span the space of complex-valued functions on the circle. Unlike the case of polynomials on $\mathbf{C}^2$, for functions on the sphere, one gets finite numbers by integrating such functions over the sphere. So an inner product on these representations for which they are unitary can be defined by simply setting

$$\langle f, g \rangle = \int_{S^2} \overline{f} g \sin\theta d\theta d\phi = \int_{\phi=0}^{2\pi} \int_{\theta=0}^{\pi} \overline{f(\theta, \phi)} g(\theta, \phi) \sin\theta d\theta d\phi \qquad (8.3)$$

We will not try and show this here, but for the allowable values of l, m the $Y_l^m(\theta, \phi)$ are mutually orthogonal with respect to this inner product.

One can derive various general formulas for the $Y_l^m(\theta, \phi)$ in terms of Legendre polynomials, but here we will just compute the first few examples, with the proper constants that give them norm 1 with respect to the chosen inner product.

- For the $l = 0$ representation,

$$Y_0^0(\theta, \phi) = \sqrt{\frac{1}{4\pi}}$$

- For the $l = 1$ representation,

$$Y_1^1 = -\sqrt{\frac{3}{8\pi}} e^{i\phi} \sin\theta, \quad Y_1^0 = \sqrt{\frac{3}{4\pi}} \cos\theta, \quad Y_1^{-1} = \sqrt{\frac{3}{8\pi}} e^{-i\phi} \sin\theta$$

(one can easily see that these have the correct eigenvalues for $\rho'(L_3) = -i\frac{\partial}{\partial\phi}$).

- For the $l = 2$ representation, one has

$$Y_2^2 = \sqrt{\frac{15}{32\pi}} e^{i2\phi} \sin^2\theta, \quad Y_2^1 = -\sqrt{\frac{15}{8\pi}} e^{i\phi} \sin\theta \cos\theta$$

$$Y_2^0 = \sqrt{\frac{5}{16\pi}} (3\cos^2\theta - 1)$$

$$Y_2^{-1} = \sqrt{\frac{15}{8\pi}} e^{-i\phi} \sin\theta \cos\theta, \quad Y_2^{-2} = \sqrt{\frac{15}{32\pi}} e^{-i2\phi} \sin^2\theta$$

We will see in chapter 21 that these functions of the angular variables in spherical coordinates are exactly the functions that give the angular dependence of wavefunctions for the physical system of a particle in a spherically symmetric potential. In such a case, the $SO(3)$ symmetry of the system implies that the state space (the wavefunctions) will provide a unitary representation π of $SO(3)$, and the action of the Hamiltonian operator H will commute with the action of the operators $L_3, L_{\pm}$. As a result, all of the states in an irreducible representation component of π will have the same energy. States are thus organized into "orbitals," with singlet states called "s" orbitals ($l = 0$), triplet states called "p" orbitals ($l = 1$), multiplicity 5 states called "d" orbitals ($l = 2$), etc.

8.4 The Casimir operator

For both $SU(2)$ and $SO(3)$, we have found that all representations can be constructed out of function spaces, with the Lie algebra acting as first-order differential operators. It turns out that there is also a very interesting second-order differential operator that comes from these Lie algebra representations, known as the Casimir operator. For the case of $SO(3)$:

Definition (Casimir operator for $SO(3)$). *The Casimir operator for the representation of $SO(3)$ on functions on S^2 is the second-order differential operator*

$$L^2 \equiv L_1^2 + L_2^2 + L_3^2$$

(the symbol L^2 is not intended to mean that this is the square of an operator L)

A straightforward calculation using the commutation relations satisfied by the L_j shows that

$$[L^2, \rho'(X)] = 0$$

for any $X \in \mathfrak{so}(3)$. Knowing this, a version of Schur's lemma says that L^2 will act on an irreducible representation as a scalar (i.e., all vectors in the representation are eigenvectors of L^2, with the same eigenvalue). This eigenvalue can be used to characterize the irreducible representation.

The easiest way to compute this eigenvalue turns out to be to act with L^2 on a highest weight vector. First one rewrites L^2 in terms of raising and lowering operators

$$\begin{aligned} L_- L_+ &= (L_1 - iL_2)(L_1 + iL_2) \\ &= L_1^2 + L_2^2 + i[L_1, L_2] \\ &= L_1^2 + L_2^2 - L_3 \end{aligned}$$

so

$$L^2 = L_1^2 + L_2^2 + L_3^2 = L_- L_+ + L_3 + L_3^2$$

For the representation ρ of $SO(3)$ on functions on S^2 constructed above, we know that on a highest weight vector of the irreducible representation ρ_l (restriction of ρ to the $2l + 1$-dimensional irreducible subspace of functions that are linear combinations of the $Y_l^m(\theta, \phi)$), we have the two eigenvalue equations

$$L_+ f = 0, \quad L_3 f = lf$$

with solution the functions proportional to $Y_l^l(\theta, \phi)$. Just from these conditions and our expression for L^2, we can immediately find the scalar eigenvalue of L^2 since

$$L^2 f = L_- L_+ f + (L_3 + L_3^2) f = (0 + l + l^2) f = l(l + 1) f$$

We have thus shown that our irreducible representation ρ_l can be characterized as the representation on which L^2 acts by the scalar $l(l + 1)$.

In summary, we have two different sets of partial differential equations whose solutions provide a highest weight vector for and thus determine the irreducible representation ρ_l:

- $$L_+ f = 0, \quad L_3 f = lf$$

  which are first-order equations, with the first using complexification and something like a Cauchy–Riemann equation, and

- $$L^2 f = l(l + 1) f, \quad L_3 f = lf$$

  where the first equation is a second-order equation, something like a Laplace equation.

That a solution of the first set of equations gives a solution of the second set is obvious. Much harder to show is that a solution of the second set gives a solution of the first set. The space of solutions to

$$L^2 f = l(l + 1) f$$

for l a nonnegative integer includes as we have seen the $2l + 1$-dimensional vector space of linear combinations of the $Y_l^m(\theta, \phi)$ (there are no other solutions, although we will not show that). Since the action of $SO(3)$ on functions commutes with the operator L^2, this $2l + 1$-dimensional space will provide a representation, the irreducible one of spin l.

The second-order differential operator L^2 in the ρ representation on functions can explicitly be computed. It is

$$L^2 = L_1^2 + L_2^2 + L_3^2$$

$$= \left(i \left(\sin\phi \frac{\partial}{\partial\theta} + \cot\theta \cos\phi \frac{\partial}{\partial\phi} \right) \right)^2$$

$$+ \left(i \left(-\cos\phi \frac{\partial}{\partial\theta} + \cot\theta \sin\phi \frac{\partial}{\partial\phi} \right) \right)^2 + \left(-i \frac{\partial}{\partial\phi} \right)^2$$

$$\doteq - \left(\frac{1}{\sin\theta} \frac{\partial}{\partial\theta} \left(\sin\theta \frac{\partial}{\partial\theta} \right) + \frac{1}{\sin^2\theta} \frac{\partial^2}{\partial\phi^2} \right) \tag{8.4}$$

We will re-encounter this operator in chapter 21 as the angular part of the Laplaceoperator on $\mathbf{R}^3$.

For the group $SU(2)$, we can also find irreducible representations as solution spaces of differential equations on functions on $\mathbf{C}^2$. In that case, the differential equation point of view is much less useful, since the solutions we are looking for are just the homogeneous polynomials, which are more easily studied by purely algebraic methods.

8.5 For further reading

The classification of $SU(2)$ representations is a standard topic in all textbooks that deal with Lie group representations. A good example is [40], which covers this material well, and from which the discussion here of the construction of representations as homogeneous polynomials is drawn (see pages 77-79). The calculation of the L_j and the derivation of expressions for spherical harmonics as Lie algebra representations of $\mathfrak{so}(3)$ appears in most quantum mechanics textbooks in one form or another (e.g., see chapter 12 of [81]). Another source used here for the explicit constructions of representations is [20], chapters 27-30.

A conventional topic in books on representation theory in physics is that of the representation theory of the group $SU(3)$, or even of $SU(n)$ for arbitrary n. The case $n = 3$ is of great historical importance, because of its use in the classification and study of strongly interacting particles, The success of these methods is now understood as due to an approximate $SU(3)$ symmetry of the strong interaction theory corresponding to the existence of three different light quarks. The highest weight theory of $SU(2)$ representations can be generalized to the case of $SU(n)$, as well as to finite dimensional representations of other Lie groups. We will not try and cover this topic here since it is a bit intricate and is already very well described in many textbooks aimed at mathematicians (e.g., [42]) and at physicists (e.g., [32]).

Chapter 9
Tensor Products, Entanglement, and Addition of Spin

If one has two independent quantum systems, with state spaces $\mathcal{H}_1$ and $\mathcal{H}_2$, the combined quantum system has a description that exploits the mathematical notion of a "tensor product," with the combined state space the tensor product $\mathcal{H}_1 \otimes \mathcal{H}_2$. Because of the ability to take linear combinations of states, this combined state space will contain much more than just products of independent states, including states that are described as "entangled" and responsible for some of the most counterintuitive behavior of quantum physical systems.

This same tensor product construction is a basic one in representation theory, allowing one to construct a new representation $(\pi_{W_1 \otimes W_2}, W_1 \otimes W_2)$ out of representations (π_{W_1}, W_1) and (π_{W_2}, W_2). When we take the tensor product of states corresponding to two irreducible representations of $SU(2)$ of spins s_1, s_2, we will get a new representation $(\pi_{V^{2s_1} \otimes V^{2s_2}}, V^{2s_1} \otimes V^{2s_2})$. It will be reducible, a direct sum of representations of various spins, a situation we will analyze in detail.

Starting with a quantum system with state space $\mathcal{H}$ that describes a single particle, a system of n particles can be described by taking an n-fold tensor product $\mathcal{H}^{\otimes n} = \mathcal{H} \otimes \mathcal{H} \otimes \cdots \otimes \mathcal{H}$. It turns out that for identical particles, we do not get the full tensor product space, but only the subspaces either symmetric or antisymmetric under the action of the permutation group by permutations of the factors, depending on whether our particles are "bosons" or "fermions." This is a separate postulate in quantum mechanics, but finds an explanation when particles are treated as quanta of quantum fields.

Digression. *When physicists refer to "tensors," they generally mean the "tensor fields" used in general relativity or other geometry-based parts of physics, not tensor products of state spaces. A tensor field is a function on a manifold, taking values in some tensor product of copies of the tangent space and its dual space. The simplest tensor fields are vector fields, functions*

© Peter Woit 2017
P. Woit, *Quantum Theory, Groups and Representations*,
DOI 10.1007/978-3-319-64612-1_9

taking values in the tangent space. A more non-trivial example is the metric tensor, which takes values in the dual of the tensor product of two copies of the tangent space.

9.1 Tensor products

Given two vector spaces V and W (over $\mathbf{R}$ or $\mathbf{C}$), the direct sum vector space $V \oplus W$ is constructed by taking pairs of elements (v, w) for $v \in V, w \in W$ and giving them a vector space structure by the obvious addition and multiplication by scalars. This space will have dimension

$$\dim(V \oplus W) = \dim V + \dim W$$

If $\{\mathbf{e}_1, \mathbf{e}_2, \ldots, \mathbf{e}_{\dim V}\}$ is a basis of V, and $\{\mathbf{f}_1, \mathbf{f}_2, \ldots, \mathbf{f}_{\dim W}\}$ a basis of W, the

$$\{\mathbf{e}_1, \mathbf{e}_2, \ldots, \mathbf{e}_{\dim V}, \mathbf{f}_1, \mathbf{f}_2, \ldots, \mathbf{f}_{\dim W}\}$$

will be a basis of $V \oplus W$.

A less trivial construction is the tensor product of the vector spaces V and W. This will be a new vector space called $V \otimes W$, of dimension

$$\dim(V \otimes W) = (\dim V)(\dim W)$$

One way to motivate the tensor product is to think of vector spaces as vector spaces of functions. Elements

$$v = v_1 \mathbf{e}_1 + v_2 \mathbf{e}_2 + \cdots + v_{\dim V} \mathbf{e}_{\dim V} \in V$$

can be thought of as functions on the $\dim V$ points $\mathbf{e}_j$, taking values v_j at $\mathbf{e}_j$. If one takes functions on the union of the sets $\{\mathbf{e}_j\}$ and $\{\mathbf{f}_k\}$, one gets elements of $V \oplus W$. The tensor product $V \otimes W$ will be what one gets by taking all functions on not the union, but the product of the sets $\{\mathbf{e}_j\}$ and $\{\mathbf{f}_k\}$. This will be the set with $(\dim V)(\dim W)$ elements, which we will write $\mathbf{e}_j \otimes \mathbf{f}_k$, and elements of $V \otimes W$ will be functions on this set, or equivalently, linear combinations of these basis vectors.

This sort of definition is less than satisfactory, since it is tied to an explicit choice of bases for V and W. We will not, however, pursue more details of this question or a better definition here. For this, one can consult pretty much any advanced undergraduate text in abstract algebra, but here we will take as given the following properties of the tensor product that we will need:

- Given vectors $v \in V, w \in W$, we get an element $v \otimes w \in V \otimes W$, satisfying bilinearity conditions (for c_1, c_2 constants)

$$v \otimes (c_1 w_1 + c_2 w_2) = c_1 (v \otimes w_1) + c_2 (v \otimes w_2)$$

$$(c_1 v_1 + c_2 v_2) \otimes w = c_1 (v_1 \otimes w) + c_2 (v_2 \otimes w)$$

- There are natural isomorphisms

$$V \otimes W \simeq W \otimes V$$

and

$$U \otimes (V \otimes W) \simeq (U \otimes V) \otimes W$$

for vector spaces U, V, W

- Given a linear operator A on V and another linear operator B on W, we can define a linear operator $A \otimes B$ on $V \otimes W$ by

$$(A \otimes B)(v \otimes w) = Av \otimes Bw$$

for $v \in V, w \in W$.

With respect to the bases $\mathbf{e}_j, \mathbf{f}_k$ of V and W, A will be a $(\dim V)$ by $(\dim V)$ matrix, B will be a $(\dim W)$ by $(\dim W)$ matrix, and $A \otimes B$ will be a $(\dim V)(\dim W)$ by $(\dim V)(\dim W)$ matrix (which can be thought of as a $(\dim V)$ by $(\dim V)$ matrix of blocks of size $(\dim W)$).

- One often wants to consider tensor products of vector spaces and dual vector spaces. An important fact is that there is an isomorphism between the tensor product $V^* \otimes W$ and linear maps from V to W. This is given by identifying $l \otimes w$ $(l \in V^*, w \in W)$ with the linear map

$$v \in V \to l(v)w \in W$$

- Given the motivation in terms of functions on a product of sets, for function spaces in general we should have an identification of the tensor product of function spaces with functions on the product set. For instance, for square-integrable functions on $\mathbf{R}$ we expect

$$\underbrace{L^2(\mathbf{R}) \otimes L^2(\mathbf{R}) \otimes \cdots \otimes L^2(\mathbf{R})}_{n \text{ times}} = L^2(\mathbf{R}^n) \tag{9.1}$$

For V a real vector space, its complexification $V_{\mathbf{C}}$ (see section 5.5) can be identified with the tensor product

$$V_{\mathbf{C}} = V \otimes_{\mathbf{R}} \mathbf{C}$$

Here, the notation $\otimes_{\mathbf{R}}$ indicates a tensor product of two real vector spaces: V of dimension $\dim V$ with basis $\{\mathbf{e}_1, \mathbf{e}_2, \ldots, \mathbf{e}_{\dim V}\}$ and $\mathbf{C} = \mathbf{R}^2$ of dimension 2 with basis $\{1, i\}$.

9.2 Composite quantum systems and tensor products

Consider two quantum systems: one defined by a state space $\mathcal{H}_1$ and a set of operators $\mathcal{O}_1$ on it, and the second given by a state space $\mathcal{H}_2$ and set of operators $\mathcal{O}_2$. One can describe the composite quantum system corresponding to considering the two quantum systems as a single one, with no interaction between them, by taking as a new state space

$$\mathcal{H}_T = \mathcal{H}_1 \otimes \mathcal{H}_2$$

with operators of the form

$$A \otimes \mathbf{Id} + \mathbf{Id} \otimes B$$

with $A \in \mathcal{O}_1, B \in \mathcal{O}_2$. The state space $\mathcal{H}_T$ can be used to describe an interacting quantum system, but with a more general class of operators.

If $\mathcal{H}$ is the state space of a quantum system, this can be thought of as describing a single particle. Then, a system of N such particles is described by the multiple tensor product

$$\mathcal{H}^{\otimes n} = \underbrace{\mathcal{H} \otimes \mathcal{H} \otimes \cdots \otimes \mathcal{H} \otimes \mathcal{H}}_{n \text{ times}}$$

The symmetric group S_n acts on this state space, and one has a representation $(\pi, \mathcal{H}^{\otimes n})$ of S_n as follows. For $\sigma \in S_n$ a permutation of the set $\{1, 2, \ldots, n\}$ of n elements, on a tensor product of vectors one has

$$\pi(\sigma)(v_1 \otimes v_2 \otimes \cdots \otimes v_n) = v_{\sigma(1)} \otimes v_{\sigma(2)} \otimes \cdots \otimes v_{\sigma(n)}$$

The representation of S_n that this gives is in general reducible, containing various components with different irreducible representations of the group S_n.

A fundamental axiom of quantum mechanics is that if $\mathcal{H}^{\otimes n}$ describes n identical particles, then all physical states occur as one-dimensional representations of S_n. These are either symmetric ("bosons") or antisymmetric ("fermions") where:

Definition. *A state $v \in \mathcal{H}^{\otimes n}$ is called*

- *symmetric, or bosonic if $\forall \sigma \in S_n$*

$$\pi(\sigma)v = v$$

The space of such states is denoted $S^n(\mathcal{H})$.

- *antisymmetric, or fermionic if $\forall \sigma \in S_n$*

$$\pi(\sigma)v = (-1)^{|\sigma|}v$$

The space of such states is denoted $\Lambda^n(\mathcal{H})$. Here, $|\sigma|$ is the minimal number of transpositions that by composition give σ.

Note that in the fermionic case, for σ a transposition interchanging two particles, π acts on the factor $\mathcal{H} \otimes \mathcal{H}$ by interchanging vectors, taking

$$w \otimes w \in \mathcal{H} \otimes \mathcal{H}$$

to itself for any vector $w \in \mathcal{H}$. Antisymmetry requires that π takes this state to its negative, so the state cannot be nonzero. As a result, one cannot have nonzero states in $\mathcal{H}^{\otimes n}$ describing two identical particles in the same state $w \in \mathcal{H}$, a fact that is known as the "Pauli principle."

While the symmetry or antisymmetry of states of multiple identical particles is a separate axiom when such particles are described in this way as tensor products, we will see later on (chapter 36) that this phenomenon instead finds a natural explanation when particles are described in terms of quantum fields.

9.3 Indecomposable vectors and entanglement

If one is given a function f on a space X and a function g on a space Y, a product function fg on the product space $X \times Y$ can be defined by taking (for $x \in X, y \in Y$)

$$(fg)(x,y) = f(x)g(y)$$

However, most functions on $X \times Y$ are not decomposable in this manner. Similarly, for a tensor product of vector spaces:

Definition (Decomposable and indecomposable vectors). *A vector in $V \otimes W$ is called decomposable if it is of the form $v \otimes w$ for some $v \in V, w \in W$. If it cannot be put in this form, it is called indecomposable.*

Note that our basis vectors of $V \otimes W$ are all decomposable since they are products of basis vectors of V and W. Linear combinations of these basis vectors, however, are in general indecomposable. If we think of an element of $V \otimes W$ as a $\dim V$ by $\dim W$ matrix, with entries the coordinates with respect to our basis vectors for $V \otimes W$, then for decomposable vectors we get a special class of matrices, those of rank one.

In the physics context, the language used is:

Definition (Entanglement). *An indecomposable state in the tensor product state space $\mathcal{H}_T = \mathcal{H}_1 \otimes \mathcal{H}_2$ is called an entangled state.*

The phenomenon of entanglement is responsible for some of the most surprising and subtle aspects of quantum mechanical systems. The Einstein–Podolsky–Rosen paradox concerns the behavior of an entangled state of two quantum systems, when one moves them far apart. Then performing a measurement on one system can give one information about what will happen if one performs a measurement on the far-removed system, introducing a sort of unexpected apparent non-locality.

Measurement theory itself involves crucially an entanglement between the state of a system being measured, thought of as in a state space $\mathcal{H}_{system}$, and the state of the measurement apparatus, thought of as lying in a state space $\mathcal{H}_{apparatus}$. The laws of quantum mechanics presumably apply to the total system $\mathcal{H}_{system} \otimes \mathcal{H}_{apparatus}$, with the counterintuitive nature of measurements appearing due to this decomposition of the world into two entangled parts: the one under study and a much larger for which only an approximate description in classical terms is possible. For much more about this, a recommended reading is chapter 2 of [75].

9.4 Tensor products of representations

Given two representations of a group, a new representation can be defined, the tensor product representation, by:

Definition (Tensor product representation of a group). *For (π_V, V) and (π_W, W) representations of a group G, there is a tensor product representation $(\pi_{V \otimes W}, V \otimes W)$ defined by*

$$(\pi_{V \otimes W}(g))(v \otimes w) = \pi_V(g)v \otimes \pi_W(g)w$$

One can easily check that $\pi_{V \otimes W}$ is a homomorphism.

To see what happens for the corresponding Lie algebra representation, compute (for X in the Lie algebra)

$$\pi'_{V \otimes W}(X)(v \otimes w) = \frac{d}{dt} \pi_{V \otimes W}(e^{tX})(v \otimes w)_{t=0}$$

$$= \frac{d}{dt}(\pi_V(e^{tX})v \otimes \pi_W(e^{tX})w)_{t=0}$$

$$= \left(\left(\frac{d}{dt} \pi_V(e^{tX})v \right) \otimes \pi_W(e^{tX})w \right)_{t=0}$$

$$+ \left(\pi_V(e^{tX})v \otimes \left(\frac{d}{dt} \pi_W(e^{tX})w \right) \right)_{t=0}$$

$$= (\pi'_V(X)v) \otimes w + v \otimes (\pi'_W(X)w)$$

which could also be written

$$\pi'_{V \otimes W}(X) = (\pi'_V(X) \otimes 1_W) + (1_V \otimes \pi'_W(X))$$

9.4.1   Tensor products of $SU(2)$ representations

Given two representations (π_V, V) and (π_W, W) of a group G, we can decompose each into irreducibles. To do the same for the tensor product of the two representations, we need to know how to decompose the tensor product of two irreducibles. This is a fundamental and non-trivial question, with the answer for $G = SU(2)$ as follows:

Theorem 9.1 (Clebsch–Gordan decomposition).
   The tensor product $(\pi_{V^{n_1} \otimes V^{n_2}}, V^{n_1} \otimes V^{n_2})$ *decomposes into irreducibles as*

$$(\pi_{n_1+n_2}, V^{n_1+n_2}) \oplus (\pi_{n_1+n_2-2}, V^{n_1+n_2-2}) \oplus \cdots \oplus (\pi_{|n_1-n_2|}, V^{|n_1-n_2|})$$

Proof. One way to prove this result is to use highest weight theory, raising and lowering operators, and the formula for the Casimir operator. We will not try and show the details of how this works out, but in the next section give a simpler argument using characters. However, in outline (for more details, see, for instance, section 5.2 of [71]), here is how one could proceed:
   One starts by noting that if $v_{n_1} \in V_{n_1}, v_{n_2} \in V_{n_2}$ are highest weight vectors for the two representations, $v_{n_1} \otimes v_{n_2}$ will be a highest weight vector in the tensor product representation (i.e., annihilated by $\pi'_{n_1+n_2}(S_+)$), of weight $n_1 + n_2$. So $(\pi_{n_1+n_2}, V^{n_1+n_2})$ will occur in the decomposition. Applying $\pi'_{n_1+n_2}(S_-)$ to $v_{n_1} \otimes v_{n_2}$, one gets a basis of the rest of the vectors in $(\pi_{n_1+n_2}, V^{n_1+n_2})$. However, at weight n_1+n_2-2, one can find another kind of vector, a highest weight vector orthogonal to the vectors in $(\pi_{n_1+n_2}, V^{n_1+n_2})$. Applying the lowering operator to this gives $(\pi_{n_1+n_2-2}, V^{n_1+n_2-2})$. As before,

at weight $n_1 + n_2 - 4$ one finds another, orthogonal highest weight vector, and gets another representation, with this process only terminating at weight $|n_1 - n_2|$.   □

9.4.2 Characters of representations

A standard tool for dealing with representations is that of associating with a representation an invariant called its character. This will be a conjugation invariant function on the group that only depends on the equivalence class of the representation. Given two representations constructed in very different ways, it is often possible to check whether they are isomorphic by seeing whether their character functions match. The problem of identifying the possible irreducible representations of a group can be attacked by analyzing the possible character functions of irreducible representations. We will not try and enter into the general theory of characters here, but will just see what the characters of irreducible representations are for the case of $G = SU(2)$. These can be used to give a simple argument for the Clebsch–Gordan decomposition of the tensor product of $SU(2)$ representations. For this, we do not need general theorems about the relations of characters and representations, but can directly check that the irreducible representations of $SU(2)$ correspond to distinct character functions which are easily evaluated.

Definition (Character). *The character of a representation (π, V) of a group G is the function on G given by*

$$\chi_V(g) = Tr(\pi(g))$$

Since the trace of a matrix is invariant under conjugation, χ_V will be a complex-valued, conjugation invariant function on G. One can easily check that it will satisfy the relations

$$\chi_{V \oplus W} = \chi_V + \chi_W, \quad \chi_{V \otimes W} = \chi_V \chi_W$$

For the case of $G = SU(2)$, any element can be conjugated to be in the $U(1)$ subgroup of diagonal matrices. Knowing the weights of the irreducible representations (π_n, V^n) of $SU(2)$, we know the characters to be the functions

$$\chi_{V^n}\left(\begin{pmatrix} e^{i\theta} & 0 \\ 0 & e^{-i\theta} \end{pmatrix}\right) = e^{in\theta} + e^{i(n-2)\theta} + \cdots + e^{-i(n-2)\theta} + e^{-in\theta} \qquad (9.2)$$

As n gets large, this becomes an unwieldy expression, but one has

Theorem (Weyl character formula).

$$\chi_{V^n}\left(\begin{pmatrix} e^{i\theta} & 0 \\ 0 & e^{-i\theta} \end{pmatrix}\right) = \frac{e^{i(n+1)\theta} - e^{-i(n+1)\theta}}{e^{i\theta} - e^{-i\theta}} = \frac{\sin((n+1)\theta)}{\sin(\theta)}$$

Proof. One just needs to use the identity

$$(e^{in\theta} + e^{i(n-2)\theta} + \cdots + e^{-i(n-2)\theta} + e^{-in\theta})(e^{i\theta} - e^{-i\theta}) = e^{i(n+1)\theta} - e^{-i(n+1)\theta}$$

and equation 9.2 for the character. □

To get a proof of 9.1, compute the character of the tensor product on the diagonal matrices using the Weyl character formula for the second factor (ordering things so that $n_2 > n_1$)

$$\chi_{V^{n_1} \otimes V^{n_2}} = \chi_{V^{n_1}} \chi_{V^{n_2}}$$

$$= (e^{in_1\theta} + e^{i(n_1-2)\theta} + \cdots + e^{-i(n_1-2)\theta} + e^{-in_1\theta}) \frac{e^{i(n_2+1)\theta} - e^{-i(n_2+1)\theta}}{e^{i\theta} - e^{-i\theta}}$$

$$= \frac{(e^{i(n_1+n_2+1)\theta} - e^{-i(n_1+n_2+1)\theta}) + \cdots + (e^{i(n_2-n_1+1)\theta} - e^{-i(n_2-n_1+1)\theta})}{e^{i\theta} - e^{-i\theta}}$$

$$= \chi_{V^{n_1+n_2}} + \chi_{V^{n_1+n_2-2}} + \cdots + \chi_{V^{n_2-n_1}}$$

So, when we decompose the tensor product of irreducibles into a direct sum of irreducibles, the ones that must occur are exactly those of theorem 9.1.

9.4.3 Some examples

Some simple examples of how this works are:

- Tensor product of two spinors:

$$V^1 \otimes V^1 = V^2 \oplus V^0$$

  This says that the four-complex dimensional tensor product of two spinor representations (which are each two-complex dimensional) decomposes into irreducibles as the sum of a three-dimensional vector representation and a one-dimensional trivial (scalar) representation. Using the basis $\begin{pmatrix} 1 \\ 0 \end{pmatrix}, \begin{pmatrix} 0 \\ 1 \end{pmatrix}$ for V^1, the tensor product $V^1 \otimes V^1$ has a basis

$$\begin{pmatrix} 1 \\ 0 \end{pmatrix} \otimes \begin{pmatrix} 1 \\ 0 \end{pmatrix}, \quad \begin{pmatrix} 1 \\ 0 \end{pmatrix} \otimes \begin{pmatrix} 0 \\ 1 \end{pmatrix}, \quad \begin{pmatrix} 0 \\ 1 \end{pmatrix} \otimes \begin{pmatrix} 1 \\ 0 \end{pmatrix}, \quad \begin{pmatrix} 0 \\ 1 \end{pmatrix} \otimes \begin{pmatrix} 0 \\ 1 \end{pmatrix}$$

The vector

$$\frac{1}{\sqrt{2}}(\begin{pmatrix}1\\0\end{pmatrix} \otimes \begin{pmatrix}0\\1\end{pmatrix} - \begin{pmatrix}0\\1\end{pmatrix} \otimes \begin{pmatrix}1\\0\end{pmatrix}) \in V^1 \otimes V^1$$

is clearly antisymmetric under permutation of the two factors of $V^1 \otimes V^1$. One can show that this vector is invariant under $SU(2)$, by computing either the action of $SU(2)$ or of its Lie algebra $\mathfrak{su}(2)$. So, this vector is a basis for the component V^0 in the decomposition of $V^1 \otimes V^1$ into irreducibles.

The other component, V^2, is three-dimensional and has a basis

$$\begin{pmatrix}1\\0\end{pmatrix} \otimes \begin{pmatrix}1\\0\end{pmatrix}, \quad \frac{1}{\sqrt{2}}(\begin{pmatrix}1\\0\end{pmatrix} \otimes \begin{pmatrix}0\\1\end{pmatrix} + \begin{pmatrix}0\\1\end{pmatrix} \otimes \begin{pmatrix}1\\0\end{pmatrix}), \quad \begin{pmatrix}0\\1\end{pmatrix} \otimes \begin{pmatrix}0\\1\end{pmatrix}$$

These three vectors span one-dimensional complex subspaces of weights $q = 2, 0, -2$ under the $U(1) \subset SU(2)$ subgroup

$$\begin{pmatrix}e^{i\theta} & 0\\0 & e^{-i\theta}\end{pmatrix}$$

They are symmetric under permutation of the two factors of $V^1 \otimes V^1$. We see that if we take two identical quantum systems with $\mathcal{H} = V^1 = \mathbf{C}^2$ and make a composite system out of them, if they were bosons we would get a three-dimensional state space $V^2 = S^2(V^1)$, transforming as a vector (spin one) under $SU(2)$. If they were fermions, we would get a one-dimensional state space $V^0 = \Lambda^2(V^1)$ of spin zero (invariant under $SU(2)$). Note that in this second case, we automatically get an entangled state, one that cannot be written as a decomposable product.

- Tensor product of three or more spinors:

$$V^1 \otimes V^1 \otimes V^1 = (V^2 \oplus V^0) \otimes V^1 = (V^2 \otimes V^1) \oplus (V^0 \otimes V^1) = V^3 \oplus V^1 \oplus V^1$$

This says that the tensor product of three spinor representations decomposes as a four-dimensional ("spin 3/2") representation plus two copies of the spinor representation. This can be generalized by considering N-fold tensor products $(V^1)^{\otimes N}$ of the spinor representation. This will be a sum of irreducible representations, including one copy of the irreducible V^N, giving an alternative to the construction using homogeneous polynomials. Doing this, however, gives the irreducible as just one component of something larger, and a method is needed to project out the desired component. This can be done using the action of the symmetric group S_N on $(V^1)^{\otimes N}$ and an understanding of the irreducible representations of S_N. This relationship between irreducible representations of $SU(2)$ and those of S_N coming from looking at how both groups act on $(V^1)^{\otimes N}$

is known as "Schur-Weyl duality." This generalizes to the case of $SU(n)$ for arbitrary n, where one can consider N-fold tensor products of the defining representation of $SU(n)$ matrices on $\mathbf{C}^n$. For $SU(n)$, this provides perhaps the most straightforward construction of all irreducible representations of the group.

9.5 Bilinear forms and tensor products

A different sort of application of tensor products that will turn out to be important is to the description of bilinear forms, which generalize the dual space V^* of linear forms on V. We have:

Definition. (Bilinear forms). *A bilinear form B on a vector space V over a field k (for us, $k = \mathbf{R}$ or $\mathbf{C}$) is a map*

$$B : (v, v') \in V \times V \to B(v, v') \in k$$

that is bilinear in both entries, i.e.,

$$B(v + v', v'') = B(v, v'') + B(v', v''), \quad B(cv, v') = cB(v, v')$$

$$B(v, v' + v'') = B(v, v') + B(v, v''), \quad B(v, cv') = cB(v, v')$$

where $c \in k$.

If $B(v', v) = B(v, v')$, the bilinear form is called symmetric; if $B(v', v) = -B(v, v')$, it is antisymmetric.

The relation to tensor products is

Theorem 9.2. *The space of bilinear forms on V is isomorphic to $V^* \otimes V^*$.*

Proof. The map

$$\alpha_1 \otimes \alpha_2 \in V^* \otimes V^* \to B : B(v, v') = \alpha_1(v)\alpha_2(v')$$

provides, in a basis-independent way, the isomorphism we are looking for. One can show this is an isomorphism using a basis. Choosing a basis $\mathbf{e}_j$ of V, the coordinate functions $v_j = \mathbf{e}_j^*$ provide a basis of V^*, so the $v_j \otimes v_k$ will be a basis of $V^* \otimes V^*$. The map above takes linear combinations of these to bilinear forms and is easily seen to be one-to-one and surjective for such linear combinations. $\square$

Given a basis $\mathbf{e}_j$ of V and dual basis v_j of V^* (the coordinates), the element of $V^* \otimes V^*$ corresponding to B can be written as the sum

$$\sum_{j,k} B_{jk} v_j \otimes v_k$$

This expresses the bilinear form B in terms of a matrix $\mathbf{B}$ with entries B_{jk}, which can be computed as

$$B_{jk} = B(\mathbf{e}_j, \mathbf{e}_k)$$

In terms of the matrix $\mathbf{B}$, the bilinear form is computed as

$$B(v, v') = \begin{pmatrix} v_1 & \cdots & v_d \end{pmatrix} \begin{pmatrix} B_{11} & \cdots & B_{1d} \\ \vdots & \vdots & \vdots \\ B_{d1} & \cdots & B_{dd} \end{pmatrix} \begin{pmatrix} v'_1 \\ \vdots \\ v'_d \end{pmatrix} = \mathbf{v} \cdot \mathbf{B} v'$$

9.6 Symmetric and antisymmetric multilinear forms

The symmetric bilinear forms lie in $S^2(V^*) \subset V^* \otimes V^*$ and correspond to symmetric matrices. Elements of V^* give linear functions on V, and one can get quadratic functions on V from elements $B \in S^2(V^*)$ by taking

$$v \in V \rightarrow B(v, v) = \mathbf{v} \cdot \mathbf{B} v$$

Equivalently, in terms of tensor products, one gets quadratic functions as the product of linear functions by taking

$$(\alpha_1, \alpha_2) \in V^* \times V^* \rightarrow \frac{1}{2}(\alpha_1 \otimes \alpha_2 + \alpha_2 \otimes \alpha_1) \in S^2(V^*)$$

and then evaluating at $v \in V$ to get the number

$$\frac{1}{2}(\alpha_1(v)\alpha_2(v) + \alpha_2(v)\alpha_1(v)) = \alpha_1(v)\alpha_2(v)$$

This multiplication can be extended to a product on the space

$$S^*(V^*) = \oplus_n S^n(V^*)$$

(called the space of symmetric multilinear forms) by defining

$$(\alpha_1 \otimes \cdots \otimes \alpha_j)(\alpha_{j+1} \otimes \cdots \otimes \alpha_n) = P^+(\alpha_1 \otimes \cdots \otimes \alpha_n)$$
$$\equiv \frac{1}{n!} \sum_{\sigma \in S_n} \alpha_{\sigma(1)} \otimes \cdots \otimes \alpha_{\sigma(n)} \qquad (9.3)$$

One can show that $S^*(V^*)$ with this product is isomorphic to the algebra of polynomials on V. For a simple example of how this works, take $v_j \in V^*$ to be the jth coordinate function. Then, the correspondence between monomials in v_j and elements of $S^*(V^*)$ is given by

$$v_j^n \leftrightarrow \underbrace{(v_j \otimes v_j \otimes \cdots \otimes v_j)}_{\text{n-times}} \tag{9.4}$$

Both sides can be thought of as the same function on V, given by evaluating the jth coordinate of $v \in V$ and multiplying it by itself n-times.

We will later find useful the fact that $S^*(V^*)$ and $(S^*(V))^*$ are isomorphic, with the tensor product

$$(\alpha_1 \otimes \cdots \otimes \alpha_j) \in S^*(V^*)$$

corresponding to the linear map

$$\mathbf{e}_{i_1} \otimes \cdots \otimes \mathbf{e}_{i_j} \to \alpha_1(\mathbf{e}_{i_1}) \cdots \alpha_j(\mathbf{e}_{i_j})$$

Antisymmetric bilinear forms lie in $\Lambda^2(V^*) \subset V^* \otimes V^*$ and correspond to antisymmetric matrices. A multiplication (called the "wedge product") can be defined on V^* that takes values in $\Lambda^2(V^*)$ by

$$(\alpha_1, \alpha_2) \in V^* \times V^* \to \alpha_1 \wedge \alpha_2 = \frac{1}{2}(\alpha_1 \otimes \alpha_2 - \alpha_2 \otimes \alpha_1) \in \Lambda^2(V^*) \tag{9.5}$$

This multiplication can be extended to a product on the space

$$\Lambda^*(V^*) = \oplus_n \Lambda^n(V^*)$$

(called the space of antisymmetric multilinear forms) by defining

$$(\alpha_1 \otimes \cdots \otimes \alpha_j) \wedge (\alpha_{j+1} \otimes \cdots \otimes \alpha_n) = P^-(\alpha_1 \otimes \cdots \otimes \alpha_n)$$
$$\equiv \frac{1}{n!} \sum_{\sigma \in S_n} (-1)^{|\sigma|} \alpha_{\sigma(1)} \otimes \cdots \otimes \alpha_{\sigma(n)} \tag{9.6}$$

This can be used to get a product on the space of antisymmetric multilinear forms of different degrees, giving something in many ways analogous to the algebra of polynomials (although without a notion of evaluation at a point v). This plays a role in the description of fermions and will be considered in more detail in chapter 30. Much like in the symmetric case, there is an isomorphism between $\Lambda^*(V^*)$ and $(\Lambda^*(V))^*$.

9.7 For further reading

For more about the tensor product and tensor product of representations, see section 6 of [95], or appendix B of [84]. Almost every quantum mechanics textbook will contain an extensive discussion of the Clebsch–Gordan decomposition for the tensor product of two irreducible $SU(2)$ representations.

A complete discussion of bilinear forms, together with the algebra of symmetric and antisymmetric multilinear forms, can be found in [36].

Chapter 10
Momentum and the Free Particle

We will now turn to the problem that conventional quantum mechanics courses generally begin with: that of the quantum system describing a free particle moving in physical space $\mathbf{R}^3$. This is something quite different from the classical mechanical description of a free particle, which will be reviewed in chapter 14. A common way of motivating this is to begin with the 1924 suggestion by de Broglie that just as photons may behave like either particles or waves, the same should be true for matter particles. Photons were known to carry an energy given by $E = \hbar\omega$, where ω is the angular frequency of the wave. De Broglie's proposal was that matter particles with momentum $\mathbf{p} = \hbar\mathbf{k}$ can also behave like a wave, with dependence on the spatial position $\mathbf{q}$ given by

$$e^{i\mathbf{k}\cdot\mathbf{q}}$$

This proposal was realized in Schrödinger's early 1926 discovery of a version of quantum mechanics in which the state space is

$$\mathcal{H} = L^2(\mathbf{R}^3)$$

which is the space of square-integrable complex-valued functions on $\mathbf{R}^3$, called "wavefunctions." The operator

$$\mathbf{P} = -i\hbar\boldsymbol{\nabla}$$

will have eigenvalues $\hbar\mathbf{k}$, the de Broglie momentum, so it can be identified as the momentum operator.

In this chapter, our discussion will emphasize the central role of the momentum operator. This operator will have the same relationship to spatial translations as the Hamiltonian operator does to time translations. In both cases, the operators are the Lie algebra representation operators corresponding to a unitary representation on the quantum state space $\mathcal{H}$ of groups of

© Peter Woit 2017
P. Woit, *Quantum Theory, Groups and Representations*,
DOI 10.1007/978-3-319-64612-1_10

translations (translations in the three space and one time directions, respectively).

One way to motivate the quantum theory of a free particle is that whatever it is, it should have analogous behavior to that of the classical case under translations in space and time. In chapter 14, we will see that in the Hamiltonian form of classical mechanics, the components of the momentum vector give a basis of the Lie algebra of the spatial translation group $\mathbf{R}^3$, the energy a basis of the Lie algebra of the time translation group $\mathbf{R}$. Invoking the classical relationship between energy and momentum

$$E = \frac{|\mathbf{p}|^2}{2m}$$

used in non-relativistic mechanics relates the Hamiltonian and momentum operators by

$$H = \frac{|\mathbf{P}|^2}{2m}$$

On wavefunctions, for this choice of H, the abstract Schrödinger equation 1.1 becomes the partial differential equation

$$i\hbar\frac{\partial}{\partial t}\psi(\mathbf{q},t) = \frac{-\hbar^2}{2m}\nabla^2\psi(\mathbf{q},t)$$

for the wavefunction of a free particle.

10.1   The group R and its representations

Some of the most fundamental symmetries of nature are translational symmetries, and the basic example of these involves the Lie group $\mathbf{R}$, with the group law given by addition. Note that $\mathbf{R}$ can be treated as a matrix group with a multiplicative group law by identifying it with the group of matrices of the form

$$\begin{pmatrix} 1 & a \\ 0 & 1 \end{pmatrix}$$

for $a \in \mathbf{R}$. Since

$$\begin{pmatrix} 1 & a \\ 0 & 1 \end{pmatrix}\begin{pmatrix} 1 & b \\ 0 & 1 \end{pmatrix} = \begin{pmatrix} 1 & a+b \\ 0 & 1 \end{pmatrix}$$

multiplication of matrices corresponds to addition of elements of $\mathbf{R}$. Using the matrix exponential, one finds that

$$e^{\begin{pmatrix} 0 & a \\ 0 & 0 \end{pmatrix}} = \begin{pmatrix} 1 & a \\ 0 & 1 \end{pmatrix}$$

so the Lie algebra of the matrix group $\mathbf{R}$ is matrices of the form

$$\begin{pmatrix} 0 & a \\ 0 & 0 \end{pmatrix}$$

with Lie bracket the matrix commutator (which is zero here). Such a Lie algebra can be identified with the Lie algebra $\mathbf{R}$ (with trivial Lie bracket).

We will sometimes find this way of expressing elements of $\mathbf{R}$ as matrices useful, but will often instead label elements of the group by scalars a, and use the additive group law. The same scalars a are also used to label elements of the Lie algebra, with the exponential map from the Lie algebra to the Lie group now just the identity map. Recall that the Lie algebra of a Lie group can be thought of as the tangent space to the group at the identity. For examples of Lie groups like $\mathbf{R}$ that are linear spaces, the space and its tangent space can be identified, and this is what we are doing here.

The irreducible representations of the group $\mathbf{R}$ are the following:

Theorem 10.1. *Irreducible representations of* $\mathbf{R}$ *are labeled by* $c \in \mathbf{C}$ *and given by*

$$\pi_c(a) = e^{ca}$$

Such representations are unitary (in $U(1)$*) when* c *is purely imaginary.*

The proof of this theorem is the same as for the $G = U(1)$ case (theorem 2.3), dropping the final part of the argument, which shows that periodicity ($U(1)$ is just $\mathbf{R}$ with a and $a + N2\pi$ identified) requires c to be i times an integer.

The representations of $\mathbf{R}$ that we are interested in are on spaces of wavefunctions, and thus infinite dimensional. The simplest case is the representation induced on functions on $\mathbf{R}$ by the action of $\mathbf{R}$ on itself by translation. Here, $a \in \mathbf{R}$ acts on $q \in \mathbf{R}$ (where q is a coordinate on $\mathbf{R}$) by

$$q \to a \cdot q = q + a$$

and the induced representation π on functions (see equation 1.3) is

$$\pi(g)f(q) = f(g^{-1} \cdot q)$$

which for this case will be

$$\pi(a)f(q) = f(q - a) \tag{10.1}$$

To get the Lie algebra version of this representation, the above can be differentiated, finding

$$\pi'(a) = -a\frac{d}{dq} \tag{10.2}$$

In the other direction, knowing the Lie algebra representation, exponentiation gives

$$\pi(a)f = e^{\pi'(a)}f = e^{-a\frac{d}{dq}}f(q) = f(q) - a\frac{df}{dq} + \frac{a^2}{2!}\frac{d^2f}{dq^2} + \cdots = f(q-a)$$

which is just Taylor's formula.[1]

In chapter 5, for finite dimensional unitary representations of a Lie group G, we found corresponding Lie algebra representations in terms of self-adjoint matrices. For the case of $G = \mathbf{R}$, even for infinite dimensional representations on $\mathcal{H} = L^2(\mathbf{R}^3)$, one gets an equivalence of unitary representations and self-adjoint operators,[2] although now this is a non-trivial theorem in analysis, not just a fact about matrices.

10.2   Translations in time and space

10.2.1   Energy and the group R of time translations

We have seen that it is a basic axiom of quantum mechanics that the observable operator responsible for infinitesimal time translations is the Hamiltonian operator H, a fact that is expressed as the Schrödinger equation

$$i\hbar\frac{d}{dt}|\psi\rangle = H|\psi\rangle$$

When H is time-independent, this equation can be understood as reflecting the existence of a unitary representation $(U(t), \mathcal{H})$ of the group $\mathbf{R}$ of time translations on the state space $\mathcal{H}$.

When $\mathcal{H}$ is finite dimensional, the fact that a differentiable unitary representation $U(t)$ of $\mathbf{R}$ on $\mathcal{H}$ is of the form

$$U(t) = e^{-\frac{i}{\hbar}tH}$$

---

[1] This requires restricting attention to a specific class of functions for which the Taylor series converges to the function.

[2] For the case of $\mathcal{H}$ infinite dimensional, this is known as Stone's theorem for one-parameter unitary groups, see, for instance, section 10.2 of [41] for details.

for H a self-adjoint matrix follows from the same sort of argument as in theorem 2.3. Such a $U(t)$ provides solutions of the Schrödinger equation by

$$|\psi(t)\rangle = U(t)|\psi(0)\rangle$$

The Lie algebra of $\mathbf{R}$ is also $\mathbf{R}$, and we get a Lie algebra representation of $\mathbf{R}$ by taking the time derivative of $U(t)$, which gives us

$$\hbar\frac{d}{dt}U(t)_{|t=0} = -iH$$

Because this Lie algebra representation comes from taking the derivative of a unitary representation, $-iH$ will be skew-adjoint, so H will be self-adjoint.

10.2.2   Momentum and the group $\mathbf{R}^3$ of space translations

Since we now want to describe quantum systems that depend not just on time, but on space variables $\mathbf{q} = (q_1, q_2, q_3)$, we will have an action by unitary transformations of not just the group $\mathbf{R}$ of time translations, but also the group $\mathbf{R}^3$ of spatial translations. We will define the corresponding Lie algebra representations using self-adjoint operators P_1, P_2, P_3 that play the same role for spatial translations that the Hamiltonian plays for time translations:

Definition (Momentum operators). *For a quantum system with state space* $\mathcal{H} = L^2(\mathbf{R}^3)$ *given by complex-valued functions of position variables* q_1, q_2, q_3, *momentum operators* P_1, P_2, P_3 *are defined by*

$$P_1 = -i\hbar\frac{\partial}{\partial q_1}, P_2 = -i\hbar\frac{\partial}{\partial q_2}, P_3 = -i\hbar\frac{\partial}{\partial q_3}$$

These are given the name "momentum operators" since we will see that their eigenvalues have an interpretation as the components of the momentum vector for the system, just as the eigenvalues of the Hamiltonian have an interpretation as the energy. Note that while in the case of the Hamiltonian, the factor of $\hbar$ kept track of the relative normalization of energy and time units; here, it plays the same role for momentum and length units. It can be set to one if appropriate choices of units of momentum and length are made.

The differentiation operator is skew-adjoint since, using integration by parts,[3] one has for each variable, for $\psi_1, \psi_2 \in \mathcal{H}$

---

[3] We are here neglecting questions of whether these integrals are well defined, which require more care in specifying the class of functions involved.

$$\int_{-\infty}^{+\infty} \overline{\psi}_1 \left(\frac{d}{dq} \psi_2 \right) dq = \int_{-\infty}^{+\infty} \left(\frac{d}{dq} (\overline{\psi}_1 \psi_2) - \left(\frac{d}{dq} \overline{\psi}_1 \right) \psi_2 \right) dq$$

$$= -\int_{-\infty}^{+\infty} \left(\frac{d}{dq} \overline{\psi}_1 \right) \psi_2 dq$$

(assuming that the $\psi_j(q)$ go to 0 at $\pm\infty$). The P_j are thus self-adjoint operators, with real eigenvalues as expected for an observable operator. Multiplying by $-i$ to get the corresponding skew-adjoint operator of a unitary Lie algebra representation, we find

$$-iP_j = -\hbar \frac{\partial}{\partial q_j}$$

Up to the $\hbar$ factor that depends on units, these are exactly the Lie algebra representation operators on basis elements of the Lie algebra, for the action of $\mathbf{R}^3$ on functions on $\mathbf{R}^3$ induced from translation:

$$\pi(a_1, a_2, a_3) f(q_1, q_2, q_3) = f(q_1 - a_1, q_2 - a_2, q_3 - a_3)$$

$$\pi'(a_1, a_2, a_3) = a_1(-iP_1) + a_2(-iP_2) + a_3(-iP_3) = -\hbar \left(a_1 \frac{\partial}{\partial q_1} + a_2 \frac{\partial}{\partial q_2} + a_3 \frac{\partial}{\partial q_3} \right)$$

Note that the convention for the sign choice here is the opposite from the case of the Hamiltonian ($-iP = -\hbar \frac{d}{dq}$ vs. $-iH = \hbar \frac{d}{dt}$). This means that the conventional sign choice we have been using for the Hamiltonian makes it minus the generator of translations in the time direction. The reason for this comes from considerations of special relativity (which will be discussed in chapter 40), where the inner product on space–time has opposite signs for the space and time dimensions.

10.3 The energy–momentum relation and the Schrödinger equation for a free particle

We will review this subject in chapter 40, but for now we just need the relationship special relativity posits between energy and momentum. Space and time are put together in "Minkowski space," which is $\mathbf{R}^4$ with indefinite inner product

$$((u_0, u_1, u_2, u_3), (v_0, v_1, v_2, v_3)) = -u_0 v_0 + u_1 v_1 + u_2 v_2 + u_3 v_3$$

Energy and momentum are the components of a Minkowski space vector $(p_0 = E, p_1, p_2, p_3)$ with norm-squared given by minus the mass-squared:

$$((E, p_1, p_2, p_3), (E, p_1, p_2, p_3)) = -E^2 + |\mathbf{p}|^2 = -m^2$$

This is the formula for a choice of space and time units such that the speed of light is 1. Putting in factors of the speed of light c to get the units right, one has

$$E^2 - |\mathbf{p}|^2 c^2 = m^2 c^4$$

Two special cases of this are:

- For photons, $m = 0$, and one has the energy–momentum relation $E = |\mathbf{p}|c$

- For velocities $\mathbf{v}$ small compared to c (and thus momenta $|\mathbf{p}|$ small compared to mc), one has

$$E = \sqrt{|\mathbf{p}|^2 c^2 + m^2 c^4} = c\sqrt{|\mathbf{p}|^2 + m^2 c^2} \approx \frac{c|\mathbf{p}|^2}{2mc} + mc^2 = \frac{|\mathbf{p}|^2}{2m} + mc^2$$

In the non-relativistic limit, we use this energy–momentum relation to describe particles with velocities small compared to c, typically dropping the momentum-independent constant term mc^2.

In later chapters, we will discuss quantum systems that describe photons, as well as other possible ways of constructing quantum systems for relativistic particles. For now though, we will just consider the non-relativistic case. To describe a quantum non-relativistic particle, we choose a Hamiltonian operator H such that its eigenvalues (the energies) will be related to the momentum operator eigenvalues (the momenta) by the classical energy–momentum relation $E = \frac{|\mathbf{p}|^2}{2m}$:

$$H = \frac{1}{2m}(P_1^2 + P_2^2 + P_3^2) = \frac{1}{2m}|\mathbf{P}|^2 = \frac{-\hbar^2}{2m}\left(\frac{\partial^2}{\partial q_1^2} + \frac{\partial^2}{\partial q_2^2} + \frac{\partial^2}{\partial q_3^2}\right)$$

The Schrödinger equation then becomes:

$$i\hbar\frac{\partial}{\partial t}\psi(\mathbf{q}, t) = \frac{-\hbar^2}{2m}\left(\frac{\partial^2}{\partial q_1^2} + \frac{\partial^2}{\partial q_2^2} + \frac{\partial^2}{\partial q_3^2}\right)\psi(\mathbf{q}, t) = \frac{-\hbar^2}{2m}\nabla^2\psi(\mathbf{q}, t)$$

This is an easily solved simple constant coefficient second-order partial differential equation. One method of solution is to separate out the time-dependence, by first finding solutions ψ_E to the time-independent equation

$$H\psi_E(\mathbf{q}) = \frac{-\hbar^2}{2m}\nabla^2\psi_E(\mathbf{q}) = E\psi_E(\mathbf{q}) \tag{10.3}$$

with eigenvalue E for the Hamiltonian operator. Then

$$\psi(\mathbf{q}, t) = \psi_E(\mathbf{q})e^{-\frac{i}{\hbar}tE}$$

will give solutions to the full time-dependent equation

$$i\hbar\frac{\partial}{\partial t}\psi(\mathbf{q}, t) = H\psi(\mathbf{q}, t)$$

The solutions $\psi_E(\mathbf{q})$ to the time-independent equation 10.3 are complex exponentials proportional to

$$e^{i(k_1 q_1 + k_2 q_2 + k_3 q_3)} = e^{i\mathbf{k}\cdot\mathbf{q}}$$

satisfying

$$\frac{-\hbar^2}{2m}i^2|\mathbf{k}|^2 = \frac{\hbar^2|\mathbf{k}|^2}{2m} = E$$

We have thus found that solutions to the Schrödinger equation are given by linear combinations of states $|\mathbf{k}\rangle$ labeled by a vector $\mathbf{k}$, which are eigenstates of the momentum and Hamiltonian operators with

$$P_j|\mathbf{k}\rangle = \hbar k_j|\mathbf{k}\rangle, \quad H|\mathbf{k}\rangle = \frac{\hbar^2}{2m}|\mathbf{k}|^2|\mathbf{k}\rangle$$

These are states with well-defined momentum and energy

$$p_j = \hbar k_j, E = \frac{|\mathbf{p}|^2}{2m}$$

so they satisfy exactly the same energy–momentum relations as those for a classical non-relativistic particle.

While the quantum mechanical state space $\mathcal{H}$ contains states with the classical energy–momentum relation, it also contains much, much more since it includes linear combinations of such states. At $t = 0$, the state can be a sum

$$|\psi\rangle = \sum_{\mathbf{k}} c_{\mathbf{k}}e^{i\mathbf{k}\cdot\mathbf{q}}$$

where $c_{\mathbf{k}}$ are complex numbers. This state will in general not have a well-defined momentum, but measurement theory says that an apparatus measuring the momentum will observe value $\hbar\mathbf{k}$ with probability

$$\frac{|c_{\mathbf{k}}|^2}{\sum_{\mathbf{k}'}|c_{\mathbf{k}'}|^2}$$

The time-dependent state will be

$$|\psi(t)\rangle = \sum_{\mathbf{k}} c_{\mathbf{k}} e^{i\mathbf{k}\cdot\mathbf{q}} \, e^{-it\hbar\frac{|\mathbf{k}|^2}{2m}}$$

Since each momentum eigenstate evolves in time by the phase factor

$$e^{-it\hbar\frac{|\mathbf{k}|^2}{2m}}$$

the probabilities of observing a momentum value stay constant in time.

10.4   For further reading

Every book about quantum mechanics covers this example of the free quantum particle somewhere very early on, in detail. Our discussion here is unusual just in emphasizing the role of the spatial translation group and its unitary representations.

Chapter 11
Fourier Analysis and the Free Particle

The quantum theory of a free particle requires not just a state space $\mathcal{H}$, but also an inner product on $\mathcal{H}$, which should be translation invariant so that translations act as unitary transformations. Such an inner product will be given by the integral

$$\langle \psi_1, \psi_2 \rangle = C \int_{\mathbf{R}^3} \overline{\psi_1(\mathbf{q})} \psi_2(\mathbf{q}) d^3\mathbf{q}$$

for some choice of normalization constant C, usually taken to be $C = 1$. $\mathcal{H}$ will be the space $L^2(\mathbf{R}^3)$ of square-integrable complex-valued functions on $\mathbf{R}^3$.

A problem arises though if we try and compute the norm-squared of one of our momentum eigenstates $|\mathbf{k}\rangle$. We find

$$\langle \mathbf{k}|\mathbf{k}\rangle = C \int_{\mathbf{R}^3} (e^{-i\mathbf{k}\cdot\mathbf{q}})(e^{i\mathbf{k}\cdot\mathbf{q}}) d^3\mathbf{q} = C \int_{\mathbf{R}^3} 1 \, d^3\mathbf{q} = \infty$$

As a result, there is no value of C which will give these states a finite norm, and they are not in the expected state space. The finite dimensional spectral theorem 4.1 assuring us that, given a self-adjoint operator, we can find an orthonormal basis of its eigenvectors, will no longer hold. Other problems arise because our momentum operators P_j may take states in $\mathcal{H}$ to states that are not in $\mathcal{H}$ (i.e., not square-integrable).

We will consider two different ways of dealing with these problems, for simplicity treating the case of just one spatial dimension. In the first, we impose periodic boundary conditions, effectively turning space into a circle of finite extent, leaving for later the issue of taking the size of the circle to infinity. The translation group action then becomes the $U(1)$ group action of rotation about the circle. This acts on the state space $\mathcal{H} = L^2(S^1)$, a situation which can be analyzed using the theory of Fourier series. Momentum eigenstates are now in $\mathcal{H}$ and labeled by an integer.

© Peter Woit 2017
P. Woit, *Quantum Theory, Groups and Representations*,
DOI 10.1007/978-3-319-64612-1_11

While this deals with the problem of eigenvectors not being in $\mathcal{H}$, it ruins an important geometrical structure of the free particle quantum system, by treating positions (taking values in the circle) and momenta (taking values in the integers) quite differently. In later chapters, we will see that physical systems like the free particle are best studied by treating positions and momenta as real-valued coordinates on a single vector space, called phase space. To do this, a formalism is needed that treats momenta as real-valued variables on a par with position variables.

The theory of Fourier analysis provides the required formalism, with the Fourier transform interchanging a state space $L^2(\mathbf{R})$ of wavefunctions depending on position with a unitarily equivalent one using wavefunctions that depend on momenta. The problems of the domain of the momentum operator P and its eigenfunctions not being in $L^2(\mathbf{R})$ still need to be addressed. This can be done by introducing

- a space $\mathcal{S}(\mathbf{R}) \subset L^2(\mathbf{R})$ of sufficiently well-behaved functions on which P is well defined, and

- a space $\mathcal{S}'(\mathbf{R}) \supset L^2(\mathbf{R})$ of "generalized functions," also known as distributions, which will include the eigenvectors of P.

Solutions to the Schrödinger equation can be studied in any of the three

$$S(\mathbf{R}) \subset L^2(\mathbf{R}) \subset \mathcal{S}'(\mathbf{R})$$

contexts, each of which will be preserved by the Fourier transform and allow one to treat position and momentum variables on the same footing.

11.1 Periodic boundary conditions and the group $U(1)$

In this section, we will describe one way to deal with the problems caused by non-normalizable eigenstates, considering first the simplified case of a single spatial dimension. In this one-dimensional case, the space $\mathbf{R}$ is replaced by the circle S^1. This is equivalent to the physicist's method of imposing "periodic boundary conditions," meaning to define the theory on an interval, and then identify the ends of the interval. The position variable q can then be thought of as an angle ϕ and one can define the inner product as

$$\langle \psi_1, \psi_2 \rangle = \frac{1}{2\pi} \int_0^{2\pi} \overline{\psi_1(\phi)} \psi_2(\phi) d\phi$$

The state space is then

$$\mathcal{H} = L^2(S^1)$$

the space of complex-valued square-integrable functions on the circle.

Instead of the group $\mathbf{R}$ acting on itself by translations, we have the standard rotation action of the group $SO(2)$ on the circle. Elements $g(\theta)$ of the group are rotations of the circle counterclockwise by an angle θ, or if we parametrize the circle by an angle ϕ, just shifts

$$\phi \to \phi + \theta$$

By the same argument as in the case $G = \mathbf{R}$, we can use the representation on functions given by equation 1.3 to get a representation on $\mathcal{H}$

$$\pi(g(\theta))\psi(\phi) = \psi(\phi - \theta)$$

If X is a basis of the Lie algebra $\mathfrak{so}(2)$ (for instance, taking the circle as the unit circle in $\mathbf{R}^2$, rotations 2 by 2 matrices, $X = \begin{pmatrix} 0 & -1 \\ 1 & 0 \end{pmatrix}$, $g(\theta) = e^{\theta X}$), then the Lie algebra representation is given by taking the derivative

$$\pi'(aX)f(\phi) = \frac{d}{d\theta}f(\phi - a\theta)_{|\theta=0} = -a\frac{d}{d\phi}f(\phi)$$

so we have (as in the $\mathbf{R}$ case, see equation 10.2)

$$\pi'(aX) = -a\frac{d}{d\phi}$$

This operator is defined on a dense subspace of $\mathcal{H} = L^2(S^1)$ and is skew-adjoint, since (using integration by parts)

$$\langle \psi_1, \frac{d}{d\phi}\psi_2 \rangle = \frac{1}{2\pi}\int_0^{2\pi} \overline{\psi_1}\frac{d}{d\phi}\psi_2 d\phi$$
$$= \frac{1}{2\pi}\int_0^{2\pi} \left(\frac{d}{d\phi}(\overline{\psi_1}\psi_2) - \left(\frac{d}{d\phi}\overline{\psi_1}\right)\psi_2 \right) d\phi$$
$$= -\langle \frac{d}{d\phi}\psi_1, \psi_2 \rangle$$

The eigenfunctions of $\pi'(X)$ are the $e^{in\phi}$, for $n \in \mathbf{Z}$, which we will also write as state vectors $|n\rangle$. These are orthonormal

$$\langle n|m \rangle = \delta_{nm} \tag{11.1}$$

and provide a countable basis for the space $L^2(S^1)$. This basis corresponds to the decomposition into irreducibles of $\mathcal{H}$ as a representation of $SO(2)$ described above. One has

$$(\pi, L^2(S^1)) = \oplus_{n \in \mathbf{Z}}(\pi_n, \mathbf{C}) \tag{11.2}$$

where π_n are the irreducible one-dimensional representations given by the multiplication action

$$\pi_n(g(\theta)) = e^{in\theta}$$

The theory of Fourier series for functions on S^1 says that any function $\psi \in L^2(S^1)$ can be expanded in terms of this basis:

Theorem 11.1 (Fourier series). *If $\psi \in L^2(S^1)$, then*

$$|\psi\rangle = \psi(\phi) = \sum_{n=-\infty}^{+\infty} c_n e^{in\phi} = \sum_{n=-\infty}^{+\infty} c_n|n\rangle$$

where

$$c_n = \langle n|\psi\rangle = \frac{1}{2\pi}\int_0^{2\pi} e^{-in\phi}\psi(\phi)d\phi$$

This is an equality in the sense of the norm on $L^2(S^1)$, i.e.,

$$\lim_{N\to\infty} ||\psi - \sum_{n=-N}^{+N} c_n e^{in\phi}|| = 0$$

The condition that $\psi \in L^2(S^1)$ corresponds to the condition

$$\sum_{n=-\infty}^{+\infty} |c_n|^2 < \infty$$

on the coefficients c_n.

One can easily derive the formula for c_n using orthogonality of the $|n\rangle$. For a detailed proof of the theorem, see, for instance, [27] and [83]. The theorem gives an equivalence (as complex vector spaces with a Hermitian inner product) between square-integrable functions on S^1 and square-summable functions on $\mathbf{Z}$. As unitary $SO(2)$ representations, this is the equivalence of equation 11.2.

The Lie algebra of the group S^1 is the same as that of the additive group $\mathbf{R}$, and the $\pi'(X)$ we have found for the S^1 action on functions is related to the momentum operator in the same way as in the $\mathbf{R}$ case. So, we can use the same momentum operator

$$P = -i\hbar \frac{d}{d\phi}$$

which satisfies

$$P|n\rangle = \hbar n|n\rangle$$

By changing space from the non-compact $\mathbf{R}$ to the compact S^1, we now have momenta that, instead of taking on any real value, can only be integral numbers times $\hbar$.

Solving the Schrödinger equation

$$i\hbar \frac{\partial}{\partial t}\psi(\phi, t) = \frac{P^2}{2m}\psi(\phi, t) = \frac{-\hbar^2}{2m}\frac{\partial^2}{\partial \phi^2}\psi(\phi, t)$$

as before, we find

$$E\psi_E(\phi) = \frac{-\hbar^2}{2m}\frac{d^2}{d\phi^2}\psi_E(\phi)$$

as the eigenvector equation. This has an orthonormal basis of solutions $|n\rangle$, with

$$E = \frac{\hbar^2 n^2}{2m}$$

The Schrödinger equation is first order in time, and the space of possible solutions can be identified with the space of possible initial values at a fixed time. Elements of this space of solutions can be characterized by

- the complex-valued square-integrable function $\psi(\phi, 0) \in L^2(S^1)$, a function on the circle S^1.

- the square-summable sequence c_n of complex numbers, a function on the integers $\mathbf{Z}$.

The c_n can be determined from the $\psi(\phi, 0)$ using the Fourier coefficient formula

$$c_n = \frac{1}{2\pi}\int_0^{2\pi} e^{-in\phi}\psi(\phi, 0)d\phi$$

Given the c_n, the corresponding solution to the Schrödinger equation will be

$$\psi(\phi, t) = \sum_{n=-\infty}^{+\infty} c_n e^{in\phi} e^{-i\frac{\hbar n^2}{2m}t}$$

To get something more realistic, we need to take our circle to have an arbitrary circumference L, and we can study our original problem with space $\mathbf{R}$ by considering the limit $L \to \infty$. To do this, we just need to change variables from ϕ to ϕ_L, where

$$\phi_L = \frac{L}{2\pi}\phi$$

The momentum operator will now be

$$P = -i\hbar\frac{d}{d\phi_L}$$

and its eigenvalues will be quantized in units of $\frac{2\pi\hbar}{L}$. The energy eigenvalues will be

$$E = \frac{2\pi^2\hbar^2 n^2}{mL^2}$$

Note that these values are discrete (as long as the size L of the circle is finite) and nonnegative.

11.2 The group R and the Fourier transform

In the previous section, we imposed periodic boundary conditions, replacing the group $\mathbf{R}$ of translations by the circle group S^1, and then used the fact that unitary representations of this group are labeled by integers. This made the analysis relatively easy, with $\mathcal{H} = L^2(S^1)$ and the self-adjoint operator $P = -i\hbar\frac{\partial}{\partial\phi}$ behaving much the same as in the finite dimensional case: The eigenvectors of P give a countable orthonormal basis of $\mathcal{H}$ and P can be thought of as an infinite dimensional matrix.

Unfortunately, in order to understand many aspects of quantum mechanics, one cannot get away with this trick, but needs to work with $\mathbf{R}$ itself. One reason for this is that the unitary representations of $\mathbf{R}$ are labeled by the same group, $\mathbf{R}$, and it will turn out (see the discussion of the Heisenberg group in Chapter 13) to be important to be able to exploit this and treat positions and momenta on the same footing. What plays the role then of $|n\rangle = e^{in\phi}$, $n \in \mathbf{Z}$ will be the $|k\rangle = e^{ikq}$, $k \in \mathbf{R}$. These are functions on $\mathbf{R}$ that are one-dimensional irreducible representations under the translation action on functions (as usual using equation 1.3)

$$\pi(a)e^{ikq} = e^{ik(q-a)} = e^{-ika}e^{ikq}$$

One can try and mimic the Fourier series decomposition, with the coefficients c_n that depend on the labels of the irreducibles replaced by a function $\widetilde{f}(k)$ depending on the label k of the irreducible representation of $\mathbf{R}$:

Definition (Fourier transform). *The Fourier transform of a function ψ is given by a function denoted $\mathcal{F}\psi$ or $\widetilde{\psi}$, where*

$$(\mathcal{F}\psi)(k) = \widetilde{\psi}(k) = \frac{1}{\sqrt{2\pi}} \int_{-\infty}^{\infty} e^{-ikq} \psi(q) dq \tag{11.3}$$

This integral is not well defined for all elements of $L^2(\mathbf{R})$, so one needs to specify a subspace of $L^2(\mathbf{R})$ to study for which it is well defined, and then extend the definition to $L^2(\mathbf{R})$ by considering limits of sequences. In our case, a good choice of such a subspace is the Schwartz space $S(\mathbf{R})$ of functions ψ such that the function and its derivatives fall off faster than any power at infinity. We will not try and give a more precise definition of $S(\mathbf{R})$ here, but a good class of examples of elements of $S(\mathbf{R})$ to keep in mind are products of polynomials and a Gaussian function. The Schwartz space has the useful property that we can apply the momentum operator P an indefinite number of times without leaving the space.

Just as a function on S^1 can be recovered from its Fourier series coefficients c_n by taking a sum, given the Fourier transform $\widetilde{\psi}(k)$ of ψ, ψ itself can be recovered by an integral, with the following theorem

Theorem (Fourier Inversion). *For $\psi \in S(\mathbf{R})$, one has $\widetilde{\psi} \in S(\mathbf{R})$ and*

$$\psi(q) = \widetilde{\mathcal{F}}\widetilde{\psi} = \frac{1}{\sqrt{2\pi}} \int_{-\infty}^{+\infty} e^{ikq} \widetilde{\psi}(k) dk \tag{11.4}$$

Note that $\widetilde{\mathcal{F}}$ is the same linear operator as $\mathcal{F}$, with a change in sign of the argument of the function it is applied to. Note also that we are choosing one of various popular ways of normalizing the definition of the Fourier transform. In others, the factor of 2π may appear instead in the exponent of the complex exponential, or just in one of $\mathcal{F}$ or $\widetilde{\mathcal{F}}$ and not the other.

The operators $\mathcal{F}$ and $\widetilde{\mathcal{F}}$ are thus inverses of each other on $S(\mathbf{R})$. One has

Theorem (Plancherel). *$\mathcal{F}$ and $\widetilde{\mathcal{F}}$ extend to unitary isomorphisms of $L^2(\mathbf{R})$ with itself. In particular*

$$\int_{-\infty}^{\infty} |\psi(q)|^2 dq = \int_{-\infty}^{\infty} |\widetilde{\psi}(k)|^2 dk \tag{11.5}$$

Note that we will be using the same inner product on functions on $\mathbf{R}$

$$\langle \psi_1, \psi_2 \rangle = \int_{-\infty}^{\infty} \overline{\psi_1(q)} \psi_2(q) dq$$

both for functions of q and their Fourier transforms, functions of k, with our normalizations chosen so that the Fourier transform is a unitary transformation.

An important example is the case of Gaussian functions, where

$$
\begin{aligned}
\mathcal{F}e^{-\alpha\frac{q^2}{2}} &= \frac{1}{\sqrt{2\pi}} \int_{-\infty}^{+\infty} e^{-ikq} e^{-\alpha\frac{q^2}{2}} dq \\
&= \frac{1}{\sqrt{2\pi}} \int_{-\infty}^{+\infty} e^{-\frac{\alpha}{2}((q+i\frac{k}{\alpha})^2 - (\frac{ik}{\alpha})^2)} dq \\
&= \frac{1}{\sqrt{2\pi}} e^{-\frac{k^2}{2\alpha}} \int_{-\infty}^{+\infty} e^{-\frac{\alpha}{2}q'^2} dq' \\
&= \frac{1}{\sqrt{\alpha}} e^{-\frac{k^2}{2\alpha}}
\end{aligned}
\tag{11.6}
$$

A crucial property of the unitary operator $\mathcal{F}$ on $\mathcal{H}$ is that it diagonalizes the differentiation operator and thus the momentum operator P. Under the Fourier transform, constant coefficient differential operators become just multiplication by a polynomial, giving a powerful technique for solving differential equations. Computing the Fourier transform of the differentiation operator using integration by parts, we find

$$
\begin{aligned}
\widetilde{\frac{d\psi}{dq}} &= \frac{1}{\sqrt{2\pi}} \int_{-\infty}^{+\infty} e^{-ikq} \frac{d\psi}{dq} dq \\
&= \frac{1}{\sqrt{2\pi}} \int_{-\infty}^{+\infty} \left(\frac{d}{dq}(e^{-ikq}\psi) - \left(\frac{d}{dq}e^{-ikq}\right)\psi \right) dq \\
&= ik \frac{1}{\sqrt{2\pi}} \int_{-\infty}^{+\infty} e^{-ikq}\psi dq \\
&= ik\widetilde{\psi}(k)
\end{aligned}
\tag{11.7}
$$

So, under Fourier transform, differentiation by q becomes multiplication by ik. This is the infinitesimal version of the fact that translation becomes multiplication by a phase under the Fourier transform, which can be seen as follows. If $\psi_a(q) = \psi(q+a)$, then

$$
\begin{aligned}
\widetilde{\psi_a}(k) &= \frac{1}{\sqrt{2\pi}} \int_{-\infty}^{+\infty} e^{-ikq}\psi(q+a) dq \\
&= \frac{1}{\sqrt{2\pi}} \int_{-\infty}^{+\infty} e^{-ik(q'-a)}\psi(q') dq' \\
&= e^{ika}\widetilde{\psi}(k)
\end{aligned}
$$

Since $p = \hbar k$, one can easily change variables and work with p instead of k. As with the factors of 2π, there is a choice of where to put the factors of $\hbar$ in the normalization of the Fourier transform. A common choice preserving symmetry between the formulas for Fourier transform and inverse Fourier transform is

$$\widetilde{\psi}(p) = \frac{1}{\sqrt{2\pi\hbar}} \int_{-\infty}^{+\infty} e^{-i\frac{pq}{\hbar}} \psi(q) dq$$

$$\psi(q) = \frac{1}{\sqrt{2\pi\hbar}} \int_{-\infty}^{+\infty} e^{i\frac{pq}{\hbar}} \widetilde{\psi}(p) dp$$

We will, however, mostly continue to set $\hbar = 1$, in which case the distinction between k and p vanishes.

11.3   Distributions

While the use of the subspace $\mathcal{S}(\mathbf{R}) \subset L^2(\mathbf{R})$ as state space gives a well-behaved momentum operator P and a formalism symmetric between positions and momenta, it still has the problem that eigenfunctions of P are not in the state space. Another problem is that unlike the case of $L^2(\mathbf{R})$ where the Riesz representation theorem provides an isomorphism between this space and its dual just like in the finite dimensional case (see 4.3), such a duality no longer holds for $\mathcal{S}(\mathbf{R})$.

For a space dual to $\mathcal{S}(\mathbf{R})$, one can take the space of linear functionals on $\mathcal{S}(\mathbf{R})$ called the Schwartz space of tempered distributions (a certain continuity condition on the functionals is needed, see, for instance, [88]), which is denoted by $\mathcal{S}'(\mathbf{R})$. An element of this space is a linear map

$$T : f \in \mathcal{S}(\mathbf{R}) \to T[f] \in \mathbf{C}$$

$\mathcal{S}(\mathbf{R})$ can be identified with a subspace of $\mathcal{S}'(\mathbf{R})$, by taking $\psi \in \mathcal{S}(\mathbf{R})$ to the linear functional T_ψ given by

$$T_\psi : f \in \mathcal{S}(\mathbf{R}) \to T_\psi[f] = \int_{-\infty}^{+\infty} \psi(q) f(q) dq \tag{11.8}$$

Note that taking $\psi \in \mathcal{S}(\mathbf{R})$ to $T_\psi \in \mathcal{S}'(\mathbf{R})$ is a complex linear map.

There are, however, elements of $\mathcal{S}'(\mathbf{R})$ that are not of this form, with three important examples

- The linear functional that takes a function to its Fourier transform at k:

$$f \in \mathcal{S}(\mathbf{R}) \to \widetilde{f}(k)$$

- The linear functional that takes a function to its value at q:

$$f \in \mathcal{S}(\mathbf{R}) \to f(q)$$

- The linear functional that takes a function to the value of its derivative at q:

$$f \in \mathcal{S}(\mathbf{R}) \to f'(q)$$

We would like to think of these as "generalized functions," corresponding to T_ψ given by the integral in equation 11.8, for some ψ which is a generalization of a function.

From the formula 11.3 for the Fourier transform, we have

$$\widetilde{f}(k) = T_{\frac{1}{\sqrt{2\pi}}e^{-ikq}}[f]$$

so the first of the above linear functionals corresponds to

$$\psi(q) = \frac{1}{\sqrt{2\pi}}e^{-ikq}$$

which is a function, but "generalized" in the sense that it is not in $\mathcal{S}(\mathbf{R})$ (or even in $L^2(\mathbf{R})$). This is an eigenfunction for the operator P, and we see that such eigenfunctions, while not in $\mathcal{S}(\mathbf{R})$ or $L^2(\mathbf{R})$, do have a meaning as elements of $\mathcal{S}'(\mathbf{R})$.

The second linear functional described above can be written as T_δ with the corresponding generalized function the "δ-function," denoted by the symbol $\delta(q - q')$, which is taken to have the property that

$$\int_{-\infty}^{+\infty} \delta(q - q')f(q')dq' = f(q)$$

$\delta(q - q')$ is manipulated in some ways like a function, although such a function does not exist. It can, however, be made sense of as a limit of actual functions. Consider the limit as $\epsilon \to 0$ of functions

$$g_\epsilon = \frac{1}{\sqrt{2\pi\epsilon}}e^{-\frac{(q-q')^2}{2\epsilon}}$$

These satisfy

$$\int_{-\infty}^{+\infty} g_\epsilon(q')dq' = 1$$

for all $\epsilon > 0$ (using equation 11.6).

Heuristically (ignoring problems of interchange of integrals that do not make sense), the Fourier inversion formula can be written as follows:

$$\psi(q) = \frac{1}{\sqrt{2\pi}} \int_{-\infty}^{+\infty} e^{ikq} \widetilde{\psi}(k) dk$$

$$= \frac{1}{\sqrt{2\pi}} \int_{-\infty}^{+\infty} e^{ikq} \left(\frac{1}{\sqrt{2\pi}} \int_{-\infty}^{+\infty} e^{-ikq'} \psi(q') dq' \right) dk$$

$$= \frac{1}{2\pi} \int_{-\infty}^{+\infty} \left(\int_{-\infty}^{+\infty} e^{ik(q-q')} \psi(q') dk \right) dq'$$

$$= \int_{-\infty}^{+\infty} \delta(q - q') \psi(q') dq'$$

Physicists interpret the above calculation as justifying the formula

$$\delta(q - q') = \frac{1}{2\pi} \int_{-\infty}^{+\infty} e^{ik(q-q')} dk$$

and then go on to consider the eigenvectors

$$|k\rangle = \frac{1}{\sqrt{2\pi}} e^{ikq}$$

of the momentum operator as satisfying a replacement for the Fourier series orthonormality relation (equation 11.1), with the δ-function replacing the δ_{nm}:

$$\langle k'|k\rangle = \int_{-\infty}^{+\infty} \overline{\left(\frac{1}{\sqrt{2\pi}} e^{ik'q} \right)} \left(\frac{1}{\sqrt{2\pi}} e^{ikq} \right) dq = \frac{1}{2\pi} \int_{-\infty}^{+\infty} e^{i(k-k')q} dq = \delta(k - k')$$

11.4 Linear transformations and distributions

The definition of distributions as linear functionals on the vector space $\mathcal{S}(\mathbf{R})$ means that for any linear transformation A acting on $\mathcal{S}(\mathbf{R})$, we can get a linear transformation on $\mathcal{S}'(\mathbf{R})$ as the transpose of A (see equation 4.1), which takes T to

$$A^t T : f \in \mathcal{S}(\mathbf{R}) \rightarrow A^t[f] = T[Af] \in \mathbf{C}$$

This gives a definition of the Fourier transform on $\mathcal{S}'(\mathbf{R})$ as

$$\mathcal{F}^t T[f] \equiv T[\mathcal{F}f]$$

and one can show that, as for $\mathcal{S}(\mathbf{R})$ and $L^2(\mathbf{R})$, the Fourier transform provides an isomorphism of $\mathcal{S}'(\mathbf{R})$ with itself. Identifying functions ψ with distributions T_ψ, one has

$$\mathcal{F}^t T_\psi[f] \equiv T_\psi[\mathcal{F}f]$$

$$= T_\psi\left[\frac{1}{\sqrt{2\pi}} \int_{-\infty}^{+\infty} e^{-ikq} f(q)dq\right]$$

$$= \frac{1}{\sqrt{2\pi}} \int_{-\infty}^{+\infty} \psi(k)\left(\int_{-\infty}^{+\infty} e^{-ikq} f(q)dq\right)dk$$

$$= \int_{-\infty}^{+\infty} \left(\frac{1}{\sqrt{2\pi}} \int_{-\infty}^{+\infty} e^{-ikq}\psi(k)dk\right) f(q)dq$$

$$= T_{\mathcal{F}\psi}[f]$$

showing that the Fourier transform is compatible with this identification.

As an example, the Fourier transform of the distribution $\frac{1}{\sqrt{2\pi}}e^{ika}$ is the δ-function $\delta(q-a)$ since

$$\mathcal{F}^t T_{\frac{1}{\sqrt{2\pi}}e^{ika}}[f] = \frac{1}{\sqrt{2\pi}} \int_{-\infty}^{+\infty} e^{ika}\left(\frac{1}{\sqrt{2\pi}} \int_{-\infty}^{+\infty} e^{-ikq} f(q)dq\right)dk$$

$$= \int_{-\infty}^{+\infty} \left(\frac{1}{2\pi} \int_{-\infty}^{+\infty} e^{-ik(q-a)}dk\right) f(q)dq$$

$$= \int_{-\infty}^{+\infty} \delta(q-a)f(q)dq$$

$$= T_{\delta(q-a)}[f]$$

For another example of a linear transformation acting on $\mathcal{S}(\mathbf{R})$, consider the translation action on functions $f \rightarrow A_a f$, where

$$(A_a f)(q) = f(q-a)$$

The transpose action on distributions is

$$A_a^t T_{\psi(q)} = T_{\psi(q+a)}$$

since

$$A_a^t T_{\psi(q)}[f] = T_{\psi(q)}[f(q-a)] = \int_{-\infty}^{+\infty} \psi(q)f(q-a)dq = \int_{-\infty}^{+\infty} \psi(q'+a)f(q')dq'$$

The derivative is an infinitesimal version of this, and one sees (using integration by parts) that

$$
\begin{aligned}
\left(\frac{d}{dq}\right)^t T_{\psi(q)}[f] &= T_{\psi(q)}\left[\frac{d}{dq}f\right] \\
&= \int_{-\infty}^{+\infty} \psi(q)\frac{d}{dq}f(q)dq \\
&= \int_{-\infty}^{+\infty}\left(-\frac{d}{dq}\psi(q)\right)f(q)dq \\
&= T_{-\frac{d}{dq}\psi(q)}[f]
\end{aligned}
$$

In order to have the standard derivative when one identifies functions and distributions, one defines the derivative on distributions by

$$
\frac{d}{dq}T[f] = T\left[-\frac{d}{dq}f\right]
$$

This allows one to define derivatives of a δ-function, with, for instance, the first derivative $\delta'(q)$ of $\delta(q)$ satisfying

$$
T_{\delta'(q)}[f] = -f'(0)
$$

11.5   Solutions of the Schrödinger equation in momentum space

Equation 11.7 shows that under Fourier transformation, the derivative operator $\frac{d}{dq}$ becomes the multiplication operator ik, and this property will extend to distributions. The Fourier transform takes constant coefficient differential equations in q to polynomial equations in k, which can often much more readily be solved, including the possibility of solutions that are distributions. The free particle Schrödinger equation

$$
i\frac{\partial}{\partial t}\psi(q,t) = -\frac{1}{2m}\frac{\partial^2}{\partial q^2}\psi(q,t)
$$

becomes after Fourier transformation in the q variable the simple ordinary differential equation

$$
i\frac{d}{dt}\widetilde{\psi}(k,t) = \frac{1}{2m}k^2\widetilde{\psi}(k,t)
$$

with solutions
$$\widetilde{\psi}(k,t) = e^{-i\frac{1}{2m}k^2 t}\widetilde{\psi}(k,0)$$

Solutions that are momentum and energy eigenstates will be distributions, with initial value
$$\widetilde{\psi}(k,0) = \delta(k - k')$$

These will have momentum k' and energy $E = \frac{k'^2}{2m}$. The space of solutions can be identified with the space of initial value data $\widetilde{\psi}(k,0)$, which can be taken to be in $\mathcal{S}(\mathbf{R})$, $L^2(\mathbf{R})$ or $\mathcal{S}'(\mathbf{R})$.

Instead of working with time-dependent momentum space solutions $\widetilde{\psi}(k,t)$, one can Fourier transform in the time variable, defining

$$\widehat{\psi}(k,\omega) = \frac{1}{\sqrt{2\pi}}\int_{-\infty}^{\infty} e^{i\omega t}\widetilde{\psi}(k,t)dt$$

Just as the Fourier transform in q takes $\frac{d}{dq}$ to multiplication by ik, here the Fourier transform in t takes $\frac{d}{dt}$ to multiplication by $-i\omega$. Note the opposite sign convention in the phase factor from the spatial Fourier transform chosen to agree with the opposite sign conventions for spatial and time translations in the definitions of momentum and energy.

One finds for free particle solutions

$$\widehat{\psi}(k,\omega) = \frac{1}{\sqrt{2\pi}}\int_{-\infty}^{\infty} e^{i\omega t}e^{-i\frac{1}{2m}k^2 t}\widetilde{\psi}(k,0)dt$$

$$= \delta(\omega - \frac{1}{2m}k^2)\sqrt{2\pi}\widetilde{\psi}(k,0)$$

so $\widehat{\psi}(k,\omega)$ will be a distribution on $k-\omega$ space that is *nonzero* only on the parabola $\omega = \frac{1}{2m}k^2$. The space of solutions can be identified with the space of functions (or distributions) supported on this parabola. Energy eigenstates of energy E will be distributions with a dependence on ω of the form

$$\widehat{\psi}_E(k,\omega) = \delta(\omega - E)\widetilde{\psi}_E(k)$$

For free particle solutions, one has $E = \frac{k^2}{2m}$. $\widetilde{\psi}_E(k)$ will be a distribution in k with a factor $\delta(E - \frac{k^2}{2m})$.

For any function $f(k)$, the delta-function distribution $\delta(f(k))$ depends only on the behavior of f near its zeros. If $f' \neq 0$ at such zeros, one has (using linear approximation near zeros of f)

$$\delta(f(k)) = \sum_{k_j:f(k_j)=0}\delta(f'(k_j)(k-k_j)) = \sum_{k_j:f(k_j)=0}\frac{1}{|f'(k_j)|}\delta(k-k_j) \quad (11.9)$$

Applying this to the case of $f(k) = E - \frac{k^2}{2m}$, with a graph that has two zeros, at $k = \pm\sqrt{2mE}$ and looks like

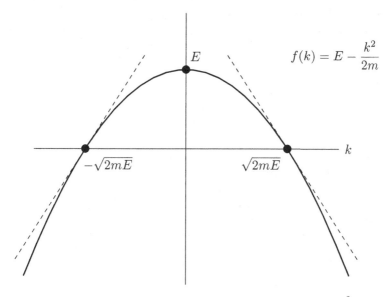

Figure 11.1: Linear approximations near zeros of $f(k) = E - \frac{k^2}{2m}$

we find that

$$\delta(E - \frac{k^2}{2m}) = \sqrt{\frac{m}{2E}}(\delta(k - \sqrt{2mE}) + \delta(k + \sqrt{2mE}))$$

and

$$\widetilde{\psi}_E(k) = c_+\delta(k - \sqrt{2mE}) + c_-\delta(k + \sqrt{2mE}) \qquad (11.10)$$

The two complex numbers c_+, c_- give the amplitudes for a free particle solution of energy E to have momentum $\pm\sqrt{2mE}$.

In the physical case of three spatial dimensions, one gets solutions

$$\widetilde{\psi}(\mathbf{k}, t) = e^{-i\frac{1}{2m}|\mathbf{k}|^2 t}\widetilde{\psi}(\mathbf{k}, 0)$$

and the space of solutions is a space of functions (or distributions) $\widetilde{\psi}(\mathbf{k}, 0)$ on $\mathbf{R}^3$. Energy eigenstates with energy E will be given by distributions that are *nonzero* only on the sphere $|\mathbf{k}|^2/2m = E$ of radius $\sqrt{2mE}$ in momentum space (these will be studied in detail in Chapter 19).

11.6 For further reading

The use of periodic boundary conditions, or "putting the system in a box," thus reducing the problem to that of Fourier series, is a conventional topic in quantum mechanics textbooks. Two good sources for the mathematics of Fourier series are [83] and [27]. The use of the Fourier transform to solve the free particle Schrödinger equation is a standard topic in physics textbooks, although the function space used is often not specified and distributions are not explicitly defined (although some discussion of the δ-function is always present). Standard mathematics textbooks discussing the Fourier transform and the theory of distributions are [88] and [27]. Lecture 6 in the notes on physics by Dolgachev [18] contains a more mathematically careful discussion of the sort of calculations with the δ-function described in this chapter and common in the physics literature.

For some insight into, and examples of, the problems that can appear when one ignores (as we generally do) the question of domains of operators such as the momentum operator, see [34]. Section 2.1 of [90] or Chapter 10 of [41] provides a formalism that includes a spectral theorem for unbounded self-adjoint operators, generalizing appropriately the spectral theorem of the finite dimensional state space case.

Chapter 12
Position and the Free Particle

Our discussion of the free particle has so far been largely in terms of one observable, the momentum operator. The free particle Hamiltonian is given in terms of this operator $(H = P^2/2m)$, and we have seen in section 11.5 that solutions of the Schrödinger equation behave very simply in momentum space. Since $[P, H] = 0$, momentum is a conserved quantity, and momentum eigenstates will remain momentum eigenstates under time evolution.

The Fourier transform interchanges momentum and position space, and a position operator Q can be defined that will play the role of the Fourier transform of the momentum operator. Position eigenstates will be position space δ-functions, but $[Q, H] \neq 0$ and the position will not be a conserved quantity. The time evolution of a state initially in a position eigenstate can be calculated in terms of a quantity called the propagator, which we will compute and study.

12.1 The position operator

On a state space $\mathcal{H}$ of functions (or distributions) of a position variable q, one can define:

Definition (Position operator). *The position operator Q is given by*

$$Q\psi(q) = q\psi(q)$$

Note that this operator has similar problems of definition to those of the momentum operator P: it can take a function in $L^2(\mathbf{R})$ to one that is no longer square-integrable. Like P, it is well defined on the Schwartz space $\mathcal{S}(\mathbf{R})$, as well as on the distributions $\mathcal{S}'(\mathbf{R})$. Also like P, it has no eigenfunc-

© Peter Woit 2017
P. Woit, *Quantum Theory, Groups and Representations*,
DOI 10.1007/978-3-319-64612-1_12

tions in $\mathcal{S}(\mathbf{R})$ or $L^2(\mathbf{R})$, but it does have eigenfunctions in $\mathcal{S}'(\mathbf{R})$. Since

$$\int_{-\infty}^{+\infty} q\delta(q - q')f(q)dq = q'f(q') = \int_{-\infty}^{+\infty} q'\delta(q - q')f(q)dq$$

one has the equality of distributions

$$q\delta(q - q') = q'\delta(q - q')$$

so $\delta(q - q')$ is an eigenfunction of Q with eigenvalue q'.

The operators Q and P do not commute, since

$$[Q, P]f = -iq\frac{d}{dq}f + i\frac{d}{dq}(qf) = if$$

and we get (reintroducing $\hbar$ for a moment) the fundamental operator commutation relation

$$[Q, P] = i\hbar\mathbf{1}$$

the Heisenberg commutation relation. This implies that Q and the free particle Hamiltonian $H = \frac{1}{2m}P^2$ also do not commute, so the position, unlike the momentum, is not a conserved quantity.

For a finite dimensional state space, recall that the spectral theorem (4.1) for a self-adjoint operator implied that any state could be written as a linear combination of eigenvectors of the operator. In this infinite dimensional case, the formula

$$\psi(q) = \int_{-\infty}^{+\infty} \delta(q - q')\psi(q')dq' \qquad (12.1)$$

can be interpreted as an expansion of an arbitrary state in terms of a continuous linear combination of eigenvectors of Q with eigenvalue q', the δ-functions $\delta(q - q')$. The Fourier inversion formula (11.4)

$$\psi(q) = \frac{1}{\sqrt{2\pi}} \int_{-\infty}^{+\infty} e^{ikq}\widetilde{\psi}(k)dk$$

similarly gives an expansion in terms of eigenvectors $\frac{1}{\sqrt{2\pi}}e^{ikq}$ of P, with eigenvalue k.

12.2 Momentum space representation

We began our discussion of the state space $\mathcal{H}$ of a free particle by taking states to be wavefunctions $\psi(q)$ defined on position space, thought of variously as being in $\mathcal{S}(\mathbf{R}), L^2(\mathbf{R})$ or $\mathcal{S}'(\mathbf{R})$. Using the Fourier transform, which takes such functions to their Fourier transforms

$$\widetilde{\psi}(k) = \mathcal{F}\psi = \frac{1}{\sqrt{2\pi}} \int_{-\infty}^{+\infty} e^{-ikq}\psi(q)dq$$

in the same sort of function space, we saw in section 11.5 that the state space $\mathcal{H}$ can instead be taken to be a space of functions $\widetilde{\psi}(k)$ on momentum space. We will call such a choice of $\mathcal{H}$, with the operator P now acting as

$$P\widetilde{\psi}(k) = k\widetilde{\psi}(k)$$

the momentum space representation, as opposed to the previous position space representation. By the Plancherel theorem (11.2), these are unitarily equivalent representations of the group $\mathbf{R}$, which acts in the position space case by translation by a in the position variable, in the momentum space case by multiplication by a phase factor e^{ika}.

In the momentum space representation, the eigenfunctions of P are the distributions $\delta(k - k')$, with eigenvalue k', and the expansion of a state in terms of eigenvectors is

$$\widetilde{\psi}(k) = \int_{-\infty}^{+\infty} \delta(k - k')\widetilde{\psi}(k')dk' \tag{12.2}$$

The position operator is

$$Q = i\frac{d}{dk}$$

which has eigenfunctions

$$\frac{1}{\sqrt{2\pi}}e^{-ikq'}$$

and the expansion of a state in terms of eigenvectors of Q is just the Fourier transform formula 11.3.

12.3   Dirac notation

In the Dirac bra-ket formalism, position and momentum eigenstates will be denoted $|q\rangle$ and $|k\rangle$, respectively, with

$$Q|q\rangle = q|q\rangle, \quad P|k\rangle = k|k\rangle$$

Arbitrary states $|\psi\rangle$ can be thought of as determined by coefficients

$$\langle q|\psi\rangle = \psi(q), \quad \langle k|\psi\rangle = \widetilde{\psi}(k) \tag{12.3}$$

with respect to either the $|q\rangle$ or $|k\rangle$ basis. The use of the bra-ket formalism requires some care, however, since states like $|q\rangle$ or $|k\rangle$ are elements of $\mathcal{S}'(\mathbf{R})$ that do not correspond to any element of $\mathcal{S}(\mathbf{R})$. Given elements $|\psi\rangle$ in $\mathcal{S}(\mathbf{R})$, they can be paired with elements of $\mathcal{S}'(\mathbf{R})$ like $\langle q|$ and $\langle k|$ as in equation 12.3 to get numbers. When working with states like $|q'\rangle$ and $|k'\rangle$, one has to invoke and properly interpret distributional relations such as

$$\langle q|q'\rangle = \delta(q - q'), \quad \langle k|k'\rangle = \delta(k - k')$$

Equation 12.1 is written in Dirac notation as

$$|\psi\rangle = \int_{-\infty}^{\infty} |q\rangle\langle q|\psi\rangle dq$$

and 12.2 as

$$|\psi\rangle = \int_{-\infty}^{\infty} |k\rangle\langle k|\psi\rangle dk$$

The resolution of the identity operator of equation 4.7 here is written

$$1 = \int_{-\infty}^{\infty} |q\rangle\langle q|dq = \int_{-\infty}^{\infty} |k\rangle\langle k|dk$$

The transformation between the $|q\rangle$ and $|k\rangle$ bases is given by the Fourier transform, which in this notation is

$$\langle k|\psi\rangle = \int_{-\infty}^{\infty} \langle k|q\rangle\langle q|\psi\rangle dq$$

where

$$\langle k|q\rangle = \frac{1}{\sqrt{2\pi}} e^{-ikq}$$

and the inverse Fourier transform

$$\langle q|\psi\rangle = \int_{-\infty}^{\infty} \langle q|k\rangle\langle k|\psi\rangle dk$$

where

$$\langle q|k\rangle = \frac{1}{\sqrt{2\pi}} e^{ikq}$$

12.4 Heisenberg uncertainty

We have seen that, describing the state of a free particle at a fixed time, one has δ-function states corresponding to a well-defined position (in the position representation) or a well-defined momentum (in the momentum representation). But Q and P do not commute, and states with both well-defined position and well-defined momentum do not exist. An example of a state peaked at $q = 0$ will be given by the Gaussian wavefunction

$$\psi(q) = e^{-\alpha \frac{q^2}{2}}$$

which becomes narrowly peaked for α large. By equation 11.6, the corresponding state in the momentum space representation is

$$\frac{1}{\sqrt{\alpha}} e^{-\frac{k^2}{2\alpha}}$$

which becomes uniformly spread out as α gets large. Similarly, as α goes to zero, one gets a state narrowly peaked at $k = 0$ in momentum space, but uniformly spread out as a position space wavefunction.

For states with expectation value of Q and P equal to zero, the width of the state in position space can be quantified by the expectation value of Q^2, and its width in momentum space by the expectation value of P^2. One has the following theorem, which makes precise the limit on simultaneously localizability of a state in position and momentum space

Theorem (Heisenberg uncertainty).

$$\frac{\langle \psi | Q^2 | \psi \rangle}{\langle \psi | \psi \rangle} \frac{\langle \psi | P^2 | \psi \rangle}{\langle \psi | \psi \rangle} \geq \frac{1}{4}$$

Proof For any real λ one has

$$\langle (Q + i\lambda P)\psi | (Q + i\lambda P)\psi \rangle \geq 0$$

but using self-adjointness of Q and P, as well as the relation $[Q, P] = i$, one has

$$\begin{aligned}
\langle (Q + i\lambda P)\psi | (Q + i\lambda P)\psi \rangle &= \lambda^2 \langle \psi | P^2 \psi \rangle + i\lambda \langle \psi | QP\psi \rangle - i\lambda \langle \psi | PQ\psi \rangle + \langle \psi | Q^2 \psi \rangle \\
&= \lambda^2 \langle \psi | P^2 \psi \rangle + \lambda(-\langle \psi | \psi \rangle) + \langle \psi | Q^2 \psi \rangle
\end{aligned}$$

This will be nonnegative for all λ if

$$\langle\psi|\psi\rangle^2 \leq 4\langle\psi|P^2\psi\rangle\langle\psi|Q^2\psi\rangle$$

$\square$

12.5 The propagator in position space

Free particle states with the simplest physical interpretation are momentum eigenstates. They describe a single quantum particle with a fixed momentum k', and this momentum is a conserved quantity that will not change. In the momentum space representation (see section 11.5), such a time-dependent state will be given by

$$\widetilde{\psi}(k,t) = e^{-i\frac{1}{2m}k'^2 t}\delta(k-k')$$

In the position space representation, such a state will be given by

$$\psi(q,t) = \frac{1}{\sqrt{2\pi}}e^{-i\frac{1}{2m}k'^2 t}e^{ik'q}$$

a wave with (restoring temporarily factors of $\hbar$ and using $p = \hbar k$) wavelength $\frac{2\pi\hbar}{p'}$ and angular frequency $\frac{p'^2}{2m\hbar}$.

As for any quantum system, time evolution of a free particle from time 0 to time t is given by a unitary operator $U(t) = e^{-itH}$. In the momentum space representation, this is just the multiplication operator

$$U(t) = e^{-i\frac{1}{2m}k^2 t}$$

In the position space representation, it is given by an integral kernel called the "propagator":

Definition (Position space propagator) *The position space propagator is the kernel $U(t, q_t, q_0)$ of the time evolution operator acting on position space wavefunctions. It determines the time evolution of wavefunctions for all times t by*

$$\psi(q_t, t) = \int_{-\infty}^{+\infty} U(t, q_t, q_0)\psi(q_0, 0)dq_0 \qquad (12.4)$$

where $\psi(q_0, 0)$ is the initial value of the wavefunction at time 0.

In the Dirac notation, one has

$$\psi(q_t, t) = \langle q_t | \psi(t) \rangle = \langle q_t | e^{-iHt} | \psi(0) \rangle = \langle q_t | e^{-iHt} \int_{-\infty}^{\infty} | q_0 \rangle \langle q_0 | \psi(0) \rangle dq_0$$

and the propagator can be written as

$$U(t, q_t, q_0) = \langle q_t | e^{-iHt} | q_0 \rangle$$

$U(t, q_t, q_0)$ can be computed for the free particle case by Fourier transform of the momentum space multiplication operator:

$$\psi(q_t, t) = \frac{1}{\sqrt{2\pi}} \int_{-\infty}^{+\infty} e^{ikq_t} \widetilde{\psi}(k, t) dk$$

$$= \frac{1}{\sqrt{2\pi}} \int_{-\infty}^{+\infty} e^{ikq_t} e^{-i\frac{1}{2m}k^2 t} \widetilde{\psi}(k, 0) dk$$

$$= \frac{1}{2\pi} \int_{-\infty}^{+\infty} e^{ikq_t} e^{-i\frac{1}{2m}k^2 t} \left(\int_{-\infty}^{+\infty} e^{-ikq_0} \psi(q_0, 0) dq_0 \right) dk$$

$$= \int_{-\infty}^{+\infty} \left(\frac{1}{2\pi} \int_{-\infty}^{+\infty} e^{ik(q_t - q_0)} e^{-i\frac{1}{2m}k^2 t} dk \right) \psi(q_0, 0) dq_0$$

so

$$U(t, q_t, q_0) = U(t, q_t - q_0) = \frac{1}{2\pi} \int_{-\infty}^{+\infty} e^{ik(q_t - q_0)} e^{-i\frac{1}{2m}k^2 t} dk \qquad (12.5)$$

Note that (as expected due to translation invariance of the Hamiltonian operator) this only depends on the difference $q_t - q_0$. Equation 12.5 can be rewritten as an inverse Fourier transform with respect to this difference

$$U(t, q_t - q_0) = \frac{1}{\sqrt{2\pi}} \int_{-\infty}^{+\infty} e^{ik(q_t - q_0)} \widetilde{U}(t, k) dk$$

where

$$\widetilde{U}(t, k) = \frac{1}{\sqrt{2\pi}} e^{-i\frac{1}{2m}k^2 t} \qquad (12.6)$$

To make sense of the integral 12.5, the product it can be replaced by a complex variable $z = \tau + it$. The integral becomes well defined when $\tau = \mathrm{Re}(z)$ ("imaginary time") is positive and then defines a holomorphic function in z. Doing the integral by the same method as in equation 11.6, one finds

$$U(z = \tau + it, q_t - q_0) = \sqrt{\frac{m}{2\pi z}} e^{-\frac{m}{2z}(q_t - q_0)^2} \tag{12.7}$$

For $z = \tau$ real and positive, this is the kernel function for solutions to the partial differential equation

$$\frac{\partial}{\partial \tau} \psi(q, \tau) = \frac{1}{2m} \frac{\partial^2}{\partial q^2} \psi(q, \tau)$$

known as the "heat equation." This equation models the way temperature diffuses in a medium; it also models the way probability of a given position diffuses in a random walk. Note that here it is $\psi(q)$ that gives the probability density, something quite different from the way probability occurs in measurement theory for the free particle quantum system. There it is $|\psi|^2$ that gives the probability density for the particle to have position observable eigenvalue q.

Taking as initial condition

$$\psi(q_0, 0) = \delta(q_0 - q')$$

the heat equation will have as solution at later times

$$\psi(q_\tau, \tau) = \sqrt{\frac{m}{2\pi \tau}} e^{-\frac{m}{2\tau}(q' - q_\tau)^2} \tag{12.8}$$

This is physically reasonable: at times $\tau > 0$, an initial source of heat localized at a point q' diffuses as a Gaussian about q' with increasing width. For $\tau < 0$, one gets something that grows exponentially at $\pm\infty$, and so is not in $L^2(\mathbf{R})$ or even $\mathcal{S}'(\mathbf{R})$.

In real time t as opposed to imaginary time τ (i.e., $z = it$, interpreted as the limit $\lim_{\epsilon \to 0^+}(\epsilon + it)$), equation 12.7 becomes

$$U(t, q_t - q_0) = \sqrt{\frac{m}{i2\pi t}} e^{-\frac{m}{i2t}(q_t - q_0)^2} \tag{12.9}$$

Unlike the case of imaginary time, this expression needs to be interpreted as a distribution, and as such equation 12.4 makes sense for $\psi(q_0, 0) \in \mathcal{S}(\mathbf{R})$. One can show that, for $\psi(q_0, 0)$ with amplitude peaked around a position q' and with amplitude of its Fourier transform peaked around a momentum k', at later times $\psi(q, t)$ will become less localized, but with a maximum amplitude at $q' + \frac{k'}{m}t$.

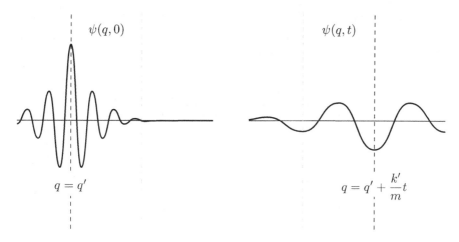

Figure 12.1: Time evolution of an initially localized wavefunction.

This is what one expects physically, since $\frac{p'}{m}$ is the velocity corresponding to momentum p' for a classical particle.

Note that the choice of square root of i in 12.9 is determined by the condition that one get an analytic continuation from the imaginary time version for $\tau > 0$, so one should take in 12.9

$$\sqrt{\frac{m}{i2\pi t}} = e^{-i\frac{\pi}{4}} \sqrt{\frac{m}{2\pi t}}$$

We have seen that an initial momentum eigenstate

$$\psi(q_0, 0) = \frac{1}{\sqrt{2\pi}} e^{ik'q}$$

evolves in time by multiplication by a phase factor. An initial position eigenstate

$$\psi(q_0, 0) = \delta(q_0 - q')$$

evolves to

$$\psi(q_t, t) = \int_{-\infty}^{+\infty} U(t, q_t - q_0)\delta(q_0 - q')dq_0 = \sqrt{\frac{m}{i2\pi t}} e^{-\frac{m}{i2t}(q_t - q')^2}$$

Near $t = 0$ this function has a rather peculiar behavior. It starts out localized at q_0 at $t = 0$, but at any later time $t > 0$, no matter how small, the wavefunction will have constant amplitude extending out to infinity in position space. Here one sees clearly the necessity of interpreting such a wavefunction as a distribution.

For a physical interpretation of this calculation, note that while a momentum eigenstate is a good approximation to a stable state one can create and then study, an approximate position eigenstate is quite different. Its creation requires an interaction with an apparatus that exchanges a very large momentum (involving a very short wavelength to resolve the position). By the Heisenberg uncertainty principle, a precisely known position corresponds to a completely unknown momentum, which may be arbitrarily large. Such arbitrarily large momenta imply arbitrarily large velocities, reaching arbitrarily far away in arbitrarily short time periods. In later chapters, we will see how relativistic quantum theories provide a more physically realistic description of what happens when one attempts to localize a quantum particle, with quite different phenomena (including possible particle production) coming into play.

12.6 Propagators in frequency–momentum space

The propagator defined by equation 12.4 will take a wavefunction at time 0 and give the wavefunction at any other time t, positive or negative. We will find it useful to define a version of the propagator that takes into account causality, only giving a nonzero result for $t > 0$:

Definition (Retarded propagator). *The retarded propagator is given by*

$$U_+(t, q_t - q_0) = \begin{cases} 0 & t < 0 \\ U(t, q_t - q_0) & t > 0 \end{cases}$$

This can also be written

$$U_+(t, q_t - q_0) = \theta(t)U(t, q_t - q_0)$$

where $\theta(t)$ is the step-function

$$\theta(t) = \begin{cases} 1 & t > 0 \\ 0 & t < 0 \end{cases}$$

We will use an integral representation of $\theta(t)$ given by

$$\theta(t) = \lim_{\epsilon \to 0+} \frac{i}{2\pi} \int_{-\infty}^{+\infty} \frac{1}{\omega + i\epsilon} e^{-i\omega t} d\omega \qquad (12.10)$$

To derive this, note that as a distribution, $\theta(t)$ has a Fourier transform given by

$$\lim_{\epsilon \to 0^+} \frac{i}{\sqrt{2\pi}} \frac{1}{\omega + i\epsilon}$$

since the calculation

$$\frac{1}{\sqrt{2\pi}} \int_{-\infty}^{+\infty} \theta(t) e^{i\omega t} d\omega = \frac{1}{\sqrt{2\pi}} \int_{0}^{+\infty} e^{i\omega t} d\omega$$

$$= \frac{1}{\sqrt{2\pi}} \left(-\frac{1}{i\omega} \right)$$

makes sense for ω replaced by $\lim_{\epsilon \to 0^+} (\omega + i\epsilon)$ (or, for real boundary values of ω complex, taking values in the upper half-plane). Fourier inversion then gives equation 12.10.

Digression. *The integral 12.10 can also be computed using methods of complex analysis in the variable ω. Cauchy's integral formula says that the integral about a closed curve of a meromorphic function with simple poles is given by $2\pi i$ times the sum of the residues at the poles. For $t < 0$, since $e^{-i\omega t}$ falls off exponentially if ω has a nonzero positive imaginary part, the integral along the real ω axis will be the same as for the semi-circle C_+ closed in the upper half-plane (with the radius of the semi-circle taken to infinity). C_+ encloses no poles so the integral is 0.*

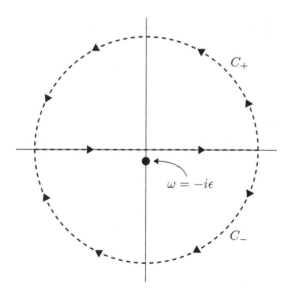

Figure 12.2: Evaluating $\theta(t)$ via contour integration.

For t > 0, one instead closes the path using C_- in the lower half-plane, and finds that the integral can be evaluated in terms of the residue of the pole at $\omega = -i\epsilon$ (with the minus sign coming from orientation of the curve), giving

$$\theta(t) = -2\pi i \left(\frac{i}{2\pi}\right) = 1$$

By similar arguments, one can show that $\theta(-t)$ has (as a distribution) Fourier transform

$$\lim_{\epsilon \to 0^+} -\frac{i}{\sqrt{2\pi}} \frac{1}{\omega - i\epsilon}$$

and the integral representation

$$\theta(-t) = \lim_{\epsilon \to 0^+} -\frac{i}{2\pi} \int_{-\infty}^{+\infty} \frac{1}{\omega - i\epsilon} e^{-i\omega t} d\omega$$

Taking $1/\sqrt{2\pi}$ times the sum of the Fourier transforms for $\theta(t)$ and $\theta(-t)$ gives the distribution

$$\lim_{\epsilon \to 0^+} \frac{i}{2\pi} \left(\frac{1}{\omega + i\epsilon} - \frac{1}{\omega - i\epsilon}\right) = \lim_{\epsilon \to 0^+} \frac{i}{2\pi} \frac{-2i\epsilon}{\omega^2 + \epsilon^2}$$

$$= \lim_{\epsilon \to 0^+} \frac{1}{\pi} \frac{\epsilon}{\omega^2 + \epsilon^2}$$

$$= \delta(\omega) \qquad\qquad (12.11)$$

as one expects since the delta-function $\delta(\omega)$ is the Fourier transform of

$$\frac{1}{\sqrt{2\pi}} (\theta(t) + \theta(-t)) = \frac{1}{\sqrt{2\pi}}$$

Returning to the propagator, as in section 11.5, one can Fourier transform with respect to time, and thus get a propagator that depends on the frequency ω. The Fourier transform of equation 12.6 with respect to time is

$$\widehat{U}(\omega, k) = \frac{1}{\sqrt{2\pi}} \int_{-\infty}^{+\infty} \left(\frac{1}{\sqrt{2\pi}} e^{-i\frac{1}{2m}k^2 t}\right) e^{i\omega t} d\omega = \delta(\omega - \frac{1}{2m}k^2)$$

Using equations 12.5 and 12.10 the retarded propagator in position space is given by

$$U_+(t, q_t - q_0) = \lim_{\epsilon \to 0^+} \left(\frac{1}{2\pi}\right)^2 \int_{-\infty}^{+\infty} \int_{-\infty}^{+\infty} \frac{i}{\omega + i\epsilon} e^{-i\omega t} e^{ik(q_t - q_0)} e^{-i\frac{1}{2m}k^2 t} \, d\omega dk$$

$$= \lim_{\epsilon \to 0^+} \left(\frac{1}{2\pi}\right)^2 \int_{-\infty}^{+\infty} \int_{-\infty}^{+\infty} \frac{i}{\omega + i\epsilon} e^{-i(\omega + \frac{1}{2m}k^2)t} e^{ik(q_t - q_0)} \, d\omega dk$$

Shifting the integration variable by

$$\omega \to \omega' = \omega + \frac{1}{2m}k^2$$

one finds

$$U_+(t, q_t - q_0) = lim_{\epsilon \to 0^+} \left(\frac{1}{2\pi}\right)^2 \int_{-\infty}^{+\infty} \int_{-\infty}^{+\infty} \frac{i}{\omega' - \frac{1}{2m}k^2 + i\epsilon} e^{-i\omega' t} e^{ik(q_T - q_0)} \, d\omega' dk$$

but this is the Fourier transform

$$U_+(t, q_t - q_0) = \frac{1}{2\pi} \int_{-\infty}^{+\infty} \int_{-\infty}^{+\infty} \widehat{U}_+(\omega, k) e^{-i\omega t} e^{ik(q_t - q_0)} \, d\omega dk \qquad (12.12)$$

where

$$\widehat{U}_+(\omega, k) = \lim_{\epsilon \to 0^+} \frac{i}{2\pi} \frac{1}{\omega - \frac{1}{2m}k^2 + i\epsilon} \qquad (12.13)$$

Digression. *By the same argument as the one above for the integral representation of $\theta(t)$, but with pole now at*

$$\omega = \frac{1}{2m}k^2 - i\epsilon$$

the ω integral in equation 12.12 can be evaluated by the Cauchy integral formula, recovering formula 12.9 for $U(t, q_t - q_0)$.

12.7   Green's functions and solutions to the Schrödinger equations

The method of Green's functions provides solutions ψ to differential equations

$$D\psi = J \qquad (12.14)$$

where D is a differential operator and J is a fixed function, by finding an inverse D^{-1} to D and then setting $\psi = D^{-1}J$. For D a constant coefficient differential operator, the Fourier transform will take D to multiplication by

a polynomial $\widehat{D}$ and we define the Green's function of D to be the function
(or distribution) with Fourier transform

$$\widehat{G} = \frac{1}{\widehat{D}} \tag{12.15}$$

Since

$$\widehat{D}\widehat{G}\widehat{J} = \widehat{J}$$

the inverse Fourier transform of $\widehat{G}\widehat{J}$ will be a solution to 12.14.

Note that $\widehat{G}$ and G are not uniquely determined by the condition 12.15 since D may have a kernel, and then solutions to 12.14 are only determined up to a solution ψ_0 of the homogeneous equation $D\psi_0 = 0$. In terms of Fourier transforms, $\widehat{D}$ may have zeros, and then $\widehat{G}$ is ambiguous up to functions on the zero set.

For the case of the Schrödinger equation, we take

$$D = i\frac{\partial}{\partial t} + \frac{1}{2m}\frac{\partial^2}{\partial q^2}$$

and then (Fourier transforming in q and t as above)

$$\widehat{D} = i(-i\omega) + \frac{1}{2m}(ik)^2 = \omega - \frac{k^2}{2m}$$

and

$$\widehat{G} = \frac{1}{\omega - \frac{k^2}{2m}}$$

A solution ψ of 12.14 will be given by computing the inverse Fourier transform of $\widehat{G}\widehat{J}$

$$\psi(q,t) = \frac{1}{2\pi}\int_{-\infty}^{+\infty}\int_{-\infty}^{+\infty}\frac{1}{\omega - \frac{k^2}{2m}}\widehat{J}(\omega,k)e^{-i\omega t}e^{ikq}d\omega dk \tag{12.16}$$

Here $\widehat{G}$ is zero on the set $\omega = \frac{k^2}{2m}$ and the non-uniqueness of the solution to $D\psi = J$ is reflected in the ambiguity of how to treat the integration through the points $\omega = \frac{k^2}{2m}$.

For solutions $\psi(q,t)$ of the Schrödinger equation with initial data $\psi(q,0)$ at time $t = 0$, if we define $\psi_+(q,t) = \theta(t)\psi(q,t)$ we get the "retarded" solution

$$\psi_+(q,t) = \int_{-\infty}^{+\infty} U_+(t, q - q_0)\psi(q_0, 0)dq_0$$

where $U_+(t, q - q_0)$ is the retarded propagator given by equations 12.12 and 12.13. Since

$$D\psi_+(q, t) = (D\theta(t))\psi(q, t) + \theta(t)D\psi(q, t) = i\delta(t)\psi(q, t) = i\delta(t)\psi(q, 0)$$

$\psi_+(q, t)$ is a solution of 12.14 with

$$J(q, t) = i\delta(t)\psi(q, 0), \quad \widehat{J}(\omega, k) = \frac{i}{\sqrt{2\pi}}\widetilde{\psi}(k, 0)$$

Using 12.16 to get an expression for $\psi_+(q, t)$ in terms of the Green's function, we have

$$\psi_+(q, t) = \frac{1}{2\pi}\int_{-\infty}^{+\infty}\int_{-\infty}^{+\infty}\widehat{G}(\omega, k)\frac{i}{\sqrt{2\pi}}\widetilde{\psi}(k, 0)e^{-i\omega t}e^{ikq}\,d\omega dk$$

$$= \left(\frac{1}{2\pi}\right)^2\int_{-\infty}^{+\infty}\left(\int_{-\infty}^{+\infty}\int_{-\infty}^{+\infty}i\widehat{G}(\omega, k)e^{-i\omega t}e^{ik(q-q')}\,d\omega dk\right)\psi(q', 0)dq'$$

Comparing this to equations 12.12 and 12.13, we find that the Green's function that will give the retarded solution $\psi_+(q, t)$ is

$$\widehat{G}_+(\omega, k) = \lim_{\epsilon \to 0^+}\frac{1}{\omega - \frac{k^2}{2m} + i\epsilon}$$

and is related to the retarded propagator by

$$\widehat{U}(\omega, k) = i\widehat{G}_+(\omega, k)$$

One can also define an "advanced" Green's function by

$$\widehat{G}_- = \lim_{\epsilon \to 0^+}\frac{1}{\omega - \frac{k^2}{2m} - i\epsilon}$$

and the inverse Fourier transform of $\widehat{G}_-\widehat{J}$ will also be a solution to 12.14. Taking the difference between retarded and advanced Green's functions gives an operator

$$\widehat{\Delta} = \frac{i}{2\pi}(\widehat{G}_+ - \widehat{G}_-)$$

with the property that, for any choice of J, ΔJ will be a solution to the Schrödinger equation (since it is the difference between two solutions of the inhomogeneous equation 12.14). The properties of Δ can be understood by using 12.11 to show that

$$\widehat{\Delta} = \delta(\omega - \frac{k^2}{2m})$$

12.8 For further reading

The topics of this chapter are covered in every quantum mechanics textbook, with a discussion providing more physical motivation. For a mathematics textbook that covers distributional solutions of the Schrödinger equation in detail, see [72]. In the textbook [41] for mathematicians, see chapter 4 for a more detailed mathematically rigorous treatment of the free particle and chapter 12 for a careful treatment of the subtleties of Heisenberg uncertainty.

Chapter 13
The Heisenberg group and the Schrödinger Representation

In our discussion of the free particle, we used just the actions of the groups $\mathbf{R}^3$ of spatial translations and the group $\mathbf{R}$ of time translations, finding corresponding observables, the self-adjoint momentum and Hamiltonian operators $\mathbf{P}$ and H. We have seen though that the Fourier transform allows a perfectly symmetrical treatment of position and momentum variables and the corresponding non-commuting position and momentum operators Q_j and P_j.

The P_j and Q_j operators satisfy relations known as the Heisenberg commutation relations, which first appeared in the earliest work of Heisenberg and collaborators on a full quantum-mechanical formalism in 1925. These were quickly recognized by Hermann Weyl as the operator relations of a Lie algebra representation, for a Lie algebra now known as the Heisenberg Lie algebra. The corresponding group is called the Heisenberg group by mathematicians, with physicists sometimes using the terminology "Weyl group" (which means something else to mathematicians). The state space of a quantum particle, either free or moving in a potential, will be a unitary representation of this group, with the group of spatial translations a subgroup.

Note that this particular use of a group and its representation theory in quantum mechanics is both at the core of the standard axioms and much more general than the usual characterization of the significance of groups as "symmetry groups." The Heisenberg group does not in any sense correspond to a group of invariances of the physical situation (there are no states invariant under the group), and its action does not commute with any *nonzero* Hamiltonian operator. Instead it plays a much deeper role, with its unique unitary representation determining much of the structure of quantum mechanics.

© Peter Woit 2017
P. Woit, *Quantum Theory, Groups and Representations*,
DOI 10.1007/978-3-319-64612-1_13

13.1   The Heisenberg Lie algebra

In either the position or momentum space representation, the operators P_j and Q_j satisfy the relation

$$[Q_j, P_k] = i\delta_{jk}\mathbf{1}$$

Soon after this commutation relation appeared in early work on quantum mechanics, Weyl realized that it can be interpreted as the relation between operators one would get from a representation of a $2d + 1$ dimensional Lie algebra, now called the Heisenberg Lie algebra. Treating first the $d = 1$ case, we define:

Definition (Heisenberg Lie algebra, $d = 1$). *The Heisenberg Lie algebra $\mathfrak{h}_3$ is the vector space $\mathbf{R}^3$ with the Lie bracket defined by its values on a basis (X, Y, Z) by*

$$[X, Y] = Z, \quad [X, Z] = [Y, Z] = 0$$

Writing a general element of $\mathfrak{h}_3$ in terms of this basis as $xX + yY + zZ$, and grouping the x, y coordinates together (we will see that it is useful to think of the vector space $\mathfrak{h}_3$ as $\mathbf{R}^2 \oplus \mathbf{R}$), the Lie bracket is given in terms of the coordinates by

$$\left[\left(\begin{pmatrix} x \\ y \end{pmatrix}, z\right), \left(\begin{pmatrix} x' \\ y' \end{pmatrix}, z'\right)\right] = \left(\begin{pmatrix} 0 \\ 0 \end{pmatrix}, xy' - yx'\right)$$

Note that this is a non-trivial Lie algebra, but only minimally so. All Lie brackets of Z with anything else are zero. All Lie brackets of Lie brackets are also zero (as a result, this is an example of what is known as a "nilpotent" Lie algebra).

The Heisenberg Lie algebra is isomorphic to the Lie algebra of 3 by 3 strictly upper triangular real matrices, with Lie bracket the matrix commutator, by the following isomorphism:

$$X \leftrightarrow \begin{pmatrix} 0 & 1 & 0 \\ 0 & 0 & 0 \\ 0 & 0 & 0 \end{pmatrix}, \quad Y \leftrightarrow \begin{pmatrix} 0 & 0 & 0 \\ 0 & 0 & 1 \\ 0 & 0 & 0 \end{pmatrix}, \quad Z \leftrightarrow \begin{pmatrix} 0 & 0 & 1 \\ 0 & 0 & 0 \\ 0 & 0 & 0 \end{pmatrix}$$

$$\left(\begin{pmatrix} x \\ y \end{pmatrix}, z\right) \leftrightarrow \begin{pmatrix} 0 & x & z \\ 0 & 0 & y \\ 0 & 0 & 0 \end{pmatrix}$$

and one has

$$\left[\begin{pmatrix} 0 & x & z \\ 0 & 0 & y \\ 0 & 0 & 0 \end{pmatrix}, \begin{pmatrix} 0 & x' & z' \\ 0 & 0 & y' \\ 0 & 0 & 0 \end{pmatrix}\right] = \begin{pmatrix} 0 & 0 & xy' - x'y \\ 0 & 0 & 0 \\ 0 & 0 & 0 \end{pmatrix}$$

The generalization of this to higher dimensions is:

Definition (Heisenberg Lie algebra). *The Heisenberg Lie algebra* $\mathfrak{h}_{2d+1}$ *is the vector space* $\mathbf{R}^{2d+1} = \mathbf{R}^{2d} \oplus \mathbf{R}$ *with the Lie bracket defined by its values on a basis* X_j, Y_j, Z $(j = 1, \ldots d)$ *by*

$$[X_j, Y_k] = \delta_{jk} Z, \quad [X_j, Z] = [Y_j, Z] = 0$$

Writing a general element as $\sum_{j=1}^{d} x_j X_j + \sum_{k=1}^{d} y_k Y_k + zZ$, in terms of coordinates the Lie bracket is

$$\left[\left(\begin{pmatrix} \mathbf{x} \\ \mathbf{y} \end{pmatrix}, z\right), \left(\begin{pmatrix} \mathbf{x}' \\ \mathbf{y}' \end{pmatrix}, z\right)\right] = \left(\begin{pmatrix} \mathbf{0} \\ \mathbf{0} \end{pmatrix}, \mathbf{x} \cdot \mathbf{y}' - \mathbf{y} \cdot \mathbf{x}'\right) \qquad (13.1)$$

This Lie algebra can be written as a Lie algebra of matrices for any d. For instance, in the physical case of $d = 3$, elements of the Heisenberg Lie algebra can be written

$$\begin{pmatrix} 0 & x_1 & x_2 & x_3 & z \\ 0 & 0 & 0 & 0 & y_1 \\ 0 & 0 & 0 & 0 & y_2 \\ 0 & 0 & 0 & 0 & y_3 \\ 0 & 0 & 0 & 0 & 0 \end{pmatrix}$$

13.2   The Heisenberg group

Exponentiating matrices in $\mathfrak{h}_3$ gives

$$\exp \begin{pmatrix} 0 & x & z \\ 0 & 0 & y \\ 0 & 0 & 0 \end{pmatrix} = \begin{pmatrix} 1 & x & z + \frac{1}{2}xy \\ 0 & 1 & y \\ 0 & 0 & 1 \end{pmatrix}$$

so the group with Lie algebra $\mathfrak{h}_3$ will be the group of upper triangular 3 by 3 real matrices with 1 on the diagonal, and this group will be the Heisenberg group H_3. For our purposes though, it is better to work in exponential coordinates (i.e., labeling a group element with the Lie algebra element that exponentiates to it). In these coordinates, the exponential map relating the Heisenberg Lie algebra $\mathfrak{h}_{2d+1}$ and the Heisenberg Lie group H_{2d+1} is just the identity map, and we will use the same notation

$$\left(\begin{pmatrix} x \\ y \end{pmatrix}, z \right)$$

for both Lie algebra and corresponding Lie group elements.

Matrix exponentials in general satisfy the Baker–Campbell–Hausdorff formula, which says

$$e^A e^B = e^{A+B+\frac{1}{2}[A,B]+\frac{1}{12}[A,[A,B]]-\frac{1}{12}[B,[A,B]]+\cdots}$$

where the higher terms can all be expressed as repeated commutators. This provides one way of showing that the Lie group structure is determined (for group elements expressible as exponentials) by knowing the Lie bracket. For the full formula and a detailed proof, see chapter 5 of [42]. One can easily check the first few terms in this formula by expanding the exponentials, but the difficulty of the proof is that it is not at all obvious why all the terms can be organized in terms of commutators.

For the case of the Heisenberg Lie algebra, since all multiple commutators vanish, the Baker–Campbell–Hausdorff formula implies for exponentials of elements of $\mathfrak{h}_3$

$$e^A e^B = e^{A+B+\frac{1}{2}[A,B]}$$

(a proof of this special case of Baker–Campbell–Hausdorff is in section 5.2 of [42]). We can use this to explicitly write the group law in exponential coordinates:

Definition (Heisenberg group, $d = 1$). *The Heisenberg group H_3 is the space* $\mathbf{R}^3 = \mathbf{R}^2 \oplus \mathbf{R}$ *with the group law*

$$\left(\begin{pmatrix} x \\ y \end{pmatrix}, z \right) \left(\begin{pmatrix} x' \\ y' \end{pmatrix}, z' \right) = \left(\begin{pmatrix} x + x' \\ y + y' \end{pmatrix}, z + z' + \frac{1}{2}(xy' - yx') \right) \qquad (13.2)$$

The isomorphism between $\mathbf{R}^2 \oplus \mathbf{R}$ with this group law and the matrix form of the group is given by

$$\left(\begin{pmatrix} x \\ y \end{pmatrix}, z \right) \leftrightarrow \begin{pmatrix} 1 & x & z + \frac{1}{2}xy \\ 0 & 1 & y \\ 0 & 0 & 1 \end{pmatrix}$$

Note that the Lie algebra basis elements X, Y, Z each generate subgroups of H_3 isomorphic to $\mathbf{R}$. Elements of the first two of these subgroups generate the full group, and elements of the third subgroup are "central," meaning they commute with all group elements. Also notice that the non-commutative nature of the Lie algebra (equation 13.1) or group (equation 13.2) depends purely on the factor $xy' - yx'$.

The generalization of this to higher dimensions is:

Definition (Heisenberg group). *The Heisenberg group H_{2d+1} is the space* $\mathbf{R}^{2d+1}$ *with the group law*

$$\left(\begin{pmatrix} \mathbf{x} \\ \mathbf{y} \end{pmatrix}, z\right)\left(\begin{pmatrix} \mathbf{x}' \\ \mathbf{y}' \end{pmatrix}, z'\right) = \left(\begin{pmatrix} \mathbf{x} + \mathbf{x}' \\ \mathbf{y} + \mathbf{y}' \end{pmatrix}, z + z' + \frac{1}{2}(\mathbf{x} \cdot \mathbf{y}' - \mathbf{y} \cdot \mathbf{x}')\right)$$

where $\mathbf{x}, \mathbf{x}' \ \mathbf{y}, \mathbf{y}' \in \mathbf{R}^d$.

13.3 The Schrödinger representation

Since it can be defined in terms of 3 by 3 matrices, the Heisenberg group H_3 has an obvious representation on $\mathbf{C}^3$, but this representation is not unitary and not of physical interest. What is of great interest is the infinite dimensional representation on functions of q for which the Lie algebra version is given by the Q, P, and unit operators:

Definition (Schrödinger representation, Lie algebra version). *The Schrödinger representation of the Heisenberg Lie algebra* $\mathfrak{h}_3$ *is the representation* $(\Gamma'_S, L^2(\mathbf{R}))$ *satisfying*

$$\Gamma'_S(X)\psi(q) = -iQ\psi(q) = -iq\psi(q), \quad \Gamma'_S(Y)\psi(q) = -iP\psi(q) = -\frac{d}{dq}\psi(q)$$

$$\Gamma'_S(Z)\psi(q) = -i\psi(q)$$

Factors of i have been chosen to make these operators skew-adjoint and the representation thus unitary. They can be exponentiated, giving in the exponential coordinates on H_3 of equation 13.2

$$\Gamma_S\left(\begin{pmatrix} x \\ 0 \end{pmatrix}, 0\right)\psi(q) = e^{-xiQ}\psi(q) = e^{-ixq}\psi(q)$$

$$\Gamma_S\left(\begin{pmatrix} 0 \\ y \end{pmatrix}, 0\right)\psi(q) = e^{-yiP}\psi(q) = e^{-y\frac{d}{dq}}\psi(q) = \psi(q - y)$$

$$\Gamma_S\left(\begin{pmatrix} 0 \\ 0 \end{pmatrix}, z\right)\psi(q) = e^{-iz}\psi(q)$$

For general group elements of H_3, one has:

Definition (Schrödinger representation, Lie group version). *The Schrödinger representation of the Heisenberg Lie group* H_3 *is the representation* $(\Gamma_S, L^2(\mathbf{R}))$ *satisfying*

$$\Gamma_S\left(\left(\begin{pmatrix} x \\ y \end{pmatrix}\right), z\right)\right)\psi(q) = e^{-iz}e^{i\frac{xy}{2}}e^{-ixq}\psi(q-y) \qquad (13.3)$$

To check that this defines a representation, one computes

$$\Gamma_S\left(\left(\begin{pmatrix} x \\ y \end{pmatrix}\right), z\right)\right)\Gamma_S\left(\left(\begin{pmatrix} x' \\ y' \end{pmatrix}\right), z'\right)\right)\psi(q)$$

$$=\Gamma_S\left(\left(\begin{pmatrix} x \\ y \end{pmatrix}\right), z\right)\right)e^{-iz'}e^{i\frac{x'y'}{2}}e^{-ix'q}\psi(q-y')$$

$$=e^{-i(z+z')}e^{i\frac{xy+x'y'}{2}}e^{-ixq}e^{-ix'(q-y)}\psi(q-y-y')$$

$$=e^{-i(z+z'+\frac{1}{2}(xy'-yx'))}e^{i\frac{(x+x')(y+y')}{2}}e^{-i(x+x')q}\psi(q-(y+y'))$$

$$=\Gamma_S\left(\left(\begin{pmatrix} x+x' \\ y+y' \end{pmatrix}\right), z+z'+\frac{1}{2}(xy'-yx')\right)\right)\psi(q)$$

The group analog of the Heisenberg commutation relations (often called the "Weyl form" of the commutation relations) is the relation

$$e^{-ixQ}e^{-iyP} = e^{-ixy}e^{-iyP}e^{-ixQ}$$

This can be derived by using the explicit representation operators in equation 13.3 (or the Baker–Campbell–Hausdorff formula and the Heisenberg commutation relations) to compute

$$e^{-ixQ}e^{-iyP} = e^{-i(xQ+yP)+\frac{1}{2}[-ixQ,-iyP]} = e^{-i\frac{xy}{2}}e^{-i(xQ+yP)}$$

as well as the same product in the opposite order, and then comparing the results.

Note that, for the Schrödinger representation, we have

$$\Gamma_S\left(\left(\begin{pmatrix} 0 \\ 0 \end{pmatrix}\right), z+2\pi\right)\right) = \Gamma_S\left(\left(\begin{pmatrix} 0 \\ 0 \end{pmatrix}\right), z\right)\right)$$

so the representation operators are periodic with period 2π in the z-coordinate. Some authors choose to define the Heisenberg group H_3 as not $\mathbf{R}^2 \oplus \mathbf{R}$, but $\mathbf{R}^2 \times S^1$, building this periodicity automatically into the definition of the group, rather than the representation.

We have seen that the Fourier transform $\mathcal{F}$ takes the Schrödinger representation to a unitarily equivalent representation of H_3, in terms of functions of p (the momentum space representation). The equivalence is given by a change

$$\Gamma_S(g) \rightarrow \widetilde{\Gamma}_S(g) = \mathcal{F}\,\Gamma_S(g)\widetilde{\mathcal{F}}$$

in the representation operators, with the Plancherel theorem (equation 11.5 ensuring that $\mathcal{F}$ and $\widetilde{\mathcal{F}} = \mathcal{F}^{-1}$ are unitary operators.

In typical physics quantum mechanics textbooks, one often sees calculations made just using the Heisenberg commutation relations, without picking a specific representation of the operators that satisfy these relations. This turns out to be justified by the remarkable fact that, for the Heisenberg group, once one picks the constant with which Z acts, all irreducible representations are unitarily equivalent. By unitarity, this constant is $-ic, c \in \mathbf{R}$. We have chosen $c = 1$, but other values of c would correspond to different choices of units.

In a sense, the representation theory of the Heisenberg group is very simple: there's only one irreducible representation. This is very different from the theory for even the simplest compact Lie groups ($U(1)$ and $SU(2)$) which have an infinity of inequivalent irreducibles labeled by weight or by spin. Representations of a Heisenberg group will appear in different guises (we have seen two and will see another in the discussion of the harmonic oscillator, and there are yet others that appear in the theory of theta-functions), but they are all unitarily equivalent, a statement known as the Stone–von Neumann theorem. Some good references for this material are [90] and [41]. In-depth discussions devoted to the mathematics of the Heisenberg group and its representations can be found in [51], [26], and [94].

In these references can be found a proof of the (not difficult)

Theorem. *The Schrödinger representation Γ_S described above is irreducible.*

and the much more difficult

Theorem (Stone–von Neumann). *Any irreducible representation π of the group H_3 on a Hilbert space, satisfying*

$$\pi'(Z) = -i\mathbf{1}$$

is unitarily equivalent to the Schrödinger representation $(\Gamma_S, L^2(\mathbf{R}))$.

Note that all of this can easily be generalized to the case of d spatial dimensions, for d finite, with the Heisenberg group now H_{2d+1} and the Stone–von Neumann theorem still true. In the case of an infinite number of degrees of freedom, which is the case of interest in quantum field theory, the Stone–von Neumann theorem no longer holds and one has an infinity of inequivalent irreducible representations, leading to quite different phenomena. For more on this topic, see chapter 39.

It is also important to note that the Stone–von Neumann theorem is formulated for Heisenberg group representations, not for Heisenberg Lie algebra representations. For infinite dimensional representations in cases like this, there are representations of the Lie algebra that are "non-integrable": they are not the derivatives of Lie group representations. For such non-integrable representations of the Heisenberg Lie algebra (i.e., operators satisfying the

Heisenberg commutation relations), there are counter-examples to the analog of the Stone–von Neumann theorem. It is only for integrable representations that the theorem holds, and one has a unique sort of irreducible representation.

13.4 For further reading

For a lot more detail about the mathematics of the Heisenberg group, its Lie algebra and the Schrödinger representation, see [8], [51], [26], and [94]. An excellent historical overview of the Stone–von Neumann theorem [74] by Jonathan Rosenberg is well worth reading. For not just a proof of Stone–von Neumann, but some motivation, see the discussion in chapter 14 of [41].

Chapter 14
The Poisson Bracket and Symplectic Geometry

We have seen that the quantum theory of a free particle corresponds to the construction of a representation of the Heisenberg Lie algebra in terms of operators Q and P, together with a choice of Hamiltonian $H = \frac{1}{2m}P^2$. One would like to use this to produce quantum systems with a similar relation to more non-trivial classical mechanical systems than the free particle. During the earliest days of quantum mechanics, it was recognized by Dirac that the commutation relations of the Q and P operators somehow corresponded to the Poisson bracket relations between the position and momentum coordinates on phase space in the Hamiltonian formalism for classical mechanics. In this chapter, we will give an outline of the topic of Hamiltonian mechanics and the Poisson bracket, including an introduction to the symplectic geometry that characterizes phase space.

The Heisenberg Lie algebra $\mathfrak{h}_{2d+1}$ is usually thought of as quintessentially quantum in nature, but it is already present in classical mechanics, as the Lie algebra of degree zero and one polynomials on phase space, with Lie bracket the Poisson bracket. In chapter 16, we will see that degree two polynomials on phase space also provide an important finite dimensional Lie algebra.

The full Lie algebra of all functions on phase space (with Lie bracket the Poisson bracket) is infinite dimensional, so not the sort of finite dimensional Lie algebra given by matrices that we have studied so far. Historically though, it is this kind of infinite dimensional Lie algebra that motivated the discovery of the theory of Lie groups and Lie algebras by Sophus Lie during the 1870s. It also provides the fundamental mathematical structure of the Hamiltonian form of classical mechanics.

© Peter Woit 2017
P. Woit, *Quantum Theory, Groups and Representations*,
DOI 10.1007/978-3-319-64612-1_14

14.1   Classical mechanics and the Poisson bracket

In classical mechanics in the Hamiltonian formalism, the space $M = \mathbf{R}^{2d}$ that one gets by putting together positions and the corresponding momenta is known as "phase space." Points in phase space can be thought of as uniquely parametrizing possible initial conditions for classical trajectories, so another interpretation of phase space is that it is the space that uniquely parametrizes solutions of the equations of motion of a given classical mechanical system. The basic axioms of Hamiltonian mechanics can be stated in a way that parallels the ones for quantum mechanics.

Axiom (States). *The state of a classical mechanical system is given by a point in the phase space $M = \mathbf{R}^{2d}$, with coordinates q_j, p_j, for $j = 1, \ldots, d$.*

Axiom (Observables). *The observables of a classical mechanical system are the functions on phase space.*

Axiom (Dynamics). *There is a distinguished observable, the Hamiltonian function h, and states evolve according to Hamilton's equations*

$$\dot{q}_j = \frac{\partial h}{\partial p_j}$$

$$\dot{p}_j = -\frac{\partial h}{\partial q_j}$$

Specializing to the case $d = 1$, for any observable function f, Hamilton's equations imply

$$\frac{df}{dt} = \frac{\partial f}{\partial q}\frac{dq}{dt} + \frac{\partial f}{\partial p}\frac{dp}{dt} = \frac{\partial f}{\partial q}\frac{\partial h}{\partial p} - \frac{\partial f}{\partial p}\frac{\partial h}{\partial q}$$

We can define:

Definition (Poisson bracket). *There is a bilinear operation on functions on the phase space $M = \mathbf{R}^2$ (with coordinates (q, p)) called the Poisson bracket, given by*

$$(f_1, f_2) \rightarrow \{f_1, f_2\} = \frac{\partial f_1}{\partial q}\frac{\partial f_2}{\partial p} - \frac{\partial f_1}{\partial p}\frac{\partial f_2}{\partial q}$$

An observable f evolves in time according to

$$\frac{df}{dt} = \{f, h\}$$

This relation is equivalent to Hamilton's equations since it implies them by taking $f = q$ and $f = p$

$$\dot{q} = \{q, h\} = \frac{\partial h}{\partial p}$$

$$\dot{p} = \{p, h\} = -\frac{\partial h}{\partial q}$$

For a non-relativistic free particle, $h = \frac{p^2}{2m}$ and these equations become

$$\dot{q} = \frac{p}{m}, \quad \dot{p} = 0$$

which says that the momentum is the mass times the velocity and is conserved. For a particle subject to a potential $V(q)$ one has

$$h = \frac{p^2}{2m} + V(q)$$

and the trajectories are the solutions to

$$\dot{q} = \frac{p}{m}, \quad \dot{p} = -\frac{\partial V}{\partial q}$$

This adds Newton's second law

$$F = -\frac{\partial V}{\partial q} = ma = m\ddot{q}$$

to the relation between momentum and velocity.

One can easily check that the Poisson bracket has the properties

- Antisymmetry

$$\{f_1, f_2\} = -\{f_2, f_1\}$$

- Jacobi identity

$$\{\{f_1, f_2\}, f_3\} + \{\{f_3, f_1\}, f_2\} + \{\{f_2, f_3\}, f_1\} = 0$$

These two properties, together with the bilinearity, show that the Poisson bracket fits the definition of a Lie bracket, making the space of functions on phase space into an infinite dimensional Lie algebra. This Lie algebra is responsible for much of the structure of the subject of Hamiltonian mechanics, and it was historically the first sort of Lie algebra to be studied.

From the fundamental dynamical equation

$$\frac{df}{dt} = \{f, h\}$$

we see that

$$\{f, h\} = 0 \implies \frac{df}{dt} = 0$$

and in this case the function f is called a "conserved quantity," since it does not change under time evolution. Note that if we have two functions f_1 and f_2 on phase space such that

$$\{f_1, h\} = 0, \quad \{f_2, h\} = 0$$

then using the Jacobi identity we have

$$\{\{f_1, f_2\}, h\} = -\{\{h, f_1\}, f_2\} - \{\{f_2, h\}, f_1\} = 0$$

This shows that if f_1 and f_2 are conserved quantities, so is $\{f_1, f_2\}$. As a result, functions f such that $\{f, h\} = 0$ make up a Lie subalgebra. It is this Lie subalgebra that corresponds to "symmetries" of the physics, commuting with the time translation determined by the dynamical law given by h.

14.2   The Poisson bracket and the Heisenberg Lie algebra

A third fundamental property of the Poisson bracket that can easily be checked is the

- Leibniz rule

$$\{f_1 f_2, f\} = \{f_1, f\} f_2 + f_1 \{f_2, f\}, \quad \{f, f_1 f_2\} = \{f, f_1\} f_2 + f_1 \{f, f_2\}$$

This property says that taking Poisson bracket with a function f acts on a product of functions in a way that satisfies the Leibniz rule for what happens when you take the derivative of a product. Unlike antisymmetry and the Jacobi identity, which reflect the Lie algebra structure on functions, the Leibniz property describes the relation of the Lie algebra structure to multiplication of functions. At least for polynomial functions, it allows one to inductively reduce the calculation of Poisson brackets to the special case of Poisson brackets of the coordinate functions q and p, for instance:

$$\{q^2, qp\} = q\{q^2, p\} + \{q^2, q\}p = q^2\{q, p\} + q\{q, p\}q = 2q^2\{q, p\} = 2q^2$$

The Poisson bracket is thus determined by its values on linear functions (thus by the relations $\{q,q\} = \{p,p\} = 0, \{q,p\} = 1$). We will define:

Definition. $\Omega(\cdot,\cdot)$ *is the restriction of the Poisson bracket to M^*, the linear functions on M. Taking as basis vectors of M^* the coordinate functions q and p, Ω is given on basis vectors by*

$$\Omega(q,q) = \Omega(p,p) = 0, \quad \Omega(q,p) = -\Omega(p,q) = 1$$

A general element of M^* will be a linear combination $c_q q + c_p p$ for some constants c_q, c_p. For general pairs of elements in M^*, Ω will be given by

$$\Omega(c_q q + c_p p, c'_q q + c'_p p) = c_q c'_p - c_p c'_q \tag{14.1}$$

We will often write elements of M^* as the column vector of their coefficients c_q, c_p, identifying

$$c_q q + c_p p \leftrightarrow \begin{pmatrix} c_q \\ c_p \end{pmatrix}$$

Then one has

$$\Omega\left(\begin{pmatrix} c_q \\ c_p \end{pmatrix}, \begin{pmatrix} c'_q \\ c'_p \end{pmatrix} \right) = c_q c'_p - c_p c'_q$$

Taking together linear functions on M and the constant function, one gets a three-dimensional space with basis elements $q, p, 1$, and this space is closed under Poisson bracket. This space is thus a Lie algebra and is isomorphic to the Heisenberg Lie algebra $\mathfrak{h}_3$ (see section 13.1), with the isomorphism given on basis elements by

$$X \leftrightarrow q, \quad Y \leftrightarrow p, \quad Z \leftrightarrow 1$$

This isomorphism preserves the Lie bracket relations since

$$[X,Y] = Z \leftrightarrow \{q,p\} = 1$$

It is convenient to choose its own notation for the dual phase space, so we will often write $M^* = \mathcal{M}$. The three-dimensional space we have identified with the Heisenberg Lie algebra is then

$$\mathcal{M} \oplus \mathbf{R}$$

We will denote elements of this space in two different ways

- As functions $c_q q + c_p p + c$, with Lie bracket the Poisson bracket

$$\{c_q q + c_p p + c, c'_q q + c'_p p + c'\} = c_q c'_p - c_p c'_q$$

- As pairs of an element of $\mathcal{M}$ and a real number

$$\left(\begin{pmatrix} c_q \\ c_p \end{pmatrix}, c\right)$$

In this second notation, the Lie bracket is

$$\left[\left(\begin{pmatrix} c_q \\ c_p \end{pmatrix}, c\right), \left(\begin{pmatrix} c'_q \\ c'_p \end{pmatrix}, c'\right)\right] = \left(\begin{pmatrix} 0 \\ 0 \end{pmatrix}, \Omega\left(\begin{pmatrix} c_q \\ c_p \end{pmatrix}, \begin{pmatrix} c'_q \\ c'_p \end{pmatrix}\right)\right)$$

which is identical to the Lie bracket for $\mathfrak{h}_3$ of equation 13.1. Notice that the Lie bracket structure is determined purely by Ω.

In higher dimensions, coordinate functions $q_1, \cdots, q_d, p_1, \cdots, p_d$ on M provide a basis for the dual space $\mathcal{M}$. Taking as an additional basis element the constant function 1, we have a $2d+1$-dimensional space with basis

$$q_1, \cdots, q_d, p_1, \cdots, p_d, 1$$

The Poisson bracket relations

$$\{q_j, q_k\} = \{p_j, p_k\} = 0, \quad \{q_j, p_k\} = \delta_{jk}$$

turn this space into a Lie algebra, isomorphic to the Heisenberg Lie algebra $\mathfrak{h}_{2d+1}$. On general functions, the Poisson bracket will be given by the obvious generalization of the $d = 1$ case

$$\{f_1, f_2\} = \sum_{j=1}^{d}\left(\frac{\partial f_1}{\partial q_j}\frac{\partial f_2}{\partial p_j} - \frac{\partial f_1}{\partial p_j}\frac{\partial f_2}{\partial q_j}\right) \tag{14.2}$$

Elements of $\mathfrak{h}_{2d+1}$ are functions on $M = \mathbf{R}^{2d}$ of the form

$$c_{q_1} q_1 + \cdots + c_{q_d} q_d + c_{p_1} p_1 + \cdots + c_{p_d} p_d + c = \mathbf{c}_q \cdot \mathbf{q} + \mathbf{c}_p \cdot \mathbf{p} + c$$

(using the notation $\mathbf{c}_q = (c_{q_1}, \ldots, c_{q_d})$, $\mathbf{c}_p = (c_{p_1}, \ldots, c_{p_d})$). We will often denote these by

$$\left(\begin{pmatrix} \mathbf{c}_q \\ \mathbf{c}_p \end{pmatrix}, c\right)$$

This Lie bracket on $\mathfrak{h}_{2d+1}$ is given by

$$\left[\left(\begin{pmatrix} \mathbf{c}_q \\ \mathbf{c}_p \end{pmatrix}, c\right), \left(\begin{pmatrix} \mathbf{c}'_q \\ \mathbf{c}'_p \end{pmatrix}, c'\right)\right] = \left(\begin{pmatrix} \mathbf{0} \\ \mathbf{0} \end{pmatrix}, \Omega\left(\begin{pmatrix} \mathbf{c}_q \\ \mathbf{c}_p \end{pmatrix}, \begin{pmatrix} \mathbf{c}'_q \\ \mathbf{c}'_p \end{pmatrix}\right)\right) \tag{14.3}$$

which depends just on the antisymmetric bilinear form

$$\Omega\left(\begin{pmatrix} \mathbf{c}_q \\ \mathbf{c}_p \end{pmatrix}, \begin{pmatrix} \mathbf{c}'_q \\ \mathbf{c}'_p \end{pmatrix}\right) = \mathbf{c}_q \cdot \mathbf{c}'_p - \mathbf{c}_p \cdot \mathbf{c}'_q \tag{14.4}$$

14.3 Symplectic geometry

We saw in chapter 4 that given a basis $\mathbf{e}_j$ of a vector space V, a dual basis $\mathbf{e}^*_j$ of V^* is given by taking $\mathbf{e}^*_j = v_j$, where v_j are the coordinate functions. If one instead is initially given the coordinate functions v_j, a dual basis of $V = (V^*)^*$ can be constructed by taking as basis vectors the first-order linear differential operators given by differentiation with respect to the v_j, in other words by taking

$$\mathbf{e}_j = \frac{\partial}{\partial v_j}$$

Elements of V are then identified with linear combinations of these operators. In effect, one is identifying vectors $\mathbf{v}$ with the directional derivative along the vector

$$\mathbf{v} \leftrightarrow \mathbf{v} \cdot \nabla$$

We also saw in chapter 4 that an inner product $(\cdot, \cdot)$ on V provides an isomorphism of V and V^* by

$$v \in V \leftrightarrow l_v(\cdot) = (v, \cdot) \in V^* \tag{14.5}$$

Such an inner product is the fundamental structure in Euclidean geometry, giving a notion of length of a vector and angle between two vectors, as well as a group; the orthogonal group of linear transformations preserving the inner product. It is a symmetric, non-degenerate bilinear form on V.

A phase space M does not usually come with a choice of inner product. Instead, we have seen that the Poisson bracket gives us not a symmetric bilinear form, but an antisymmetric bilinear form Ω, defined on the dual space $\mathcal{M}$. We will define an analog of an inner product, with symmetry replaced by antisymmetry:

Definition (Symplectic form). *A symplectic form ω on a vector space V is a bilinear map*

$$\omega : V \times V \to \mathbf{R}$$

such that

- *ω is antisymmetric: $\omega(v, v') = -\omega(v, v')$*
- *ω is non-degenerate: if $v \neq 0$, then $\omega(v, \cdot) \in V^*$ is nonzero.*

A vector space V with a symplectic form ω is called a symplectic vector space. The analog of Euclidean geometry, replacing the inner product by a symplectic form, is called symplectic geometry. In this sort of geometry, there is no notion of length (since antisymmetry implies $\omega(v,v) = 0$). There is an analog of the orthogonal group, called the symplectic group, which consists of linear transformations preserving ω, a group we will study in detail in chapter 16.

Just as an inner product gives an identification of V and V^*, a symplectic form can be used in a similar way, giving an identification of M and $\mathcal{M}$. Using the symplectic form Ω on $\mathcal{M}$, we can define an isomorphism by identifying basis vectors by

$$q_j \in \mathcal{M} \leftrightarrow \Omega(\cdot, q_j) = -\Omega(q_j, \cdot) = -\frac{\partial}{\partial p_j} \in M$$

$$p_j \in \mathcal{M} \leftrightarrow \Omega(\cdot, p_j) = -\Omega(p_j, \cdot) = \frac{\partial}{\partial q_j} \in M$$

and in general

$$u \in \mathcal{M} \leftrightarrow \Omega(\cdot, u) = -\Omega(u, \cdot) \in M \tag{14.6}$$

Note that unlike the inner product case, a choice of convention of minus sign must be made and is done here.

Recalling the discussion of bilinear forms from section 9.5, a bilinear form on a vector space V can be identified with an element of $V^* \otimes V^*$. Taking $V = M^*$ we have $V^* = (M^*)^* = M$, and the bilinear form Ω on M^* is an element of $M \otimes M$ given by

$$\Omega = \sum_{j=1}^{d} \left(\frac{\partial}{\partial q_j} \otimes \frac{\partial}{\partial p_j} - \frac{\partial}{\partial p_j} \otimes \frac{\partial}{\partial q_j} \right)$$

Under the identification 14.6 of M and M^*, $\Omega \in M \otimes M$ corresponds to

$$\omega = \sum_{j=1}^{d} (q_j \otimes p_j - p_j \otimes q_j) \in M^* \otimes M^* \tag{14.7}$$

Another version of the identification of M and $\mathcal{M}$ is then given by

$$v \in M \to \omega(v, \cdot) \in \mathcal{M}$$

In the case of Euclidean geometry, one can show by Gram–Schmidt orthogonalization that a basis $\mathbf{e}_j$ can always be found that puts the inner product (which is a symmetric element of $V^* \otimes V^*$) in the standard form

$$\sum_{j=1}^{n} v_j \otimes v_j$$

in terms of basis elements of V^*, the coordinate functions v_j. There is an analogous theorem in symplectic geometry (for a proof, see, for instance, Proposition 1.1 of [8]). This says that a basis of a symplectic vector space V can always be found so that the dual basis coordinate functions come in pairs q_j, p_j, with the symplectic form ω, the same one we have found based on the Poisson bracket, that given by equation 14.7. Note that one difference between Euclidean and symplectic geometry is that a symplectic vector space will always be even-dimensional.

Digression. *For those familiar with differential manifolds, vector fields and differential forms, the notion of a symplectic vector space can be extended to:*

Definition (Symplectic manifold). *A symplectic manifold M is a manifold with a differential two-form $\omega(\cdot, \cdot)$ (called a symplectic two-form) satisfying the conditions*

- *ω is non-degenerate (i.e., for a nowhere zero vector field X, $\omega(X, \cdot)$ is a nowhere zero one-form).*

- *$d\omega = 0$, in which case ω is said to be closed.*

The cotangent bundle T^*N of a manifold N (i.e., the space of pairs of a point on N together with a linear function on the tangent space at that point) provides one class of symplectic manifolds, generalizing the linear case $N = \mathbf{R}^d$, and corresponding physically to a particle moving on N. A simple example that is neither linear nor a cotangent bundle is the sphere $M = S^2$, with ω the area two-form. The Darboux theorem says that, by an appropriate choice of local coordinates q_j, p_j on M, symplectic two-forms ω can always be written in such local coordinates as

$$\omega = \sum_{j=1}^{d} dq_j \wedge dp_j$$

Unlike the linear case though, there will in general be no global choice of coordinates for which this true. Later on, our discussion of quantization will rely crucially on having a linear structure on phase space, so will not apply to general symplectic manifolds.

Note that there is no assumption here that M has a metric (i.e., it may not be a Riemannian manifold). A symplectic two-form ω is a structure on a manifold analogous to a metric but with opposite symmetry properties. Whereas a metric is a symmetric non-degenerate bilinear form on the tangent space at each point, a symplectic form is an antisymmetric non-degenerate bilinear form on the tangent space.

14.4 For further reading

Some good sources for discussions of symplectic geometry and the geometrical formulation of Hamiltonian mechanics are [2, 8, 13].

Chapter 15
Hamiltonian Vector Fields and the Moment Map

A basic feature of Hamiltonian mechanics is that, for any function f on phase space M, there are parametrized curves in phase space that solve Hamilton's equations

$$\dot{q}_j = \frac{\partial f}{\partial p_j} \quad \dot{p}_j = -\frac{\partial f}{\partial q_j}$$

and the tangent vectors of these parametrized curves provide a vector field on phase space. Such vector fields are called Hamiltonian vector fields. There is a distinguished choice of f, the Hamiltonian function h, which gives the velocity vector fields for time evolution trajectories in phase space.

More generally, when a Lie group G acts on phase space M, the infinitesimal action of the group associates to each element $L \in \mathfrak{g}$ a vector field X_L on phase space. When these are Hamiltonian vector fields, there is (up to a constant) a corresponding function μ_L. The map from $L \in \mathfrak{g}$ to the function μ_L on M is called the moment map, and such functions play a central role in both classical and quantum mechanics. For the case of the action of $G = \mathbf{R}^3$ on $M = \mathbf{R}^6$ by spatial translations, the components of the momentum arise in this way, for the action of $G = SO(3)$ by rotations, the angular momentum.

Conventional physics discussions of symmetry in quantum mechanics focus on group actions on configuration space that preserve the Lagrangian, using Noether's theorem to provide corresponding conserved quantities (see chapter 35). In the Hamiltonian formalism described here, these same conserved quantities appear as moment map functions. The operator quantizations of these functions provide quantum observables and (modulo the problem of indeterminacy up to a constant) a unitary representation of G on the state space $\mathcal{H}$. The use of moment map functions rather than Lagrangian-derived conserved quantities allows one to work with cases where G acts not on

© Peter Woit 2017
P. Woit, *Quantum Theory, Groups and Representations*,
DOI 10.1007/978-3-319-64612-1_15

configuration space, but on phase space, mixing position and momentum co-ordinates. It also applies to cases where the group action is not a "symmetry" (i.e., does not commute with time evolution), with the functions μ_L having nonzero Poisson brackets with the Hamiltonian function.

15.1 Vector fields and the exponential map

A vector field on $M = \mathbf{R}^2$ can be thought of as a choice of a two-dimensional vector at each point in $\mathbf{R}^2$, so given by a vector-valued function

$$\mathbf{F}(q,p) = \begin{pmatrix} F_q(q,p) \\ F_p(q,p) \end{pmatrix}$$

Such a vector field determines a system of differential equations

$$\frac{dq}{dt} = F_q, \quad \frac{dp}{dt} = F_p$$

Once initial conditions

$$q(0) = q_0, \quad p(0) = p_0$$

are specified, if F_q and F_p are differentiable functions these differential equa-tions have a unique solution $q(t), p(t)$, at least for some neighborhood of $t = 0$ (from the existence and uniqueness theorem that can be found, for instance, in [48]). These solutions $q(t), p(t)$ describe trajectories in $\mathbf{R}^2$ with velocity vector $\mathbf{F}(q(t), p(t))$, and such trajectories can be used to define the "flow" of the vector field: for each t, this is the map that takes the initial point $(q(0), p(0)) \in \mathbf{R}^2$ to the point $(q(t), p(t)) \in \mathbf{R}^2$.

Another equivalent way to define vector fields on $\mathbf{R}^2$ is to use instead the directional derivative along the vector field, identifying

$$\mathbf{F}(q,p) \leftrightarrow \mathbf{F}(q,p) \cdot \boldsymbol{\nabla} = F_q(q,p)\frac{\partial}{\partial q} + F_p(q,p)\frac{\partial}{\partial p}$$

The case of $\mathbf{F}$ a constant vector is just our previous identification of the vector space M with linear combinations of $\frac{\partial}{\partial q}$ and $\frac{\partial}{\partial p}$.

An advantage of defining vector fields in this way as first-order linear differential operators is that it shows that vector fields form a Lie algebra, where one takes as Lie bracket of vector fields X_1, X_2 the commutator

$$[X_1, X_2] = X_1 X_2 - X_2 X_1 \tag{15.1}$$

of the differential operators. The commutator of two first-order differential operators is another first-order differential operator since second-order derivatives will cancel, using equality of mixed partial derivatives. In addition, such a commutator will satisfy the Jacobi identity.

Given this Lie algebra of vector fields, one can ask what the corresponding group might be. This is not a finite dimensional matrix Lie algebra, so exponentiation of matrices will not give the group. The flow of the vector field X can be used to define an analog of the exponential of a parameter t times X:

Definition (Flow of a vector field and the exponential map). *The flow of the vector field X on M is the map*

$$\Phi_X : (t, m) \in \mathbf{R} \times M \rightarrow \Phi_X(t, m) \in M$$

satisfying

$$\frac{d}{dt}\Phi_X(t, m) = X(\Phi_X(t, m))$$

$$\Phi_X(0, m) = m$$

In words, $\Phi_X(t, m)$ is the trajectory in M that passes through $m \in M$ at $t = 0$, with velocity vector given by the vector field X evaluated along the trajectory.

The flow can be written as a map

$$\exp(tX) : m \in M \rightarrow \Phi_X(t, m) \in M$$

called the exponential map.

If the vector field X is differentiable (with bounded derivative), $\exp(tX)$ will be a well-defined map for some neighborhood of $t = 0$ and satisfy

$$\exp(t_1 X)\exp(t_2 X) = \exp((t_1 + t_2)X)$$

thus providing a one-parameter group of maps from M to itself, with derivative X at the identity.

Digression. *For any manifold M, there is an infinite dimensional Lie group, the group of invertible maps from M to itself, such that the maps and their inverses are both differentiable. This group is called the diffeomorphism group of M and written Diff(M). Its Lie algebra is the Lie algebra of vector fields.*

The Lie algebra of vector fields acts on functions on M by differentiation. This is the differential of the representation of Diff*(M) on functions induced in the usual way (see equation 1.3) from the action of* Diff*(M) on the space M. This representation, however, is not one of relevance to quantum mechanics, since it acts on functions on phase space, whereas the quantum state space is given by functions on just half the phase space coordinates (positions or momenta).*

15.2 Hamiltonian vector fields and canonical transformations

Our interest is not in general vector fields but in vector fields corresponding to Hamilton's equations for some Hamiltonian function f, e.g., the case

$$F_q = \frac{\partial f}{\partial p}, \quad F_p = -\frac{\partial f}{\partial q}$$

We call such vector fields Hamiltonian vector fields, defining:

Definition (Hamiltonian Vector Field). *A vector field on $M = \mathbf{R}^2$ given by*

$$\frac{\partial f}{\partial p}\frac{\partial}{\partial q} - \frac{\partial f}{\partial q}\frac{\partial}{\partial p} = -\{f, \cdot\}$$

for some function f on $M = \mathbf{R}^2$ is called a Hamiltonian vector field and will be denoted by X_f. In higher dimensions, Hamiltonian vector fields will be those of the form

$$X_f = \sum_{j=1}^{d} \left(\frac{\partial f}{\partial p_j}\frac{\partial}{\partial q_j} - \frac{\partial f}{\partial q_j}\frac{\partial}{\partial p_j} \right) = -\{f, \cdot\} \tag{15.2}$$

for some function f on $M = \mathbf{R}^{2d}$.

The simplest nonzero Hamiltonian vector fields are those for f a linear function. For c_q, c_p constants, if

$$f = c_q q + c_p p$$

then

$$X_f = c_p \frac{\partial}{\partial q} - c_q \frac{\partial}{\partial p}$$

and the map

$$f \to X_f$$

is the isomorphism of $\mathcal{M}$ and M of equation 14.6.

For example, taking $f = p$, we have $X_p = \frac{\partial}{\partial q}$. The exponential map for this vector field satisfies

$$q(\exp(tX_p)(m)) = q(m) + t, \quad p(\exp(tX_p)(m)) = p(m) \tag{15.3}$$

Similarly, for $f = q$ one has $X_q = -\frac{\partial}{\partial p}$ and

$$q(\exp(tX_q)(m)) = q(m), \quad p(\exp(tX_q)(m)) = p(m) - t \tag{15.4}$$

Quadratic functions f give vector fields X_f with components linear in the coordinates. An important example is the case of the quadratic function

$$h = \frac{1}{2}(q^2 + p^2)$$

which is the Hamiltonian function for a harmonic oscillator, a system that will be treated in much more detail beginning in chapter 22. The Hamiltonian vector field for this function is

$$X_h = p\frac{\partial}{\partial q} - q\frac{\partial}{\partial p}$$

The trajectories satisfy

$$\frac{dq}{dt} = p, \quad \frac{dp}{dt} = -q$$

and are given by

$$q(t) = q(0)\cos t + p(0)\sin t, \quad p(t) = p(0)\cos t - q(0)\sin t$$

The exponential map is given by clockwise rotation through an angle t

$$q(\exp(tX_h)(m)) = q(m)\cos t + p(m)\sin t$$
$$p(\exp(tX_h)(m)) = -q(m)\sin t + p(m)\cos t$$

The vector field X_h and the trajectories in the qp plane look like this

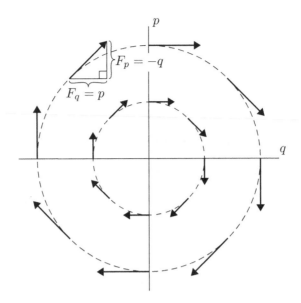

Figure 15.1: Hamiltonian vector field for a simple harmonic oscillator.

and describe a periodic motion in phase space.

The relation of vector fields to the Poisson bracket is given by (see equation 15.2)

$$\{f_1, f_2\} = X_{f_2}(f_1) = -X_{f_1}(f_2)$$

and, in particular,

$$\{q, f\} = \frac{\partial f}{\partial p}, \quad \{p, f\} = -\frac{\partial f}{\partial q}$$

The definition we have given here of X_f (equation 15.2) carries with it a choice of how to deal with a confusing sign issue. Recall that vector fields on M form a Lie algebra with Lie bracket the commutator of differential operators. A natural question is that of how this Lie algebra is related to the Lie algebra of functions on M (with Lie bracket the Poisson bracket).

The Jacobi identity implies

$$\{f, \{f_1, f_2\}\} = \{\{f, f_1\}, f_2\} + \{\{f_2, f\}, f_1\} = \{\{f, f_1\}, f_2\} - \{\{f, f_2\}, f_1\}$$

so

$$X_{\{f_1, f_2\}} = X_{f_2} X_{f_1} - X_{f_1} X_{f_2} = -[X_{f_1}, X_{f_2}] \tag{15.5}$$

This shows that the map $f \to X_f$ of equation 15.2 that we defined between these Lie algebras is not quite a Lie algebra homomorphism because of the - sign in equation 15.5 (it is called a Lie algebra "antihomomorphism"). The map that is a Lie algebra homomorphism is

$$f \to -X_f \tag{15.6}$$

To keep track of the minus sign here, one needs to keep straight the difference between

- The functions on phase space M are a Lie algebra, with a function f acting on the function space by the adjoint action

$$ad(f)(\cdot) = \{f, \cdot\}$$

and

- The functions f provide vector fields X_f acting on functions on M, where

$$X_f(\cdot) = \{\cdot, f\} = -\{f, \cdot\}$$

As a simple example, the function p satisfies

$$\{p, F(q, p)\} = -\frac{\partial F}{\partial q}$$

so

$$\{p, \cdot\} = -\frac{\partial(\cdot)}{\partial q} = -X_p$$

Note that acting on functions with p in this way is the Lie algebra version of the representation of the translation group on functions induced from the translation action on the position (see equations 10.1 and 10.2).

It is important to note that the Lie algebra homomorphism 15.6 from functions to vector fields is not an isomorphism, for two reasons:

- It is not injective (one-to-one), since functions f and $f + C$ for any constant C correspond to the same X_f.

- It is not surjective since not all vector fields are Hamiltonian vector fields (i.e., of the form X_f for some f). One property that a vector field X must satisfy in order to possibly be a Hamiltonian vector field is

$$X\{g_1, g_2\} = \{Xg_1, g_2\} + \{g_1, Xg_2\} \tag{15.7}$$

for g_1 and g_2 on M. This is the Jacobi identity for f, g_1, g_2, when $X = X_f$.

Digression. *For a general symplectic manifold M, the symplectic two-form ω gives us an analog of Hamilton's equations. This is the following equality of one-forms, relating a Hamiltonian function h and a vector field X_h determining time evolution of trajectories in M*

$$i_{X_h}\omega = \omega(X_h, \cdot) = dh$$

(here, i_X is interior product with the vector field X). The Poisson bracket in this context can be defined as follows:

$$\{f_1, f_2\} = \omega(X_{f_1}, X_{f_2})$$

Recall that a symplectic two-form is defined to be closed, satisfying the equation $d\omega = 0$, which is then a condition on a three-form $d\omega$. Standard differential form computations allow one to express $d\omega(X_{f_1}, X_{f_2}, X_{f_3})$ in terms of Poisson brackets of functions f_1, f_2, f_3, and one finds that $d\omega = 0$ is the Jacobi identity for the Poisson bracket.

The theory of "prequantization" (see [52], [41]) enlarges the phase space M to a $U(1)$ bundle with connection, where the curvature of the connection is the symplectic form ω. Then, the problem of lack of injectivity of the Lie algebra homomorphism

$$f \to -X_f$$

is resolved by instead using the map

$$f \to -\nabla_{X_f} + if \tag{15.8}$$

where ∇_X is the covariant derivative with respect to the connection. For details of this, see [52] or [41].

In our treatment of functions on phase space M, we have always been taking such functions to be time-independent. M can be thought of as the space of trajectories of a classical mechanical system, with coordinates q, p having the interpretation of initial conditions $q(0), p(0)$ of the trajectories. The exponential maps $\exp(tX_h)$ give an action on the space of trajectories for the Hamiltonian function h, taking the trajectory with initial conditions given by $m \in M$ to the time-translated one with initial conditions given by $\exp(tX_h)(m)$. One should really interpret the formula for Hamilton's equations

$$\frac{df}{dt} = \{f, h\}$$

as meaning

$$\frac{d}{dt}f(\exp(tX_h)(m))|_{t=0} = \{f(m), h(m)\}$$

for each $m \in M$.

Given a Hamiltonian vector field X_f, the maps

$$\exp(tX_f) : M \to M$$

are known to physicists as "canonical transformations," and to mathematicians as "symplectomorphisms." We will not try and work out in any more detail how the exponential map behaves in general. In chapter 16, we will see what happens for f an order-two homogeneous polynomial in the q_j, p_j. In that case, the vector field X_f will take linear functions on M to linear functions, thus acting on $\mathcal{M}$, in which case, its behavior can be studied using the matrix for the linear transformation with respect to the basis elements q_j, p_j.

Digression. *The exponential map* $\exp(tX)$ *can be defined as above on a general manifold. For a symplectic manifold* M, *Hamiltonian vector fields* X_f *will have the property that they preserve the symplectic form, in the sense that*

$$\exp(tX_f)^*\omega = \omega$$

This is because

$$\mathcal{L}_{X_f}\omega = (di_{X_f} + i_{X_f}d)\omega = di_{X_f}\omega = d\omega(X_f, \cdot) = ddf = 0 \qquad (15.9)$$

where $\mathcal{L}_{X_f}$ *is the Lie derivative along* X_f.

15.3   Group actions on M and the moment map

Our fundamental interest is in studying the implications of Lie group actions on physical systems. In classical Hamiltonian mechanics, with a Lie group G acting on phase space M, such actions are characterized by their derivative, which takes elements of the Lie algebra to vector fields on M. When these are Hamiltonian vector fields, equation 15.6 can often be used to instead take elements of the Lie algebra to functions on M. This is known as the moment map of the group action, and such functions on M will provide our central tool to understand the implications of a Lie group action on a physical system. Quantization then takes such functions to operators which will turn out to be the important observables of the quantum theory.

Given an action of a Lie group G on a space M, there is a map

$$L \in \mathfrak{g} \to X_L$$

from $\mathfrak{g}$ to vector fields on M. This takes L to the vector field X_L which acts on functions on M by

$$X_L F(m) = \frac{d}{dt} F(e^{tL} \cdot m)_{|t=0} \qquad (15.10)$$

This map, however, is not a homomorphism (for the Lie bracket 15.1 on vector fields), but an antihomomorphism. To see why this is, recall that when a group G acts on a space, we get a representation π on functions F on the space by

$$\pi(g)F(m) = F(g^{-1} \cdot m)$$

The derivative of this representation will be the Lie algebra representation

$$\pi'(L)F(m) = \frac{d}{dt}F(e^{-tL} \cdot m)_{|t=0} = -X_L F(m)$$

so we see that it is the map

$$L \to \pi'(L) = -X_L$$

that will be a homomorphism.

When the vector field X_L is a Hamiltonian vector field, we can define:

Definition (Moment map). *Given an action of G on phase space M, a Lie algebra homomorphism*

$$L \to \mu_L$$

from $\mathfrak{g}$ to functions on M is said to be a moment map if

$$X_L = X_{\mu_L}$$

Equivalently, for functions F on M, μ_L satisfies

$$\{\mu_L, F\}(m) = -X_L F = \frac{d}{dt}F(e^{-tL} \cdot m)_{|t=0} \qquad (15.11)$$

This is sometimes called a "co-moment map," with the term "moment map" referring to a repackaged form of the same information, the map

$$\mu : M \to \mathfrak{g}^*$$

where

$$(\mu(m))(L) = \mu_L(m)$$

A conventional physical terminology for the statement 15.11 is that "the function μ_L generates the symmetry L," giving its infinitesimal action on functions.

Only for certain actions of G on M will the X_L be Hamiltonian vector fields and an identity $X_L = X_{\mu_L}$ possible. A necessary condition is that X_L satisfy equation 15.7

$$X_L\{g_1, g_2\} = \{X_Lg_1, g_2\} + \{g_1, X_Lg_2\}$$

Even when a function μ_L exists such that $X_{\mu_L} = X_L$, it is only unique up to a constant, since μ_L and $\mu_L + C$ will give the same vector field. To get a moment map, we need to be able to choose these constants in such a way that the map

$$L \to \mu_L$$

is a Lie algebra homomorphism from $\mathfrak{g}$ to the Lie algebra of functions on M. When this is possible, the G-action is said to be a Hamiltonian G-action. When such a choice of constants is not possible, the G-action on the classical phase space is said to have an "anomaly."

Digression. *For the case of M a general symplectic manifold, the moment map can still be defined, whenever one has a Lie group G acting on M, preserving the symplectic form ω. The infinitesimal condition for such a G-action is (see equation 15.9)*

$$\mathcal{L}_X\omega = 0$$

Using the formula

$$\mathcal{L}_X = (d + i_X)^2 = di_X + i_Xd$$

for the Lie derivative acting on differential forms (i_X is interior product with the vector field X), one has

$$(di_X + i_Xd)\omega = 0$$

and since $d\omega = 0$ we have

$$di_X\omega = 0$$

When M is simply connected, one-forms $i_X\omega$ whose differential is 0 (called "closed") will be the differentials of a function (and called "exact"). So there will be a function μ such that

$$i_X\omega(\cdot) = \omega(X, \cdot) = d\mu(\cdot)$$

although such a μ is only unique up to a constant.

Given an element $L \in \mathfrak{g}$, a G-action on M gives a vector field X_L by equation 15.10. When we can choose the constants appropriately and find functions μ_L satisfying

$$i_{X_L}\omega(\cdot) = d\mu_L(\cdot)$$

such that the map

$$L \to \mu_L$$

taking Lie algebra elements to functions on M (with Lie bracket the Poisson bracket) is a Lie algebra homomorphism, then this is called the moment map. One can equivalently work with

$$\mu : M \to \mathfrak{g}^*$$

by defining

$$(\mu(m))(L) = \mu_L(m)$$

15.4 Examples of Hamiltonian group actions

Some examples of Hamiltonian group actions are the following:

- For $d = 3$, an element $\mathbf{a}$ of the translation group $G = \mathbf{R}^3$ acts on the phase space $M = \mathbf{R}^6$ by translation

$$m \in M \to \mathbf{a} \cdot m \in M$$

  such that the coordinates satisfy

$$\mathbf{q}(\mathbf{a} \cdot m) = \mathbf{q} + \mathbf{a}, \quad \mathbf{p}(\mathbf{a} \cdot m) = \mathbf{p}(m)$$

  Taking $\mathbf{a}$ to be the corresponding element in the Lie algebra of $G = \mathbf{R}^3$, the vector field on M corresponding to this action (by 15.10) is

$$X_{\mathbf{a}} = a_1 \frac{\partial}{\partial q_1} + a_2 \frac{\partial}{\partial q_2} + a_3 \frac{\partial}{\partial q_3}$$

  and the moment map is given by

$$\mu_{\mathbf{a}}(m) = \mathbf{a} \cdot \mathbf{p}(m) \qquad (15.12)$$

  This can be interpreted as a function $\mathbf{a} \cdot \mathbf{p}$ on M for each element $\mathbf{a}$ of the Lie algebra or as an element $\mathbf{p}(m)$ of the dual of the Lie algebra $\mathbf{R}^3$ for each point $m \in M$.

- For another example, consider the action of the group $G = SO(3)$ of rotations on phase space $M = \mathbf{R}^6$ given by performing the same $SO(3)$ rotation on position and momentum vectors. This gives a map from $\mathfrak{so}(3)$ to vector fields on $\mathbf{R}^6$, taking for example

$$l_1 \in \mathfrak{so}(3) \to X_{l_1} = -q_3 \frac{\partial}{\partial q_2} + q_2 \frac{\partial}{\partial q_3} - p_3 \frac{\partial}{\partial p_2} + p_2 \frac{\partial}{\partial p_3}$$

(this is the vector field for an infinitesimal counterclockwise rotation in the $q_2 - q_3$ and $p_2 - p_3$ planes, in the opposite direction to the case of the vector field $X_{\frac{1}{2}(q^2+p^2)}$ in the qp plane of section 15.2). The moment map here gives the usual expression for the 1-component of the angular momentum

$$\mu_{l_1} = q_2 p_3 - q_3 p_2$$

since one can check from equation 15.2 that $X_{l_1} = X_{\mu_{l_1}}$. On basis elements of $\mathfrak{so}(3)$, one has

$$\mu_{l_j}(m) = (\mathbf{q}(m) \times \mathbf{p}(m))_j$$

Formulated as a map from M to $\mathfrak{so}(3)^*$, the moment map is

$$\mu(m)(\mathbf{l}) = (\mathbf{q}(m) \times \mathbf{p}(m)) \cdot \mathbf{l}$$

where $\mathbf{l} \in \mathfrak{so}(3)$.

- While most of the material of this chapter also applies to the case of a general symplectic manifold M, the case of M a vector space has the feature that G can be taken to be a group of linear transformations of M, and the moment map will give quadratic polynomials. The previous example is a special case of this, and more general linear transformations will be studied in great detail in later chapters. In this linear case, it turns out that it is generally best to work not with M but with its dual space $\mathcal{M}$.

15.5 The dual of a Lie algebra and symplectic geometry

We have been careful to keep track of the difference between phase space $M = \mathbf{R}^{2d}$ and its dual $\mathcal{M} = M^*$, even though the symplectic form provides an isomorphism between them (see equation 14.6). One reason for this is that it is $\mathcal{M} = M^*$ that is related to the Heisenberg Lie algebra by

$$\mathfrak{h}_{2d+1} = \mathcal{M} \oplus \mathbf{R}$$

with $\mathcal{M}$ the linear functions on phase space, $\mathbf{R}$ the constant functions, and the Poisson bracket the Lie bracket. It is this Lie algebra that we want to use in chapter 17 when we define the quantization of a classical system.

Another reason to carefully keep track of the difference between M and $\mathcal{M}$ is that they carry two different actions of the Heisenberg group, coming from the fact that the group acts quite differently on its Lie algebra (the adjoint action) and on the dual of its Lie algebra (the "co-adjoint" action). On M and $\mathcal{M}$ these actions become:

- The Heisenberg group H_{2d+1} acts on its Lie algebra $\mathfrak{h}_{2d+1} = \mathcal{M} \oplus \mathbf{R}$ by the adjoint action, with the differential of this action given as usual by the Lie bracket (see equation 5.4). Here, this means

$$ad\left(\left(\begin{pmatrix}\mathbf{c}_q\\\mathbf{c}_p\end{pmatrix},c\right)\right)\left(\left(\begin{pmatrix}\mathbf{c}'_q\\\mathbf{c}'_p\end{pmatrix},c'\right)\right) = \left[\left(\begin{pmatrix}\mathbf{c}_q\\\mathbf{c}_p\end{pmatrix},c\right),\left(\begin{pmatrix}\mathbf{c}'_q\\\mathbf{c}'_p\end{pmatrix},c'\right)\right]$$
$$= \left(\begin{pmatrix}\mathbf{0}\\\mathbf{0}\end{pmatrix},\Omega\left(\begin{pmatrix}\mathbf{c}_q\\\mathbf{c}_p\end{pmatrix},\begin{pmatrix}\mathbf{c}'_q\\\mathbf{c}'_p\end{pmatrix}\right)\right)$$

This action is trivial on the subspace $\mathcal{M}$, taking

$$\begin{pmatrix}\mathbf{c}'_q\\\mathbf{c}'_p\end{pmatrix} \rightarrow \begin{pmatrix}\mathbf{0}\\\mathbf{0}\end{pmatrix}$$

- The simplest way to define the "co-adjoint" action in this case is to define it as the Hamiltonian action of H_{2d+1} on $M = \mathbf{R}^{2d}$ such that its moment map μ_L is just the identification of $\mathfrak{h}_{2d+1}$ with functions on M. For the case $d = 1$, one has

$$\mu_L = L = c_q q + c_p p + c \in \mathfrak{h}_3$$

and

$$X_{\mu_L} = -c_q \frac{\partial}{\partial p} + c_p \frac{\partial}{\partial q}$$

This is the action described in equations 15.3 and 15.4, satisfying

$$q\left(\left(\begin{pmatrix}x\\y\end{pmatrix},z\right)\cdot m\right) = q(m) + y,\quad p\left(\left(\begin{pmatrix}x\\y\end{pmatrix},z\right)\cdot m\right) = p(m) - x \quad (15.13)$$

Here, the subgroup of elements of H_3 with $x = z = 0$ acts as the usual translations in position q.

It is a general phenomenon that for any Lie algebra $\mathfrak{g}$, a Poisson bracket on functions on the dual space $\mathfrak{g}^*$ can be defined. This is because the Leibniz property ensures that the Poisson bracket only depends on Ω, its restriction to linear functions, and linear functions on $\mathfrak{g}^*$ are elements of $\mathfrak{g}$. So a Poisson bracket on functions on $\mathfrak{g}^*$ is given by first defining

$$\Omega(X, X') = [X, X'] \tag{15.14}$$

for $X, X' \in \mathfrak{g} = (\mathfrak{g}^*)^*$ and then extending this to all functions on $\mathfrak{g}^*$ by the Leibniz property.

Such a Poisson bracket on functions on the vector space $\mathfrak{g}^*$ is said to provide a "Poisson structure" on $\mathfrak{g}^*$. In general, it will not provide a symplectic structure on $\mathfrak{g}^*$, since it will not be non-degenerate. For example, in the case of the Heisenberg Lie algebra

$$\mathfrak{g}^* = M \oplus \mathbf{R}$$

and Ω will be non-degenerate only on the subspace M, the phase space, which it will give a symplectic structure.

Digression. *The Poisson structure on $\mathfrak{g}^*$ can often be used to get a symplectic structure on submanifolds of $\mathfrak{g}^*$. As an example, take $\mathfrak{g} = \mathfrak{so}(3)$, in which case $\mathfrak{g}^* = \mathbf{R}^3$, with antisymmetric bilinear form ω given by the vector cross-product. In this case, it turns out that if one considers spheres of fixed radius in $\mathbf{R}^3$, ω provides a symplectic form proportional to the area two-form, giving such spheres the structure of a symplectic manifold.*

This is a special case of a general construction. Taking the dual of the adjoint representation Ad on $\mathfrak{g}$, there is an action of $g \in G$ on $\mathfrak{g}^*$ by the representation Ad^*, satisfying

$$(Ad^*(g) \cdot l)(X) = l(Ad(g^{-1})X)$$

This is called the "co-adjoint" action on $\mathfrak{g}^*$. Picking an element $l \in \mathfrak{g}^*$, the orbit $\mathcal{O}_l$ of the co-adjoint action turns out to be a symplectic manifold. It comes with an action of G preserving the symplectic structure (the restriction of the co-adjoint action on $\mathfrak{g}^*$ to the orbit). In such a case, the moment map

$$\mu : \mathcal{O}_l \to \mathfrak{g}^*$$

is just the inclusion map. Two simple examples are

- *For $\mathfrak{g} = \mathfrak{h}_3$, phase space $M = \mathbf{R}^2$ with the Heisenberg group action of equation 15.13 is given by a co-adjoint orbit, taking $l \in \mathfrak{h}_3^*$ to be the dual basis vector to the basis vector of $\mathfrak{h}_3$ given by the constant function 1 on $\mathbf{R}^2$.*

- *For $\mathfrak{g} = \mathfrak{so}(3)$ the nonzero co-adjoint orbits are spheres, with radius the length of l, the symplectic form described above, and an action of $G = SO(3)$ preserving the symplectic form.*

Note that in the second example, the standard inner product on $\mathbf{R}^3$ provides a $SO(3)$ invariant identification of the Lie algebra with its dual, and as a result, the adjoint and co-adjoint actions are much the same. In the case of H_3, there is no invariant inner product on $\mathfrak{h}_3$, so the adjoint and co-adjoint actions are rather different, explaining the different actions of H_3 on M and $\mathcal{M}$ described earlier.

15.6 For further reading

For a general discussion of vector fields on $\mathbf{R}^n$, see [48]. See [2], [8] and [13] for more on Hamiltonian vector fields and the moment map. For more on the duals of Lie algebras and co-adjoint orbits, see [9] and [51].

Chapter 16
Quadratic Polynomials
and the Symplectic Group

In Chapters 14 and 15, we studied in detail the Heisenberg Lie algebra as the Lie algebra of linear functions on phase space. After quantization, such functions will give operators Q_j and P_j on the state space $\mathcal{H}$. In this chapter, we will begin to investigate what happens for quadratic functions with the symplectic Lie algebra now the one of interest.

The existence of non-trivial Poisson brackets between homogeneous order two and order one polynomials reflects the fact that the symplectic group acts by automorphisms on the Heisenberg group. The significance of this phenomenon will only become clear in later chapters, where examples will appear of interesting observables coming from the symplectic Lie algebra that are quadratic in the Q_j and P_j and act not just on states, but non-trivially on the Q_j and P_j observables.

The identification of elements L of the Lie algebra $\mathfrak{sp}(2d, \mathbf{R})$ with order two polynomials μ_L on phase space M is just the moment map for the action of the symplectic group $Sp(2d, \mathbf{R})$ on M. Quantization of these quadratic functions will provide quantum observables corresponding to any Lie subgroup $G \subset Sp(2d, \mathbf{R})$ (any Lie group G that acts linearly on M preserving the symplectic form). Such quantum observables may or may not be "symmetries," with the term "symmetry" usually meaning that they arise by quantization of a μ_L such that $\{\mu_L, h\} = 0$ for h the Hamiltonian function.

The reader should be warned that the discussion here is not at this stage physically very well motivated, with much of the motivation only appearing in later chapters, especially in the case of the observables of quantum field theory, which will be quadratic in the fields, and act by automorphisms on the fields themselves.

© Peter Woit 2017
P. Woit, *Quantum Theory, Groups and Representations*,
DOI 10.1007/978-3-319-64612-1_16

16.1 The symplectic group

Recall that the orthogonal group can be defined as the group of linear transformations preserving an inner product, which is a symmetric bilinear form.
We now want to study the analog of the orthogonal group that comes from
replacing the inner product by the antisymmetric bilinear form Ω that determines the symplectic geometry of phase space. We will define:

Definition (Symplectic Group). *The symplectic group $Sp(2d, \mathbf{R})$ is the subgroup of linear transformations g of $\mathcal{M} = \mathbf{R}^{2d}$ that satisfy*

$$\Omega(gv_1, gv_2) = \Omega(v_1, v_2)$$

for $v_1, v_2 \in \mathcal{M}$

While this definition uses the dual phase space $\mathcal{M}$ and Ω, it would have
been equivalent to have made the definition using M and ω, since these
transformations preserve the isomorphism between M and $\mathcal{M}$ given by Ω
(see equation 14.6). For an action on $\mathcal{M}$

$$u \in \mathcal{M} \to gu \in \mathcal{M}$$

the action on elements of M (such elements correspond to linear functions
$\Omega(u, \cdot)$ on $\mathcal{M}$) is given by

$$\Omega(u, \cdot) \in M \to g \cdot \Omega(u, \cdot) = \Omega(u, g^{-1}(\cdot)) = \Omega(gu, \cdot) \in M \qquad (16.1)$$

Here, the first equality uses the definition of the dual representation (see 4.2)
to get a representation on linear functions on $\mathcal{M}$ given a representation on
$\mathcal{M}$, and the second uses the invariance of Ω.

16.1.1 The symplectic group for $d = 1$

In order to study symplectic groups as groups of matrices, we will begin with
the case $d = 1$ and the group $Sp(2, \mathbf{R})$. We can write Ω as

$$\Omega\left(\begin{pmatrix} c_q \\ c_p \end{pmatrix}, \begin{pmatrix} c'_q \\ c'_p \end{pmatrix}\right) = c_q c'_p - c_p c'_q = \begin{pmatrix} c_q & c_p \end{pmatrix} \begin{pmatrix} 0 & 1 \\ -1 & 0 \end{pmatrix} \begin{pmatrix} c'_q \\ c'_p \end{pmatrix} \qquad (16.2)$$

A linear transformation g of $\mathcal{M}$ will be given by

$$\begin{pmatrix} c_q \\ c_p \end{pmatrix} \to \begin{pmatrix} \alpha & \beta \\ \gamma & \delta \end{pmatrix} \begin{pmatrix} c_q \\ c_p \end{pmatrix} \qquad (16.3)$$

The condition for Ω to be invariant under such a transformation is

$$\begin{pmatrix} \alpha & \beta \\ \gamma & \delta \end{pmatrix}^T \begin{pmatrix} 0 & 1 \\ -1 & 0 \end{pmatrix} \begin{pmatrix} \alpha & \beta \\ \gamma & \delta \end{pmatrix} = \begin{pmatrix} 0 & 1 \\ -1 & 0 \end{pmatrix} \qquad (16.4)$$

or

$$\begin{pmatrix} 0 & \alpha\delta - \beta\gamma \\ -\alpha\delta + \beta\gamma & 0 \end{pmatrix} = \begin{pmatrix} 0 & 1 \\ -1 & 0 \end{pmatrix}$$

so

$$\det \begin{pmatrix} \alpha & \beta \\ \gamma & \delta \end{pmatrix} = \alpha\delta - \beta\gamma = 1$$

This says that we can have any linear transformation with unit determinant. In other words, we find that $Sp(2, \mathbf{R}) = SL(2, \mathbf{R})$. This isomorphism with a special linear group occurs only for $d = 1$.

Now turning to the Lie algebra, for group elements $g \in GL(2, \mathbf{R})$ near the identity, g can be written in the form $g = e^{tL}$ where L is in the Lie algebra $\mathfrak{gl}(2, \mathbf{R})$. The condition that g acts on $\mathcal{M}$ preserving Ω implies that (differentiating 16.4)

$$\frac{d}{dt}\left((e^{tL})^T \begin{pmatrix} 0 & 1 \\ -1 & 0 \end{pmatrix} e^{tL} \right) = (e^{tL})^T \left(L^T \begin{pmatrix} 0 & 1 \\ -1 & 0 \end{pmatrix} + \begin{pmatrix} 0 & 1 \\ -1 & 0 \end{pmatrix} L \right) e^{tL} = 0$$

Setting $t = 0$, the condition on L is

$$L^T \begin{pmatrix} 0 & 1 \\ -1 & 0 \end{pmatrix} + \begin{pmatrix} 0 & 1 \\ -1 & 0 \end{pmatrix} L = 0 \qquad (16.5)$$

This requires that L must be of the form

$$L = \begin{pmatrix} a & b \\ c & -a \end{pmatrix} \qquad (16.6)$$

which is what one expects: L is in the Lie algebra $\mathfrak{sl}(2, \mathbf{R})$ of 2 by 2 real matrices with zero trace.

The homogeneous degree two polynomials in p and q form a three-dimensional sub-Lie algebra of the Lie algebra of functions on phase space, since the nonzero Poisson bracket relations on a basis $\frac{q^2}{2}, \frac{p^2}{2}, qp$ are

$$\{\frac{q^2}{2}, \frac{p^2}{2}\} = qp \quad \{qp, p^2\} = 2p^2 \quad \{qp, q^2\} = -2q^2$$

We have

Theorem 16.1. *The Lie algebra of degree two homogeneous polynomials on* $M = \mathbf{R}^2$ *is isomorphic to the Lie algebra* $\mathfrak{sp}(2, \mathbf{R}) = \mathfrak{sl}(2, \mathbf{R})$, *with the isomorphism given explicitly by*

$$- aqp + \frac{bq^2}{2} - \frac{cp^2}{2} = \frac{1}{2} \begin{pmatrix} q & p \end{pmatrix} L \begin{pmatrix} 0 & -1 \\ 1 & 0 \end{pmatrix} \begin{pmatrix} q \\ p \end{pmatrix} \leftrightarrow L = \begin{pmatrix} a & b \\ c & -a \end{pmatrix} \quad (16.7)$$

Proof. One can identify basis elements as follows:

$$\frac{q^2}{2} \leftrightarrow E = \begin{pmatrix} 0 & 1 \\ 0 & 0 \end{pmatrix} \quad -\frac{p^2}{2} \leftrightarrow F = \begin{pmatrix} 0 & 0 \\ 1 & 0 \end{pmatrix} \quad -qp \leftrightarrow G = \begin{pmatrix} 1 & 0 \\ 0 & -1 \end{pmatrix} \quad (16.8)$$

The commutation relations among these matrices are

$$[E, F] = G \quad [G, E] = 2E \quad [G, F] = -2F$$

which are the same as the Poisson bracket relations between the corresponding quadratic polynomials. ☐

The moment map for the $SL(2, \mathbf{R})$ action on $\mathcal{M} = \mathbf{R}^2$ of equation 16.3 is given by

$$\mu_L = -aqp + \frac{bq^2}{2} - \frac{cp^2}{2} \quad (16.9)$$

To check this, first compute using the definition of the Poisson bracket

$$-X_{\mu_L} F(q, p) = \{\mu_L, F\} = (bq - ap) \frac{\partial F}{\partial p} + (aq + cp) \frac{\partial F}{\partial q}$$

Elements $e^{tL} \in SL(2, \mathbf{R})$ act on functions on M by

$$e^{tL} \cdot F(q(m), p(m)) = F(q(e^{-tL} \cdot m), p(e^{-tL} \cdot m))$$

where (for $m \in M$ written as column vectors) $e^{-tL} \cdot m$ is multiplication by the matrix e^{-tL}. On linear functions $l \in \mathcal{M}$ written as column vectors, the same group action takes l to $e^{tL} l$ and acts on basis vectors q, p of $\mathcal{M}$ by

$$\begin{pmatrix} q \\ p \end{pmatrix} \to (e^{tL})^T \begin{pmatrix} q \\ p \end{pmatrix}$$

The vector field X_L is then given by

$$-X_L F(q,p) = \frac{d}{dt} F((q(e^{-tL} \cdot m), p(e^{-tL} \cdot m))_{t=0}$$

$$= \left(\frac{\partial F}{\partial q}, \frac{\partial F}{\partial p} \right) \cdot \frac{d}{dt} (e^{tL})^T \begin{pmatrix} q \\ p \end{pmatrix}_{t=0}$$

$$= \left(\frac{\partial F}{\partial q}, \frac{\partial F}{\partial p} \right) \cdot L^T \begin{pmatrix} q \\ p \end{pmatrix}$$

$$= \left(\frac{\partial F}{\partial q}, \frac{\partial F}{\partial p} \right) \cdot \begin{pmatrix} aq + cp \\ bq - ap \end{pmatrix}$$

and one sees that $X_L = X_{\mu_L}$ as required. The isomorphism of the theorem is the statement that μ_L has the Lie algebra homomorphism property characterizing moment maps:

$$\{\mu_L, \mu_{L'}\} = \mu_{[L,L']}$$

Two important subgroups of $SL(2, \mathbf{R})$ are

- The subgroup of elements one gets by exponentiating G, which is isomorphic to the multiplicative group of positive real numbers

$$e^{tG} = \begin{pmatrix} e^t & 0 \\ 0 & e^{-t} \end{pmatrix}$$

Here, one can explicitly see that this group has elements going off to infinity.

- Exponentiating the Lie algebra element $E - F$ gives rotations of the plane

$$e^{\theta(E-F)} = \begin{pmatrix} \cos\theta & \sin\theta \\ -\sin\theta & \cos\theta \end{pmatrix}$$

Note that the Lie algebra element being exponentiated here is

$$E - F \leftrightarrow \frac{1}{2}(p^2 + q^2)$$

the function studied in section 15.2, which we will later re-encounter as the Hamiltonian function for the harmonic oscillator in Chapter 22.

The group $SL(2, \mathbf{R})$ is non-compact, and its representation theory is quite unlike the case of $SU(2)$. In particular, all of its non-trivial irreducible unitary representations are infinite dimensional, forming an important topic in mathematics, but one that is beyond our scope. We will be studying just one such irreducible representation (the one provided by the quantum mechanical state space), and it is a representation only of a double cover of $SL(2, \mathbf{R})$, not of $SL(2, \mathbf{R})$ itself.

16.1.2   The symplectic group for arbitrary d

For general d, the symplectic group $Sp(2d, \mathbf{R})$ is the group of linear transformations g of $\mathcal{M}$ that leave Ω (see 14.4) invariant, i.e., satisfy

$$\Omega\left(g\begin{pmatrix}\mathbf{c}_q\\\mathbf{c}_p\end{pmatrix}, g\begin{pmatrix}\mathbf{c}'_q\\\mathbf{c}'_p\end{pmatrix}\right) = \Omega\left(\begin{pmatrix}\mathbf{c}_q\\\mathbf{c}_p\end{pmatrix}, \begin{pmatrix}\mathbf{c}'_q\\\mathbf{c}'_p\end{pmatrix}\right)$$

where $\mathbf{c}_q, \mathbf{c}_p$ are d-dimensional vectors. By essentially the same calculation as in the $d = 1$ case, we find the d-dimensional generalization of equation 16.4. This says that $Sp(2d, \mathbf{R})$ is the group of real $2d$ by $2d$ matrices g satisfying

$$g^T\begin{pmatrix}\mathbf{0} & \mathbf{1}\\-\mathbf{1} & \mathbf{0}\end{pmatrix}g = \begin{pmatrix}\mathbf{0} & \mathbf{1}\\-\mathbf{1} & \mathbf{0}\end{pmatrix} \tag{16.10}$$

where $\mathbf{0}$ is the d by d zero matrix, $\mathbf{1}$ the d by d unit matrix.

Again by a similar argument to the $d = 1$ case where the Lie algebra $\mathfrak{sp}(2, \mathbf{R})$ was determined by the condition 16.5, $\mathfrak{sp}(2d, \mathbf{R})$ is the Lie algebra of $2d$ by $2d$ matrices L satisfying

$$L^T\begin{pmatrix}\mathbf{0} & \mathbf{1}\\-\mathbf{1} & \mathbf{0}\end{pmatrix} + \begin{pmatrix}\mathbf{0} & \mathbf{1}\\-\mathbf{1} & \mathbf{0}\end{pmatrix}L = 0 \tag{16.11}$$

Such matrices will be those with the block-diagonal form

$$L = \begin{pmatrix}A & B\\C & -A^T\end{pmatrix} \tag{16.12}$$

where A, B, C are d by d real matrices, with B and C symmetric, i.e.,

$$B = B^T, \quad C = C^T$$

Note that, replacing the block antisymmetric matrix by the unit matrix, in 16.10 one recovers the definition of an orthogonal matrix, in 16.11 the definition of the Lie algebra of the orthogonal group.

The generalization of 16.7 is

Theorem 16.2. *The Lie algebra* $\mathfrak{sp}(2d, \mathbf{R})$ *is isomorphic to the Lie algebra of order two homogeneous polynomials on* $M = \mathbf{R}^{2d}$ *by the isomorphism (using a vector notation for the coefficient functions* $q_1, \cdots, q_d, p_1, \cdots, p_d$)

$$L \leftrightarrow \mu_L$$

where

$$\mu_L = \frac{1}{2} \begin{pmatrix} \mathbf{q} & \mathbf{p} \end{pmatrix} L \begin{pmatrix} 0 & -1 \\ 1 & 0 \end{pmatrix} \begin{pmatrix} \mathbf{q} \\ \mathbf{p} \end{pmatrix}$$

$$= \frac{1}{2} \begin{pmatrix} \mathbf{q} & \mathbf{p} \end{pmatrix} \begin{pmatrix} B & -A \\ -A^T & -C \end{pmatrix} \begin{pmatrix} \mathbf{q} \\ \mathbf{p} \end{pmatrix}$$

$$= \frac{1}{2} (\mathbf{q} \cdot B\mathbf{q} - 2\mathbf{q} \cdot A\mathbf{p} - \mathbf{p} \cdot C\mathbf{p}) \qquad (16.13)$$

We will postpone the proof of this theorem until section 16.2, since it is easier to first study Poisson brackets between order two and order one polynomials, then use this to prove the theorem about Poisson brackets between order two polynomials. As in $d = 1$, the function μ_L is the moment map function for L.

The Lie algebra $\mathfrak{sp}(2d, \mathbf{R})$ has a subalgebra $\mathfrak{gl}(d, \mathbf{R})$ consisting of matrices of the form

$$L = \begin{pmatrix} A & 0 \\ 0 & -A^T \end{pmatrix}$$

or, in terms of quadratic functions, the functions

$$- \mathbf{q} \cdot A\mathbf{p} = -(A^T \mathbf{p}) \cdot \mathbf{q} \qquad (16.14)$$

where A is any real d by d matrix. This shows that one way to get symplectic transformations is to take any linear transformation of the position coordinates, together with the dual linear transformation (see definition 4.2) on momentum coordinates. In this way, any linear group acting on position space gives a subgroup of the symplectic transformations of phase space.

An example of this is the group $SO(d)$ of spatial rotations, with Lie algebra $\mathfrak{so}(d) \subset \mathfrak{gl}(d, \mathbf{R})$, the antisymmetric d by d matrices, for which $-A^T = A$. The special case $d = 3$ was an example already worked out earlier, in section 15.4, where μ_L gives the standard expression for the angular momentum as a function of the q_j, p_j coordinates on phase space.

Another important special case comes from taking $A = 0, B = 1, C = -1$ in equation 16.12, which by equation 16.13 gives

$$\mu_L = \frac{1}{2} (|\mathbf{q}|^2 + |\mathbf{p}|^2)$$

This generalizes the case of $d = 1$ described earlier and will be the Hamiltonian function for a d-dimensional harmonic oscillator. Note that exponentiating L gives a symplectic action on phase space that mixes position and momentum coordinates, so this an example that cannot be understood just in terms of a group action on configuration space.

16.2   The symplectic group and automorphisms of the Heisenberg group

Returning to the $d = 1$ case, we have found two three-dimensional Lie algebras ($\mathfrak{h}_3$ and $\mathfrak{sl}(2,\mathbf{R})$) as subalgebras of the infinite dimensional Lie algebra of functions on phase space:

- $\mathfrak{h}_3$, the Lie algebra of linear polynomials on M, with basis $1, q, p$.
- $\mathfrak{sl}(2,\mathbf{R})$, the Lie algebra of order two homogeneous polynomials on M, with basis q^2, p^2, qp.

Taking all quadratic polynomials, we get a six-dimensional Lie algebra with basis elements $1, q, p, qp, q^2, p^2$. This is not the direct product of $\mathfrak{h}_3$ and $\mathfrak{sl}(2,\mathbf{R})$ since there are nonzero Poisson brackets

$$\{qp, q\} = \ -q, \ \ \{qp, p\} = p$$
$$\{\frac{p^2}{2}, q\} = \ -p, \ \ \{\frac{q^2}{2}, p\} = q \tag{16.15}$$

These relations show that operating on a basis of linear functions on M by taking the Poisson bracket with something in $\mathfrak{sl}(2,\mathbf{R})$ (a quadratic function) provides a linear transformation on $M^* = \mathcal{M}$.

In this section, we will see that this is the infinitesimal version of the fact that $SL(2,\mathbf{R})$ acts on the Heisenberg group H_3 by automorphisms. We will begin with a general discussion of what happens when a Lie group G acts by automorphisms on a Lie group H, then turn to two examples: the conjugation action of G on itself and the action of $SL(2,\mathbf{R})$ on H_3.

An action of one group on another by automorphisms means the following:

Definition (Group automorphisms). *If an action of elements g of a group G on a group H*

$$h \in H \to \Phi_g(h) \in H$$

satisfies

$$\Phi_g(h_1)\Phi_g(h_2) = \Phi_g(h_1 h_2)$$

for all $g \in G$ and $h_1, h_2 \in H$, the group G is said to act on H by automorphisms. Each map Φ_g is an automorphism of H. Note that since Φ_g is an action of G, we have $\Phi_{g_1 g_2} = \Phi_{g_1}\Phi_{g_2}$.

When the groups are Lie groups, taking the derivative $\phi_g : \mathfrak{h} \to \mathfrak{h}$ of the map $\Phi_g : H \to H$ at the identity of H gives a Lie algebra automorphism, defined by

Definition (Lie algebra automorphisms). *If an action of elements g of a group G on a Lie algebra $\mathfrak{h}$*

$$X \in \mathfrak{h} \to \phi_g(X) \in \mathfrak{h}$$

satisfies

$$[\phi_g(X), \phi_g(Y)] = \phi_g([X, Y])$$

for all $g \in G$ and $X, Y \in \mathfrak{h}$, the group is said to act on $\mathfrak{h}$ by automorphisms.

Given an action ϕ_g of a Lie group G on $\mathfrak{h}$, we get an action of elements $Z \in \mathfrak{g}$ on $\mathfrak{h}$ by linear maps:

$$X \to Z \cdot X = \frac{d}{dt}(\phi_{e^{tZ}}(X))_{|t=0} \tag{16.16}$$

that we will often refer to as the infinitesimal version of the action ϕ_g of G on $\mathfrak{h}$. These maps satisfy

$$
\begin{aligned}
Z \cdot [X, Y] &= \frac{d}{dt}(\phi_{e^{tZ}}([X, Y]))_{|t=0} \\
&= \frac{d}{dt}([\phi_{e^{tZ}}(X), \phi_{e^{tZ}}(Y)])_{|t=0} \\
&= [Z \cdot X, Y] + [X, Z \cdot Y]
\end{aligned}
$$

and one can define

Definition (Lie algebra derivations). *If an action of a Lie algebra $\mathfrak{g}$ on a Lie algebra $\mathfrak{h}$ by linear maps*

$$X \in \mathfrak{h} \to Z \cdot X \in \mathfrak{h}$$

satisfies

$$[Z \cdot X, Y] + [X, Z \cdot Y] = Z \cdot [X, Y] \tag{16.17}$$

for all $Z \in \mathfrak{g}$ and $X, Y \in \mathfrak{h}$, the Lie algebra $\mathfrak{g}$ is said to act on $\mathfrak{h}$ by derivations. The action of an element Z on $\mathfrak{h}$ is a derivation of $\mathfrak{h}$.

16.2.1 The adjoint representation and inner automorphisms

Any group G acts on itself by conjugation, with

$$\Phi_g(g') = gg'g^{-1}$$

giving an action by automorphisms (these are called "inner automorphisms"). The derivative at the identity of the map Φ_g is the linear map on $\mathfrak{g}$ given by the adjoint representation operators $Ad(g)$ discussed in Chapter 5. So, in this case the corresponding action by automorphisms on the Lie algebra $\mathfrak{g}$ is the adjoint action

$$X \in \mathfrak{g} \rightarrow \phi_g(X) = Ad(g)(X) = gXg^{-1}$$

The infinitesimal version of the Lie group adjoint representation by $Ad(g)$ on $\mathfrak{g}$ is the Lie algebra adjoint representation by operators $ad(Z)$ on $\mathfrak{g}$

$$X \in \mathfrak{g} \rightarrow Z \cdot X = ad(Z)(X) = [Z, X]$$

This is an action of $\mathfrak{g}$ on itself by derivations.

16.2.2   The symplectic group as automorphism group

Recall the definition 13.2 of the Heisenberg group H_3 as elements

$$\left(\begin{pmatrix} x \\ y \end{pmatrix}, z \right) \in \mathbf{R}^2 \oplus \mathbf{R}$$

with the group law

$$\left(\begin{pmatrix} x \\ y \end{pmatrix}, z \right) \left(\begin{pmatrix} x' \\ y' \end{pmatrix}, z' \right) = \left(\begin{pmatrix} x + x' \\ y + y' \end{pmatrix}, z + z' + \frac{1}{2}\Omega \left(\begin{pmatrix} x \\ y \end{pmatrix}, \begin{pmatrix} x' \\ y' \end{pmatrix} \right) \right)$$

Elements $g \in SL(2, \mathbf{R})$ act on H_3 by

$$\left(\begin{pmatrix} x \\ y \end{pmatrix}, z \right) \rightarrow \Phi_g \left(\left(\begin{pmatrix} x \\ y \end{pmatrix}, z \right) \right) = \left(g \begin{pmatrix} x \\ y \end{pmatrix}, z \right) \qquad (16.18)$$

Here, $G = SL(2, \mathbf{R})$, $H = H_3$ and Φ_g given above is an action by automorphisms since

$$\Phi_g \left(\begin{pmatrix} x \\ y \end{pmatrix}, z \right) \Phi_g \left(\begin{pmatrix} x' \\ y' \end{pmatrix}, z' \right) = \left(g \begin{pmatrix} x \\ y \end{pmatrix}, z \right) \left(g \begin{pmatrix} x' \\ y' \end{pmatrix}, z' \right)$$

$$= \left(g \begin{pmatrix} x + x' \\ y + y' \end{pmatrix}, z + z' + \frac{1}{2}\Omega \left(g \begin{pmatrix} x \\ y \end{pmatrix}, g \begin{pmatrix} x' \\ y' \end{pmatrix} \right) \right)$$

$$= \left(g \begin{pmatrix} x+x' \\ y+y' \end{pmatrix}, z+z' + \frac{1}{2}\Omega \left(\begin{pmatrix} x \\ y \end{pmatrix}, \begin{pmatrix} x' \\ y' \end{pmatrix} \right) \right)$$

$$= \Phi_g \left(\left(\begin{pmatrix} x \\ y \end{pmatrix}, z \right) \left(\begin{pmatrix} x' \\ y' \end{pmatrix}, z' \right) \right) \tag{16.19}$$

Recall that, in the exponential coordinates we use, the exponential map between the Lie algebra $\mathfrak{h}_3$ and the Lie group H_3 is the identity map, with both $\mathfrak{h}_3$ and H_3 identified with $\mathbf{R}^2 \oplus \mathbf{R}$. As in section 14.2, we will explicitly identify $\mathfrak{h}_3$ with functions $c_q q + c_p p + c$ on M, writing these as

$$\left(\begin{pmatrix} c_q \\ c_p \end{pmatrix}, c \right)$$

with Lie bracket

$$\left[\left(\begin{pmatrix} c_q \\ c_p \end{pmatrix}, c \right), \left(\begin{pmatrix} c_q' \\ c_p' \end{pmatrix}, c' \right) \right] = \left(\begin{pmatrix} 0 \\ 0 \end{pmatrix}, c_q c_p' - c_p c_q' \right) = \left(\begin{pmatrix} 0 \\ 0 \end{pmatrix}, \Omega \left(\begin{pmatrix} c_q \\ c_p \end{pmatrix}, \begin{pmatrix} c_q' \\ c_p' \end{pmatrix} \right) \right)$$

The linearized action of Φ_g at the identity of H_3 gives the action ϕ_g on $\mathfrak{h}_3$, but since the exponential map is the identity, ϕ_g acts on $\mathbf{R}^2 \oplus \mathbf{R} = \mathcal{M} \oplus \mathbf{R}$ in the same way as Φ_g, by

$$\left(\begin{pmatrix} c_q \\ c_p \end{pmatrix}, c \right) \in \mathfrak{h}_3 \rightarrow \phi_g \left(\left(\begin{pmatrix} c_q \\ c_p \end{pmatrix}, c \right) \right) = \left(g \begin{pmatrix} c_q \\ c_p \end{pmatrix}, c \right)$$

Since the Lie bracket just depends on Ω, which is $SL(2,\mathbf{R})$ invariant, ϕ_g preserves the Lie bracket and so acts by automorphisms on $\mathfrak{h}_3$.

The infinitesimal version of the $SL(2,\mathbf{R})$ action ϕ_g on $\mathfrak{h}_3$ is an action of $\mathfrak{sl}(2,\mathbf{R})$ on $\mathfrak{h}_3$ by derivations. This action can be found by computing (for $L \in \mathfrak{sl}(2,\mathbf{R})$ and $X \in \mathfrak{h}_3$) using equation 16.16 to get

$$L \cdot \left(\begin{pmatrix} c_q \\ c_p \end{pmatrix}, c \right) = \frac{d}{dt} \left(\phi_{e^{tL}} \left(\begin{pmatrix} c_q \\ c_p \end{pmatrix}, c \right) \right)_{|t=0} = \left(L \begin{pmatrix} c_q \\ c_p \end{pmatrix}, 0 \right) \tag{16.20}$$

The Poisson brackets between degree two and degree one polynomials discussed at the beginning of this section give an alternate way of calculating this action of $\mathfrak{sl}(2,\mathbf{R})$ on $\mathfrak{h}_3$ by derivations. For a general $L \in \mathfrak{sl}(2,\mathbf{R})$ (see equation 16.6) and $c_q q + c_p p + C \in \mathfrak{h}_3$, we have

$$\{ \mu_L, c_q q + c_p p + C \} = c_q' q + c_p' p, \quad \begin{pmatrix} c_q' \\ c_p' \end{pmatrix} = \begin{pmatrix} ac_q + bc_p \\ cc_q - ac_p \end{pmatrix} = L \begin{pmatrix} c_q \\ c_p \end{pmatrix} \tag{16.21}$$

(here, μ_L is given by 16.9). We see that this is the action of $\mathfrak{sl}(2,\mathbf{R})$ by derivations on $\mathfrak{h}_3$ of equation 16.20, the infinitesimal version of the action of $SL(2,\mathbf{R})$ on $\mathfrak{h}_3$ by automorphisms.

Note that in the larger Lie algebra of all polynomials on M of order two or less, the action of $\mathfrak{sl}(2, \mathbf{R})$ on $\mathfrak{h}_3$ by derivations is part of the adjoint action of the Lie algebra on itself, since it is given by the Poisson bracket (which is the Lie bracket), between order two and order one polynomials.

16.3 The case of arbitrary d

For the general case of arbitrary d, the group $Sp(2d, \mathbf{R})$ will act by automorphisms on H_{2d+1} and $\mathfrak{h}_{2d+1}$, both of which can be identified with $\mathcal{M} \oplus \mathbf{R}$. The group acts by linear transformations on the $\mathcal{M}$ factor, preserving Ω. The infinitesimal version of this action is computed as in the $d = 1$ case to be

$$L \cdot \left(\begin{pmatrix} \mathbf{c}_q \\ \mathbf{c}_p \end{pmatrix}, c \right) = \frac{d}{dt} \left(\phi_{e^{tL}} \left(\begin{pmatrix} \mathbf{c}_q \\ \mathbf{c}_p \end{pmatrix}, c \right) \right)_{|t=0} = \left(L \begin{pmatrix} \mathbf{c}_q \\ \mathbf{c}_p \end{pmatrix}, 0 \right)$$

where $L \in \mathfrak{sp}(2d, \mathbf{R})$. This action, as in the $d = 1$ case, is given by taking Poisson brackets of a quadratic function with a linear function:

Theorem 16.3. *The $\mathfrak{sp}(2d, \mathbf{R})$ action on $\mathfrak{h}_{2d+1} = \mathcal{M} \oplus \mathbf{R}$ by derivations is*

$$L \cdot (\mathbf{c}_q \cdot \mathbf{q} + \mathbf{c}_p \cdot \mathbf{p} + c) = \{\mu_L, \mathbf{c}_q \cdot \mathbf{q} + \mathbf{c}_p \cdot \mathbf{p} + c\} = \mathbf{c}'_q \cdot \mathbf{q} + \mathbf{c}'_p \cdot \mathbf{p} \quad (16.22)$$

where

$$\begin{pmatrix} \mathbf{c}'_q \\ \mathbf{c}'_p \end{pmatrix} = L \begin{pmatrix} \mathbf{c}_q \\ \mathbf{c}_p \end{pmatrix}$$

or, equivalently (see section 4.1), on basis vectors of $\mathcal{M}$ one has

$$\left\{ \mu_L, \begin{pmatrix} \mathbf{q} \\ \mathbf{p} \end{pmatrix} \right\} = L^T \begin{pmatrix} \mathbf{q} \\ \mathbf{p} \end{pmatrix}$$

Proof. One can first prove 16.22 for the cases when only one of A, B, C is nonzero, and then the general case follows by linearity. For instance, taking the special case

$$L = \begin{pmatrix} 0 & B \\ 0 & 0 \end{pmatrix}, \quad \mu_L = \frac{1}{2} \mathbf{q} \cdot B\mathbf{q}$$

the action on coordinate functions (the basis vectors of $\mathcal{M}$) is

$$\{\frac{1}{2} \mathbf{q} \cdot B\mathbf{q}, \begin{pmatrix} \mathbf{q} \\ \mathbf{p} \end{pmatrix}\} = L^T \begin{pmatrix} \mathbf{q} \\ \mathbf{p} \end{pmatrix} = \begin{pmatrix} 0 \\ B\mathbf{q} \end{pmatrix}$$

since

$$\{\frac{1}{2}\sum_{j,k} q_j B_{jk} q_k, p_l\} = \frac{1}{2}\sum_{j,k}(q_j\{B_{jk}q_k, p_l\} + \{q_j B_{jk}, p_l\}q_k)$$

$$= \frac{1}{2}(\sum_j q_j B_{jl} + \sum_k B_{lk}q_k)$$

$$= \sum_j B_{lj}q_j \quad (\text{since } B = B^T)$$

Repeating for A and C gives in general

$$\left\{\mu_L, \begin{pmatrix} \mathbf{q} \\ \mathbf{p} \end{pmatrix}\right\} = L^T \begin{pmatrix} \mathbf{q} \\ \mathbf{p} \end{pmatrix}$$

$\square$

We can now prove theorem 16.2 as follows:

Proof.

$$L \to \mu_L$$

is clearly a vector space isomorphism of matrices and of quadratic polynomials. To show that it is a Lie algebra isomorphism, the Jacobi identity for the Poisson bracket can be used to show

$$\{\mu_L, \{\mu_{L'}, \mathbf{c}_q \cdot \mathbf{q} + \mathbf{c}_p \cdot \mathbf{p}\}\} - \{\mu_{L'}, \{\mu_L, \mathbf{c}_q \cdot \mathbf{q} + \mathbf{c}_p \cdot \mathbf{p}\}\} = \{\{\mu_L, \mu_{L'}\}, \mathbf{c}_q \cdot \mathbf{q} + \mathbf{c}_p \cdot \mathbf{p}\}$$

The left-hand side of this equation is $\mathbf{c}''_q \cdot \mathbf{q} + \mathbf{c}''_p \cdot \mathbf{p}$, where

$$\begin{pmatrix} \mathbf{c}''_q \\ \mathbf{c}''_p \end{pmatrix} = (LL' - L'L)\begin{pmatrix} \mathbf{c}_q \\ \mathbf{c}_p \end{pmatrix}$$

As a result, the right-hand side is the linear map given by

$$\{\mu_L, \mu_{L'}\} = \mu_{[L,L']}$$

$\square$

16.4   For further reading

For more on symplectic groups and the isomorphism between $\mathfrak{sp}(2d, \mathbf{R})$ and homogeneous degree two polynomials, see chapter 14 of [37] or chapter 4 of [26]. Chapter 15 of [37] and chapter 1 of [26] discuss the action of the symplectic group on the Heisenberg group and Lie algebra by automorphisms.

Chapter 17
Quantization

Given any Hamiltonian classical mechanical system with phase space $\mathbf{R}^{2d}$, physics textbooks have a standard recipe for producing a quantum system, by a method known as "canonical quantization." We will see that for linear functions on phase space, this is just the construction we have already seen of a unitary representation Γ'_S of the Heisenberg Lie algebra, the Schrödinger representation. The Stone–von Neumann theorem assures us that this is the unique such construction, up to unitary equivalence. We will also see that this recipe can only ever be partially successful: the Schrödinger representation gives us a representation of a subalgebra of the Lie algebra of all functions on phase space (the polynomials of degree two and below), but a no-go theorem shows that this cannot be extended to a representation of the full infinite dimensional Lie algebra. Recipes for quantizing higher-order polynomials will always suffer from a lack of uniqueness, a phenomenon known to physicists as the existence of "operator ordering ambiguities."

In later chapters, we will see that this quantization prescription does give unique quantum systems corresponding to some Hamiltonian systems (in particular the harmonic oscillator and the hydrogen atom), and does so in a manner that allows a description of the quantum system purely in terms of representation theory.

17.1 Canonical quantization

Very early on in the history of quantum mechanics, when Dirac first saw the Heisenberg commutation relations, he noticed an analogy with the Poisson bracket. One has

$$\{q,p\} = 1 \quad \text{and} \quad -\frac{i}{\hbar}[Q,P] = 1$$

© Peter Woit 2017
P. Woit, *Quantum Theory, Groups and Representations*,
DOI 10.1007/978-3-319-64612-1_17

as well as

$$\frac{df}{dt} = \{f, h\} \quad \text{and} \quad \frac{d}{dt}\mathcal{O}(t) = -\frac{i}{\hbar}[\mathcal{O}, H]$$

where the last of these equations is the equation for the time-dependence of a Heisenberg picture observable $\mathcal{O}(t)$ in quantum mechanics. Dirac's suggestion was that given any classical Hamiltonian system, one could "quantize" it by finding a rule that associates to a function f on phase space a self-adjoint operator O_f (in particular $O_h = H$), acting on a state space $\mathcal{H}$ such that

$$O_{\{f,g\}} = -\frac{i}{\hbar}[O_f, O_g]$$

This is completely equivalent to asking for a unitary representation $(\pi', \mathcal{H})$ of the infinite dimensional Lie algebra of functions on phase space (with the Poisson bracket as Lie bracket). To see this, note that units for momentum p and position q can be chosen such that $\hbar = 1$. Then, as usual getting a skew-adjoint Lie algebra representation operator by multiplying a self-adjoint operator by $-i$, setting

$$\pi'(f) = -iO_f$$

the Lie algebra homomorphism property

$$\pi'(\{f, g\}) = [\pi'(f), \pi'(g)]$$

corresponds to

$$-iO_{\{f,g\}} = [-iO_f, -iO_g] = -[O_f, O_g]$$

so one has Dirac's suggested relation.

Recall that the Heisenberg Lie algebra is isomorphic to the three-dimensional subalgebra of functions on phase space given by linear combinations of the constant function, the function q and the function p. The Schrödinger representation Γ'_S provides a unitary representation not of the Lie algebra of all functions on phase space, but of these polynomials of degree at most one, as follows

$$O_1 = \mathbf{1}, \quad O_q = Q, \quad O_p = P$$

so

$$\Gamma'_S(1) = -i\mathbf{1}, \quad \Gamma'_S(q) = -iQ = -iq, \quad \Gamma'_S(p) = -iP = -\frac{d}{dq}$$

Moving on to quadratic polynomials, these can also be quantized, as follows

$$O_{\frac{p^2}{2}} = \frac{P^2}{2}, \quad O_{\frac{q^2}{2}} = \frac{Q^2}{2}$$

For the function pq, one can no longer just replace p by P and q by Q since the operators P and Q do not commute, so the ordering matters. In addition, neither PQ nor QP is self-adjoint. What does work, satisfying all the conditions to give a Lie algebra homomorphism, is the self-adjoint combination

$$O_{pq} = \frac{1}{2}(PQ + QP)$$

This shows that the Schrödinger representation Γ'_S that was defined as a representation of the Heisenberg Lie algebra $\mathfrak{h}_3$ extends to a unitary Lie algebra representation of a larger Lie algebra, that of all quadratic polynomials on phase space, a representation that we will continue to denote by Γ'_S and refer to as the Schrödinger representation. On a basis of homogeneous-order two polynomials, we have

$$\Gamma'_S\left(\frac{q^2}{2}\right) = -i\frac{Q^2}{2} = -\frac{i}{2}q^2$$

$$\Gamma'_S\left(\frac{p^2}{2}\right) = -i\frac{P^2}{2} = \frac{i}{2}\frac{d^2}{dq^2}$$

$$\Gamma'_S(pq) = -\frac{i}{2}(PQ + QP)$$

Restricting Γ'_S to linear combinations of these homogeneous-order two polynomials (which give the Lie algebra $\mathfrak{sl}(2,\mathbf{R})$, see theorem 16.1), we get a Lie algebra representation of $\mathfrak{sl}(2,\mathbf{R})$ called the metaplectic representation.

Restricted to the Heisenberg Lie algebra, the Schrödinger representation Γ'_S exponentiates to give a representation Γ_S of the corresponding Heisenberg Lie group (recall section 13.3). As an $\mathfrak{sl}(2,\mathbf{R})$ representation however, it turns out that Γ'_S has the same sort of problem as the spinor representation of $\mathfrak{su}(2) = \mathfrak{so}(3)$, which was not a representation of $SO(3)$, but only of its double cover $SU(2) = Spin(3)$. To get a group representation, one must go to a double cover of the group $SL(2,\mathbf{R})$, which will be called the metaplectic group and denoted $Mp(2,\mathbf{R})$.

For an indication of the problem, consider the element

$$\frac{1}{2}(q^2 + p^2) \leftrightarrow E - F = \begin{pmatrix} 0 & 1 \\ -1 & 0 \end{pmatrix}$$

in $\mathfrak{sl}(2,\mathbf{R})$. Exponentiating this gives a subgroup $SO(2) \subset SL(2,\mathbf{R})$ of clockwise rotations in the qp plane. The Lie algebra representation operator is

$$\Gamma'_S\left(\frac{1}{2}(q^2+p^2)\right) = -\frac{i}{2}(Q^2+P^2) = -\frac{i}{2}\left(q^2 - \frac{d^2}{dq^2}\right)$$

which is a second-order differential operator in both the position space and momentum space representations. As a result, it is not obvious how to exponentiate this operator.

One can, however, see what happens on the state

$$\psi_0(q) = e^{-\frac{q^2}{2}} \subset \mathcal{H} = L^2(\mathbf{R})$$

where one has

$$-\frac{i}{2}\left(q^2 - \frac{d^2}{dq^2}\right)\psi_0(q) = -\frac{i}{2}\psi_0(q)$$

so $\psi_0(q)$ is an eigenvector of $\Gamma'_S(\frac{1}{2}(q^2+p^2))$ with eigenvalue $-\frac{i}{2}$. Exponentiating $\Gamma'_S(\frac{1}{2}(q^2+p^2))$, the representation Γ_S acts on this state by multiplication by a phase. As one goes around the group $SO(2)$ once (rotating the qp plane by an angle from 0 to 2π), the phase angle only goes from 0 to π, demonstrating the same problem that occurs in the case of the spinor representation.

When we study the Schrödinger representation using its action on the quantum harmonic oscillator state space $\mathcal{H}$ in Chapter 22, we will see that the operator

$$\frac{1}{2}(Q^2 + P^2)$$

is the Hamiltonian operator for the quantum harmonic oscillator, and all of its eigenvectors (not just $\psi_0(q)$) have half-integer eigenvalues. In Chapter 24, we will go on to discuss in more detail the construction of the metaplectic representation, using methods developed to study the harmonic oscillator.

17.2 The Groenewold–van Hove no-go theorem

If one wants to quantize polynomial functions on phase space of degree greater than two, it quickly becomes clear that the problem of "operator ordering ambiguities" is a significant one. Different prescriptions involving different ways of ordering the P and Q operators lead to different O_f for the same function f, with physically different observables (although the differences involve the commutator of P and Q, so higher-order terms in $\hbar$).

When physicists first tried to find a consistent prescription for producing an operator O_f corresponding to a polynomial function on phase space of degree greater than two, they found that there was no possible way to do this consistent with the relation

$$O_{\{f,g\}} = -\frac{i}{\hbar}[O_f, O_g]$$

for polynomials of degree greater than two. Whatever method one devises for quantizing higher-degree polynomials, it can only satisfy that relation to lowest order in $\hbar$, and there will be higher-order corrections, which depend upon one's choice of quantization scheme. Equivalently, it is only for the six-dimensional Lie algebra of polynomials of degree up to two that the Schrödinger representation gives one a Lie algebra representation, and this cannot be consistently extended to a representation of a larger subalgebra of the functions on phase space. This problem is made precise by the following no-go theorem

Theorem (Groenewold–van Hove). *There is no map $f \to O_f$ from polynomials on $\mathbf{R}^2$ to self-adjoint operators on $L^2(\mathbf{R})$ satisfying*

$$O_{\{f,g\}} = -\frac{i}{\hbar}[O_f, O_g]$$

and

$$O_p = P, \quad O_q = Q$$

for any Lie subalgebra of the functions on $\mathbf{R}^2$ for which the subalgebra of polynomials of degree less than or equal to two is a proper subalgebra.

Proof. For a detailed proof, see section 5.4 of [8], section 4.4 of [26], or chapter 16 of [37]. In outline, the proof begins by showing that taking Poisson brackets of polynomials of degree three leads to higher-order polynomials and that furthermore for degree three and above there will be no finite dimensional subalgebras of polynomials of bounded degree. The assumptions of the theorem force certain specific operator ordering choices in degree three. These are then used to get a contradiction in degree four, using the fact that the same degree four polynomial has two different expressions as a Poisson bracket:

$$q^2 p^2 = \frac{1}{3}\{q^2 p, p^2 q\} = \frac{1}{9}\{q^3, p^3\}$$

□

17.3 Canonical quantization in d dimensions

The above can easily be generalized to the case of d dimensions, with the Schrödinger representation Γ'_S now giving a unitary representation of the Heisenberg Lie algebra $\mathfrak{h}_{2d+1}$ determined by

$$\Gamma'_S(q_j) = -iQ_j, \quad \Gamma'_S(p_j) = -iP_j$$

which satisfy the Heisenberg relations

$$[Q_j, P_k] = i\delta_{jk}$$

Generalizing to quadratic polynomials in the phase space coordinate functions, we have

$$\Gamma'_S(q_j q_k) = -iQ_j Q_k, \quad \Gamma'_S(p_j p_k) = -iP_j P_k, \quad \Gamma'_S(q_j p_k) = -\frac{i}{2}(Q_j P_k + P_k Q_j)$$
$$(17.1)$$

These operators can be exponentiated to get a representation on the same $\mathcal{H}$ of $Mp(2d, \mathbf{R})$, a double cover of the symplectic group $Sp(2d, \mathbf{R})$. This phenomenon will be examined carefully in later chapters, starting with Chapter 20 and the calculation in Section 20.3.2, followed by a discussion in Chapters 24 and 25 using a different (but unitarily equivalent) representation that appears in the quantization of the harmonic oscillator. The Groenewold–van Hove theorem implies that we cannot find a unitary representation of a larger group of canonical transformations extending this one of the Heisenberg and metaplectic groups.

17.4 Quantization and symmetries

The Schrödinger representation is thus a Lie algebra representation providing observables corresponding to elements of the Lie algebras $\mathfrak{h}_{2d+1}$ (linear combinations of Q_j and P_k) and $\mathfrak{sp}(2d, \mathbf{R})$ (linear combinations of degree two combinations of Q_j and P_k). The observables that commute with the Hamiltonian operator H will make up a Lie algebra of symmetries of the quantum system and will take energy eigenstates to energy eigenstates of the same energy. Some examples for the physical case of $d = 3$ are:

- The group $\mathbf{R}^3$ of translations in coordinate space is a subgroup of the Heisenberg group and has a Lie algebra representation as linear combinations of the operators $-iP_j$. If the Hamiltonian is position-independent, for instance the free particle case of

$$H = \frac{1}{2m}(P_1^2 + P_2^2 + P_3^2)$$

then the momentum operators correspond to symmetries. Note that the position operators Q_j do not commute with this Hamiltonian, and so do not correspond to a symmetry of the dynamics.

- The group $SO(3)$ of spatial rotations is a subgroup of $Sp(6, \mathbf{R})$, with $\mathfrak{so}(3) \subset \mathfrak{sp}(6, \mathbf{R})$ given by the quadratic polynomials in equation 16.14 for A an antisymmetric matrix. Quantizing, the operators

$$-i(Q_2 P_3 - Q_3 P_2), \quad -i(Q_3 P_1 - Q_1 P_3), \quad -i(Q_1 P_2 - Q_2 P_1)$$

provide a basis for a Lie algebra representation of $\mathfrak{so}(3)$. This phenomenon will be studied in detail in Chapter 19.2 where we will find that for the Schrödinger representation on position space wavefunctions, these are the same operators that were studied in Chapter 8 under the name $\rho'(l_j)$. They will be symmetries of rotationally invariant Hamiltonians, for instance the free particle as above, or the particle in a potential

$$H = \frac{1}{2m}(P_1^2 + P_2^2 + P_3^2) + V(Q_1, Q_2, Q_3)$$

when the potential only depends on the combination $Q_1^2 + Q_2^2 + Q_3^2$.

17.5   More general notions of quantization

The definition given here of quantization using the Schrödinger representation of $\mathfrak{h}_{2d+1}$ only allows the construction of a quantum system based on a classical phase space for the linear case of $M = \mathbf{R}^{2d}$. For other sorts of classical systems, one needs other methods to get a corresponding quantum system. One possible approach is the path integral method, which starts with a choice of configuration space and Lagrangian and will be discussed in Chapter 35.

Digression. *The name "geometric quantization" refers to attempt to generalize quantization to the case of any symplectic manifold M, starting with the idea of prequantization (see equation 15.8). This gives a representation of the Lie algebra of functions on M on a space of sections of a line bundle with connection ∇, with ∇ a connection with curvature ω, where ω is the symplectic form on M. One then has to deal with two problems:*

- *The space of all functions on M is far too big, allowing states localized in both position and coordinate variables in the case $M = \mathbf{R}^{2d}$. One needs some way to cut down this space to something like a space of functions depending on only half the variables (e.g., just the positions,* *or just the momenta). This requires finding an appropriate choice of a so-called "polarization" that will accomplish this.*

- *To get an inner product on the space of states, one needs to introduce a twist by a "square root" of a certain line bundle, something called the "metaplectic correction."*

For more details, see, for instance, [41] or [103].

Geometric quantization focuses on finding an appropriate state space. Another general method, the method of "deformation quantization," focuses instead on the algebra of operators, with a quantization given by finding an appropriate non-commutative algebra that is in some sense a deformation of a commutative algebra of functions. To first order the deformation in the product law is determined by the Poisson bracket.

Starting with any Lie algebra $\mathfrak{g}$, in principle 15.14 can be used to get a Poisson bracket on functions on the dual space $\mathfrak{g}^*$, and then one can take the quantization of this to be the algebra of operators known as the universal enveloping algebra $U(\mathfrak{g})$. This will in general have many different irreducible representations and corresponding possible quantum state spaces. The co-adjoint orbit philosophy posits an approximate matching between orbits in $\mathfrak{g}^*$ under the dual of the adjoint representation (which are symplectic manifolds) and irreducible representations. Geometric quantization provides one possible method for trying to associate representations to orbits. For more details, see [51].

None of the general methods of quantization are fully satisfactory, with each running into problems in certain cases, or not providing a construction with all the properties that one would want.

17.6 For further reading

Just about all quantum mechanics textbooks contain some version of the discussion here of canonical quantization starting with classical mechanical systems in Hamiltonian form. For discussions of quantization from the point of view of representation theory, see [8] and chapters 14–16 of [37]. For a detailed discussion of the Heisenberg group and Lie algebra, together with their representation theory, also see chapter 2 of [51].

Chapter 18
Semi-direct Products

The theory of a free particle is largely determined by its group of symmetries, the group of symmetries of three-dimensional space, a group which includes a subgroup $\mathbf{R}^3$ of spatial translations, and a subgroup $SO(3)$ of rotations. The second subgroup acts non-trivially on the first, since the direction of a translation is rotated by an element of $SO(3)$. In later chapters dealing with special relativity, these groups get enlarged to include a fourth dimension, time, and the theory of a free particle will again be determined by the action of these groups, now on space–time, not just space. In chapters 15 and 16, we studied two groups acting on phase space: the Heisenberg group H_{2d+1} and the symplectic group $Sp(2d, \mathbf{R})$. In this situation also, the second group acts non-trivially on the first by automorphisms (see 16.19).

This situation of two groups, with one acting on the other by automorphisms, allows one to construct a new sort of product of the two groups, called the semi-direct product, and this will be the topic for this chapter. The general theory of such a construction will be given, but our interest will be in certain specific examples: the semi-direct product of $\mathbf{R}^3$ and $SO(3)$, the semi-direct product of H_{2d+1} and $Sp(2d, \mathbf{R})$, and the Poincaré group (which will be discussed later, in chapter 42). This chapter will just be concerned with the groups and their Lie algebras, with their representations the topics of later chapters (19, 20, and 42).

18.1 An example: the Euclidean group

Given two groups G' and G'', a product group is formed by taking pairs of elements $(g', g'') \in G' \times G''$. However, when the two groups act on the same space, but elements of G' and G'' do not commute, a different sort of product

© Peter Woit 2017
P. Woit, *Quantum Theory, Groups and Representations*,
DOI 10.1007/978-3-319-64612-1_18

group is needed to describe the group action. As an example, consider the case of pairs $(\mathbf{a}_2, R_2)$ of elements $\mathbf{a}_2 \in \mathbf{R}^3$ and $R_2 \in SO(3)$, acting on $\mathbf{R}^3$ by translation and rotation

$$\mathbf{v} \rightarrow (\mathbf{a}_2, R_2) \cdot \mathbf{v} = \mathbf{a}_2 + R_2 \mathbf{v}$$

If one then acts on the result with $(\mathbf{a}_1, R_1)$, one gets

$$(\mathbf{a}_1, R_1) \cdot ((\mathbf{a}_2, R_2) \cdot \mathbf{v}) = (\mathbf{a}_1, R_1) \cdot (\mathbf{a}_2 + R_2 \mathbf{v}) = \mathbf{a}_1 + R_1 \mathbf{a}_2 + R_1 R_2 \mathbf{v}$$

Note that this is not what one would get if one took the product group law on $\mathbf{R}^3 \times SO(3)$, since then the action of $(\mathbf{a}_1, R_1)(\mathbf{a}_2, R_2)$ on $\mathbf{R}^3$ would be

$$\mathbf{v} \rightarrow \mathbf{a}_1 + \mathbf{a}_2 + R_1 R_2 \mathbf{v}$$

To get the correct group action on $\mathbf{R}^3$, one needs to take $\mathbf{R}^3 \times SO(3)$ not with the product group law, but instead with the group law

$$(\mathbf{a}_1, R_1)(\mathbf{a}_2, R_2) = (\mathbf{a}_1 + R_1 \mathbf{a}_2, R_1 R_2)$$

This group law differs from the standard product law by a term $R_1 \mathbf{a}_2$, which is the result of $R_1 \in SO(3)$ acting non-trivially on $\mathbf{a}_2 \in \mathbf{R}^3$. We will denote the set $\mathbf{R}^3 \times SO(3)$ with this group law by

$$\mathbf{R}^3 \rtimes SO(3)$$

This is the group of orientation-preserving transformations of $\mathbf{R}^3$ preserving the standard inner product.

The same construction works in arbitrary dimensions, where one has:

Definition (Euclidean group). *The Euclidean group $E(d)$ (sometimes written $ISO(d)$ for "inhomogeneous" rotation group) in dimension d is the product of the translation and rotation groups of $\mathbf{R}^d$ as a set, with multiplication law*

$$(\mathbf{a}_1, R_1)(\mathbf{a}_2, R_2) = (\mathbf{a}_1 + R_1 \mathbf{a}_2, R_1 R_2)$$

(where $\mathbf{a}_j \in \mathbf{R}^d, R_j \in SO(d)$) and can be denoted by

$$\mathbf{R}^d \rtimes SO(d)$$

$E(d)$ can also be written as a matrix group, taking it to be the subgroup of $GL(d+1, \mathbf{R})$ of matrices of the form (R is a d by d orthogonal matrix, $\mathbf{a}$ a d-dimensional column vector)

$$\begin{pmatrix} R & \mathbf{a} \\ \mathbf{0} & 1 \end{pmatrix}$$

One gets the multiplication law for $E(d)$ from matrix multiplication since

$$\begin{pmatrix} R_1 & \mathbf{a}_1 \\ \mathbf{0} & 1 \end{pmatrix}\begin{pmatrix} R_2 & \mathbf{a}_2 \\ \mathbf{0} & 1 \end{pmatrix} = \begin{pmatrix} R_1 R_2 & \mathbf{a}_1 + R_1\mathbf{a}_2 \\ \mathbf{0} & 1 \end{pmatrix}$$

18.2   Semi-direct product groups

The Euclidean group example of the previous section can be generalized to the following:

Definition (Semi-direct product group). *Given a group K, a group N, and an action Φ of K on N by automorphisms*

$$\Phi_k : n \in N \rightarrow \Phi_k(n) \in N$$

the semi-direct product $N \rtimes K$ is the set of pairs $(n, k) \in N \times K$ with group law

$$(n_1, k_1)(n_2, k_2) = (n_1\Phi_{k_1}(n_2), k_1 k_2)$$

One can easily check that this satisfies the group axioms. The inverse is

$$(n, k)^{-1} = (\Phi_{k^{-1}}(n^{-1}), k^{-1})$$

Checking associativity, one finds

$$\begin{aligned}
((n_1, k_1)(n_2, k_2))(n_3, k_3) &= (n_1\Phi_{k_1}(n_2), k_1 k_2)(n_3, k_3)\\
&= (n_1\Phi_{k_1}(n_2)\Phi_{k_1 k_2}(n_3), k_1 k_2 k_3)\\
&= (n_1\Phi_{k_1}(n_2)\Phi_{k_1}(\Phi_{k_2}(n_3)), k_1 k_2 k_3)\\
&= (n_1\Phi_{k_1}(n_2\Phi_{k_2}(n_3)), k_1 k_2 k_3)\\
&= (n_1, k_1)(n_2\Phi_{k_2}(n_3), k_2 k_3)\\
&= (n_1, k_1)((n_2, k_2)(n_3, k_3))
\end{aligned}$$

The notation $N \rtimes K$ for this construction has the weakness of not explicitly indicating the automorphism Φ which it depends on. There may be multiple possible choices for Φ, and these will always include the trivial choice $\Phi_k = 1$ for all $k \in K$, which will give the standard product of groups.

Digression. *For those familiar with the notion of a normal subgroup, N is a normal subgroup of $N \rtimes K$. A standard notation for "N is a normal subgroup* *of G" is $N \lhd G$. The symbol $N \rtimes K$ is supposed to be a mixture of the $\times$ and $\lhd$ symbols (note that some authors define it to point in the other direction).*

The Euclidean group $E(d)$ is an example with $N = \mathbf{R}^d, K = SO(d)$. For $\mathbf{a} \in \mathbf{R}^d, R \in SO(d)$, one has

$$\Phi_R(\mathbf{a}) = R\mathbf{a}$$

In chapter 42, we will see another important example, the Poincaré group which generalizes $E(3)$ to include a time dimension, treating space and time according to the principles of special relativity.

The most important example for quantum theory is:

Definition (Jacobi group). *The Jacobi group in d dimensions is the semidirect product group*

$$G^J(d) = H_{2d+1} \rtimes Sp(2d, \mathbf{R})$$

If we write elements of the group as

$$\left(\left(\begin{pmatrix} \mathbf{c}_q \\ \mathbf{c}_p \end{pmatrix}, c \right), k \right)$$

where $k \in Sp(2d, \mathbf{R})$, then the automorphism Φ_k that defines the Jacobi group is given by the one studied in section 16.2

$$\Phi_k \left(\left(\begin{pmatrix} \mathbf{c}_q \\ \mathbf{c}_p \end{pmatrix}, c \right) \right) = \left(k \begin{pmatrix} \mathbf{c}_q \\ \mathbf{c}_p \end{pmatrix}, c \right) \tag{18.1}$$

Note that the Euclidean group $E(d)$ is a subgroup of the Jacobi group $G^J(d)$, the subgroup of elements of the form

$$\left(\left(\begin{pmatrix} \mathbf{0} \\ \mathbf{c}_p \end{pmatrix}, 0 \right), \begin{pmatrix} R & \mathbf{0} \\ \mathbf{0} & R \end{pmatrix} \right)$$

where $R \in SO(d)$. The

$$\left(\begin{pmatrix} \mathbf{0} \\ \mathbf{c}_p \end{pmatrix}, 0 \right) \subset H_{2d+1}$$

make up the group $\mathbf{R}^d$ of translations in the q_j coordinates, and the

$$k = \begin{pmatrix} R & \mathbf{0} \\ \mathbf{0} & R \end{pmatrix} \subset Sp(2d, \mathbf{R})$$

are symplectic transformations since

$$\Omega\left(k\begin{pmatrix}\mathbf{c}_q\\\mathbf{c}_p\end{pmatrix},k\begin{pmatrix}\mathbf{c}'_q\\\mathbf{c}'_p\end{pmatrix}\right) = R\mathbf{c}_q \cdot R\mathbf{c}'_p - R\mathbf{c}_p \cdot R\mathbf{c}'_q$$

$$= \mathbf{c}_q \cdot \mathbf{c}'_p - \mathbf{c}_p \cdot \mathbf{c}'_q$$

$$= \Omega\left(\begin{pmatrix}\mathbf{c}_q\\\mathbf{c}_p\end{pmatrix},\begin{pmatrix}\mathbf{c}'_q\\\mathbf{c}'_p\end{pmatrix}\right)$$

(R is orthogonal so preserves dot products).

18.3 Semi-direct product Lie algebras

We have seen that semi-direct product Lie groups can be constructed by taking a product $N \times K$ of Lie groups as a set, and imposing a group multiplication law that uses an action of K on N by automorphisms. In a similar manner, semi-direct product Lie algebras $\mathfrak{n} \rtimes \mathfrak{k}$ can be constructed by taking the direct sum of $\mathfrak{n}$ and $\mathfrak{k}$ as vector spaces, and defining a Lie bracket that uses an action of $\mathfrak{k}$ on $\mathfrak{n}$ by derivations (the infinitesimal version of automorphisms, see equation 16.17).

Considering first the example $E(d) = \mathbf{R}^d \rtimes SO(d)$, recall that elements $E(d)$ can be written in the form

$$\begin{pmatrix} R & \mathbf{a} \\ \mathbf{0} & 1 \end{pmatrix}$$

for $R \in SO(d)$ and $\mathbf{a} \in \mathbf{R}^d$. The tangent space to this group at the identity will be given by matrices of the form

$$\begin{pmatrix} X & \mathbf{a} \\ \mathbf{0} & 0 \end{pmatrix}$$

where X is an antisymmetric d by d matrix and $\mathbf{a} \in \mathbf{R}^d$. Exponentiating such matrices will give elements of $E(d)$.

The Lie bracket is then given by the matrix commutator

$$\left[\begin{pmatrix} X_1 & \mathbf{a}_1 \\ \mathbf{0} & 0 \end{pmatrix},\begin{pmatrix} X_2 & \mathbf{a}_2 \\ \mathbf{0} & 0 \end{pmatrix}\right] = \begin{pmatrix} [X_1,X_2] & X_1\mathbf{a}_2 - X_2\mathbf{a}_1 \\ \mathbf{0} & 0 \end{pmatrix} \qquad (18.2)$$

We see that the Lie algebra of $E(d)$ will be given by taking the sum of $\mathbf{R}^d$ (the Lie algebra of $\mathbf{R}^d$) and $\mathfrak{so}(d)$, with elements pairs $(\mathbf{a}, X)$ with $\mathbf{a} \in \mathbf{R}^d$ and X an antisymmetric d by d matrix. The infinitesimal version of the rotation action of $SO(d)$ on $\mathbf{R}^d$ by automorphisms

$$\Phi_R(a) = Ra$$

is

$$\frac{d}{dt}\Phi_{e^{tX}}(\mathbf{a})_{|t=0} = \frac{d}{dt}(e^{tX}\mathbf{a})_{|t=0} = X\mathbf{a}$$

Just in terms of such pairs, the Lie bracket can be written

$$[(\mathbf{a}_1, X_1), (\mathbf{a}_2, X_2)] = (X_1\mathbf{a}_2 - X_2\mathbf{a}_1, [X_1, X_2])$$

We can define in general:

Definition (Semi-direct product Lie algebra). *Given Lie algebras $\mathfrak{k}$ and $\mathfrak{n}$, and an action of elements $Y \in \mathfrak{k}$ on $\mathfrak{n}$ by derivations*

$$X \in \mathfrak{n} \to Y \cdot X \in \mathfrak{n}$$

the semi-direct product $\mathfrak{n} \rtimes \mathfrak{k}$ is the set of pairs $(X, Y) \in \mathfrak{n} \oplus \mathfrak{k}$ with the Lie bracket

$$[(X_1, Y_1), (X_2, Y_2)] = ([X_1, X_2] + Y_1 \cdot X_2 - Y_2 \cdot X_1, [Y_1, Y_2])$$

One can easily see that in the special case of the Lie algebra of $E(d)$ this agrees with the construction above.

In section 16.1.2 we studied the Lie algebra of all polynomials of degree at most two in d-dimensional phase space coordinates q_j, p_j, with the Poisson bracket as Lie bracket. There we found two Lie subalgebras, the degree zero and one polynomials (isomorphic to $\mathfrak{h}_{2d+1}$), and the homogeneous degree two polynomials (isomorphic to $\mathfrak{sp}(2d, \mathbf{R})$) with the second subalgebra acting on the first by derivations as in equation 16.22.

Recall from chapter 16 that elements of this Lie algebra can also be written as pairs

$$\left(\left(\begin{pmatrix}\mathbf{c}_q \\ \mathbf{c}_p\end{pmatrix}, c\right), L\right)$$

of elements in $\mathfrak{h}_{2d+1}$ and $\mathfrak{sp}(2d, \mathbf{R})$, with this pair corresponding to the polynomial

$$\mu_L + \mathbf{c}_q \cdot \mathbf{q} + \mathbf{c}_p \cdot \mathbf{p} + c$$

In terms of such pairs, the Lie bracket is given by

$$\left[\left(\left(\begin{pmatrix}\mathbf{c}_q \\ \mathbf{c}_p\end{pmatrix}, c\right), L\right), \left(\left(\begin{pmatrix}\mathbf{c}_q' \\ \mathbf{c}_p'\end{pmatrix}, c\right), L'\right)\right] =$$
$$\left(\left(L\begin{pmatrix}\mathbf{c}_q' \\ \mathbf{c}_p'\end{pmatrix} - L'\begin{pmatrix}\mathbf{c}_q \\ \mathbf{c}_p\end{pmatrix}, \Omega\left(\begin{pmatrix}\mathbf{c}_q \\ \mathbf{c}_p\end{pmatrix}, \begin{pmatrix}\mathbf{c}_q' \\ \mathbf{c}_p'\end{pmatrix}\right)\right), [L, L']\right)$$

which satisfies the definition above and defines the semi-direct product Lie algebra

$$\mathfrak{g}^J(d) = \mathfrak{h}_{2d+1} \rtimes \mathfrak{sp}(2d, \mathbf{R})$$

The fact that this is the Lie algebra of the semi-direct product group

$$G^J(d) = H_{2d+1} \rtimes Sp(2d, \mathbf{R})$$

follows from the discussion in section 16.2.

The Lie algebra of $E(d)$ will be a sub-Lie algebra of $\mathfrak{g}^J(d)$, consisting of elements of the form

$$\left(\left(\begin{pmatrix} \mathbf{0} \\ \mathbf{c}_p \end{pmatrix}, 0 \right), \begin{pmatrix} X & \mathbf{0} \\ \mathbf{0} & X \end{pmatrix} \right)$$

where X is an antisymmetric d by d matrix.

Digression. *Just as $E(d)$ can be identified with a group of $d+1$ by $d+1$ matrices, the Jacobi group $G_J(d)$ is also a matrix group and one can in principle work with it and its Lie algebra using usual matrix methods. The construction is slightly complicated and represents elements of $G_J(d)$ as matrices in $Sp(2d+1, \mathbf{R})$. See section 8.5 of [9] for details of the $d = 1$ case.*

18.4 For further reading

Semi-direct products are not commonly covered in detail in either physics or mathematics textbooks, with the exception of the case of the Poincaré group of special relativity, which will be discussed in chapter 42. Some textbooks that do cover the subject include section 3.8 of [84], chapter 6 of [39] and [9].

Chapter 19
The Quantum Free Particle as a Representation of the Euclidean Group

The quantum theory of a free particle is intimately connected to the representation theory of the group of symmetries of space and time. This is well known for relativistic theories, where it is the representation theory of the Poincaré group that is relevant, a topic that will be discussed in chapter 42. It is less well known that even in the non-relativistic case, the Euclidean group $E(3)$ of symmetries of space plays a similar role, with irreducible representations of $E(3)$ corresponding to free particle quantum theories for a fixed value of the energy. In this chapter we will examine this phenomenon, for both two and three spatial dimensions.

The Euclidean groups $E(2)$ and $E(3)$ in two and three dimensions act on phase space by a Hamiltonian group action. The corresponding moment maps (momenta p_j for translations, angular momenta l_j for rotations) Poisson-commute with the free particle Hamiltonian giving symmetries of the theory. The quantum free particle theory then provides a construction of unitary representations of the Euclidean group, with the space of states of a fixed energy giving an irreducible representation. The momentum operators P_j give the infinitesimal action of translations on the state space, while angular momentum operators L_j give the infinitesimal rotation action (there will be only one angular momentum operator L in two dimensions since the dimension of $SO(2)$ is one, three in three dimensions since the dimension of $SO(3)$ is three).

The Hamiltonian of the free particle is proportional to the operator $|\mathbf{P}|^2$. This is a quadratic operator that commutes with the action of all the elements of the Lie algebra of the Euclidean group, and so is a Casimir operator playing an analogous role to that of the $SO(3)$ Casimir operator $|\mathbf{L}|^2$ of section 8.4. Irreducible representations will be labeled by the eigenvalue of this operator, which in this case will be proportional to the energy. In the Schrödinger representation, where the P_j are differentiation operators, this will be a second-order differential operator, and the eigenvalue equation will be a second-order differential equation (the time-independent Schrödinger equation).

© Peter Woit 2017
P. Woit, *Quantum Theory, Groups and Representations*,
DOI 10.1007/978-3-319-64612-1_19

Using the Fourier transform, the space of solutions of the Schrödinger equation of fixed energy becomes something much easier to analyze, the space of functions (or, more generally, distributions) on momentum space supported only on the subspace of momenta of a fixed length. In the case of $E(2)$ this is just a circle, whereas for $E(3)$ it is a sphere. In both cases, for each radius one gets an irreducible representation in this manner.

In the case of $E(3)$, other classes of irreducible representations can be constructed. This can be done by introducing multicomponent wavefunctions, with a new action of the rotation group $SO(3)$. A second Casimir operator is available in this case, and irreducible representations are eigenfunctions of this operator in the space of wavefunctions of fixed energy. The eigenvalues of this second Casimir operator turn out to be proportional to an integer, the "helicity" of the representation.

19.1 The quantum free particle and representations of $E(2)$

We will begin for simplicity with the case of two spatial dimensions. Recall from chapter 18 that the Euclidean group $E(d)$ is a subgroup of the Jacobi group $G^J(d) = H_{2d+1} \rtimes Sp(2d, \mathbf{R})$. For the case $d = 2$, the translations $\mathbf{R}^2$ are a subgroup of the Heisenberg group H_5 (translations in q_1, q_2) and the rotations are a subgroup $SO(2) \subset Sp(4, \mathbf{R})$ (simultaneous rotations of q_1, q_2 and p_1, p_2). The Lie algebra of $E(2)$ is a sub-Lie algebra of the Lie algebra $\mathfrak{g}^J(2)$ of polynomials in q_1, q_2, p_1, p_2 of degree at most two.

More specifically, a basis for the Lie algebra of $E(2)$ is given by the functions

$$l = q_1 p_2 - q_2 p_1, \quad p_1, \quad p_2$$

on the $d = 2$ phase space $M = \mathbf{R}^4$, where l is a basis for the Lie algebra $\mathfrak{so}(2)$ of rotations, p_1, p_2 a basis for the Lie algebra $\mathbf{R}^2$ of translations. The nonzero Lie bracket relations are given by the Poisson brackets

$$\{l, p_1\} = p_2, \quad \{l, p_2\} = -p_1$$

which are the infinitesimal version of the rotation action of $SO(2)$ on $\mathbf{R}^2$. There is an isomorphism of this Lie algebra with a matrix Lie algebra of 3 by 3 matrices given by

$$l \leftrightarrow \begin{pmatrix} 0 & -1 & 0 \\ 1 & 0 & 0 \\ 0 & 0 & 0 \end{pmatrix}, \quad p_1 \leftrightarrow \begin{pmatrix} 0 & 0 & 1 \\ 0 & 0 & 0 \\ 0 & 0 & 0 \end{pmatrix}, \quad p_2 \leftrightarrow \begin{pmatrix} 0 & 0 & 0 \\ 0 & 0 & 1 \\ 0 & 0 & 0 \end{pmatrix}$$

Since we have realized the Lie algebra of $E(2)$ as a sub-Lie algebra of the Jacobi Lie algebra $\mathfrak{g}^J(2)$, quantization via the Schrödinger representation

Γ'_S provides a unitary Lie algebra representation on the state space $\mathcal{H}$ of functions of the position variables q_1, q_2. This will be given by the operators

$$\Gamma'_S(p_1) = -iP_1 = -\frac{\partial}{\partial q_1}, \quad \Gamma'_S(p_2) = -iP_1 = -\frac{\partial}{\partial q_2} \tag{19.1}$$

and

$$\Gamma'_S(l) = -iL = -i(Q_1 P_2 - Q_2 P_1) = -\left(q_1 \frac{\partial}{\partial q_2} - q_2 \frac{\partial}{\partial q_1}\right) \tag{19.2}$$

The Hamiltonian operator for the free particle is

$$H = \frac{1}{2m}(P_1^2 + P_2^2) = -\frac{1}{2m}\left(\frac{\partial^2}{\partial q_1^2} + \frac{\partial^2}{\partial q_2^2}\right)$$

and solutions to the Schrödinger equation can be found by solving the eigenvalue equation

$$H\psi(q_1, q_2) = -\frac{1}{2m}\left(\frac{\partial^2}{\partial q_1^2} + \frac{\partial^2}{\partial q_2^2}\right)\psi(q_1, q_2) = E\psi(q_1, q_2)$$

The operators L, P_1, P_2 commute with H and so provide a representation of the Lie algebra of $E(2)$ on the space of wavefunctions of energy E.

This construction of irreducible representations of $E(2)$ is similar in spirit to the construction of irreducible representations of $SO(3)$ in section 8.4. There the Casimir operator L^2 commuted with the $SO(3)$ action, and gave a differential operator on functions on the sphere whose eigenfunctions were spaces of dimension $2l + 1$ with eigenvalue $l(l + 1)$, for l nonnegative and integral. For $E(2)$ the quadratic function $p_1^2 + p_2^2$ Poisson commutes with l, p_1, p_2. After quantization,

$$|\mathbf{P}|^2 = P_1^2 + P_2^2$$

is a second-order differential operator which commutes with L, P_1, P_2. This operator has infinite dimensional eigenspaces that each carry an irreducible representation of $E(2)$. They are characterized by a nonnegative eigenvalue that has physical interpretation as $2mE$ where m, E are the mass and energy of a free quantum particle moving in two spatial dimensions.

From our discussion of the free particle in chapter 11 we see that, in momentum space, solutions of the Schrödinger equation are given by

$$\widetilde{\psi}(\mathbf{p}, t) = e^{-i\frac{1}{2m}|\mathbf{p}|^2 t}\widetilde{\psi}(\mathbf{p}, 0)$$

and are parametrized by distributions

$$\widetilde{\psi}(\mathbf{p}, 0) \equiv \widetilde{\psi}(\mathbf{p})$$

on $\mathbf{R}^2$. These will have well-defined momentum $\mathbf{p}_0$ when

$$\widetilde{\psi}(\mathbf{p}) = \delta(\mathbf{p} - \mathbf{p}_0)$$

The position space wavefunctions can be recovered from the Fourier inversion formula

$$\psi(\mathbf{q}, t) = \frac{1}{2\pi} \int_{\mathbf{R}^2} e^{i\mathbf{p}\cdot\mathbf{q}} \widetilde{\psi}(\mathbf{p}, t) d^2\mathbf{p}$$

Since, in the momentum space representation, the momentum operator is the multiplication operator

$$\mathbf{P}\widetilde{\psi}(\mathbf{p}) = \mathbf{p}\widetilde{\psi}(\mathbf{p})$$

an eigenfunction for the Hamiltonian with eigenvalue E will satisfy

$$\left(\frac{|\mathbf{p}|^2}{2m} - E\right) \widetilde{\psi}(\mathbf{p}) = 0$$

$\widetilde{\psi}(\mathbf{p})$ can only be nonzero if $E = \frac{|\mathbf{p}|^2}{2m}$, so free particle solutions of energy E will thus be parametrized by distributions that are supported on the circle

$$|\mathbf{p}|^2 = 2mE$$

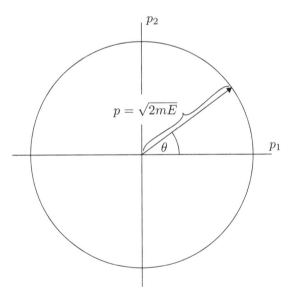

Figure 19.1: Parametrizing free particle solutions of Schrödinger's equation via distributions supported on a circle in momentum space.

Going to polar coordinates $\mathbf{p} = (p\cos\theta, p\sin\theta)$, such solutions are given by distributions $\widetilde{\psi}(\mathbf{p})$ of the form

$$\widetilde{\psi}(\mathbf{p}) = \widetilde{\psi}_E(\theta)\delta(p^2 - 2mE)$$

depending on two variables θ, p. To put this delta-function in a more useful form, recall the discussion leading to equation 11.9 and note that for $p \approx \sqrt{2mE}$ one has the linear approximation

$$p^2 - 2mE \approx 2\sqrt{2mE}(p - \sqrt{2mE})$$

so one has the equality of distributions

$$\delta(p^2 - 2mE) = \frac{1}{2\sqrt{2mE}}\delta(p - \sqrt{2mE})$$

In the one-dimensional case (see equation 11.10), we found that the space of solutions of energy E was parametrized by two complex numbers, corresponding to the two possible momenta $\pm\sqrt{2mE}$. In this two-dimensional case, the space of such solutions will be infinite dimensional, parametrized by distributions $\widetilde{\psi}_E(\theta)$ on the circle.

It is this space of distributions $\widetilde{\psi}_E(\theta)$ on the circle of radius $\sqrt{2mE}$ that will provide an infinite dimensional representation of the group $E(2)$, one that turns out to be irreducible, although we will not show that here. The position space wavefunction corresponding to $\widetilde{\psi}_E(\theta)$ will be

$$\begin{aligned}
\psi(\mathbf{q}) &= \frac{1}{2\pi}\iint e^{i\mathbf{p}\cdot\mathbf{q}}\widetilde{\psi}_E(\theta)\delta(p^2 - 2mE)p\,dp\,d\theta \\
&= \frac{1}{2\pi}\iint e^{i\mathbf{p}\cdot\mathbf{q}}\widetilde{\psi}_E(\theta)\frac{1}{2\sqrt{2mE}}\delta(p - \sqrt{2mE})p\,dp\,d\theta \\
&= \frac{1}{4\pi}\int_0^{2\pi} e^{i\sqrt{2mE}(q_1\cos\theta + q_2\sin\theta)}\widetilde{\psi}_E(\theta)\,d\theta
\end{aligned}$$

Functions $\widetilde{\psi}_E(\theta)$ with simple behavior in θ will correspond to wavefunctions with more complicated behavior in position space. For instance, taking $\widetilde{\psi}_E(\theta) = e^{-in\theta}$ one finds that the wavefunction along the q_2 direction is given by

$$\begin{aligned}
\psi(0, q) &= \frac{1}{4\pi}\int_0^{2\pi} e^{i\sqrt{2mE}(q\sin\theta)}e^{-in\theta}\,d\theta \\
&= \frac{1}{2}J_n(\sqrt{2mE}q)
\end{aligned}$$

where J_n is the n'th Bessel function.

Equations 19.1 and 19.2 give the representation of the Lie algebra of $E(2)$ on wavefunctions $\psi(\mathbf{q})$. The representation of this Lie algebra on the $\tilde{\psi}_E(\theta)$ is given by the Fourier transform, and we will denote this by $\tilde{\Gamma}'_S$. Using the formula for the Fourier transform, we find that

$$\tilde{\Gamma}'_S(p_1) = -\widetilde{\frac{\partial}{\partial q_1}} = -ip_1 = -i\sqrt{2mE}\cos\theta$$

$$\tilde{\Gamma}'_S(p_2) = -\widetilde{\frac{\partial}{\partial q_2}} = -ip_2 = -i\sqrt{2mE}\sin\theta$$

are multiplication operators and taking the Fourier transform of 19.2 gives the differentiation operator

$$\tilde{\Gamma}'_S(l) = -\left(p_1\frac{\partial}{\partial p_2} - p_2\frac{\partial}{\partial p_1}\right)$$

$$= -\frac{\partial}{\partial\theta}$$

(use integration by parts to show $q_j = i\frac{\partial}{\partial p_j}$ and thus the first equality, then the chain rule for functions $f(p_1(\theta), p_2(\theta))$ for the second).

This construction of a representation of $E(2)$ starting with the Schrödinger representation gives the same result as starting with the action of $E(2)$ on configuration space, and taking the induced action on functions on $\mathbf{R}^2$ (the wavefunctions). To see this, note that $E(2)$ has elements $(\mathbf{a}, R(\phi))$ which can be written as a product $(\mathbf{a}, R(\phi)) = (\mathbf{a}, \mathbf{1})(\mathbf{0}, R(\phi))$ or in terms of matrices

$$\begin{pmatrix} \cos\phi & -\sin\phi & a_1 \\ \sin\phi & \cos\phi & a_2 \\ 0 & 0 & 1 \end{pmatrix} = \begin{pmatrix} 1 & 0 & a_1 \\ 0 & 1 & a_2 \\ 0 & 0 & 1 \end{pmatrix} \begin{pmatrix} \cos\phi & -\sin\phi & 0 \\ \sin\phi & \cos\phi & 0 \\ 0 & 0 & 1 \end{pmatrix}$$

The group has a unitary representation

$$(\mathbf{a}, R(\phi)) \rightarrow u(\mathbf{a}, R(\phi))$$

on the position space wavefunctions $\psi(\mathbf{q})$, given by the induced action on functions from the action of $E(2)$ on position space $\mathbf{R}^2$

$$u(\mathbf{a}, R(\phi))\psi(\mathbf{q}) = \psi((\mathbf{a}, R(\phi))^{-1} \cdot \mathbf{q})$$
$$= = \psi((-R(-\phi)\mathbf{a}, R(-\phi)) \cdot \mathbf{q})$$
$$= \psi(R(-\phi)(\mathbf{q} - \mathbf{a}))$$

This representation of $E(2)$ is the same as the exponentiated version of the Schrödinger representation Γ'_S of the Jacobi Lie algebra $\mathfrak{g}^J(2)$, restricted to the Lie algebra of $E(2)$. This can be seen by considering the action of translations as the exponential of the Lie algebra representation operators $\Gamma'_S(p_j) = -iP_j$

$$u(\mathbf{a}, 1)\psi(\mathbf{q}) = e^{-i(a_1 P_1 + a_2 P_2)}\psi(\mathbf{q}) = \psi(\mathbf{q} - \mathbf{a})$$

and the action of rotations as the exponential of the $\Gamma'_S(l) = -iL$

$$u(\mathbf{0}, R(\phi))\psi(\mathbf{q}) = e^{-i\phi L}\psi(\mathbf{q}) = \psi(R(-\phi)\mathbf{q})$$

One also has a Fourier-transformed version $\widetilde{u}$ of this representation, with translations now acting by multiplication operators on the $\widetilde{\psi}_E$

$$\widetilde{u}(\mathbf{a}, 1)\widetilde{\psi}_E(\theta) = e^{-i(\mathbf{a}\cdot\mathbf{p})}\widetilde{\psi}_E(\theta) = e^{-i\sqrt{2mE}(a_1 \cos\theta + a_2 \sin\theta)}\widetilde{\psi}_E(\theta) \qquad (19.3)$$

and rotations acting by rotating the circle in momentum space

$$\widetilde{u}(\mathbf{0}, R(\phi))\widetilde{\psi}_E(\theta) = \widetilde{\psi}_E(\theta - \phi) \qquad (19.4)$$

Although we will not prove it here, the representations constructed in this way provide essentially all the unitary irreducible representations of $E(2)$, parametrized by a real number $E > 0$. The only other ones are those on which the translations act trivially, corresponding to $E = 0$, with $SO(2)$ acting as an irreducible representation. We have seen that such $SO(2)$ representations are one dimensional, and characterized by an integer, the weight. We thus get another class of $E(2)$ irreducible representations, labeled by an integer, but they are just one-dimensional representations on $\mathbf{C}$.

19.2 The case of $E(3)$

In the physical case of three spatial dimensions, the state space of the theory of a quantum free particle is again a Euclidean group representation, with the same relationship to the Schrödinger representation as in two spatial dimensions. The main difference is that the rotation group is now three dimensional and non-commutative, so instead of the single Lie algebra basis element l we have three of them, satisfying Poisson bracket relations that are the Lie algebra relations of $\mathfrak{so}(3)$

$$\{l_1, l_2\} = l_3, \ \{l_2, l_3\} = l_1, \ \{l_3, l_1\} = l_2$$

The p_j give the other three basis elements of the Lie algebra of $E(3)$. They commute among themselves, and the action of rotations on vectors provides the rest of the non-trivial Poisson bracket relations

$$\{l_1, p_2\} = p_3, \quad \{l_1, p_3\} = -p_2$$

$$\{l_2, p_1\} = -p_3, \quad \{l_2, p_3\} = p_1$$

$$\{l_3, p_1\} = p_2, \quad \{l_3, p_2\} = -p_1$$

An isomorphism of this Lie algebra with a Lie algebra of matrices is given by

$$l_1 \leftrightarrow \begin{pmatrix} 0 & 0 & 0 & 0 \\ 0 & 0 & -1 & 0 \\ 0 & 1 & 0 & 0 \\ 0 & 0 & 0 & 0 \end{pmatrix} \quad l_2 \leftrightarrow \begin{pmatrix} 0 & 0 & 1 & 0 \\ 0 & 0 & 0 & 0 \\ -1 & 0 & 0 & 0 \\ 0 & 0 & 0 & 0 \end{pmatrix} \quad l_3 \leftrightarrow \begin{pmatrix} 0 & -1 & 0 & 0 \\ 1 & 0 & 0 & 0 \\ 0 & 0 & 0 & 0 \\ 0 & 0 & 0 & 0 \end{pmatrix}$$

$$p_1 \leftrightarrow \begin{pmatrix} 0 & 0 & 0 & 1 \\ 0 & 0 & 0 & 0 \\ 0 & 0 & 0 & 0 \\ 0 & 0 & 0 & 0 \end{pmatrix} \quad p_2 \leftrightarrow \begin{pmatrix} 0 & 0 & 0 & 0 \\ 0 & 0 & 0 & 1 \\ 0 & 0 & 0 & 0 \\ 0 & 0 & 0 & 0 \end{pmatrix} \quad p_3 \leftrightarrow \begin{pmatrix} 0 & 0 & 0 & 0 \\ 0 & 0 & 0 & 0 \\ 0 & 0 & 0 & 1 \\ 0 & 0 & 0 & 0 \end{pmatrix}$$

The l_j are quadratic functions in the q_j, p_j, given by the classical mechanical expression for the angular momentum

$$\mathbf{l} = \mathbf{q} \times \mathbf{p}$$

or in components

$$l_1 = q_2 p_3 - q_3 p_2, \quad l_2 = q_3 p_1 - q_1 p_3, \quad l_3 = q_1 p_2 - q_2 p_1$$

The Euclidean group $E(3)$ is a subgroup of the Jacobi group $G^J(3)$ in the same way as in two dimensions, and, just as in the $E(2)$ case, exponentiating the Schrödinger representation Γ'_S

$$\Gamma'_S(l_1) = -iL_1 = -i(Q_2 P_3 - Q_3 P_2) = -\left(q_2 \frac{\partial}{\partial q_3} - q_3 \frac{\partial}{\partial q_2} \right)$$

$$\Gamma'_S(l_2) = -iL_2 = -i(Q_3 P_1 - Q_1 P_3) = -\left(q_3 \frac{\partial}{\partial q_1} - q_1 \frac{\partial}{\partial q_3} \right)$$

$$\Gamma'_S(l_3) = -iL_3 = -i(Q_1 P_2 - Q_2 P_1) = -\left(q_1 \frac{\partial}{\partial q_2} - q_2 \frac{\partial}{\partial q_1} \right)$$

$$\Gamma'_S(p_j) = -iP_j = -\frac{\partial}{\partial q_j}$$

provides a representation of $E(3)$.

As in the $E(2)$ case, the above Lie algebra representation is just the infinitesimal version of the action of $E(3)$ on functions induced from its action on position space $\mathbf{R}^3$. Given an element $g = (\mathbf{a}, R) \in E(3)$, we have a unitary transformation on wavefunctions

$$u(\mathbf{a}, R)\psi(\mathbf{q}) = \psi(g^{-1} \cdot \mathbf{q}) = \psi(R^{-1}(\mathbf{q} - \mathbf{a}))$$

Such group elements g will be a product of a translation and a rotation, and treating these separately, the unitary transformations u are exponentials of the Lie algebra actions above, with

$$u(\mathbf{a}, 1)\psi(\mathbf{q}) = e^{-i(a_1 P_1 + a_2 P_2 + a_3 P_3)}\psi(\mathbf{q}) = \psi(\mathbf{q} - \mathbf{a})$$

for a translation by $\mathbf{a}$, and

$$u(\mathbf{0}, R(\phi, \mathbf{e}_j))\psi(\mathbf{q}) = e^{-i\phi L_j}\psi(\mathbf{q}) = \psi(R(-\phi, \mathbf{e}_j)\mathbf{q})$$

for $R(\phi, \mathbf{e}_j)$ a rotation about the j-axis by angle ϕ.

This representation of $E(3)$ on wavefunctions is reducible, since in terms of momentum eigenstates, rotations will only take eigenstates with one value of the momentum to those with another value of the same norm-squared. We can get an irreducible representation by using the Casimir operator $P_1^2 + P_2^2 + P_3^2$, which commutes with all elements in the Lie algebra of $E(3)$. The Casimir operator will act on an irreducible representation as a scalar, and the representation will be characterized by that scalar. The Casimir operator is just $2m$ times the Hamiltonian

$$H = \frac{1}{2m}(P_1^2 + P_2^2 + P_3^2)$$

and so the constant characterizing an irreducible will be the energy $2mE$. Our irreducible representation will be on the space of solutions of the time-independent Schrödinger equation

$$\frac{1}{2m}(P_1^2 + P_2^2 + P_3^2)\psi(\mathbf{q}) = -\frac{1}{2m}\left(\frac{\partial^2}{\partial q_1^2} + \frac{\partial^2}{\partial q_2^2} + \frac{\partial^2}{\partial q_3^2}\right)\psi(\mathbf{q}) = E\psi(\mathbf{q})$$

Using the Fourier transform

$$\psi(\mathbf{q}) = \frac{1}{(2\pi)^{\frac{3}{2}}} \int_{\mathbf{R}^3} e^{i\mathbf{p}\cdot\mathbf{q}}\widetilde{\psi}(\mathbf{p})d^3\mathbf{p}$$

the time-independent Schrödinger equation becomes

$$\left(\frac{|\mathbf{p}|^2}{2m} - E\right)\widetilde{\psi}(\mathbf{p}) = 0$$

and we have distributional solutions

$$\widetilde{\psi}(\mathbf{p}) = \widetilde{\psi}_E(\mathbf{p})\delta(|\mathbf{p}|^2 - 2mE)$$

characterized by distributions $\widetilde{\psi}_E(\mathbf{p})$ defined on the sphere $|\mathbf{p}|^2 = 2mE$.

Such complex-valued distributions on the sphere of radius $\sqrt{2mE}$ provide a Fourier-transformed version $\widetilde{u}$ of the irreducible representation of $E(3)$. Here the action of the group $E(3)$ is by

$$\widetilde{u}(\mathbf{a}, 1)\widetilde{\psi}_E(\mathbf{p}) = e^{-i(\mathbf{a}\cdot\mathbf{p})}\widetilde{\psi}_E(\mathbf{p})$$

for translations, by

$$\widetilde{u}(\mathbf{0}, R)\widetilde{\psi}_E(\mathbf{p}) = \widetilde{\psi}_E(R^{-1}\mathbf{p})$$

for rotations, and by

$$\widetilde{u}(\mathbf{a}, R)\widetilde{\psi}_E(\mathbf{p}) = \widetilde{u}(\mathbf{a}, 1)\widetilde{u}(\mathbf{0}, R)\widetilde{\psi}_E(\mathbf{p}) = e^{-i\mathbf{a}\cdot R^{-1}\mathbf{p}}\widetilde{\psi}_E(R^{-1}\mathbf{p})$$

for a general element.

19.3   Other representations of $E(3)$

For the case of $E(3)$, besides the representations parametrized by $E > 0$ constructed above, as in the $E(2)$ case there are finite dimensional representations where the translation subgroup of $E(3)$ acts trivially. Such irreducible representations are just the spin-s representations $(\rho_s, \mathbf{C}^{2s+1})$ of $SO(3)$ for $s = 0, 1, 2, \ldots$.

$E(3)$ has some structure not seen in the $E(2)$ case, which can be used to construct new classes of infinite dimensional irreducible representations. This can be seen from two different points of view:

- There is a second Casimir operator which one can show commutes with the $E(3)$ action, given by

$$\mathbf{L} \cdot \mathbf{P} = L_1 P_1 + L_2 P_2 + L_3 P_3$$

- The group $SO(3)$ acts on momentum vectors by rotation, with orbit of the group action the sphere of momentum vectors of fixed energy $E > 0$.

This is the sphere on which the Fourier transform of the wavefunctions in the representation is supported. Unlike the corresponding circle in the $E(2)$ case, here there is a non-trivial subgroup of the rotation group $SO(3)$ which leaves a given momentum vector invariant. This is the $SO(2) \subset SO(3)$ subgroup of rotations about the axis determined by the momentum vector, and it is different for different points in momentum space.

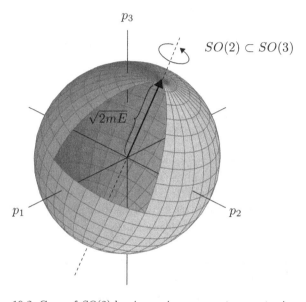

Figure 19.2: Copy of $SO(2)$ leaving a given momentum vector invariant.

For single-component wavefunctions, a straightforward computation shows that the second Casimir operator $\mathbf{L} \cdot \mathbf{P}$ acts as zero. By introducing wavefunctions with several components, together with an action of $SO(3)$ that mixes the components, it turns out that one can get new irreducible representations, with a nonzero value of the second Casimir corresponding to a non-trivial weight of the action of the $SO(2)$ of rotations about the momentum vector.

Such multiple-component wavefunctions can be constructed as representations of $E(3)$ by taking the tensor product of our irreducible representation on wavefunctions of energy E (call this $\mathcal{H}_E$) and the finite dimensional irreducible representation $\mathbf{C}^{2s+1}$

$$\mathcal{H}_E \otimes \mathbf{C}^{2s+1}$$

The Lie algebra representation operators for the translation part of $E(3)$ act as momentum operators on $\mathcal{H}_E$ and as 0 on $\mathbf{C}^{2s+1}$. For the $SO(3)$ part of

$E(3)$, we get angular momentum operators that can be written as

$$J_j = L_j + S_j \equiv L_j \otimes \mathbf{1} + \mathbf{1} \otimes S_j$$

where L_j acts on $\mathcal{H}_E$ and $S_j = \rho'(l_j)$ acts on $\mathbf{C}^{2s+1}$.

This tensor product representation will not be irreducible, but its irreducible components can be found by taking the eigenspaces of the second Casimir operator, which will now be

$$\mathbf{J} \cdot \mathbf{P}$$

We will not work out the details of this here (although details can be found in chapter 34 for the case $s = \frac{1}{2}$, where the half-integrality corresponds to replacing $SO(3)$ by its double cover $Spin(3)$). What happens is that the tensor product breaks up into irreducibles as

$$\mathcal{H}_E \otimes \mathbf{C}^{2s+1} = \bigoplus_{n=-s}^{n=s} \mathcal{H}_{E,n}$$

where n is an integer taking values from $-s$ to s that is called the "helicity." $\mathcal{H}_{E,n}$ is the subspace of the tensor product on which the first Casimir $|\mathbf{P}|^2$ takes the value $2mE$, and the second Casimir $\mathbf{J} \cdot \mathbf{P}$ takes the value np, where $p = \sqrt{2mE}$. The physical interpretation of the helicity is that it is the component of angular momentum along the axis given by the momentum vector. The helicity can also be thought of as the weight of the action of the $SO(2)$ subgroup of $SO(3)$ corresponding to rotations about the axis of the momentum vector.

Choosing $E > 0$ and $n \in \mathbf{Z}$, the representations on $\mathcal{H}_{E,n}$ (which we have constructed using some s such that $s \geq |n|$) give all possible irreducible representations of $E(3)$. The representation spaces have a physical interpretation as the state space for a free quantum particle of energy E which carries an "internal" quantized angular momentum about its direction of motion, given by the helicity.

19.4   For further reading

The angular momentum operators are a standard topic in every quantum mechanics textbook, see, for example, chapter 12 of [81]. The characterization here of free particle wavefunctions at fixed energy as giving irreducible representations of the Euclidean group is not so conventional, but it is just a non-relativistic version of the conventional description of relativistic quantum particles in terms of representations of the Poincaré group (see chapter 42).

In the Poincaré group case the analog of the $E(3)$ irreducible representations of nonzero energy E and helicity n considered here will be irreducible representations labeled by a nonzero mass and an irreducible representation of $SO(3)$ (the spin). In the Poincaré group case, for massless particles one will again see representations labeled by an integral helicity (an irreducible representation of $SO(2)$), but there is no analog of such massless particles in the $E(3)$ case.

For more details about representations of $E(2)$ and $E(3)$, see [94] or [97] (which is based on [92]).

Chapter 20
Representations of Semi-direct Products

In this chapter, we will examine some aspects of representations of semi-direct products, in particular for the case of the Jacobi group and its Lie algebra, as well as the case of $N \rtimes K$, for N commutative. The latter case includes the Euclidean groups $E(d)$, as well as the Poincaré group which will come into play once we introduce special relativity.

The Schrödinger representation provides a unitary representation of the Heisenberg group, one that carries extra structure arising from the fact that the symplectic group acts on the Heisenberg group by automorphisms. Each such automorphism takes a given construction of the Schrödinger representation to a unitarily equivalent one, providing an operator on the state space called an "intertwining operator." These intertwining operators will give (up to a phase factor) a representation of the symplectic group. Up to the problem of the phase factor, the Schrödinger representation in this way extends to a representation of the full Jacobi group. To explicitly find the phase factor, one can start with the Lie algebra representation, where the $\mathfrak{sp}(2d, \mathbf{R})$ action is given by quantizing quadratic functions on phase space. It turns out that, for a finite dimensional phase space, exponentiating the Lie algebra representation gives a group representation up to sign, which can be turned into a true representation by taking a double cover (called $Mp(2d, \mathbf{R})$) of $Sp(2d, \mathbf{R})$.

In later chapters, we will find that many groups acting on quantum systems can be understood as subgroups of this $Mp(2d, \mathbf{R})$, with the corresponding observables arising as the quadratic combinations of momentum and position operators determined by the moment map.

The Euclidean group $E(d)$ is a subgroup of the Jacobi group, and we saw in Chapter 19 how some of its representations can be understood by restricting the Schrödinger representation to this subgroup. More generally, this is an example of a semi-direct product $N \rtimes K$ with N commutative. In such cases, irreducible representations can be characterized in terms of the action of K on irreducible representations of N, together with the irreducible representations of certain subgroups of K.

© Peter Woit 2017

P. Woit, *Quantum Theory, Groups and Representations*,
DOI 10.1007/978-3-319-64612-1_20

The reader should be warned that much of the material included in this chapter is motivated not by its applications to non-relativistic quantum mechanics, a context in which such an abstract point of view is not particularly helpful. The motivation for this material is provided by more complicated cases in relativistic quantum field theory, but it seems worthwhile to first see how these ideas work in a simpler context. In particular, the discussion of representations of $N \rtimes K$ for N commutative is motivated by the case of the Poincaré group (see chapter 42). The treatment of intertwining operators is motivated by the way symmetry groups act on quantum fields (a topic which will first appear in chapter 38).

20.1 Intertwining operators and the metaplectic representation

For a general semi-direct product $N \rtimes K$ with non-commutative N, the representation theory can be quite complicated. For the Jacobi group case though, it turns out that things simplify dramatically because of the Stone–von Neumann theorem which says that, up to unitary equivalence, we only have one irreducible representation of $N = H_{2d+1}$.

In the general case, recall that for each $k \in K$ the definition of the semidirect product comes with an automorphism $\Phi_k : N \to N$ satisfying $\Phi_{k_1 k_2} = \Phi_{k_1} \Phi_{k_2}$. Given a representation π of N, for each k we can define a new representation π_k of N by first acting with Φ_k:

$$\pi_k(n) = \pi(\Phi_k(n))$$

In the special case of the Heisenberg group and Schrödinger representation Γ_S, we can do this for each $k \in K = Sp(2d, \mathbf{R})$, defining a new representation by

$$\Gamma_{S,k}(n) = \Gamma_S(\Phi_k(n))$$

The Stone–von Neumann theorem assures us that these must all be unitarily equivalent, so there must exist unitary operators U_k satisfying

$$\Gamma_{S,k} = U_k \Gamma_S U_k^{-1} = \Gamma_S(\Phi_k(n))$$

We will generally work with the Lie algebra version Γ'_S of the Schrödinger representation, for which the same argument applies: We expect to be able to find unitary operators U_k relating Lie algebra representations Γ'_S and $\Gamma'_{S,k}$ by

$$\Gamma'_{S,k}(X) = U_k \Gamma'_S(X) U_k^{-1} = \Gamma'_S(\Phi_k(X)) \tag{20.1}$$

where X is in the Heisenberg Lie algebra, and k acts by automorphism Φ_k on this Lie algebra.

Operators like U_k that relate two representations are called "intertwining operators":

Definition (Intertwining operator). *If $(\pi_1, V_1), (\pi_2, V_2)$ are two representations of a group G, an intertwining operator between these two representations is an operator U such that*

$$\pi_2(g)U = U\pi_1(g) \quad \forall g \in G$$

In our case, $V_1 = V_2$ is the Schrödinger representation state space $\mathcal{H}$ and $U_k : \mathcal{H} \to \mathcal{H}$ is an intertwining operator between Γ_S and $\Gamma_{S,k}$ for each $k \in Sp(2d, \mathbf{R})$.

Since

$$\Gamma_{S,k_1 k_2} = U_{k_1 k_2} \Gamma_S U_{k_1 k_2}^{-1}$$

one might expect that the U_k should satisfy the group homomorphism property

$$U_{k_1 k_2} = U_{k_1} U_{k_2}$$

and give us a representation of the group $Sp(2d, \mathbf{R})$ on $\mathcal{H}$. This is what would follow from the general principle that a group action on the classical phase space after quantization becomes a unitary representation on the quantum state space.

The problem with this argument is that the U_k are not uniquely defined. Schur's lemma tells us that since the representation on $\mathcal{H}$ is irreducible, the operators commuting with the representation operators are just the complex scalars. These give a phase ambiguity in the definition of the unitary operators U_k, which then give a representation of $Sp(2d, \mathbf{R})$ on $\mathcal{H}$ only up to a phase, i.e.,

$$U_{k_1 k_2} = U_{k_1} U_{k_2} e^{i\varphi(k_1, k_2)}$$

for some real-valued function φ of pairs of group elements. In terms of corresponding Lie algebra representation operators U'_L, this ambiguity appears as an unknown constant times the identity operator.

The question then arises whether the phases of the U_k can be chosen so as to satisfy the homomorphism property (i.e., can phases be chosen so that $\varphi(k_1, k_2) = N 2\pi$ for N integral?). It turns out that this cannot quite be done, since N may have to be half-integral, giving the homomorphism property only up to a sign. Just as in the $SO(d)$ case where a similar sign ambiguity showed the need to go to a double cover $Spin(d)$ to get a true representation, here one needs to go to a double cover of $Sp(2d, \mathbf{R})$, called the metaplectic group $Mp(2d, \mathbf{R})$. The nature of this sign ambiguity and double cover is quite subtle, and unlike for the $Spin(d)$ case, we will not provide an actual construction of $Mp(2d, \mathbf{R})$. For more details on this, see [56] or [37]. In section 20.3.2, we will show by computation one aspect of the double cover.

Since this is just a sign ambiguity, it does not appear infinitesimally: The ambiguous constants in the Lie algebra representation operators can be chosen so that the Lie algebra homomorphism property is satisfied. However, this will no longer necessarily be true for infinite dimensional phase spaces, a situation that is described as an "anomaly" in the symmetry. This phenomenon will be examined in more detail in chapter 39.

20.2 Constructing intertwining operators

The method we will use to construct the intertwining operators U_k is to find a solution to the differentiated version of equation 20.1 and then get U_k by exponentiation. Differentiating 20.1 for $k = e^{tL}$ at $t = 0$ gives

$$[U'_L, \Gamma'_S(X)] = \Gamma'_S(L \cdot X) \tag{20.2}$$

where

$$L \cdot X = \frac{d}{dt} \Phi_{e^{tL}}(X)_{|t=0}$$

and we have used equation 5.1 on the left-hand side.

In terms of Q_j and P_j operators, which are i times the $\Gamma'_S(X)$ for X a basis vector q_j, p_j, equation 20.2 is

$$\left[U'_L, \begin{pmatrix} \mathbf{Q} \\ \mathbf{P} \end{pmatrix} \right] = L^T \begin{pmatrix} \mathbf{Q} \\ \mathbf{P} \end{pmatrix} \tag{20.3}$$

We can find U'_L by quantizing the moment map function μ_L, which satisfies

$$\left\{ \mu_L, \begin{pmatrix} \mathbf{q} \\ \mathbf{p} \end{pmatrix} \right\} = L^T \begin{pmatrix} \mathbf{q} \\ \mathbf{p} \end{pmatrix} \tag{20.4}$$

Recall from 16.1.2 that the μ_L are quadratic polynomials in the q_j, p_j. We saw in section 17.3 that the Schrödinger representation Γ'_S could be extended from the Heisenberg Lie algebra to the symplectic Lie algebra, by taking a product of Q_j, P_j operators corresponding to the product in μ_L. The ambiguity in ordering for non-commuting operators is resolved by quantizing $q_j p_j$ using

$$\Gamma'_S(q_j p_j) = -\frac{i}{2}(Q_j P_j + P_j Q_j)$$

We thus take

$$U'_L = \Gamma'_S(\mu_L)$$

and this will satisfy 20.3 as desired. It will also satisfy the Lie algebra homo-morphism property

$$[U'_{L_1}, U'_{L_2}] = U'_{[L_1, L_2]} \tag{20.5}$$

If one shifts U'_L by a constant operator, it will still satisfy 20.3, but in general will no longer satisfy 20.5. Exponentiating this U'_L will give us our U_k, and thus the intertwining operators that we want.

This method will be our fundamental way of producing observable opera-tors. They come from an action of a Lie group on phase space preserving the Poisson bracket. For an element L of the Lie algebra, we first use the moment map to find μ_L, the classical observable, then quantize to get the quantum observable U'_L.

20.3 Explicit calculations

As a balance to the abstract discussion so far in this chapter, in this section we will work out explicitly what happens for some simple examples of subgroups of $Sp(2d, \mathbf{R})$ acting on phase space. They are chosen because of important later applications, but also because the calculations are quite simple, while demonstrating some of the phenomena that occur. The general story of how to explicitly construct the full metaplectic representation is quite a bit more complex. These calculations will also make clear the conventions being chosen and show the basic structure of what the quadratic operators corresponding to actions of subgroups of the symplectic group look like, a structure that will reappear in the much more complicated infinite dimensional quantum field theory examples we will come to later.

20.3.1 The $SO(2)$ action by rotations of the plane for $d = 2$

In the case $d = 2$, one can consider the $SO(2)$ group which acts as the group of rotations of the configuration space $\mathbf{R}^2$, with a simultaneous rotation of the momentum space. This leaves invariant the Poisson bracket and so is a subgroup of $Sp(4, \mathbf{R})$ (this is just the $SO(2)$ subgroup of $E(2)$ studied in section 19.1).

From the discussion in section 16.2, this $SO(2)$ acts by automorphisms on the Heisenberg group H_5 and Lie algebra $\mathfrak{h}_5$, both of which can be identified with $\mathcal{M} \oplus \mathbf{R}$, by an action leaving invariant the $\mathbf{R}$ component. The group $SO(2)$ acts on $c_{q_1} q_1 + c_{q_2} q_2 + c_{p_1} p_1 + c_{p_2} p_2 \in \mathcal{M}$ by

$$
\begin{pmatrix} c_{q_1} \\ c_{q_2} \\ c_{p_1} \\ c_{p_2} \end{pmatrix} \to g \begin{pmatrix} c_{q_1} \\ c_{q_2} \\ c_{p_1} \\ c_{p_2} \end{pmatrix} = \begin{pmatrix} \cos\theta & -\sin\theta & 0 & 0 \\ \sin\theta & \cos\theta & 0 & 0 \\ 0 & 0 & \cos\theta & -\sin\theta \\ 0 & 0 & \sin\theta & \cos\theta \end{pmatrix} \begin{pmatrix} c_{q_1} \\ c_{q_2} \\ c_{p_1} \\ c_{p_2} \end{pmatrix}
$$

so $g = e^{\theta L}$ where $L \in \mathfrak{sp}(4, \mathbf{R})$ is given by

$$
L = \begin{pmatrix} 0 & -1 & 0 & 0 \\ 1 & 0 & 0 & 0 \\ 0 & 0 & 0 & -1 \\ 0 & 0 & 1 & 0 \end{pmatrix}
$$

L acts on phase space coordinate functions by

$$
\begin{pmatrix} q_1 \\ q_2 \\ p_1 \\ p_2 \end{pmatrix} \to L^T \begin{pmatrix} q_1 \\ q_2 \\ p_1 \\ p_2 \end{pmatrix} = \begin{pmatrix} q_2 \\ -q_1 \\ p_2 \\ -p_1 \end{pmatrix}
$$

By equation 16.22, with

$$
A = \begin{pmatrix} 0 & -1 \\ 1 & 0 \end{pmatrix}, \quad B = C = 0
$$

the quadratic function μ_L that satisfies

$$
\left\{ \mu_L, \begin{pmatrix} q_1 \\ q_2 \\ p_1 \\ p_2 \end{pmatrix} \right\} = L^T \begin{pmatrix} q_1 \\ q_2 \\ p_1 \\ p_2 \end{pmatrix} = \begin{pmatrix} 0 & 1 & 0 & 0 \\ -1 & 0 & 0 & 0 \\ 0 & 0 & 0 & 1 \\ 0 & 0 & -1 & 0 \end{pmatrix} \begin{pmatrix} q_1 \\ q_2 \\ p_1 \\ p_2 \end{pmatrix} = \begin{pmatrix} q_2 \\ -q_1 \\ p_2 \\ -p_1 \end{pmatrix}
$$

is

$$
\mu_L = -\mathbf{q} \cdot \begin{pmatrix} 0 & -1 \\ 1 & 0 \end{pmatrix} \mathbf{p} = q_1 p_2 - q_2 p_1
$$

which is just the formula for the angular momentum

$$
l = q_1 p_2 - q_2 p_1
$$

in $d = 2$.

Quantization gives a representation of the Lie algebra $\mathfrak{so}(2)$ with

$$U'_L = -i(Q_1 P_2 - Q_2 P_1)$$

satisfying

$$\left[U'_L, \begin{pmatrix} Q_1 \\ Q_2 \end{pmatrix}\right] = \begin{pmatrix} Q_2 \\ -Q_1 \end{pmatrix}, \quad \left[U'_L, \begin{pmatrix} P_1 \\ P_2 \end{pmatrix}\right] = \begin{pmatrix} P_2 \\ -P_1 \end{pmatrix}$$

Exponentiating gives a representation of $SO(2)$

$$U_{e^{\theta L}} = e^{-i\theta(Q_1 P_2 - Q_2 P_1)}$$

with conjugation by $U_{e^{\theta L}}$ rotating linear combinations of the Q_1, Q_2 (or the P_1, P_2) each by an angle θ.

$$U_{e^{\theta L}}(c_{q_1} Q_1 + c_{q_2} Q_2) U_{e^{\theta L}}^{-1} = c'_{q_1} Q_1 + c'_{q_2} Q_2$$

where

$$\begin{pmatrix} c'_{q_1} \\ c'_{q_2} \end{pmatrix} = \begin{pmatrix} \cos\theta & -\sin\theta \\ \sin\theta & \cos\theta \end{pmatrix} \begin{pmatrix} c_{q_1} \\ c_{q_2} \end{pmatrix}$$

These representation operators are exactly the ones found in (section 19.1) the discussion of the representation of $E(2)$ corresponding to the quantum free particle in two dimensions. There we saw that on position space wavefunctions this is just the representation induced from rotations of the position space. It also comes from the Schrödinger representation, by taking a specific quadratic combination of the Q_j, P_j operators, the one corresponding to the quadratic function l. Note that there is no ordering ambiguity in this case since one does not multiply q_j and p_j with the same value of j. Also note that for this $SO(2)$ the double cover is trivial: As one goes around the circle in $SO(2)$ once, the operator $U_{e^{\theta L}}$ is well defined and returns to its initial value. As far as this subgroup of $Sp(4, \mathbf{R})$ is concerned, there is no need to consider the double cover $Mp(4, \mathbf{R})$ to get a well-defined representation.

The case of the group $SO(2) \subset Sp(4, \mathbf{R})$ can be generalized to a larger subgroup, the group $GL(2, \mathbf{R})$ of all invertible linear transformations of $\mathbf{R}^2$, performed simultaneously on position and momentum space. Replacing the matrix L by

$$\begin{pmatrix} A & \mathbf{0} \\ \mathbf{0} & A \end{pmatrix}$$

for A any real 2 by 2 matrix

$$A = \begin{pmatrix} a_{11} & a_{12} \\ a_{21} & a_{22} \end{pmatrix}$$

we get an action of the group $GL(2, \mathbf{R}) \subset Sp(4, \mathbf{R})$ on $\mathcal{M}$, and after quantization a Lie algebra representation

$$U'_A = i \begin{pmatrix} Q_1 & Q_2 \end{pmatrix} \begin{pmatrix} a_{11} & a_{12} \\ a_{21} & a_{22} \end{pmatrix} \begin{pmatrix} P_1 \\ P_2 \end{pmatrix}$$

which will satisfy

$$\left[U'_A, \begin{pmatrix} Q_1 \\ Q_2 \end{pmatrix} \right] = -A \begin{pmatrix} Q_1 \\ Q_2 \end{pmatrix}, \quad \left[U'_A, \begin{pmatrix} P_1 \\ P_2 \end{pmatrix} \right] = A^T \begin{pmatrix} P_1 \\ P_2 \end{pmatrix}$$

Note that the action of A on the momentum operators is the dual of the action on the position operators. Only in the case of an orthogonal action (the $SO(2)$ earlier) are these the same, with $A^T = -A$.

20.3.2   An $SO(2)$ action on the $d = 1$ phase space

Another sort of $SO(2)$ action on phase space provides a $d = 1$ example that mixes position and momentum coordinates. This will lead to quite non-trivial intertwining operators, with an action on wavefunctions that does not come about as an induced action from a group action on position space. This example will be studied in much greater detail when we get to the theory of the quantum harmonic oscillator, beginning with chapter 22. Such a physical system is periodic in time, so the usual group $\mathbf{R}$ of time translations becomes this $SO(2)$, with the corresponding intertwining operators giving the time evolution of the quantum states.

In this case, $d = 1$ and one has elements $g \in SO(2) \subset Sp(2, \mathbf{R})$ acting on $c_q q + c_p p \in \mathcal{M}$ by

$$\begin{pmatrix} c_q \\ c_p \end{pmatrix} \to g \begin{pmatrix} c_q \\ c_p \end{pmatrix} = \begin{pmatrix} \cos\theta & \sin\theta \\ -\sin\theta & \cos\theta \end{pmatrix} \begin{pmatrix} c_q \\ c_p \end{pmatrix}$$

so

$$g = e^{\theta L}$$

where

$$L = \begin{pmatrix} 0 & 1 \\ -1 & 0 \end{pmatrix}$$

(Note that for such phase space rotations, we are making the opposite choice for convention of the positive direction of rotation, clockwise instead of counterclockwise).

To find the intertwining operators, we first find the quadratic function μ_L in q, p that satisfies

$$\left\{\mu_L, \begin{pmatrix} q \\ p \end{pmatrix}\right\} = L^T \begin{pmatrix} q \\ p \end{pmatrix} = \begin{pmatrix} -p \\ q \end{pmatrix}$$

By equation 16.7, this is

$$\mu_L = \frac{1}{2} \begin{pmatrix} q & p \end{pmatrix} \begin{pmatrix} 1 & 0 \\ 0 & 1 \end{pmatrix} \begin{pmatrix} q \\ p \end{pmatrix} = \frac{1}{2}(q^2 + p^2)$$

Quantizing μ_L using the Schrödinger representation Γ'_S, one has a unitary Lie algebra representation U' of $\mathfrak{so}(2)$ with

$$U'_L = -\frac{i}{2}(Q^2 + P^2)$$

satisfying

$$\left[U'_L, \begin{pmatrix} Q \\ P \end{pmatrix}\right] = \begin{pmatrix} -P \\ Q \end{pmatrix} \qquad (20.6)$$

and intertwining operators

$$U_g = e^{\theta U'_L} = e^{-i\frac{\theta}{2}(Q^2 + P^2)}$$

These give a representation of $SO(2)$ only up to a sign, for reasons mentioned in section 17.1 that will be discussed in more detail in chapter 24.

Conjugating the Heisenberg Lie algebra representation operators by the unitary operators U_g intertwines the representations corresponding to rotations of the phase space plane by an angle θ

$$e^{-i\frac{\theta}{2}(Q^2 + P^2)} \begin{pmatrix} Q \\ P \end{pmatrix} e^{i\frac{\theta}{2}(Q^2 + P^2)} = \begin{pmatrix} \cos\theta & -\sin\theta \\ \sin\theta & \cos\theta \end{pmatrix} \begin{pmatrix} Q \\ P \end{pmatrix} \qquad (20.7)$$

Note that this is a different calculation than in the spin case where we also constructed a double cover of $SO(2)$. Despite the different context ($SO(2)$ acting on an infinite dimensional state space), again one sees an aspect of the double cover here, as either U_g or $-U_g$ will give the same $SO(2)$ rotation action on the operators Q, P (while each having a different action on the states, to be worked out in chapter 24).

In our discussion here we have blithely assumed that the operator U'_L can be exponentiated, but doing so turns out to be quite non-trivial. As remarked earlier, this representation on wavefunctions does not arise as the induced action from an action on position space. U'_L is (up to a factor of i) the Hamiltonian operator for a quantum system that is not translation invariant. It involves quadratic operators in both Q and P, so neither the position space nor momentum space version of the Schrödinger representation can be used to make the operator a multiplication operator. Further details of the construction of the needed exponentiated operators will be given in section 23.4.

20.3.3 The Fourier transform as an intertwining operator

For another indication of the non-trivial nature of the intertwining operators of section 20.3.2, note that a group element g acting by a $\frac{\pi}{2}$ rotation of the $d = 1$ phase space interchanges the role of q and p. It turns out that the corresponding intertwining operator U_g is closely related to the Fourier transform $\mathcal{F}$. Up to a phase factor $e^{i\frac{\pi}{4}}$, Fourier transformation is just such an intertwining operator: We will see in section 23.4 that, acting on wave-functions,

$$U_{e^{\frac{\pi}{2}L}} = e^{i\frac{\pi}{4}}\mathcal{F}$$

Squaring this gives

$$U_{e^{\pi L}} = (U_{e^{\frac{\pi}{2}L}})^2 = i\mathcal{F}^2$$

and we know from the definition of $\mathcal{F}$ and Fourier inversion that

$$\mathcal{F}^2\psi(q) = \psi(-q)$$

The non-trivial double cover here appears because

$$U_{e^{2\pi L}} = -\mathcal{F}^4 = -\mathbf{1}$$

which takes a wavefunction $\psi(q)$ to $-\psi(q)$.

20.3.4 An R action on the $d = 1$ phase space

For another sort of example in $d = 1$, consider the action of a subgroup $\mathbf{R} \subset SL(2, \mathbf{R})$ on $d = 1$ phase space by

$$\begin{pmatrix} c_q \\ c_p \end{pmatrix} \rightarrow g\begin{pmatrix} c_q \\ c_p \end{pmatrix} = \begin{pmatrix} e^r & 0 \\ 0 & e^{-r} \end{pmatrix}\begin{pmatrix} c_q \\ c_p \end{pmatrix}$$

where

$$g = e^{rL}, \ L = \begin{pmatrix} 1 & 0 \\ 0 & -1 \end{pmatrix}$$

Now, by equation 16.7 the moment map will be

$$\mu_L = \frac{1}{2} \begin{pmatrix} q & p \end{pmatrix} \begin{pmatrix} 0 & -1 \\ -1 & 0 \end{pmatrix} \begin{pmatrix} q \\ p \end{pmatrix} = -qp$$

which satisfies

$$\left\{ \mu_L, \begin{pmatrix} q \\ p \end{pmatrix} \right\} = \begin{pmatrix} q \\ -p \end{pmatrix}$$

Quantization gives intertwining operators by

$$U'_L = -\frac{i}{2}(QP + PQ), \ U_g = e^{rU'_L} = e^{-\frac{ir}{2}(QP+PQ)}$$

These act on operators Q and P by a simple rescaling

$$e^{-\frac{ir}{2}(QP+PQ)} \begin{pmatrix} Q \\ P \end{pmatrix} e^{\frac{ir}{2}(QP+PQ)} = \begin{pmatrix} e^r & 0 \\ 0 & e^{-r} \end{pmatrix} \begin{pmatrix} Q \\ P \end{pmatrix}$$

Note that in the Schrödinger representation

$$-i\frac{1}{2}(QP + PQ) = -i(QP - \frac{i}{2})\mathbf{1} = -q\frac{d}{dq} - \frac{1}{2}\mathbf{1}$$

The operator will have as eigenfunctions

$$\psi(q) = q^c$$

with eigenvalues $-c - \frac{1}{2}$. Such states are far from square-integrable, but do have an interpretation as distributions on the Schwartz space.

20.4 Representations of $N \rtimes K$, N commutative

The representation theory of semi-direct products $N \rtimes K$ will in general be rather complicated. However, when N is commutative things simplify considerably, and in this section, we will survey some of the general features of this case. The special cases of the Euclidean groups in 2 and 3 dimensions were covered in chapter 19, and the Poincaré group case will be discussed in chapter 42.

For a general commutative group N, one does not have the simplifying feature of the Heisenberg group, the uniqueness of its irreducible representation. On the other hand, while N will have many irreducible representations, they are all one dimensional. As a result, the set of representations of N acquires its own group structure, also commutative, and one can define:

Definition (Character group). *For N a commutative group, let $\widehat{N}$ be the set of characters of N, i.e., functions*

$$\alpha : N \to \mathbf{C}$$

that satisfy the homomorphism property

$$\alpha(n_1 n_2) = \alpha(n_1)\alpha(n_2)$$

The elements of $\widehat{N}$ form a group, with multiplication

$$(\alpha_1 \alpha_2)(n) = \alpha_1(n)\alpha_2(n)$$

When N is a Lie group, we will restrict attention to characters that are differentiable functions on N. We only will actually need the case $N = \mathbf{R}^d$, where we have already seen that the differentiable irreducible representations are one dimensional and given by

$$\alpha_{\mathbf{p}}(\mathbf{a}) = e^{i\mathbf{p}\cdot\mathbf{a}}$$

where $\mathbf{a} \in N$. So the character group in this case is $\widehat{N} = \mathbf{R}^d$, with elements labeled by the vector $\mathbf{p}$.

For a semi-direct product $N \rtimes K$, we will have an automorphism Φ_k of N for each $k \in K$. From this action on N, we get an induced action on functions on N, in particular on elements of $\widehat{N}$, by

$$\widehat{\Phi}_k : \alpha \in \widehat{N} \to \widehat{\Phi}_k(\alpha) \in \widehat{N}$$

where $\widehat{\Phi}_k(\alpha)$ is the element of $\widehat{N}$ satisfying

$$\widehat{\Phi}_k(\alpha)(n) = \alpha(\Phi_k^{-1}(n))$$

For the case of $N = \mathbf{R}^d$, we have

$$\widehat{\Phi}_k(\alpha_{\mathbf{p}})(\mathbf{a}) = e^{i\mathbf{p}\cdot\Phi_k^{-1}(\mathbf{a})} = e^{i(\Phi_k^{-1})^T(\mathbf{p})\cdot\mathbf{a}}$$

so

$$\widehat{\Phi}_k(\alpha_{\mathbf{p}}) = \alpha_{(\Phi_k^{-1})^T(\mathbf{p})}$$

When K acts by orthogonal transformations on $N = \mathbf{R}^d$, $\Phi_k^T = \Phi_k^{-1}$ so

$$\widehat{\Phi}_k(\alpha_{\mathbf{p}}) = \alpha_{\Phi_k(\mathbf{p})}$$

To analyze representations (π, V) of $N \rtimes K$, one can begin by restricting attention to the N action, decomposing V into subspaces V_α where N acts according to α. $v \in V$ is in the subspace V_α when

$$\pi(n, \mathbf{1})v = \alpha(n)v$$

Acting by K will take this subspace to another one according to

Theorem.

$$v \in V_\alpha \implies \pi(\mathbf{0}, k)v \in V_{\widehat{\Phi}_k(\alpha)}$$

Proof. Using the definition of the semi-direct product in chapter 18, one can show that the group multiplication satisfies

$$(\mathbf{0}, k^{-1})(n, \mathbf{1})(\mathbf{0}, k) = (\Phi_{k^{-1}}(n), \mathbf{1})$$

Using this, one has

$$
\begin{aligned}
\pi(n, \mathbf{1})\pi(\mathbf{0}, k)v &= \pi(\mathbf{0}, k)\pi(\mathbf{0}, k^{-1})\pi(n, \mathbf{1})\pi(\mathbf{0}, k)v \\
&= \pi(\mathbf{0}, k)\pi(\Phi_{k^{-1}}(n), \mathbf{1})v \\
&= \pi(\mathbf{0}, k)\alpha(\Phi_{k^{-1}}(n))v \\
&= \widehat{\Phi}_k(\alpha)(n)\pi(\mathbf{0}, k)v
\end{aligned}
$$

$\square$

For each $\alpha \in \widehat{N}$, one can look at its orbit under the action of K by $\widehat{\Phi}_k$, which will give a subset $\mathcal{O}_\alpha \subset \widehat{N}$. From the above theorem, we see that if $V_\alpha \neq 0$, then we will also have $V_\beta \neq 0$ for $\beta \in \mathcal{O}_\alpha$, so one piece of information that characterizes a representation V is the set of orbits one gets in this way.

α also defines a subgroup $K_\alpha \subset K$ consisting of group elements whose action on $\widehat{N}$ leaves α invariant:

Definition (Stabilizer group or little group). *The subgroup $K_\alpha \subset K$ of elements $k \in K$ such that*

$$\widehat{\Phi}_k(\alpha) = \alpha$$

for a given $\alpha \in \widehat{N}$ is called the stabilizer subgroup (by mathematicians) or little subgroup (by physicists).

The group K_α will act on the subspace V_α, and this representation of K_α is a second piece of information that can be used to characterize a representation.

In the case of the Euclidean group $E(2)$, we found that the nonzero orbits $\mathcal{O}_\alpha$ were circles and the groups K_α were trivial. For $E(3)$, the nonzero orbits were spheres, with K_α an $SO(2)$ subgroup of $SO(3)$ (one that varies with α). In these cases, we found that our construction of representations of $E(2)$ or $E(3)$ on spaces of solutions of the single-component Schrödinger equation corresponded under Fourier transform to a representation on functions on the orbits $\mathcal{O}_\alpha$. We also found in the $E(3)$ case that using multiple-component wavefunctions gave new representations corresponding to a choice of orbit $\mathcal{O}_\alpha$ and a choice of irreducible representation of $K_\alpha = SO(2)$. We did not show this, but this construction gives an irreducible representation when a single orbit $\mathcal{O}_\alpha$ occurs (with a transitive K action), with an irreducible representation of K_α on V_α.

We will not further pursue the general theory here, but one can show that distinct irreducible representations of $N \rtimes K$ will occur for each choice of an orbit $\mathcal{O}_\alpha$ and an irreducible representation of K_α. One way to construct these representations is as the solution space of an appropriate wave equation, with the wave equation corresponding to the eigenvalue equation for a Casimir operator. In general, other "subsidiary conditions" then must be imposed to pick out a subspace of solutions that gives an irreducible representation of $N \rtimes K$; this corresponds to the existence of other Casimir operators. Another part of the general theory has to do with the question of the unitarity of representations produced in this way, which will require that one starts with an irreducible representation of K_α that is unitary.

20.5   For further reading

For more on representations of semi-direct products, see section 3.8 of [84], chapter 5 of [94], [9], and [39]. The general theory was developed by Mackey during the late 1940s and 1950s, and his lecture notes on representation theory [58] are a good source for the details of this. The point of view taken here that emphasizes constructing representations as solution spaces of differential equations where the differential operators are Casimir operators, is explained in more detail in [47].

The conventional derivation found in most physics textbooks of the operators U'_L coming from an infinitesimal group action uses Lagrangian methods and Noether's theorem. The purely Hamiltonian method used here treats configuration and momentum variables on the same footing and is useful especially in the case of group actions that mix them (such as the example of section 20.3.2) For another treatment of these operators along the lines of this chapter, see section 14 of [37].

For a concise but highly insightful discussion of the metaplectic representation, see chapters 16 and 17 in Graeme Segal's section of [14]. For a discussion of this topic emphasizing the role of the Fourier transform as an intertwining operator, see [49]. The issue of the phase factor in the intertwining operators and the metaplectic double cover will be discussed later in the context of the harmonic oscillator, using a different realization of the Heisenberg Lie algebra representation. For a discussion of this in terms of the Schrödinger representation, see part I of [56].

Chapter 21
Central Potentials and the Hydrogen Atom

When the Hamiltonian function is invariant under rotations, we then expect eigenspaces of the corresponding Hamiltonian operator to carry representations of $SO(3)$. These spaces of eigenfunctions of a given energy break up into irreducible representations of $SO(3)$, and we have seen that these are labeled by an integer $l = 0, 1, 2, \ldots$ and have dimension $2l + 1$. This can be used to find properties of the solutions of the Schrödinger equation whenever one has a rotation invariant potential energy. We will work out what happens for the case of the Coulomb potential describing the hydrogen atom. This specific case is exactly solvable because it has a second not-so-obvious $SO(3)$ symmetry, in addition to the one coming from rotations of $\mathbf{R}^3$.

21.1 Quantum particle in a central potential

In classical physics, to describe not free particles, but particles experiencing some sort of force, one just needs to add a "potential energy" term to the kinetic energy term in the expression for the energy (the Hamiltonian function). In one dimension, for potential energies that just depend on position, one has

$$h = \frac{p^2}{2m} + V(q)$$

for some function $V(q)$. In the physical case of three dimensions, this will be

$$h = \frac{1}{2m}(p_1^2 + p_2^2 + p_3^2) + V(q_1, q_2, q_3)$$

Quantizing and using the Schrödinger representation, the Hamiltonian operator for a particle moving in a potential $V(q_1, q_2, q_3)$ will be

© Peter Woit 2017
P. Woit, *Quantum Theory, Groups and Representations*,
DOI 10.1007/978-3-319-64612-1_21

$$H = \frac{1}{2m}(P_1^2 + P_2^2 + P_3^3) + V(Q_1, Q_2, Q_3)$$
$$= -\frac{\hbar^2}{2m}\left(\frac{\partial^2}{\partial q_1^2} + \frac{\partial^2}{\partial q_2^2} + \frac{\partial^2}{\partial q_3^2}\right) + V(q_1, q_2, q_3)$$
$$= -\frac{\hbar^2}{2m}\Delta + V(q_1, q_2, q_3)$$

We will be interested in so-called "central potentials," potential functions that are functions only of $q_1^2 + q_2^2 + q_3^2$, and thus only depend upon r, the radial distance to the origin. For such V, both terms in the Hamiltonian will be $SO(3)$ invariant, and eigenspaces of H will be representations of $SO(3)$.

Using the expressions for the angular momentum operators in spherical coordinates derived in chapter 8 (including equation 8.4 for the Casimir operator L^2), one can show that the Laplacian has the following expression in spherical coordinates

$$\Delta = \frac{\partial^2}{\partial r^2} + \frac{2}{r}\frac{\partial}{\partial r} - \frac{1}{r^2}L^2$$

The Casimir operator L^2 has eigenvalues $l(l+1)$ on irreducible representations of dimension $2l + 1$ (integral spin l). So, restricted to such an irreducible representation, we have

$$\Delta = \frac{\partial^2}{\partial r^2} + \frac{2}{r}\frac{\partial}{\partial r} - \frac{l(l+1)}{r^2}$$

To solve the Schrödinger equation, we want to find the eigenfunctions of H. The space of eigenfunctions of energy E will be a sum of irreducible representations of $SO(3)$, with the $SO(3)$ acting on the angular coordinates of the wavefunctions, leaving the radial coordinate invariant. To find eigenfunctions of the Hamiltonian

$$H = -\frac{\hbar^2}{2m}\Delta + V(r)$$

we can first look for functions $g_{lE}(r)$, depending on $l = 0, 1, 2, \ldots$ and the energy eigenvalue E, and satisfying

$$\left(-\frac{\hbar^2}{2m}\left(\frac{d^2}{dr^2} + \frac{2}{r}\frac{d}{dr} - \frac{l(l+1)}{r^2}\right) + V(r)\right)g_{lE}(r) = Eg_{lE}(r)$$

Turning to the angular coordinates, we have seen in chapter 8 that representations of $SO(3)$ on functions of angular coordinates can be explicitly expressed in terms of the spherical harmonic functions $Y_l^m(\theta, \phi)$, on which L^2 acts with eigenvalue $l(l+1)$. For each solution $g_{lE}(r)$, we will have the eigenvalue equation

$$Hg_{lE}(r)Y_l^m(\theta,\phi) = Eg_{lE}(r)Y_l^m(\theta,\phi)$$

and the

$$\psi(r,\theta,\phi) = g_{lE}(r)Y_l^m(\theta,\phi)$$

will span a $2l+1$-dimensional (since $m = -l, -l+1, \ldots, l-1, l$) space of energy eigenfunctions for H of eigenvalue E.

For a general potential function $V(r)$, exact solutions for the eigenvalues E and corresponding functions $g_{lE}(r)$ cannot be found in closed form. One special case where we can find such solutions is for the three-dimensional harmonic oscillator, where $V(r) = \frac{1}{2}m\omega^2 r^2$. These are much more easily found though using the creation and annihilation operator techniques to be discussed in chapter 22.

The other well known and physically very important such case is the case of a $\frac{1}{r}$ potential, called the Coulomb potential. This describes a light charged particle moving in the potential due to the electric field of a much heavier charged particle, a situation that corresponds closely to that of a hydrogen atom. In this case, we have

$$V = -\frac{e^2}{r}$$

where e is the charge of the electron, so we are looking for solutions to

$$\left(\frac{-\hbar^2}{2m}\left(\frac{d^2}{dr^2} + \frac{2}{r}\frac{d}{dr} - \frac{l(l+1)}{r^2}\right) - \frac{e^2}{r}\right)g_{lE}(r) = Eg_{lE}(r) \qquad (21.1)$$

Since on functions $f(r)$

$$\frac{d^2}{dr^2}(rf) = r\left(\frac{d^2}{dr^2} + \frac{2}{r}\frac{d}{dr}\right)f$$

multiplying both sides of equation 21.1 by r gives

$$\left(\frac{-\hbar^2}{2m}\left(\frac{d^2}{dr^2} - \frac{l(l+1)}{r^2}\right) - \frac{e^2}{r}\right)rg_{lE}(r) = Erg_{lE}(r)$$

The solutions to this equation can be found through a rather elaborate process described in most quantum mechanics textbooks, which involves looking for a power series solution. For $E \geq 0$, there are non-normalizable solutions that describe scattering phenomena that we will not study here. For $E < 0$, solutions correspond to an integer $n = 1, 2, 3, \ldots$, with $n \geq l+1$. So, for each n, we get n solutions, with $l = 0, 1, 2, \ldots, n-1$, all with the same energy

$$E_n = -\frac{me^4}{2\hbar^2 n^2}$$

A plot of the different energy eigenstates looks like this:

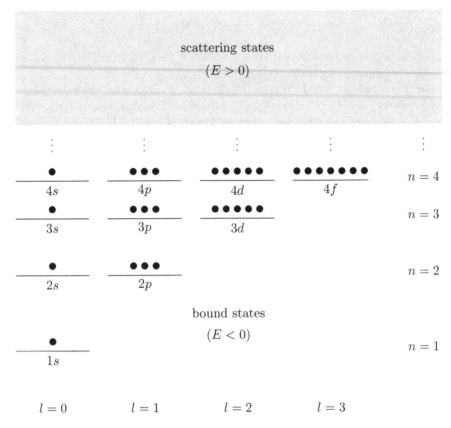

Figure 21.1: Energy eigenstates in the Coulomb potential.

    The degeneracy in the energy values leads one to suspect that there is some extra group action in the problem commuting with the Hamiltonian. If so, the eigenspaces of energy eigenfunctions will come in irreducible representations of some larger group than $SO(3)$. If the representation of the larger group is reducible when one restricts to the $SO(3)$ subgroup, giving n copies of the $SO(3)$ representation of spin l, that would explain the pattern observed here. In the next section, we will see that this is the case, and there use representation theory to derive the above formula for E_n.

    We will not go through the process of showing how to explicitly find the functions $g_{lE_n}(r)$ but just quote the result. Setting

$$a_0 = \frac{\hbar^2}{me^2}$$

(this has dimensions of length and is known as the "Bohr radius"), and defining $g_{nl}(r) = g_{lE_n}(r)$ the solutions are of the form

$$g_{nl}(r) \propto e^{-\frac{r}{na_0}} \left(\frac{2r}{na_0} \right)^l L_{n+l}^{2l+1} \left(\frac{2r}{na_0} \right)$$

where the L_{n+l}^{2l+1} are certain polynomials known as associated Laguerre polynomials.

So, finally, we have found energy eigenfunctions

$$\psi_{nlm}(r, \theta, \phi) = g_{nl}(r) Y_l^m(\theta, \phi)$$

for

$$n = 1, 2, \ldots$$

$$l = 0, 1, \ldots, n - 1$$

$$m = -l, -l + 1, \ldots, l - 1, l$$

The first few of these, properly normalized, are

$$\psi_{100} = \frac{1}{\sqrt{\pi a_0^3}} e^{-\frac{r}{a_0}}$$

(called the $1S$ state, "S" meaning $l = 0$)

$$\psi_{200} = \frac{1}{\sqrt{8\pi a_0^3}} \left(1 - \frac{r}{2a_0} \right) e^{-\frac{r}{2a_0}}$$

(called the $2S$ state), and the three-dimensional $l = 1$ (called $2P$, "P" meaning $l = 1$) states with basis elements

$$\psi_{211} = -\frac{1}{8\sqrt{\pi a_0^3}} \frac{r}{a_0} e^{-\frac{r}{2a_0}} \sin\theta e^{i\phi}$$

$$\psi_{210} = -\frac{1}{4\sqrt{2\pi a_0^3}} \frac{r}{a_0} e^{-\frac{r}{2a_0}} \cos\theta$$

$$\psi_{21-1} = \frac{1}{8\sqrt{\pi a_0^3}} \frac{r}{a_0} e^{-\frac{r}{2a_0}} \sin\theta e^{-i\phi}$$

21.2 $\mathfrak{so}(4)$ symmetry and the Coulomb potential

The Coulomb potential problem is very special in that it has an additional symmetry, of a non-obvious kind. This symmetry appears even in the classical problem, where it is responsible for the relatively simple solution one can find to the essentially identical Kepler problem. This is the problem of finding the classical trajectories for bodies orbiting around a central object exerting a gravitational force, which also has a $\frac{1}{r}$ potential.

Kepler's second law for such motion comes from conservation of angular momentum, which corresponds to the Poisson bracket relation

$$\{l_j, h\} = 0$$

Here, we will take the Coulomb version of the Hamiltonian that we need for the hydrogen atom problem

$$h = \frac{1}{2m}|\mathbf{p}|^2 - \frac{e^2}{r}$$

The relation $\{l_j, h\} = 0$ can be read in two ways:

- The Hamiltonian h is invariant under the action of the group $(SO(3))$ whose infinitesimal generators are l_j.
- The components of the angular momentum (l_j) are invariant under the action of the group ($\mathbf{R}$ of time translations) whose infinitesimal generator is h, so the angular momentum is a conserved quantity.

For this special choice of Hamiltonian, there is a different sort of conserved quantity. This quantity is, like the angular momentum, a vector, often called the Lenz (or sometimes Runge–Lenz, or even Laplace–Runge–Lenz) vector:

Definition (Lenz vector). *The Lenz vector is the vector-valued function on the phase space $\mathbf{R}^6$ given by*

$$\mathbf{w} = \frac{1}{m}(\mathbf{l} \times \mathbf{p}) + e^2 \frac{\mathbf{q}}{|\mathbf{q}|}$$

Simple manipulations of the cross product show that one has

$$\mathbf{l} \cdot \mathbf{w} = 0$$

We will not here explicitly calculate the various Poisson brackets involving the components w_j of $\mathbf{w}$, since this is a long and unilluminating calculation, but will just quote the results, which are

•
$$\{w_j, h\} = 0$$

This says that, like the angular momentum, the vector with components w_j is a conserved quantity under time evolution of the system, and its components generate symmetries of the classical system.

•
$$\{l_j, w_k\} = \epsilon_{jkl} w_l$$

These relations say that the generators of the $SO(3)$ symmetry act on w_j in the way one would expect for the components w_j of a vector in $\mathbf{R}^3$.

•
$$\{w_j, w_k\} = \epsilon_{jkl} l_l \left(\frac{-2h}{m} \right)$$

This is the most surprising relation, and it has no simple geometrical explanation (although one can change variables in the problem to try and give it one). It expresses a highly non-trivial relationship between the Hamiltonian h and the two sets of symmetries generated by the vectors $\mathbf{l}, \mathbf{w}$.

The w_j are cubic in the q and p variables, so the Groenewold–van Hove no-go theorem implies that there is no consistent way to quantize this system by finding operators W_j, L_j, H providing a representation of the Lie algebra generated by the functions w_j, l_j, h (taking Poisson brackets). Away from the locus $h = 0$ in phase space, the function h can be used to rescale the w_j, defining

$$k_j = \sqrt{\frac{-m}{2h}} w_j$$

and the functions l_j, k_j then do generate a finite dimensional Lie algebra. Quantization of the system can be performed by finding appropriate operators W_j, then rescaling them using the energy eigenvalue, giving operators L_j, K_j that provide a representation of a finite dimensional Lie algebra on energy eigenspaces.

A choice of operators W_j that will work is

$$\mathbf{W} = \frac{1}{2m}(\mathbf{L} \times \mathbf{P} - \mathbf{P} \times \mathbf{L}) + e^2 \frac{\mathbf{Q}}{|\mathbf{Q}|}$$

where the last term is the operator of multiplication by $e^2 q_j/|\mathbf{q}|$. By elaborate and unenlightening computations, the W_j can be shown to satisfy the commutation relations corresponding to the Poisson bracket relations of the w_j:

$$[W_j, H] = 0$$

$$[L_j, W_k] = i\hbar \epsilon_{jkl} W_l$$

$$[W_j, W_k] = i\hbar \epsilon_{jkl} L_l \left(-\frac{2}{m} H \right)$$

as well as

$$\mathbf{L} \cdot \mathbf{W} = \mathbf{W} \cdot \mathbf{L} = 0$$

The first of these shows that energy eigenstates will be preserved not just by the angular momentum operators L_j, but by a new set of non-trivial operators, the W_j, so will be representations of a larger Lie algebra than $\mathfrak{so}(3)$. In addition, one has the following relation between W^2, H and the Casimir operator L^2

$$W^2 = e^4 \mathbf{1} + \frac{2}{m} H(L^2 + \hbar^2 \mathbf{1}) \tag{21.2}$$

If we now restrict attention to the subspace $\mathcal{H}_E \subset \mathcal{H}$ of energy eigenstates of energy E, on this space we can define rescaled operators

$$\mathbf{K} = \sqrt{\frac{-m}{2E}} \mathbf{W}$$

On this subspace, equation 21.2 becomes the relation

$$2H(K^2 + L^2 + \hbar^2 \mathbf{1}) = -me^4 \mathbf{1}$$

and we will be able to use this to find the eigenvalues of H in terms of those of L^2 and K^2.

We will assume that $E < 0$, in which case we have the following commutation relations

$$[L_j, L_k] = i\hbar \epsilon_{jkl} L_l$$

$$[L_j, K_k] = i\hbar \epsilon_{jkl} K_l$$

$$[K_j, K_k] = i\hbar \epsilon_{jkl} L_l$$

Defining

$$\mathbf{M} = \frac{1}{2}(\mathbf{L} + \mathbf{K}), \quad \mathbf{N} = \frac{1}{2}(\mathbf{L} - \mathbf{K})$$

one has

$$[M_j, M_k] = i\hbar \epsilon_{jkl} M_l$$

$$[N_j, N_k] = i\hbar\epsilon_{jkl}N_l$$

$$[M_j, N_k] = 0$$

This shows that we have two commuting copies of $\mathfrak{so}(3)$ acting on states, spanned, respectively, by the M_j and N_j, with two corresponding Casimir operators M^2 and N^2.

Using the fact that

$$\mathbf{L}\cdot\mathbf{K} = \mathbf{K}\cdot\mathbf{L} = 0$$

one finds that

$$M^2 = N^2$$

Recall from our discussion of rotations in three dimensions that representations of $\mathfrak{so}(3) = \mathfrak{su}(2)$ correspond to representations of $Spin(3) = SU(2)$, the double cover of $SO(3)$ and the irreducible ones have dimension $2l + 1$, with l half-integral. Only for l integral does one get representations of $SO(3)$, and it is these that occur in the $SO(3)$ representation on functions on $\mathbf{R}^3$. For four dimensions, we found that $Spin(4)$, the double cover of $SO(4)$, is $SU(2) \times SU(2)$, and one thus has $\mathfrak{spin}(4) = \mathfrak{so}(4) = \mathfrak{su}(2) \times \mathfrak{su}(2) = \mathfrak{so}(3) \times \mathfrak{so}(3)$. This is exactly the Lie algebra we have found here, so one can think of the Coulomb problem at a fixed negative value of E as having an $\mathfrak{so}(4)$ symmetry. The representations that will occur can include the half-integral ones, since neither of the two $\mathfrak{so}(3)$ factors is the $\mathfrak{so}(3)$ of physical rotations in 3-space (the physical angular momentum operators are the $\mathbf{L} = \mathbf{M} + \mathbf{N}$).

The relation between the Hamiltonian and the Casimir operators M^2 and N^2 is

$$2H(K^2 + L^2 + \hbar^2\mathbf{1}) = 2H(2M^2 + 2N^2 + \hbar^2\mathbf{1}) = 2H(4M^2 + \hbar^2\mathbf{1}) = -me^4\mathbf{1}$$

On irreducible representations of $\mathfrak{so}(3)$ of spin μ, we will have

$$M^2 = \mu(\mu + 1)\hbar^2\mathbf{1}$$

for some half-integral μ, so we get the following equation for the energy eigenvalues

$$E = -\frac{-me^4}{2\hbar^2(4\mu(\mu + 1) + 1)} = -\frac{-me^4}{2\hbar^2(2\mu + 1)^2}$$

Letting $n = 2\mu + 1$, for $\mu = 0, \frac{1}{2}, 1, \ldots$ we get $n = 1, 2, 3, \ldots$ and precisely the same equation for the eigenvalues described earlier

$$E_n = -\frac{me^4}{2\hbar^2 n^2}$$

One can show that the irreducible representations of the product Lie algebra $\mathfrak{so}(3) \times \mathfrak{so}(3)$ are tensor products of irreducible representations of the factors, and in this case, the two factors in the tensor product are identical due to the equality of the Casimirs $M^2 = N^2$. The dimension of the $\mathfrak{so}(3) \times \mathfrak{so}(3)$ irreducibles is thus $(2\mu + 1)^2 = n^2$, explaining the multiplicity of states one finds at energy eigenvalue E_n.

The states with $E < 0$ are called "bound states" and correspond physically to quantized particles that remain localized near the origin. If we had chosen $E > 0$, our operators would have satisfied the relations for a different real Lie algebra, called $\mathfrak{so}(3,1)$, with quite different properties. Such states are called "scattering states," corresponding to quantized particles that behave as free particles far from the origin in the distant past, but have their momentum direction changed by the Coulomb potential (the Hamiltonian is non-translation invariant, so momentum is not conserved) as they propagate in time.

21.3 The hydrogen atom

The Coulomb potential problem provides a good description of the quantum physics of the hydrogen atom, but it is missing an important feature of that system, the fact that electrons are spin $\frac{1}{2}$ systems. To describe this, one really needs to take as space of states two-component wavefunctions

$$|\psi\rangle = \begin{pmatrix} \psi_1(\mathbf{q}) \\ \psi_2(\mathbf{q}) \end{pmatrix}$$

(or, equivalently, replace our state space $\mathcal{H}$ of wavefunctions by the tensor product $\mathcal{H} \otimes \mathbf{C}^2$) in a way that we will examine in detail in chapter 34.

The Hamiltonian operator for the hydrogen atom acts trivially on the $\mathbf{C}^2$ factor, so the only effect of the additional wavefunction component is to double the number of energy eigenstates at each energy. Electrons are fermions, so antisymmetry of multiparticle wavefunctions implies the Pauli principle that states can only be occupied by a single particle. As a result, one finds that when adding electrons to an atom described by the Coulomb potential problem, the first two fill up the lowest Coulomb energy eigenstate (the ψ_{100} or $1S$ state at $n = 1$), the next eight fill up the $n = 2$ states (two each for $\psi_{200}, \psi_{211}, \psi_{210}, \psi_{21-1}$), etc. This goes a long ways toward explaining the structure of the periodic table of elements.

When one puts a hydrogen atom in a constant magnetic field $\mathbf{B}$, for reasons that will be described in section 45.3, the Hamiltonian acquires a term that acts only on the $\mathbf{C}^2$ factor, of the form

$$\frac{2e}{mc} \mathbf{B} \cdot \boldsymbol{\sigma}$$

This is exactly the sort of Hamiltonian we began our study of quantum mechanics with for a simple two-state system. It causes a shift in energy eigenvalues proportional to $\pm|\mathbf{B}|$ for the two different components of the wavefunction, and the observation of this energy splitting makes clear the necessity of treating the electron using the two-component formalism.

21.4   For further reading

This is a standard topic in all quantum mechanics books. For example, see chapters 12 and 13 of [81]. The $\mathfrak{so}(4)$ calculation is not in [81], but is in some of the other such textbooks, a good example is chapter 7 of [5]. For extensive discussion of the symmetries of the $\frac{1}{r}$ potential problem, see [38] or [39].

Chapter 22
The Harmonic Oscillator

In this chapter, we will begin the study of the most important exactly solvable physical system, the harmonic oscillator. Later chapters will discuss extensions of the methods developed here to the case of fermionic oscillators, as well as free quantum field theories, which are harmonic oscillator systems with an infinite number of degrees of freedom.

For a finite number of degrees of freedom, the Stone–von Neumann theorem tells us that there is essentially just one way to non-trivially represent the (exponentiated) Heisenberg commutation relations as operators on a quantum mechanical state space. We have seen two unitarily equivalent constructions of these operators: the Schrödinger representation in terms of functions on either coordinate space or momentum space. It turns out that there is another class of quite different constructions of these operators, one that depends upon introducing complex coordinates on phase space and then using properties of holomorphic functions. We will refer to this as the Bargmann–Fock representation, although quite a few mathematicians have had their name attached to it for one good reason or another (some of the other names one sees are Friedrichs, Segal, Shale, Weil, as well as the descriptive terms "holomorphic" and "oscillator").

Physically, the importance of this representation is that it diagonalizes the Hamiltonian operator for a fundamental sort of quantum system: the harmonic oscillator. In the Bargmann–Fock representation, the energy eigenstates of such a system are monomials, and energy eigenvalues are (up to a half-integral constant) integers. These integers label the irreducible representations of the $U(1)$ symmetry generated by the Hamiltonian, and they can be interpreted as counting the number of "quanta" in the system. It is the ubiquity of this example that justifies the "quantum" in "quantum mechanics." The operators on the state space can be simply understood in terms of so-called "annihilation" and "creation" operators which decrease or increase by one the number of quanta.

© Peter Woit 2017
P. Woit, *Quantum Theory, Groups and Representations*,
DOI 10.1007/978-3-319-64612-1_22

22.1 The harmonic oscillator with one degree of freedom

An even simpler case of a particle in a potential than the Coulomb potential of Chapter 21 is the case of $V(q)$ quadratic in q. This is also the lowest-order approximation when one studies motion near a local minimum of an arbitrary $V(q)$, expanding $V(q)$ in a power series around this point. We will write this as

$$h = \frac{p^2}{2m} + \frac{1}{2}m\omega^2 q^2$$

with coefficients chosen so as to make ω the angular frequency of periodic motion of the classical trajectories. These satisfy Hamilton's equations

$$\dot{p} = -\frac{\partial V}{\partial q} = -m\omega^2 q, \quad \dot{q} = \frac{p}{m}$$

so

$$\ddot{q} = -\omega^2 q$$

which will have solutions with periodic motion of angular frequency ω. These solutions can be written as

$$q(t) = c_+ e^{i\omega t} + c_- e^{-i\omega t}$$

for $c_+, c_- \in \mathbf{C}$ where, since $q(t)$ must be real, we have $c_- = \bar{c}_+$. The space of solutions of the equation of motion is thus two real dimensional, and abstractly this can be thought of as the phase space of the system.

More conventionally, the phase space can be parametrized by initial values that determine the classical trajectories, for instance by the position $q(0)$ and momentum $p(0)$ at an initial time $t(0)$. Since

$$p(t) = m\dot{q} = mc_+ i\omega e^{i\omega t} - mc_- i\omega e^{-i\omega t} = im\omega(c_+ e^{i\omega t} - \bar{c}_+ e^{-i\omega t})$$

we have

$$q(0) = c_+ + c_- = 2\operatorname{Re}(c_+), \quad p(0) = im\omega(c_+ - c_-) = 2m\omega \operatorname{Im}(c_+)$$

so

$$c_+ = \frac{1}{2}q(0) + i\frac{1}{2m\omega}p(0)$$

The classical phase space trajectories are

$$q(t) = \left(\frac{1}{2}q(0) + i\frac{1}{2m\omega}p(0)\right)e^{i\omega t} + \left(\frac{1}{2}q(0) - i\frac{1}{2m\omega}p(0)\right)e^{-i\omega t}$$

$$p(t) = \left(\frac{im\omega}{2}q(0) - \frac{1}{2}p(0)\right)e^{i\omega t} + \left(\frac{-im\omega}{2}q(0) + \frac{1}{2}p(0)\right)e^{-i\omega t}$$

Instead of using two real coordinates to describe points in the phase space (and having to introduce a reality condition when using complex exponentials), one can instead use a single complex coordinate, which we will choose as

$$z(t) = \sqrt{\frac{m\omega}{2}}\left(q(t) - \frac{i}{m\omega}p(t)\right)$$

Then, the equation of motion is a first-order rather than second-order differential equation

$$\dot{z} = i\omega z$$

with solutions

$$z(t) = z(0)e^{i\omega t} \tag{22.1}$$

The classical trajectories are then realized as complex functions of t and parametrized by the complex number

$$z(0) = \sqrt{\frac{m\omega}{2}}\left(q(0) - \frac{i}{m\omega}p(0)\right)$$

Since the Hamiltonian is quadratic in the p and q, we have seen that we can construct the corresponding quantum operator uniquely using the Schrödinger representation. For $\mathcal{H} = L^2(\mathbf{R})$, we have a Hamiltonian operator

$$H = \frac{P^2}{2m} + \frac{1}{2}m\omega^2 Q^2 = -\frac{\hbar^2}{2m}\frac{d^2}{dq^2} + \frac{1}{2}m\omega^2 q^2$$

To find solutions of the Schrödinger equation, as with the free particle, one proceeds by first solving for eigenvectors of H with eigenvalue E, which means finding solutions to

$$H\psi_E = \left(-\frac{\hbar^2}{2m}\frac{d^2}{dq^2} + \frac{1}{2}m\omega^2 q^2\right)\psi_E = E\psi_E$$

Solutions to the Schrödinger equation will then be linear combinations of the functions

$$\psi_E(q)e^{-\frac{i}{\hbar}Et}$$

Standard but somewhat intricate methods for solving differential equations like this show that one gets solutions for $E = E_n = (n+\frac{1}{2})\hbar\omega$, n a nonnegative integer, and the normalized solution for a given n (which we will denote ψ_n) will be

$$\psi_n(q) = \left(\frac{m\omega}{\pi\hbar 2^{2n}(n!)^2}\right)^{\frac{1}{4}} H_n\left(\sqrt{\frac{m\omega}{\hbar}}q\right) e^{-\frac{m\omega}{2\hbar}q^2} \qquad (22.2)$$

where H_n is a family of polynomials called the Hermite polynomials. The ψ_n provide an orthonormal basis for $\mathcal{H}$ (one does not need to consider non-normalizable wavefunctions as in the free particle case), so any initial wave-function $\psi(q, 0)$ can be written in the form

$$\psi(q, 0) = \sum_{n=0}^{\infty} c_n \psi_n(q)$$

with

$$c_n = \int_{-\infty}^{+\infty} \psi_n(q)\psi(q, 0)dq$$

(note that the ψ_n are real-valued). At later times, the wavefunction will be

$$\psi(q, t) = \sum_{n=0}^{\infty} c_n \psi_n(q) e^{-\frac{i}{\hbar}E_n t} = \sum_{n=0}^{\infty} c_n \psi_n(q) e^{-i(n+\frac{1}{2})\hbar\omega t}$$

22.2 Creation and annihilation operators

It turns out that there is a quite easy method which allows one to explicitly find eigenfunctions and eigenvalues of the harmonic oscillator Hamiltonian (although it is harder to show, it gives all of them). This also leads to a new representation of the Heisenberg group (of course unitarily equivalent to the Schrödinger one by the Stone–von Neumann theorem). Instead of working with the self-adjoint operators Q and P that satisfy the commutation relation

$$[Q, P] = i\hbar\mathbf{1}$$

we define

$$a = \sqrt{\frac{m\omega}{2\hbar}}Q + i\sqrt{\frac{1}{2m\omega\hbar}}P, \quad a^\dagger = \sqrt{\frac{m\omega}{2\hbar}}Q - i\sqrt{\frac{1}{2m\omega\hbar}}P$$

which satisfy the commutation relation

$$[a, a^\dagger] = 1$$

Since

$$Q = \sqrt{\frac{\hbar}{2m\omega}}(a + a^\dagger), \quad P = -i\sqrt{\frac{m\omega\hbar}{2}}(a - a^\dagger)$$

the Hamiltonian operator is

$$
\begin{aligned}
H &= \frac{P^2}{2m} + \frac{1}{2}m\omega^2 Q^2 \\
&= \frac{1}{4}\hbar\omega\left(-(a - a^\dagger)^2 + (a + a^\dagger)^2\right) \\
&= \frac{1}{2}\hbar\omega(aa^\dagger + a^\dagger a) \\
&= \hbar\omega\left(a^\dagger a + \frac{1}{2}\right)
\end{aligned}
$$

The problem of finding eigenvectors and eigenvalues for H is seen to be equivalent to the same problem for the operator

$$N = a^\dagger a$$

Such an operator satisfies the commutation relations

$$[N, a] = [a^\dagger a, a] = a^\dagger[a, a] + [a^\dagger, a]a = -a$$

and

$$[N, a^\dagger] = a^\dagger$$

If $|c\rangle$ is a normalized eigenvector of N with eigenvalue c, one has

$$c = \langle c|a^\dagger a|c\rangle = |a|c\rangle|^2 \geq 0$$

so eigenvalues of N must be nonnegative. Using the commutation relations of $N, a, a^\dagger$ gives

$$Na|c\rangle = ([N, a] + aN)|c\rangle = a(N - 1)|c\rangle = (c - 1)a|c\rangle$$

and

$$Na^\dagger|c\rangle = ([N, a^\dagger] + a^\dagger N)|c\rangle = a^\dagger(N + 1)|c\rangle = (c + 1)a^\dagger|c\rangle$$

This shows that $a|c\rangle$ will have eigenvalue $c - 1$ for N, and a normalized eigenfunction for N will be

$$|c - 1\rangle = \frac{1}{\sqrt{c}} a|c\rangle$$

Similarly, since

$$|a^\dagger|c\rangle|^2 = \langle c|aa^\dagger|c\rangle = \langle c|(N + 1)|c\rangle = c + 1$$

we have

$$|c + 1\rangle = \frac{1}{\sqrt{c + 1}} a^\dagger|c\rangle$$

If we start off with a state $|0\rangle$ that is a nonzero eigenvector for N with eigenvalue 0, we see that the eigenvalues of N will be the nonnegative integers, and for this reason N is called the "number operator."

We can find such a state by looking for solutions to

$$a|0\rangle = 0$$

$|0\rangle$ will have energy eigenvalue $\frac{1}{2}\hbar\omega$, and this will be the lowest energy eigenstate. Acting by $a^\dagger$ n-times on $|0\rangle$ gives states with energy eigenvalue $(n + \frac{1}{2})\hbar\omega$. The equation for $|0\rangle$ is

$$a|0\rangle = \left(\sqrt{\frac{m\omega}{2\hbar}} Q + i\sqrt{\frac{1}{2m\omega\hbar}} P \right) \psi_0(q) = \sqrt{\frac{m\omega}{2\hbar}} \left(q + \frac{\hbar}{m\omega} \frac{d}{dq} \right) \psi_0(q) = 0$$

One can check that this equation has a single normalized solution

$$\psi_0(q) = (\frac{m\omega}{\pi\hbar})^{\frac{1}{4}} e^{-\frac{m\omega}{2\hbar} q^2}$$

which is the lowest energy eigenfunction.

The rest of the energy eigenfunctions can be found by computing

$$|n\rangle = \frac{a^\dagger}{\sqrt{n}} \cdots \frac{a^\dagger}{\sqrt{2}} \frac{a^\dagger}{\sqrt{1}} |0\rangle = \frac{1}{\sqrt{n!}} \left(\frac{m\omega}{2\hbar} \right)^{\frac{n}{2}} \left(q - \frac{\hbar}{m\omega} \frac{d}{dq} \right)^n \psi_0(q)$$

To show that these are the eigenfunctions of equation 22.2, one starts with the definition of Hermite polynomials as a generating function

$$e^{2qt - t^2} = \sum_{n=0}^{\infty} H_n(q) \frac{t^n}{n!} \qquad (22.3)$$

and interprets the $H_n(q)$ as the Taylor coefficients of the left-hand side at $t = 0$, deriving the identity

$$H_n(q) = \left(\frac{d^n}{dt^n} e^{2qt - t^2}\right)_{|t=0}$$

$$= e^{q^2}\left(\frac{d^n}{dt^n} e^{-(q-t)^2}\right)_{|t=0}$$

$$= (-1)^n e^{q^2}\left(\frac{d^n}{dq^n} e^{-(q-t)^2}\right)_{|t=0}$$

$$= (-1)^n e^{q^2} \frac{d^n}{dq^n} e^{-q^2}$$

$$= e^{\frac{q^2}{2}}\left(q - \frac{d}{dq}\right)^n e^{-\frac{q^2}{2}}$$

Taking q to $\sqrt{\frac{m\omega}{\hbar}}q$, this can be used to show that $|n\rangle = \psi_n(q)$ is given by 22.2.

In the physical interpretation of this quantum system, the state $|n\rangle$, with energy $\hbar\omega(n + \frac{1}{2})$, is thought of as a state describing n "quanta." The state $|0\rangle$ is the "vacuum state" with zero quanta, but still carrying a "zero-point" energy of $\frac{1}{2}\hbar\omega$. The operators $a^\dagger$ and a have somewhat similar properties to the raising and lowering operators we used for $SU(2)$ but their commutator is different (the identity operator), leading to simpler behavior. In this case, they are called "creation" and "annihilation" operators, respectively, due to the way they change the number of quanta. The relation of such quanta to physical particles like the photon is that quantization of the electromagnetic field (see Chapter 46) involves quantization of an infinite collection of oscillators, with the quantum of an oscillator corresponding physically to a photon with a specific momentum and polarization. This leads to a well-known problem of how to handle the infinite vacuum energy corresponding to adding up $\frac{1}{2}\hbar\omega$ for each oscillator.

The first few eigenfunctions are plotted below. The lowest energy eigenstate is a Gaussian centered at $q = 0$, with a Fourier transform that is also a Gaussian centered at $p = 0$. Classically, the lowest energy solution is an oscillator at rest at its equilibrium point ($q = p = 0$), but for a quantum oscillator one cannot have such a state with a well-defined position and momentum. Note that the plot gives the wavefunctions, which in this case are real and can be negative. The square of this function is what has an interpretation as the probability density for measuring a given position.

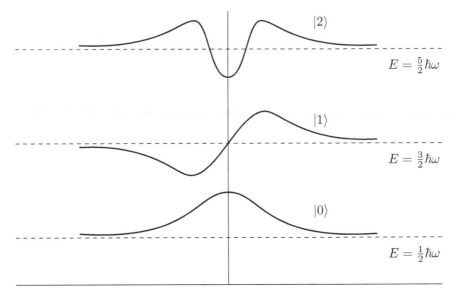

$|2\rangle$

$E = \frac{5}{2}\hbar\omega$

$|1\rangle$

$E = \frac{3}{2}\hbar\omega$

$|0\rangle$

$E = \frac{1}{2}\hbar\omega$

Figure 22.1: Harmonic oscillator energy eigenfunctions.

While we have preserved constants in our calculations in this section, in what follows we will often for simplicity set $\hbar = m = \omega = 1$, which can be done by an appropriate choice of units. Equations with the constants can be recovered by rescaling. In particular, our definition of annihilation and creation operators will be given by

$$a = \frac{1}{\sqrt{2}}(Q + iP), \quad a^\dagger = \frac{1}{\sqrt{2}}(Q - iP)$$

22.3 The Bargmann–Fock representation

Working with the operators a and $a^\dagger$ and their commutation relation

$$[a, a^\dagger] = 1$$

makes it clear that there is a simpler way to represent these operators than the Schrödinger representation as operators on position space functions that we have been using, while the Stone–von Neumann theorem assures us that this will be unitarily equivalent to the Schrödinger representation. This representation appears in the literature under a large number of different names, depending on the context, all of which refer to the same representation:

Definition (Bargmann–Fock or oscillator or holomorphic or Segal-Shale-Weil representation). *The Bargmann–Fock (etc.) representation is given by taking as state space* $\mathcal{H} = \mathcal{F}$, *where* $\mathcal{F}$ *is the space of holomorphic functions (satisfying* $\frac{d}{d\bar{z}}\psi = 0$*) on* $\mathbf{C}$ *with finite norm in the inner product*

$$\langle \psi_1 | \psi_2 \rangle = \frac{1}{\pi} \int_{\mathbf{C}} \overline{\psi_1(z)} \psi_2(z) e^{-|z|^2} d^2 z \tag{22.4}$$

where $d^2 z = dRe(z)dIm(z)$. *The space* $\mathcal{F}$ *is sometimes called "Fock space." We define the following two operators acting on this space:*

$$a = \frac{d}{dz}, \quad a^\dagger = z$$

Since

$$[a, a^\dagger] z^n = \frac{d}{dz}(z z^n) - z \frac{d}{dz} z^n = (n + 1 - n) z^n = z^n$$

the commutator is the identity operator on polynomials

$$[a, a^\dagger] = \mathbf{1}$$

One finds

Theorem. *The Bargmann–Fock representation has the following properties*

- *The elements*

$$\frac{z^n}{\sqrt{n!}}$$

  of $\mathcal{F}$ *for* $n = 0, 1, 2, \ldots$ *are orthonormal.*

- *The operators* a *and* $a^\dagger$ *are adjoints with respect to the given inner product on* $\mathcal{F}$.

- *The basis*

$$\frac{z^n}{\sqrt{n!}}$$

  of $\mathcal{F}$ *for* $n = 0, 1, 2, \ldots$ *is complete.*

Proof. The proofs of the above statements are not difficult, in outline they are

- For orthonormality, one can compute the integrals

$$\int_{\mathbf{C}} \bar{z}^m z^n e^{-|z|^2} d^2 z$$

in polar coordinates.

- To show that z and $\frac{d}{dz}$ are adjoint operators, use integration by parts.

- For completeness, assume $\langle n|\psi\rangle = 0$ for all n. The expression for the $|n\rangle$ as Hermite polynomials times a Gaussian then implies that

$$\int F(q)e^{-\frac{q^2}{2}}\psi(q)dq = 0$$

for all polynomials $F(q)$. Computing the Fourier transform of $\psi(q)e^{-\frac{q^2}{2}}$ gives

$$\int e^{-ikq}e^{-\frac{q^2}{2}}\psi(q)dq = \int \sum_{j=0}^{\infty}\frac{(-ikq)^j}{j!}e^{-\frac{q^2}{2}}\psi(q)dq = 0$$

So $\psi(q)e^{-\frac{q^2}{2}}$ has Fourier transform 0 and must be 0 itself. Alternatively, one can invoke the spectral theorem for the self-adjoint operator H, which guarantees that its eigenvectors form a complete and orthonormal set.

$\square$

Since in this representation, the number operator $N = a^\dagger a$ satisfies

$$Nz^n = z\frac{d}{dz}z^n = nz^n$$

the monomials in z diagonalize the number and energy operators, so one has

$$\frac{z^n}{\sqrt{n!}}$$

for the normalized energy eigenstate of energy $\hbar\omega(n+\frac{1}{2})$.

Note that we are here taking the state space $\mathcal{F}$ to include infinite linear combinations of the states $|n\rangle$, as long as the Bargmann–Fock norm is finite. We will sometimes want to restrict to the subspace of finite linear combinations of the $|n\rangle$, which we will denote $\mathcal{F}^{fin}$. This is the space $\mathbf{C}[z]$ of polynomials, and $\mathcal{F}$ is its completion for the Bargmann–Fock norm.

22.4 Quantization by annihilation and creation operators

The introduction of annihilation and creation operators involves allowing linear combinations of position and momentum operators with complex coefficients. These can be thought of as giving a Lie algebra representation of $\mathfrak{h}_3 \otimes \mathbf{C}$, the complexified Heisenberg Lie algebra. This is the Lie algebra of complex polynomials of degree zero and one on phase space M, with a basis $1, z, \bar{z}$. One has

$$\mathfrak{h}_3 \otimes \mathbf{C} = (\mathcal{M} \otimes \mathbf{C}) \oplus \mathbf{C}$$

with

$$z = \frac{1}{\sqrt{2}}(q - ip), \quad \bar{z} = \frac{1}{\sqrt{2}}(q + ip)$$

a basis for the complexified dual phase space $\mathcal{M} \otimes \mathbf{C}$. Note that these coordinates provide a decomposition

$$\mathcal{M} \otimes \mathbf{C} = \mathbf{C} \oplus \mathbf{C}$$

of the complexified dual phase space into subspaces spanned by z and by $\bar{z}$. The Lie bracket is the Poisson bracket, extended by complex linearity. The only nonzero bracket between basis elements is given by

$$\{z, \bar{z}\} = i$$

Quantization by annihilation and creation operators produces a Lie algebra representation by

$$\Gamma'(1) = -i\mathbf{1}, \quad \Gamma'(z) = -ia^\dagger, \quad \Gamma'(\bar{z}) = -ia \tag{22.5}$$

with the operator relation

$$[a, a^\dagger] = \mathbf{1}$$

equivalent to the Lie algebra homomorphism property

$$[\Gamma'(z), \Gamma'(\bar{z})] = \Gamma'(\{z, \bar{z}\})$$

We have now seen two different unitarily equivalent realizations of this Lie algebra representation: the Schrödinger version Γ'_S on functions of q, where

$$a = \frac{1}{\sqrt{2}}\left(q + \frac{d}{dq}\right), \quad a^\dagger = \frac{1}{\sqrt{2}}\left(q - \frac{d}{dq}\right)$$

and the Bargmann–Fock version Γ'_{BF} on functions of z, where

$$a = \frac{d}{dz}, \quad a^\dagger = z$$

Note that while annihilation and creation operators give a representation of the complexified Heisenberg Lie algebra $\mathfrak{h}_3 \otimes \mathbf{C}$, this representation is only unitary on the real Lie subalgebra $\mathfrak{h}_3$. This corresponds to the fact that general complex linear combinations of a and $a^\dagger$ are not self-adjoint, and to get something self-adjoint one must take real linear combinations of

$$a + a^\dagger \quad \text{and} \quad i(a - a^\dagger)$$

22.5   For further reading

All quantum mechanics books should have a similar discussion of the harmonic oscillator, with a good example the detailed one in chapter 7 of Shankar [81]. One source for a detailed treatment of the Bargmann–Fock representation is [26].

Chapter 23
Coherent States and the Propagator for the Harmonic Oscillator

In Chapter 22 we found the energy eigenstates for the harmonic oscillator using annihilation and creation operator methods and showed that these give a new construction of the representation of the Heisenberg group on the quantum mechanical state space, called the Bargmann–Fock representation. This representation comes with a distinguished state, the state $|0\rangle$, and the Heisenberg group action takes this state to a set of states known as "coherent states." These states are labeled by points of the phase space and provide the closest analog possible in the quantum system of classical states (i.e., those with a well-defined value of position and momentum variables).

Coherent states also evolve in time very simply, with their time evolution given just by the classical time evolution of the corresponding point in phase space. This fact can be used to calculate relatively straightforwardly the harmonic oscillator position space propagator, which gives the kernel for the action of time evolution on position space wavefunctions.

23.1 Coherent states and the Heisenberg group action

Since the Hamiltonian for the $d = 1$ harmonic oscillator does not commute with the operators a or $a^\dagger$ which give the representation of the Lie algebra $\mathfrak{h}_3$ on the state space $\mathcal{F}$, the Heisenberg Lie group and its Lie algebra are not symmetries of the system. Energy eigenstates do not break up into irreducible representations of the group but rather the entire state space makes up such an irreducible representation. The state space for the harmonic oscillator does, however, have a distinguished state, the lowest energy state $|0\rangle$, and one can ask what happens to this state under the Heisenberg group action.

Elements of the complexified Heisenberg Lie algebra $\mathfrak{h}_3 \otimes \mathbf{C}$ can be written as

© Peter Woit 2017
P. Woit, *Quantum Theory, Groups and Representations*,
DOI 10.1007/978-3-319-64612-1_23

$$i\alpha z + \beta \bar{z} + \gamma$$

for α, β, γ in $\mathbf{C}$ (this choice of α simplifies later formulas). The Lie algebra $\mathfrak{h}_3$ is the subspace of real functions, which will be those of the form

$$i\alpha z - i\bar{\alpha}\bar{z} + \gamma$$

for $\alpha \in \mathbf{C}$ and $\gamma \in \mathbf{R}$. The Lie algebra structure is given by the Poisson bracket

$$\{i\alpha_1 z - i\bar{\alpha}_1\bar{z} + \gamma_1, i\alpha_2 z - i\bar{\alpha}_2\bar{z} + \gamma_2\} = 2\mathrm{Im}(\bar{\alpha}_1\alpha_2)$$

Here, $\mathfrak{h}_3$ is identified with $\mathbf{C} \oplus \mathbf{R}$, and elements can be written as pairs (α, γ), with the Lie bracket

$$[(\alpha_1, \gamma_1), (\alpha_2, \gamma_2)] = (0, 2\mathrm{Im}(\bar{\alpha}_1\alpha_2))$$

This is just a variation on the labeling of $\mathfrak{h}_3$ elements discussed in Chapter 13, and one can again use exponential coordinates and write elements of the Heisenberg group H_3 also as such pairs, with group law

$$(\alpha_1, \gamma_1)(\alpha_2, \gamma_2) = (\alpha_1 + \alpha_2, \gamma_1 + \gamma_2 + \mathrm{Im}(\bar{\alpha}_1\alpha_2))$$

Quantizing using equation 22.5, one has a Lie algebra representation Γ', with operators for elements of $\mathfrak{h}_3$

$$\Gamma'(\alpha, \gamma) = \Gamma'(i\alpha z - i\bar{\alpha}\bar{z} + \gamma) = \alpha a^\dagger - \bar{\alpha}a - i\gamma\mathbf{1} \qquad (23.1)$$

and exponentiating these will give the unitary representation

$$\Gamma(\alpha, \gamma) = e^{\alpha a^\dagger - \bar{\alpha}a - i\gamma}$$

We define operators

$$D(\alpha) = e^{\alpha a^\dagger - \bar{\alpha}a}$$

which satisfy (using Baker-Campbell-Hausdorff)

$$D(\alpha_1)D(\alpha_2) = D(\alpha_1 + \alpha_2)e^{-i\mathrm{Im}(\bar{\alpha}_1\alpha_2)}$$

Then

$$\Gamma(\alpha, \gamma) = D(\alpha)e^{-i\gamma}$$

and the operators $\Gamma(\alpha, \gamma)$ give a representation, since they satisfy

$$\Gamma(\alpha_1, \gamma_1)\Gamma(\alpha_2, \gamma_2) = D(\alpha_1 + \alpha_2)e^{-i(\gamma_1 + \gamma_2 + \mathrm{Im}(\bar{\alpha}_1\alpha_2))} = \Gamma((\alpha_1, \gamma_1)(\alpha_2, \gamma_2))$$

Acting on $|0\rangle$ with $D(\alpha)$ gives:

Definition (Coherent states). *The coherent states in $\mathcal{H}$ are the states*

$$|\alpha\rangle = D(\alpha)|0\rangle = e^{\alpha a^\dagger - \bar{\alpha} a}|0\rangle$$

where $\alpha \in \mathbf{C}$.

Using the Baker-Campbell-Hausdorff formula

$$e^{\alpha a^\dagger - \bar{\alpha} a} = e^{\alpha a^\dagger} e^{-\bar{\alpha} a} e^{-\frac{|\alpha|^2}{2}} = e^{-\bar{\alpha} a} e^{\alpha a^\dagger} e^{\frac{|\alpha|^2}{2}}$$

so

$$|\alpha\rangle = e^{\alpha a^\dagger} e^{-\bar{\alpha} a} e^{-\frac{|\alpha|^2}{2}}|0\rangle$$

and since $a|0\rangle = 0$ this becomes

$$|\alpha\rangle = e^{-\frac{|\alpha|^2}{2}} e^{\alpha a^\dagger}|0\rangle = e^{-\frac{|\alpha|^2}{2}} \sum_{n=0}^{\infty} \frac{\alpha^n}{\sqrt{n!}}|n\rangle \tag{23.2}$$

Since $a|n\rangle = \sqrt{n}|n-1\rangle$

$$a|\alpha\rangle = e^{-\frac{|\alpha|^2}{2}} \sum_{n=1}^{\infty} \frac{\alpha^n}{\sqrt{(n-1)!}}|n-1\rangle = \alpha|\alpha\rangle$$

and this property could be used as an equivalent definition of coherent states. In a coherent state, the expectation value of a is

$$\langle \alpha|a|\alpha\rangle = \langle \alpha|\frac{1}{\sqrt{2}}(Q + iP)|\alpha\rangle = \alpha$$

so

$$\langle \alpha|Q|\alpha\rangle = \sqrt{2}\mathrm{Re}(\alpha), \quad \langle \alpha|P|\alpha\rangle = \sqrt{2}\mathrm{Im}(\alpha)$$

Note that coherent states are superpositions of different states $|n\rangle$, so are not eigenvectors of the number operator N, and do not describe states with a fixed (or even finite) number of quanta. They are eigenvectors of

$$a = \frac{1}{\sqrt{2}}(Q + iP)$$

with eigenvalue α, so one can try and think of $\sqrt{2}\alpha$ as a complex number whose real part gives the position and imaginary part the momentum. This does not lead to a violation of the Heisenberg uncertainty principle since this is not a self-adjoint operator, and thus not an observable. Such states are, however, very useful for describing certain sorts of physical phenomena, for instance, the state of a laser beam, where (for each momentum component of the electromagnetic field) one does not have a definite number of photons, but does have a definite amplitude and phase.

Digression (Spin coherent states). *One can perform a similar construction replacing the group H_3 by the group $SU(2)$, and the state $|0\rangle$ by a highest weight vector of an irreducible representation $(\pi_n, V^n = \mathbf{C}^{n+1})$ of spin $\frac{n}{2}$. Writing $|\frac{n}{2}\rangle$ for such a highest weight vector, we have*

$$\pi'_n(S_3)|\frac{n}{2}\rangle = \frac{n}{2}|\frac{n}{2}\rangle, \quad \pi'_n(S_+)|\frac{n}{2}\rangle = 0$$

and we can create a family of spin coherent states by acting on $|\frac{n}{2}\rangle$ by elements of $SU(2)$. If we identify states in this family that differ only by a phase, the states are parametrized by a sphere.

For the case $n = 1$, this is precisely the Bloch sphere construction of section 7.5, where we took as highest weight vector $|\frac{1}{2}\rangle = \begin{pmatrix} 1 \\ 0 \end{pmatrix}$. In that case, all states in the representation space $\mathbf{C}^2$ were spin coherent states (identifying states that differ only by scalar multiplication). For larger values of n, only a subset of the states in $\mathbf{C}^{n+1}$ will be spin coherent states.

23.2 Coherent states and the Bargmann–Fock state space

One thing coherent states provide is an alternate complete set of norm one vectors in $\mathcal{H}$, so any state can be written in terms of them. However, these states are not orthogonal (they are eigenvectors of a non-self-adjoint operator, so the spectral theorem for self-adjoint operators does not apply). The inner product of two coherent states is

$$\begin{aligned}
\langle\beta|\alpha\rangle &= \langle 0|e^{-\frac{1}{2}|\beta|^2}e^{\bar{\beta}a}e^{-\frac{1}{2}|\alpha|^2}e^{\alpha a^\dagger}|0\rangle \\
&= e^{-\frac{1}{2}(|\alpha|^2+|\beta|^2)}e^{\bar{\beta}\alpha}\langle 0|e^{\alpha a^\dagger}e^{\bar{\beta}a}|0\rangle \\
&= e^{\bar{\beta}\alpha-\frac{1}{2}(|\alpha|^2+|\beta|^2)}
\end{aligned} \tag{23.3}$$

and

$$|\langle\beta|\alpha\rangle|^2 = e^{-|\alpha-\beta|^2}$$

The Dirac formalism used for representing states as position space or momentum space distributions with a continuous basis $|q\rangle$ or $|p\rangle$ can also be adapted to the Bargmann–Fock case. In the position space case, with states functions of q, the delta-function distribution $\delta(q - q')$ provides an eigenvector $|q'\rangle$ for the Q operator, with eigenvalue q'. As discussed in Chapter 12, the position space wavefunction of a state $|\psi\rangle$ can be thought of as given by

$$\psi(q) = \langle q | \psi \rangle$$

with

$$\langle q | q' \rangle = \delta(q - q')$$

In the Bargmann–Fock case, there is an analog of the distributional states $|q\rangle$, given by taking states that are eigenvectors for a, but unlike the $|\alpha\rangle$, are not normalizable. We define

$$|\delta_w\rangle = e^{\overline{w} a^\dagger} |0\rangle = \sum_{n=0}^{\infty} \frac{\overline{w}^n}{\sqrt{n!}} |n\rangle = e^{\overline{w} z} = e^{\frac{|w|^2}{2}} |\overline{w}\rangle$$

Instead of equation 23.3, such states satisfy

$$\langle \delta_{w_1} | \delta_{w_2} \rangle = e^{\overline{w}_1 w_2}$$

The $|\delta_w\rangle$ behave in a manner analogous to the delta-function, since the Bargmann–Fock analog of computing $\langle q | \psi \rangle$ using the function space inner product is, writing

$$\psi(w) = \sum_{n=0}^{\infty} c_n \frac{w^n}{\sqrt{n!}}$$

the computation

$$\langle \delta_z | \psi \rangle = \frac{1}{\pi} \int_{\mathbf{C}} e^{z\overline{w}} e^{-|w|^2} \psi(w) d^2 w$$

$$= \frac{1}{\pi} \int_{\mathbf{C}} e^{z\overline{w}} e^{-|w|^2} \sum_{n=0}^{\infty} c_n \frac{w^n}{\sqrt{n!}} d^2 w$$

$$= \frac{1}{\pi} \int_{\mathbf{C}} \sum_{m=0}^{\infty} \frac{z^m}{\sqrt{m!}} \frac{\overline{w}^m}{\sqrt{m!}} e^{-|w|^2} \sum_{n=0}^{\infty} c_n \frac{w^n}{\sqrt{n!}} d^2 w$$

$$= \sum_{n=0}^{\infty} c_n \frac{z^n}{\sqrt{n!}} = \psi(z)$$

Here, we have used the orthogonality relations

$$\int_{\mathbf{C}} \overline{w}^m w^n e^{-|w|^2} d^2 w = \pi n! \delta_{n,m} \tag{23.4}$$

It is easily seen that the Bargmann–Fock wavefunction of a coherent state is given by

$$\langle \delta_z | \alpha \rangle = e^{-\frac{|\alpha|^2}{2}} e^{\alpha z} \tag{23.5}$$

while for number operator eigenvector states

$$\langle \delta_z | n \rangle = \frac{z^n}{\sqrt{n!}}$$

In section 23.5, we will compute the Bargmann–Fock wavefunction $\langle \delta_z | q \rangle$ for position eigenstates, see equation 23.13.

Like the $|\alpha\rangle$ (and unlike the $|q\rangle$ or $|p\rangle$), these states $|\delta_w\rangle$ are not orthogonal for different eigenvalues of a, but they span the state space, providing an over-complete basis, and satisfy the resolution of the identity relation

$$1 = \frac{1}{\pi} \int_{\mathbf{C}} |\delta_w\rangle\langle\delta_w| e^{-|w|^2} d^2 w \tag{23.6}$$

This can be shown using

$$|\delta_w\rangle\langle\delta_w| = \left(\sum_{n=0}^{\infty} \frac{\overline{w}^n}{\sqrt{n!}} |n\rangle \right) \left(\sum_{m=0}^{\infty} \frac{z^m}{\sqrt{m!}} \langle m| \right)$$

as well as

$$1 = \sum_{n=0}^{\infty} |n\rangle\langle n|$$

and the orthogonality relations 23.4. Note that the normalized coherent states similarly provide an over-complete basis, with

$$1 = \frac{1}{\pi} \int_{\mathbf{C}} |\alpha\rangle\langle\alpha| d^2\alpha \tag{23.7}$$

To avoid confusion over the various ways in which complex variables z and w appear here, note that this is just the analog of what happens in the position space representation, where q is variously a coordinate on classical phase space, an argument of a wavefunction, a label of a position operator eigenstate, and a multiplication operator. The analog of the position operator Q here is $a^\dagger$, which is multiplication by z (unlike Q, not self-adjoint). The conjugate complex coordinate $\overline{z}$ is analogous to the momentum coordinate, quantized to a differentiation operator. One confusing aspect of this formalism is that complex conjugation takes elements of $\mathcal{H}$ (holomorphic functions) to antiholomorphic functions, which are in a different space. The quantization of $\overline{z}$ is not the complex conjugate of z, but the adjoint operator.

23.3 The Heisenberg group action on operators

The representation operators

$$\Gamma(\alpha, \gamma) = D(\alpha)e^{-i\gamma}$$

act not just on states, but also on operators, by the conjugation action

$$D(\alpha)aD(\alpha)^{-1} = a - \alpha, \quad D(\alpha)a^\dagger D(\alpha)^{-1} = a^\dagger - \overline{\alpha}$$

(on operators the phase factors cancel). These relations follow from the fact that the commutation relations

$$[\alpha a^\dagger - \overline{\alpha}a, a] = -\alpha, \quad [\alpha a^\dagger - \overline{\alpha}a, a^\dagger] = -\overline{\alpha}$$

are the derivatives with respect to t of

$$D(t\alpha)aD(t\alpha)^{-1} = a - t\alpha, \quad D(t\alpha)a^\dagger D(t\alpha)^{-1} = a^\dagger - t\overline{\alpha} \qquad (23.8)$$

At $t = 0$, this is just equation 5.1, but it holds for all t since multiple commutators vanish.

We thus see that the Heisenberg group acts on annihilation and creation operators by shifting the operators by a constant. The Heisenberg group acts by automorphisms on its Lie algebra by the adjoint representation (see section 15.5), and one can check that the $\Gamma(\alpha, \gamma)$ are intertwining operators for this action (see Chapter 20). The constructions of this chapter can easily be generalized from $d = 1$ to general values of the dimension d. For finite values of d, the $\Gamma(\alpha, \gamma)$ act on states as an irreducible representation, as required by the Stone–von Neumann theorem. We will see in Chapter 39 that in infinite dimensions this is no longer necessarily the case.

23.4 The harmonic oscillator propagator

In section 12.5, we saw that for the free particle quantum system, energy eigenstates were momentum eigenstates, and in the momentum space representation, time evolution by a time interval T was given by a kernel (see equation 12.6)

$$\widetilde{U}(T, k) = \frac{1}{\sqrt{2\pi}}e^{-i\frac{1}{2m}k^2 T}$$

The position space propagator was found by computing the Fourier transform of this. For the harmonic oscillator, energy eigenstates are no longer momen-

tum eigenstates and different methods are needed to compute the action of
the time evolution operator e^{-iHT}.

23.4.1   The propagator in the Bargmann–Fock representation

In the Bargmann–Fock representation, the Hamiltonian is the operator

$$H = \omega \left(a^\dagger a + \frac{1}{2} \right) = \omega \left(z \frac{d}{dz} + \frac{1}{2} \right)$$

(here, we choose $\hbar = 1$ and $m = 1$, but no longer fix $\omega = 1$) and energy
eigenstates are the states

$$\frac{z^n}{\sqrt{n!}} = |n\rangle$$

with energy eigenvalues

$$\omega \left(n + \frac{1}{2} \right)$$

e^{-iHT} will be diagonal in this basis, with

$$e^{-iHT} |n\rangle = e^{-i\omega(n+\frac{1}{2})T} |n\rangle$$

Instead of the Schrödinger picture in which states evolve and operators
are constant, one can instead go to the Heisenberg picture (see section 7.3)
where states are constant and operators $\mathcal{O}$ evolve in time according to

$$\frac{d}{dt} \mathcal{O}(t) = i[H, \mathcal{O}(t)]$$

with solution

$$\mathcal{O}(t) = e^{itH} \mathcal{O}(0) e^{-itH}$$

In the harmonic oscillator problem, we can express other operators in terms
of the annihilation and creation operators, which evolve according to

$$\frac{d}{dt} a(t) = i[H, a(t)] = -i\omega a, \quad \frac{d}{dt} a^\dagger(t) = i[H, a^\dagger(t)] = i\omega a^\dagger$$

with solutions

$$a(t) = e^{-i\omega t} a(0), \quad a^\dagger(t) = e^{i\omega t} a^\dagger(0)$$

The Hamiltonian operator is time invariant.

Questions about time evolution now become questions about various products of annihilation and creation operators taken at various times, applied to various Heisenberg picture states. Since an arbitrary state is given as a linear combination of states produced by repeatedly applying $a^\dagger(0)$ to $|0\rangle$, such problems can be reduced to evaluating expressions involving just the state $|0\rangle$, with various creation and annihilation operators applied at different times. Nonzero results will come from terms involving

$$\langle 0|a(T)a^\dagger(0)|0\rangle = e^{-i\omega T}$$

which for $T > 0$ has an interpretation as an amplitude for the process of adding one quantum to the lowest energy state at time $t = 0$, then removing it at time $t = T$.

23.4.2 The coherent state propagator

One possible reason these states are given the name "coherent" is that they remain coherent states as they evolve in time (for the harmonic oscillator Hamiltonian), with α evolving in time along a classical phase space trajectory. If the state at $t = 0$ is a coherent state labeled by α_0 ($|\psi(0)\rangle = |\alpha_0\rangle$), by 23.2, at later times one has

$$
\begin{aligned}
|\psi(t)\rangle &= e^{-iHt}|\alpha_0\rangle \\
&= e^{-iHt}e^{-\frac{|\alpha_0|^2}{2}}\sum_{n=0}^{\infty}\frac{\alpha_0^n}{\sqrt{n!}}|n\rangle \\
&= e^{-i\frac{1}{2}\omega t}e^{-\frac{|\alpha_0|^2}{2}}\sum_{n=0}^{\infty}\frac{e^{-i\omega nt}\alpha_0^n}{\sqrt{n!}}|n\rangle \\
&= e^{-i\frac{1}{2}\omega t}e^{-\frac{|e^{-i\omega t}\alpha_0|^2}{2}}\sum_{n=0}^{\infty}\frac{(e^{-i\omega t}\alpha_0)^n}{\sqrt{n!}}|n\rangle \\
&= e^{-i\frac{1}{2}\omega t}|e^{-i\omega t}\alpha_0\rangle
\end{aligned}
\tag{23.9}
$$

Up to the phase factor $e^{-i\frac{1}{2}\omega t}$, this remains a coherent state, with time-dependence of the label α given by the classical time-dependence of the complex coordinate $\overline{z}(t) = \frac{1}{\sqrt{2}}(\sqrt{\omega}q(t) + \frac{i}{\sqrt{\omega}}p(t))$ for the harmonic oscillator (see 22.1) with $\overline{z}(0) = \alpha_0$.

Equations 23.3 and 23.9 can be used to calculate a propagator function in terms of coherent states, with the result

$$\langle \alpha_T|e^{-iHT}|\alpha_0\rangle = \exp(-\frac{1}{2}(|\alpha_0|^2 + |\alpha_T|^2) + \overline{\alpha}_T\alpha_0 e^{-i\omega T} - \frac{i}{2}\omega T) \tag{23.10}$$

23.4.3 The position space propagator

Coherent states can be expressed in the position space representation by calculating

$$\langle q|\alpha\rangle = \langle q|e^{-\frac{|\alpha|^2}{2}}e^{\alpha a^\dagger}|0\rangle$$
$$=e^{-\frac{|\alpha|^2}{2}}e^{\frac{\alpha}{\sqrt{2}}\left(\sqrt{\omega}q-\frac{1}{\sqrt{\omega}}\frac{d}{dq}\right)}\left(\frac{\omega}{\pi}\right)^{\frac{1}{4}}e^{-\frac{\omega}{2}q^2}$$
$$=\left(\frac{\omega}{\pi}\right)^{\frac{1}{4}}e^{-\frac{|\alpha|^2}{2}}e^{\alpha\sqrt{\frac{\omega}{2}}q}e^{-\frac{\alpha}{\sqrt{2\omega}}\frac{d}{dq}}e^{-\frac{\alpha^2}{4}}e^{-\frac{\omega}{2}q^2}$$
$$=\left(\frac{\omega}{\pi}\right)^{\frac{1}{4}}e^{-\frac{|\alpha|^2}{2}}e^{-\frac{\alpha^2}{4}}e^{\alpha\sqrt{\frac{\omega}{2}}q}e^{-\frac{\omega}{2}(q-\frac{\alpha}{\sqrt{2\omega}})^2}$$
$$=\left(\frac{\omega}{\pi}\right)^{\frac{1}{4}}e^{-\frac{|\alpha|^2}{2}}e^{-\frac{\alpha^2}{2}}e^{-\frac{\omega}{2}q^2}e^{\sqrt{2\omega}\alpha q} \tag{23.11}$$

This expression gives the transformation between the position space basis and coherent state basis. The propagator in the position space basis can then be calculated as

$$\langle q_T|e^{-iHT}|q_0\rangle = \frac{1}{\pi^2}\int_{\mathbb{C}^2}\langle q_T|\alpha_T\rangle\langle\alpha_T|e^{-iHT}|\alpha_0\rangle\langle\alpha_0|q_0\rangle d^2\alpha_T d^2\alpha_0$$

using equations 23.10, 23.11 (and its complex conjugate), as well as equation 23.7.

We will not perform this (rather difficult) calculation here, but just quote the result, which is

$$\langle q_T|e^{-iHT}|q_0\rangle = \sqrt{\frac{\omega}{i2\pi\sin(\omega T)}}\exp\left(\frac{i\omega}{2\sin(\omega T)}((q_0^2+q_T^2)\cos(\omega T)-2q_0q_T)\right) \tag{23.12}$$

One can easily see that as $T\to 0$, this will approach the free particle propagator (equation 12.9, with $m=1$)

$$\langle q_T|e^{-iHT}|q_0\rangle \approx \sqrt{\frac{1}{i2\pi T}}e^{\frac{i}{2T}(q_T-q_0)^2}$$

and as in that case becomes the distribution $\delta(q_0-q_T)$ as $T\to 0$. Without too much difficulty, one can check that 23.12 satisfies the harmonic oscillator Schrödinger equation (in $q=q_T$ and $t=T$, for any initial $\psi(q_0,0)$).

As in the free particle case, the harmonic oscillator propagator can be defined first as a function of a complex variable $s=\tau+iT$, holomorphic for $\tau>0$, then taking the boundary value as $\tau\to 0$. This fixes the branch of the square root in 23.12, and one finds (see, for instance, section 7.6.7 of [107]) that the square root factor needs to be taken to be

$$\sqrt{\frac{\omega}{i2\pi \sin(\omega T)}} = e^{-i\frac{\pi}{4}} e^{-in\frac{\pi}{2}} \sqrt{\frac{\omega}{2\pi |\sin(\omega T)|}}$$

for $\omega T \in [n\pi, (n+1)\pi]$.

23.5 The Bargmann transform

The Stone–von Neumann theorem implies the existence of:

Definition. *Bargmann transform*
 There is a unitary map called the Bargmann transform

$$\mathcal{B} : \mathcal{H}_S \to \mathcal{F}$$

intertwining the Schrödinger representation and the Bargmann–Fock representation, i.e., with operators satisfying the relation

$$\Gamma'_S(X) = \mathcal{B}^{-1} \Gamma'_{BF}(X) \mathcal{B}$$

for $X \in \mathfrak{h}_3$.

In practice, knowing $\mathcal{B}$ explicitly is often not needed, since the representation independent relation

$$a = \frac{1}{\sqrt{2}}(Q + iP)$$

can be used to express operators either purely in terms of a and $a^\dagger$, which have a simple expression

$$a = \frac{d}{dz}, \quad a^\dagger = z$$

in the Bargmann–Fock representation, or purely in terms of Q and P which have a simple expression

$$Q = q, \quad P = -i\frac{d}{dq}$$

in the Schrödinger representation.
    To compute the Bargmann transform, one uses equation 23.11, for non-normalizable continuous basis states $|\delta_u\rangle$, to get

$$\langle q|\delta_u\rangle = \left(\frac{\omega}{\pi}\right)^{\frac{1}{4}} e^{-\frac{u^2}{2}} e^{-\frac{\omega}{2}q^2} e^{\sqrt{2\omega}\overline{u}q}$$

and

$$\langle \delta_u | q \rangle = \left(\frac{\omega}{\pi}\right)^{\frac{1}{4}} e^{-\frac{u^2}{2}} e^{-\frac{\omega}{2}q^2} e^{\sqrt{2\omega}uq} \qquad (23.13)$$

The Bargmann transform is then given by

$$(\mathcal{B}\psi)(z) = \int_{-\infty}^{+\infty} \langle \delta_z | q \rangle \langle q | \psi \rangle dq$$

$$= \left(\frac{\omega}{\pi}\right)^{\frac{1}{4}} e^{-\frac{z^2}{2}} \int_{-\infty}^{+\infty} e^{-\frac{\omega}{2}q^2} e^{\sqrt{2\omega}zq} \psi(q) dq \qquad (23.14)$$

(here, $\psi(q)$ is the position space wavefunction) while the inverse Bargmann transform is given by

$$(\mathcal{B}^{-1}\phi)(q) = \int_{\mathbb{C}} \langle q | \delta_u \rangle \langle \delta_u | \phi \rangle e^{-|u|^2} d^2 u$$

$$= \left(\frac{\omega}{\pi}\right)^{\frac{1}{4}} e^{-\frac{\omega}{2}q^2} \int_{\mathbb{C}} e^{-\frac{\bar{u}^2}{2}} e^{\sqrt{2\omega}\bar{u}q} \phi(u) e^{-|u|^2} d^2 u$$

(here, $\phi(z)$ is the Bargmann–Fock wavefunction).

As a check of equation 23.14, consider the case of the lowest energy state in the Schrödinger representation, where $|0\rangle$ has coordinate space representation

$$\psi(q) = \left(\frac{\omega}{\pi}\right)^{\frac{1}{4}} e^{-\frac{\omega q^2}{2}}$$

and

$$(\mathcal{B}\psi)(z) = \left(\frac{\omega}{\pi}\right)^{\frac{1}{4}} \left(\frac{\omega}{\pi}\right)^{\frac{1}{4}} e^{-\frac{z^2}{2}} \int_{-\infty}^{+\infty} e^{-\frac{\omega}{2}q^2} e^{\sqrt{2\omega}zq} e^{-\frac{\omega q^2}{2}} dq$$

$$= \left(\frac{\omega}{\pi}\right)^{\frac{1}{2}} \int_{-\infty}^{+\infty} e^{-\frac{z^2}{2}} e^{-\omega q^2} e^{\sqrt{2\omega}zq} dq$$

$$= \left(\frac{\omega}{\pi}\right)^{\frac{1}{2}} \int_{-\infty}^{+\infty} e^{-\omega\left(q - \frac{z}{\sqrt{2\omega}}\right)^2} dq$$

$$= 1$$

which is the expression for the state $|0\rangle$ in the Bargmann–Fock representation.

For an alternate way to compute the harmonic oscillator propagator, the kernel corresponding to applying the Bargmann transform, then the time evolution operator, then the inverse Bargmann transform can be calculated. This will give

$$\langle q_T | e^{-iHT} | q_0 \rangle = \int_{\mathbf{C}} \langle q_T | \delta_u \rangle \langle \delta_u | e^{-iTH} | q_0 \rangle e^{-|u|^2} d^2 u$$

$$= e^{-i\frac{\omega T}{2}} \int_{\mathbf{C}} \langle q_T | \delta_u \rangle \langle \delta_{ue^{-i\omega T}} | q_0 \rangle e^{-|u|^2} d^2 u$$

from which 23.12 can be derived by a (difficult) manipulation of Gaussian integrals.

23.6   For further reading

Coherent states and spin coherent states are discussed in more detail in chapter 21 of [81] and in [66]. For more about the Bargmann transform, see chapter 4 of [62] for its relation to coherent states and [26] for its relation to the Heisenberg group.

Chapter 24
The Metaplectic Representation and Annihilation and Creation Operators, $d = 1$

In section 22.4, we saw that annihilation and creation operators quantize complexified coordinate functions $\overline{z}, z$ on phase space, giving a representation of the complexified Heisenberg Lie algebra $\mathfrak{h}_3 \otimes \mathbf{C}$. In this chapter, we will see what happens for quadratic combinations of the $\overline{z}, z$, which after quantization give quadratic combinations of the annihilation and creation operators. These provide a Bargmann–Fock realization of the metaplectic representation of $\mathfrak{sl}(2, \mathbf{R})$, the representation which was studied in section 17.1 using the Schrödinger realization. Using annihilation and creation operators, the fact that the exponentiated quadratic operators act with a sign ambiguity (requiring the introduction of a double cover of $SL(2, \mathbf{R})$) is easily seen.

The metaplectic representation gives intertwining operators for the $SL(2, \mathbf{R})$ action by automorphisms of the Heisenberg group. The use of annihilation and creation operators to construct these operators introduces an extra piece of structure, in particular picking out a distinguished subgroup $U(1) \subset SL(2, \mathbf{R})$. Linear transformations of the $a, a^\dagger$ preserving the commutation relations (and thus acting as automorphisms of the Heisenberg Lie algebra structure) are known to physicists as "Bogoliubov transformations." They are naturally described using a different, isomorphic, form of the group $SL(2, \mathbf{R})$, a group of complex matrices denoted $SU(1, 1)$.

24.1 The metaplectic representation for $d = 1$ in terms of a and $a^\dagger$

Poisson brackets of order two combinations of z and $\overline{z}$ can easily be computed using the basic relation $\{z, \overline{z}\} = i$ and the Leibniz rule. On basis elements $z^2, \overline{z}^2, z\overline{z}$, the nonzero brackets are

© Peter Woit 2017
P. Woit, *Quantum Theory, Groups and Representations*,
DOI 10.1007/978-3-319-64612-1_24

$$\{z\bar{z}, z^2\} = -2iz^2, \quad \{z\bar{z}, \bar{z}^2\} = 2i\bar{z}^2, \quad \{\bar{z}^2, z^2\} = 4iz\bar{z}$$

Recall from equation 16.8 that quadratic real combinations of p and q can be identified with the Lie algebra $\mathfrak{sl}(2, \mathbf{R})$ of traceless 2 by 2 real matrices with basis

$$E = \begin{pmatrix} 0 & 1 \\ 0 & 0 \end{pmatrix}, \quad F = \begin{pmatrix} 0 & 0 \\ 1 & 0 \end{pmatrix}, \quad G = \begin{pmatrix} 1 & 0 \\ 0 & -1 \end{pmatrix}$$

Since we have complexified, allowing complex linear combinations of basis elements, our quadratic combinations of z and $\bar{z}$ are in the complexification of $\mathfrak{sl}(2, \mathbf{R})$. This is the Lie algebra $\mathfrak{sl}(2, \mathbf{C})$ of traceless 2 by 2 complex matrices. We can take as a basis of $\mathfrak{sl}(2, \mathbf{C})$ over the complex numbers

$$Z = E - F, \quad X_\pm = \frac{1}{2}(G \pm i(E + F))$$

which satisfy

$$[Z, X_-] = -2iX_-, \quad [Z, X_+] = 2iX_+, \quad [X_+, X_-] = -iZ$$

and then use as our isomorphism between quadratics in $z, \bar{z}$, and $\mathfrak{sl}(2, \mathbf{C})$

$$\frac{\bar{z}^2}{2} \leftrightarrow X_+, \quad \frac{z^2}{2} \leftrightarrow X_-, \quad z\bar{z} \leftrightarrow Z$$

The element

$$z\bar{z} = \frac{1}{2}(q^2 + p^2) \leftrightarrow Z = \begin{pmatrix} 0 & 1 \\ -1 & 0 \end{pmatrix}$$

exponentiates to give a $SO(2) = U(1)$ subgroup of $SL(2, \mathbf{R})$ with elements of the form

$$e^{\theta Z} = \begin{pmatrix} \cos\theta & \sin\theta \\ -\sin\theta & \cos\theta \end{pmatrix} \tag{24.1}$$

Note that $h = \frac{1}{2}(p^2 + q^2) = z\bar{z}$ is the classical Hamiltonian function for the harmonic oscillator.

We can now quantize quadratics in z and $\bar{z}$ using annihilation and creation operators acting on the Fock space $\mathcal{F}$. There is no operator ordering ambiguity for

$$z^2 \to (a^\dagger)^2 = z^2, \quad \bar{z}^2 \to a^2 = \frac{d^2}{dz^2}$$

For the case of $z\bar{z}$ (which is real), in order to get the $\mathfrak{sl}(2, \mathbf{R})$ commutation relations to come out right (in particular, the Poisson bracket $\{\bar{z}^2, z^2\} = 4iz\bar{z}$), we must take the symmetric combination

$$z\bar{z} \to \frac{1}{2}(aa^\dagger + a^\dagger a) = a^\dagger a + \frac{1}{2} = z\frac{d}{dz} + \frac{1}{2}$$

(which of course is the standard Hamiltonian for the quantum harmonic oscillator).

Multiplying as usual by $-i$ (to get a unitary representation of the real Lie algebra $\mathfrak{sl}(2, \mathbf{R})$), an extension of the Bargmann–Fock representation Γ'_{BF} of $\mathfrak{h}_3 \otimes \mathbf{C}$ (see section 22.4) to an $\mathfrak{sl}(2, \mathbf{C})$ representation can be defined by taking

$$\Gamma'_{BF}(X_+) = -\frac{i}{2}a^2, \quad \Gamma'_{BF}(X_-) = -\frac{i}{2}(a^\dagger)^2, \quad \Gamma'_{BF}(Z) = -i\frac{1}{2}(a^\dagger a + aa^\dagger)$$

This is the right choice of $\Gamma'_{BF}(Z)$ to get an $\mathfrak{sl}(2, \mathbf{C})$ representation since

$$[\Gamma'_{BF}(X_+), \Gamma'_{BF}(X_-)] = \left[-\frac{i}{2}a^2, -\frac{i}{2}(a^\dagger)^2\right] = -\frac{1}{2}(aa^\dagger + a^\dagger a)$$
$$= -i\Gamma'_{BF}(Z) = \Gamma'_{BF}([X_+, X_-])$$

As a representation of the real sub-Lie algebra $\mathfrak{sl}(2, \mathbf{R})$ of $\mathfrak{sl}(2, \mathbf{C})$, one has (using the fact that $G, E + F, E - F$ is a real basis of $\mathfrak{sl}(2, \mathbf{R})$):

Definition (Metaplectic representation of $\mathfrak{sl}(2, \mathbf{R})$). *The representation* Γ'_{BF} *on $\mathcal{F}$ given by*

$$\Gamma'_{BF}(G) = \Gamma'_{BF}(X_+ + X_-) = -\frac{i}{2}((a^\dagger)^2 + a^2)$$

$$\Gamma'_{BF}(E + F) = \Gamma'_{BF}(-i(X_+ - X_-)) = -\frac{1}{2}((a^\dagger)^2 - a^2) \quad\quad (24.2)$$

$$\Gamma'_{BF}(E - F) = \Gamma'_{BF}(Z) = -i\frac{1}{2}(a^\dagger a + aa^\dagger)$$

is a representation of $\mathfrak{sl}(2, \mathbf{R})$, called the metaplectic representation.

Note that this is clearly a unitary representation, since all the operators are skew-adjoint (using the fact that a and $a^\dagger$ are each other's adjoints).

This representation Γ'_{BF} on $\mathcal{F}$ will be unitarily equivalent using the Bargmann transform (see section 23.5) to the Schrödinger representation Γ'_S found earlier when quantizing q^2, p^2, pq as operators on $\mathcal{H} = L^2(\mathbf{R})$. For many purposes, it is, however, much easier to work with since it can be studied as the state space of the quantum harmonic oscillator, which comes with a basis of eigenvectors of the number operator $a^\dagger a$. The Lie algebra acts simply on such eigenvectors by quadratic expressions in the annihilation and creation operators.

One thing that can now easily be seen is that this representation Γ'_{BF} does not integrate to give a representation of the group $SL(2, \mathbf{R})$. If the Lie algebra representation Γ'_{BF} comes from a Lie group representation Γ_{BF} of $SL(2, \mathbf{R})$, we have

$$\Gamma_{BF}(e^{\theta Z}) = e^{\theta \Gamma'_{BF}(Z)}$$

where

$$\Gamma'_{BF}(Z) = -i\left(a^\dagger a + \frac{1}{2}\right) = -i\left(N + \frac{1}{2}\right)$$

so

$$\Gamma_{BF}(e^{\theta Z})|n\rangle = e^{-i\theta(n+\frac{1}{2})}|n\rangle$$

Taking $\theta = 2\pi$, this gives an inconsistency

$$\Gamma_{BF}(\mathbf{1})|n\rangle = -|n\rangle$$

This is the same phenomenon first described in the context of the Schrödinger representation in section 17.1.

As remarked there, it is the same sort of problem we found when studying the spinor representation of the Lie algebra $\mathfrak{so}(3)$. Just as in that case, the problem indicates that we need to consider not the group $SL(2, \mathbf{R})$, but a double cover, the metaplectic group $Mp(2, \mathbf{R})$. The behavior here is quite a bit more subtle than in the $Spin(3)$ double cover case, where $Spin(3)$ was the group $SU(2)$, and topologically, the only non-trivial cover of $SO(3)$ was the $Spin(3)$ one since $\pi_1(SO(3)) = \mathbf{Z}_2$. Here $\pi_1(SL(2, \mathbf{R})) = \mathbf{Z}$, and each extra time one goes around the $U(1)$ subgroup we are looking at, one gets a topo- logically different non-contractible loop in the group. As a result, $SL(2, \mathbf{R})$ has lots of non-trivial covering groups, of which only one interests us, the double cover $Mp(2, \mathbf{R})$. In particular, there is an infinite-sheeted universal cover $\widetilde{SL(2, \mathbf{R})}$, but that plays no role here.

Digression. *This group $Mp(2, \mathbf{R})$ is quite unusual in that it is a finite dim- ensional Lie group, but does not have any sort of description as a group of finite dimensional matrices. This is due to the fact that all its finite dimen- sional irreducible representations are the same as those of $SL(2, \mathbf{R})$, which has the same Lie algebra (these are representations on homogeneous polyno- mials in two variables, those first studied in chapter 8, which are $SL(2, \mathbf{C})$ representations which can be restricted to $SL(2, \mathbf{R})$). These finite dimensional representations factor through $SL(2, \mathbf{R})$ so their matrices do not distinguish* *between two different elements of $Mp(2, \mathbf{R})$ that correspond as $SL(2, \mathbf{R})$ elements.*

There are no faithful finite dimensional representations of $Mp(2, \mathbf{R})$ itself which could be used to identify $Mp(2, \mathbf{R})$ with a group of matrices. The only faithful irreducible representation available is the infinite dimensional one we are studying. Note that the lack of a matrix description means that this is a case where the definition we gave of a Lie algebra in terms of the matrix exponential does not apply. The more general geometric definition of the Lie algebra of a group in terms of the tangent space at the identity of the group does apply, although to do this one really needs a construction of the double cover $Mp(2, \mathbf{R})$, which is quite non-trivial and not done here. This is not a problem for purely Lie algebra calculations, since the Lie algebras of $Mp(2, \mathbf{R})$ and $SL(2, \mathbf{R})$ can be identified.

Another aspect of the metaplectic representation that is relatively easy to see in the Bargmann–Fock construction is that the state space $\mathcal{F}$ is not an irreducible representation, but is the sum of two irreducible representations

$$\mathcal{F} = \mathcal{F}_{even} \oplus \mathcal{F}_{odd}$$

where $\mathcal{F}_{even}$ consists of the even functions of z, $\mathcal{F}_{odd}$ of odd functions of z. On the subspace $\mathcal{F}^{fin} \subset \mathcal{F}$ of finite sums of the number eigenstates, these are the even and odd degree polynomials. Since the generators of the Lie algebra representation are degree two combinations of annihilation and creation operators, they will take even functions to even functions and odd to odd. The separate irreducibility of these two pieces is due to the fact that (when n and m have the same parity) one can get from state $|n\rangle$ to any another $|m\rangle$ by repeated application of the Lie algebra representation operators.

24.2 Intertwining operators in terms of a and $a^\dagger$

Recall from the discussion in chapter 20 that the metaplectic representation of $Mp(2, \mathbf{R})$ can be understood in terms of intertwining operators that arise due to the action of the group $SL(2, \mathbf{R})$ as automorphisms of the Heisenberg group H_3. Such intertwining operators can be constructed by exponentiating quadratic operators that have the commutation relations with the Q, P operators that reflect the intertwining relations (see equation 20.3). These quadratic operators provide the Lie algebra version of the metaplectic representation, discussed in section 24.1 using the Lie algebra $\mathfrak{sl}(2, \mathbf{R})$, which is identical to the Lie algebra of $Mp(2, \mathbf{R})$. In sections 20.3.2 and 20.3.4, these representations were constructed explicitly for $SO(2)$ and $\mathbf{R}$ subgroups of $SL(2, \mathbf{R})$ using quadratic combinations of the Q and P operators. Here, we will do the same thing using annihilation and creation operators instead of Q and P operators.

For the $SO(2)$ subgroup of equation 24.1 (this is the same one discussed in section 20.3.2), in terms of z and $\overline{z}$ coordinates, the moment map will be

$$\mu_Z = z\overline{z}$$

and one has

$$\{\mu_Z, \begin{pmatrix} \overline{z} \\ z \end{pmatrix}\} = \begin{pmatrix} i & 0 \\ 0 & -i \end{pmatrix} \begin{pmatrix} \overline{z} \\ z \end{pmatrix} \tag{24.3}$$

Quantization by annihilation and creation operators gives (see 24.2)

$$\Gamma'_{BF}(z\overline{z}) = \Gamma'_{BF}(Z) = -\frac{i}{2}(aa^\dagger + a^\dagger a)$$

and the quantized analog of 24.3 is

$$\left[-\frac{i}{2}(aa^\dagger + a^\dagger a), \begin{pmatrix} a \\ a^\dagger \end{pmatrix}\right] = \begin{pmatrix} i & 0 \\ 0 & -i \end{pmatrix} \begin{pmatrix} a \\ a^\dagger \end{pmatrix} \tag{24.4}$$

For group elements, $g_\theta = e^{\theta Z} \in SO(2) \subset SL(2, \mathbf{R})$ and the representation is given by unitary operators

$$U_{g_\theta} = \Gamma_{BF}(e^{\theta Z}) = e^{-i\frac{\theta}{2}(aa^\dagger + a^\dagger a)}$$

which satisfy

$$U_{g_\theta} \begin{pmatrix} a \\ a^\dagger \end{pmatrix} U_{g_\theta}^{-1} = \begin{pmatrix} e^{i\theta} & 0 \\ 0 & e^{-i\theta} \end{pmatrix} \begin{pmatrix} a \\ a^\dagger \end{pmatrix} \tag{24.5}$$

Note that, using equation 5.1

$$\frac{d}{d\theta}\left(U_{g_\theta} \begin{pmatrix} a \\ a^\dagger \end{pmatrix} U_{g_\theta}^{-1}\right)_{|\theta=0} = \left[-\frac{i}{2}(aa^\dagger + a^\dagger a), \begin{pmatrix} a \\ a^\dagger \end{pmatrix}\right]$$

so equation 24.4 is the derivative at $\theta = 0$ of equation 24.5. We see that, on operators, conjugation by the action of this $SO(2)$ subgroup of $SL(2, \mathbf{R})$ does not mix creation and annihilation operators. On the distinguished state $|0\rangle$, U_{g_θ} acts as the phase transformation

$$U_{g_\theta}|0\rangle = e^{-\frac{i}{2}\theta}|0\rangle$$

Besides 24.3, there are also the following other Poisson bracket relations between order two and order one polynomials in $z, \bar{z}$

$$\{z^2, \bar{z}\} = 2iz, \quad \{z^2, z\} = 0, \quad \{\bar{z}^2, z\} = -2i\bar{z}, \quad \{\bar{z}^2, \bar{z}\} = 0 \qquad (24.6)$$

The function

$$\mu = \frac{i}{2}(\bar{z}^2 - z^2)$$

will provide a moment map for the $\mathbf{R} \subset SL(2, \mathbf{R})$ subgroup studied in section 20.3.4. This is the subgroup of elements g_r that for $r \in \mathbf{R}$ act on basis elements q, p by

$$\begin{pmatrix} q \\ p \end{pmatrix} \rightarrow \begin{pmatrix} e^r & 0 \\ 0 & e^{-r} \end{pmatrix} \begin{pmatrix} q \\ p \end{pmatrix} = \begin{pmatrix} e^r q \\ e^{-r} p \end{pmatrix} \qquad (24.7)$$

or on basis elements $z, \bar{z}$ by

$$\begin{pmatrix} \bar{z} \\ z \end{pmatrix} \rightarrow \begin{pmatrix} \cosh r & \sinh r \\ \sinh r & \cosh r \end{pmatrix} \begin{pmatrix} \bar{z} \\ z \end{pmatrix}$$

This moment map satisfies the relations

$$\left\{ \mu, \begin{pmatrix} \bar{z} \\ z \end{pmatrix} \right\} = \begin{pmatrix} 0 & 1 \\ 1 & 0 \end{pmatrix} \begin{pmatrix} \bar{z} \\ z \end{pmatrix} \qquad (24.8)$$

Quantization gives

$$\Gamma'_{BF}(\mu) = \frac{1}{2}(a^2 - (a^\dagger)^2)$$

which satisfies

$$\left[\frac{1}{2}(a^2 - (a^\dagger)^2), \begin{pmatrix} a \\ a^\dagger \end{pmatrix} \right] = \begin{pmatrix} 0 & 1 \\ 1 & 0 \end{pmatrix} \begin{pmatrix} a \\ a^\dagger \end{pmatrix}$$

and intertwining operators

$$U_{g_r} = e^{\Gamma'_{BF}(r\mu)} = e^{\frac{r}{2}(a^2 - (a^\dagger)^2)}$$

which satisfy

$$U_{g_r} \begin{pmatrix} a \\ a^\dagger \end{pmatrix} U_{g_r}^{-1} = e^{r \begin{pmatrix} 0 & 1 \\ 1 & 0 \end{pmatrix}} \begin{pmatrix} a \\ a^\dagger \end{pmatrix} = \begin{pmatrix} \cosh r & \sinh r \\ \sinh r & \cosh r \end{pmatrix} \begin{pmatrix} a \\ a^\dagger \end{pmatrix} \qquad (24.9)$$

The operator $\frac{1}{2}(a^2 - (a^\dagger)^2)$ does not commute with the number operator N, or the harmonic oscillator Hamiltonian H, so the transformations U_{g_r} are not "symmetry transformations," preserving energy eigenspaces. In particular, they act non-trivially on the state $|0\rangle$, taking it to a different state

$$|0\rangle_r = e^{\frac{\tau}{2}(a^2 - (a^\dagger)^2)}|0\rangle$$

24.3 Implications of the choice of $z, \overline{z}$

The definition of annihilation and creation operators requires making a specific choice, in our case

$$\overline{z} = \frac{1}{\sqrt{2}}(q + ip), \quad z = \frac{1}{\sqrt{2}}(q - ip)$$

for complexified coordinates on phase space, which after quantization becomes the choice

$$a = \frac{1}{\sqrt{2}}(Q + iP), \quad a^\dagger = \frac{1}{\sqrt{2}}(Q - iP)$$

Besides the complexification of coordinates on phase space M, the choice of z introduces a new piece of structure into the problem. In chapter 26, we will examine other possible consistent such choices, here will just point out the various different ways in which this extra structure appears.

- The Schrödinger representation of the Heisenberg group comes with no particular distinguished state. The unitarily equivalent Bargmann–Fock representation does come with a distinguished state, the constant function 1. It has zero eigenvalue for the number operator $N = a^\dagger a$, so can be thought of as the state with zero "quanta," or the "vacuum" state and can be written $|0\rangle$. Such a constant function could also be characterized (up to scalar multiplication), as the state that satisfies the condition

$$a|0\rangle = 0$$

- The choice of coordinates z and $\overline{z}$ gives a distinguished choice of Hamiltonian function, $h = z\overline{z}$. After quantization, this corresponds to a distinguished choice of Hamiltonian operator

$$H = \frac{1}{2}(a^\dagger a + a a^\dagger) = a^\dagger a + \frac{1}{2} = N + \frac{1}{2}$$

With this choice the distinguished state $|0\rangle$ will be an eigenstate of H with eigenvalue $\frac{1}{2}$.

- The choice of the coordinate z gives a decomposition

$$\mathcal{M} \otimes \mathbf{C} = \mathbf{C} \oplus \mathbf{C} \tag{24.10}$$

where the first subspace $\mathbf{C}$ has basis vector z, the second subspace has basis vector $\overline{z}$.

- The decomposition 24.10 picks out a subgroup $U(1) \subset SL(2, \mathbf{R})$, those symplectic transformations that preserve the decomposition. In terms of the coordinates $z, \overline{z}$, the Lie bracket relations 16.15 giving the action of $\mathfrak{sl}(2, \mathbf{R})$ on $\mathcal{M}$ become

$$\{z\overline{z}, z\} = -iz, \quad \{z\overline{z}, \overline{z}\} = i\overline{z}$$

$$\left\{\frac{z^2}{2}, z\right\} = 0, \quad \left\{\frac{z^2}{2}, \overline{z}\right\} = iz$$

$$\left\{\frac{\overline{z}^2}{2}, z\right\} = -i\overline{z}, \quad \left\{\frac{\overline{z}^2}{2}, \overline{z}\right\} = 0$$

The only basis element of $\mathfrak{sl}(2, \mathbf{R})$ does not mix the z and $\overline{z}$ coordinates is $z\overline{z}$. We saw (see equation 24.1) that upon exponentiation this basis element gives the subgroup of $SL(2, \mathbf{R})$ of matrices of the form

$$\begin{pmatrix} \cos\theta & \sin\theta \\ -\sin\theta & \cos\theta \end{pmatrix}$$

- Quantization of polynomials in $z, \overline{z}$ involves an operator ordering ambiguity since a and $a^\dagger$ do not commute. This can be resolved by the following specific choice, one that depends on the choice of z and $\overline{z}$:

Definition. *Normal-ordered product*
Given any product P of the a and $a^\dagger$ operators, the normal-ordered product of P, written :P: is given by reordering the product so that all factors $a^\dagger$ are on the left, all factors a on the right, for example

$$:a^2 a^\dagger a (a^\dagger)^3: = (a^\dagger)^4 a^3$$

For the case of the Hamiltonian H, the normal-ordered version

$$:H: = :\frac{1}{2}(aa^\dagger + a^\dagger a): = a^\dagger a$$

could be chosen. This has the advantage that it acts trivially on $|0\rangle$ and has integer rather than half-integer eigenvalues on $\mathcal{F}$. Upon exponentiation, one gets a representation of $U(1)$ with no sign ambiguity and thus, no need to invoke a double covering. The disadvantage is that $:H:$ gives a representation of $\mathfrak{u}(1)$ that does not extend to a representation of $\mathfrak{sl}(2, \mathbf{R})$.

24.4 $SU(1,1)$ and Bogoliubov transformations

Changing bases in complexified phase space from q, p to $z, \overline{z}$ changes the group of linear transformations preserving the Poisson bracket from the group $SL(2, \mathbf{R})$ of real 2 by 2 matrices of determinant one to an isomorphic group of complex 2 by 2 matrices. We have

Theorem. *The group $SL(2, \mathbf{R})$ is isomorphic to the group $SU(1,1)$ of complex 2 by 2 matrices*

$$\begin{pmatrix} \alpha & \beta \\ \overline{\beta} & \overline{\alpha} \end{pmatrix}$$

such that

$$|\alpha|^2 - |\beta|^2 = 1$$

Proof. The equations for $\overline{z}, z$ in terms of q, p imply that the change of basis between these two bases is

$$\begin{pmatrix} \overline{z} \\ z \end{pmatrix} = \frac{1}{\sqrt{2}} \begin{pmatrix} 1 & i \\ 1 & -i \end{pmatrix} \begin{pmatrix} q \\ p \end{pmatrix}$$

The matrix for this transformation has inverse

$$\frac{1}{\sqrt{2}} \begin{pmatrix} 1 & 1 \\ -i & i \end{pmatrix}$$

Conjugating by this change of basis matrix, one finds

$$\frac{1}{\sqrt{2}} \begin{pmatrix} 1 & 1 \\ -i & i \end{pmatrix} \begin{pmatrix} \alpha & \beta \\ \overline{\beta} & \overline{\alpha} \end{pmatrix} \frac{1}{\sqrt{2}} \begin{pmatrix} 1 & i \\ 1 & -i \end{pmatrix} = \begin{pmatrix} \mathrm{Re}(\alpha + \beta) & -\mathrm{Im}(\alpha - \beta) \\ \mathrm{Im}(\alpha + \beta) & \mathrm{Re}(\alpha - \beta) \end{pmatrix}$$

$$(24.11)$$

The right-hand side is a real matrix, with determinant one, since conjugation does not change the determinant. □

Note that the change of basis 24.11 is reflected in equations 24.3 and 24.8, where the matrices on the right-hand side are the matrix Z and G, respectively, but transformed to the $\bar{z}, z$ basis by 24.11.

Another equivalent characterization of the group $SU(1,1)$ is as the group of linear transformations of $\mathbf{C}^2$, with determinant one, preserving the indefinite Hermitian inner product

$$\left\langle \begin{pmatrix} c_1 \\ c_2 \end{pmatrix}, \begin{pmatrix} c_1' \\ c_2' \end{pmatrix} \right\rangle_{1,1} = \bar{c}_1 c_1' - \bar{c}_2 c_2'$$

One finds that

$$\left\langle \begin{pmatrix} \alpha & \beta \\ \gamma & \delta \end{pmatrix} \begin{pmatrix} c_1 \\ c_2 \end{pmatrix}, \begin{pmatrix} \alpha & \beta \\ \gamma & \delta \end{pmatrix} \begin{pmatrix} c_1' \\ c_2' \end{pmatrix} \right\rangle_{1,1} = \left\langle \begin{pmatrix} c_1 \\ c_2 \end{pmatrix}, \begin{pmatrix} c_1' \\ c_2' \end{pmatrix} \right\rangle_{1,1}$$

when $\gamma = \bar{\beta}, \delta = \bar{\alpha}$, and $|\alpha|^2 - |\beta|^2 = 1$.

Applied not to $\bar{z}, z$ but to their quantizations $a, a^\dagger$, such $SU(1,1)$ transformations are known to physicists as "Bogoliubov transformations." One can easily see that replacing the annihilation operator a by

$$a' = \alpha a + \beta a^\dagger$$

leads to operators with the same commutation relations when $|\alpha|^2 - |\beta|^2 = 1$, since

$$[a', (a')^\dagger] = [\alpha a + \beta a^\dagger, \bar{\alpha} a^\dagger + \bar{\beta} a] = (|\alpha|^2 - |\beta|^2)\mathbf{1}$$

By equation 24.11, the $SO(2) \subset SL(2, \mathbf{R})$ subgroup of equation 24.1 appears in the isomorphic $SU(1,1)$ group as the special case $\alpha = e^{i\theta}, \beta = 0$, so matrices of the form

$$\begin{pmatrix} e^{i\theta} & 0 \\ 0 & e^{-i\theta} \end{pmatrix}$$

Acting with this subgroup on the annihilation and creation operators just changes a by a phase (and $a^\dagger$ by the conjugate phase).

The subgroup 24.7 provides more non-trivial Bogoliubov transformations, with conjugation by U_{g_r} giving (see equation 24.9) annihilation and creation operators

$$a_r = a \cosh r + a^\dagger \sinh r, \quad a_r^\dagger = a \sinh r + a^\dagger \cosh r$$

For $r \neq 0$, the state

$$|0\rangle_r = e^{\frac{r}{2}(a^2 - (a^\dagger)^2)} |0\rangle$$

will be an eigenstate of neither H nor the number operator N and describes a state without a definite number of quanta. It will be the ground state for a quantum system with Hamiltonian operator

$$\begin{aligned}
H_r = a_r^\dagger a_r + \frac{1}{2} &= (\cosh^2 r + \sinh^2 r) a^\dagger a + \cosh r \sinh r (a^2 - (a^\dagger)^2) - \frac{1}{2} \\
&= (\cosh 2r) a^\dagger a + \frac{1}{2} \sinh 2r (a^2 - (a^\dagger)^2) - \frac{1}{2}
\end{aligned}$$

Such quadratic Hamiltonians that do not commute with the number operator have lowest energy states $|0\rangle_r$ with indefinite number eigenvalue. Examples of this kind occur for instance in the theory of superfluidity.

24.5   For further reading

The metaplectic representation is not usually mentioned in the physics literature, and the discussions in the mathematical literature tend to be aimed at an advanced audience. Two good examples of such detailed discussions can be found in [26] and chapters 1 and 11 of [94]. To see how Bogoliubov transformations appear in the theory of superfluidity, see, for instance, chapter 10.3 of [86].

Chapter 25
The Metaplectic Representation and Annihilation and Creation Operators, arbitrary d

In this chapter, we will turn from the $d = 1$ case of chapter 24 to the general case of arbitrary d. The choice of d annihilation and creation operators picks out a distinguished subgroup $U(d) \subset Sp(2d, \mathbf{R})$ of transformations that do not mix annihilation and creation operators, and the metaplectic representation gives one a representation of a double cover of this group. We will see that normal-ordering the products of annihilation and creation operators turns this into a representation of $U(d)$ itself (rather than the double cover). In this way, a $U(d)$ action on the finite dimensional phase space gives operators that provide an infinite dimensional representation of $U(d)$ on the state space of the d-dimensional harmonic oscillator.

This method for turning unitary symmetries of the classical phase space into unitary representations of the symmetry group on a quantum state space is elaborated in great detail here not just because of its application to simple quantum systems like the d-dimensional harmonic oscillator, but because it will turn out to be fundamental in our later study of quantum field theories. In such theories the observables of interest will be operators of a Lie algebra representation, built out of quadratic combinations of annihilation and creation operators. These arise from the construction in this chapter, applied to a unitary group action on phase space (which in the quantum field theory case will be infinite dimensional).

Studying the d-dimensional quantum harmonic oscillator using these methods, we will see in detail how in the case $d = 2$ the group $U(2) \subset Sp(4, \mathbf{R})$ commutes with the Hamiltonian, so acts as symmetries preserving energy eigenspaces on the harmonic oscillator state space. This gives the same construction of all $SU(2) \subset U(2)$ irreducible representations that we studied in chapter 8. The case $d = 3$ corresponds to the physical example of an isotropic quadratic central potential in three dimensions, with the rotation group acting on the state space as an $SO(3)$ subgroup of the subgroup $U(3) \subset Sp(6, \mathbf{R})$ of symmetries commuting with the Hamiltonian. This gives a construction of angular momentum operators in terms of annihilation and creation operators.

© Peter Woit 2017
P. Woit, *Quantum Theory, Groups and Representations*,
DOI 10.1007/978-3-319-64612-1_25

25.1 Multiple degrees of freedom

Up until now we have been working with the simple case of one physical degree of freedom, i.e., one pair (Q, P) of position and momentum operators satisfying the Heisenberg relation $[Q, P] = i\mathbf{1}$, or one pair of adjoint operators $a, a^\dagger$ satisfying $[a, a^\dagger] = \mathbf{1}$. We can easily extend this to any number d of degrees of freedom by taking tensor products of our state space $\mathcal{F}$, and d copies of our operators, each acting on a different factor of the tensor product. Our new state space will be

$$\mathcal{H} = \mathcal{F}_d = \underbrace{\mathcal{F} \otimes \cdots \otimes \mathcal{F}}_{d \text{ times}}$$

and we will have operators

$$Q_j, P_j \quad j = 1, \ldots, d$$

satisfying

$$[Q_j, P_k] = i\delta_{jk}\mathbf{1}, \quad [Q_j, Q_k] = [P_j, P_k] = 0$$

Here Q_j and P_j act on the j'th term of the tensor product in the usual way, and trivially on the other terms.

We define annihilation and creation operators then by

$$a_j = \frac{1}{\sqrt{2}}(Q_j + iP_j), \quad a_j^\dagger = \frac{1}{\sqrt{2}}(Q_j - iP_j), \quad j = 1, \ldots, d$$

These satisfy:

Definition (Canonical commutation relations). *The canonical commutation relations (often abbreviated CCR) are*

$$[a_j, a_k^\dagger] = \delta_{jk}\mathbf{1}, \quad [a_j, a_k] = [a_j^\dagger, a_k^\dagger] = 0$$

In the Bargmann–Fock representation, $\mathcal{H} = \mathcal{F}_d$ is the space of holomorphic functions in d complex variables z_j (with finite norm in the d-dimensional version of 22.4) and we have

$$a_j = \frac{\partial}{\partial z_j}, \quad a_j^\dagger = z_j$$

The harmonic oscillator Hamiltonian for d degrees of freedom will be

$$H = \frac{1}{2} \sum_{j=1}^d (P_j^2 + Q_j^2) = \sum_{j=1}^d \left(a_j^\dagger a_j + \frac{1}{2} \right) \tag{25.1}$$

where one should keep in mind that each degree of freedom can be rescaled separately, allowing different parameters ω_j for the different degrees of freedom. The energy and number operator eigenstates will be written

$$|n_1, \ldots, n_d\rangle$$

where

$$a_j^\dagger a_j |n_1, \ldots, n_d\rangle = N_j |n_1, \ldots, n_d\rangle = n_j |n_1, \ldots, n_d\rangle$$

For $d = 3$, the harmonic oscillator problem is an example of the central potential problem described in chapter 21, and will be discussed in more detail in section 25.4.2. It has an $SO(3)$ symmetry, with angular momentum operators that commute with the Hamiltonian, and spaces of energy eigenstates that can be organized into irreducible $SO(3)$ representations. In the Schrödinger representation states are in $\mathcal{H} = L^2(\mathbf{R}^3)$, described by wavefunctions that can be written in rectangular or spherical coordinates, and the Hamiltonian is a second-order differential operator. In the Bargmann–Fock representation, states in $\mathcal{F}_3$ are described by holomorphic functions of 3 complex variables, with operators given in terms of products of annihilation and creation operators. The Hamiltonian is, up to a constant, just the number operator, with energy eigenstates homogeneous polynomials (with eigenvalue of the number operator their degree).

Either the P_j, Q_j or the $a_j, a_j^\dagger$ together with the identity operator will give a representation of the Heisenberg Lie algebra $\mathfrak{h}_{2d+1}$ on $\mathcal{H}$, and by exponentiation a representation of the Heisenberg group H_{2d+1}. Quadratic combinations of these operators will give a representation of the Lie algebra $\mathfrak{sp}(2d, \mathbf{R})$, one that exponentiates to the metaplectic representation of a double cover of $Sp(2d, \mathbf{R})$.

25.2 Complex coordinates on phase space and $U(d) \subset Sp(2d, \mathbf{R})$

As in the $d = 1$ case, annihilation and creation operators can be thought of as the quantization of complexified coordinates $z_j, \overline{z}_j$ on phase space, with the standard choice given by

$$z_j = \frac{1}{\sqrt{2}}(q_j - ip_j), \quad \overline{z}_j = \frac{1}{\sqrt{2}}(q_j + ip_j)$$

Such a choice of $z_j, \overline{z}_j$ gives a decomposition of the complexified Lie algebra $\mathfrak{sp}(2d, \mathbf{C})$ (as usual, the Lie bracket is the Poisson bracket) into three Lie subalgebras as follows:

- A Lie subalgebra with basis elements $z_j z_k$. There are $\frac{1}{2}(d^2 + d)$ distinct such basis elements. This is a commutative Lie subalgebra, since the Poisson bracket of any two basis elements is zero.

- A Lie subalgebra with basis elements $\overline{z}_j \overline{z}_k$. Again, this has dimension $\frac{1}{2}(d^2 + d)$ and is a commutative Lie subalgebra.

- A Lie subalgebra with basis elements $z_j \overline{z}_k$, which has dimension d^2. Computing Poisson brackets one finds

$$
\begin{aligned}
\{z_j \overline{z}_k, z_l \overline{z}_m\} &= z_j \{\overline{z}_k, z_l \overline{z}_m\} + \overline{z}_k \{z_j, z_l \overline{z}_m\} \\
&= -i z_j \overline{z}_m \delta_{kl} + i z_l \overline{z}_k \delta_{jm}
\end{aligned}
\tag{25.2}
$$

In this chapter, we will focus on the third subalgebra and the operators that arise by quantization of its elements.

Taking all complex linear combinations, this subalgebra can be identified with the Lie algebra $\mathfrak{gl}(d, \mathbf{C})$ of all d by d complex matrices. One can see this by noting that if E_{jk} is the matrix with 1 at the j-th row and k-th column, zeros elsewhere, one has

$$
[E_{jk}, E_{lm}] = E_{jm} \delta_{kl} - E_{lk} \delta_{jm}
$$

and these provide a basis of $\mathfrak{gl}(d, \mathbf{C})$. Identifying bases by

$$
i z_j \overline{z}_k \leftrightarrow E_{jk}
$$

gives the isomorphism of Lie algebras. This $\mathfrak{gl}(d, \mathbf{C})$ is the complexification of $\mathfrak{u}(d)$, the Lie algebra of the unitary group $U(d)$. Elements of $\mathfrak{u}(d)$ will correspond to, equivalently, skew-adjoint matrices, or real linear combinations of the quadratic functions

$$
z_j \overline{z}_k + \overline{z}_j z_k, \quad i(z_j \overline{z}_k - \overline{z}_j z_k)
$$

on M.

In section 16.1.2, we saw that the moment map for the action of the symplectic group on phase space is just the identity map when we identify the Lie algebra $\mathfrak{sp}(2d, \mathbf{R})$ with order two homogeneous polynomials in the phase space coordinates q_j, p_j. We can complexify and identify $\mathfrak{sp}(2d, \mathbf{C})$ with complex-valued order two homogeneous polynomials which we write in terms of the complexified coordinates $z_j, \overline{z}_j$. The moment map is again the identity map, and on the sub-Lie algebra we are concerned with, is explicitly given by

$$
A \in \mathfrak{gl}(d, \mathbf{C}) \to \mu_A = i \sum_{j,k} z_j A_{jk} \overline{z}_k
\tag{25.3}
$$

We can at the same time consider the complexification of the Heisenberg Lie algebra, using linear functions of z_j and $\overline{z}_j$, with Poisson brackets between these and the order two homogeneous functions giving a complexified version of the derivation action of $\mathfrak{sp}(2d, \mathbf{R})$ on $\mathfrak{h}_{2d+1}$.

We have (complexifying and restricting to $\mathfrak{gl}(d, \mathbf{C}) \subset \mathfrak{sp}(2d, \mathbf{C})$) the following version of theorems 16.2 and 16.3

Theorem 25.1. *The map of equation 25.3 is a Lie algebra homomorphism, i.e.,*

$$\{\mu_A, \mu_{A'}\} = \mu_{[A,A']}$$

The μ_A satisfy (for column vectors $\mathbf{z}$ with components $z_1, \ldots, z_d$)

$$\{\mu_A, \overline{\mathbf{z}}\} = -A\overline{\mathbf{z}}, \quad \{\mu_A, \mathbf{z}\} = A^T \mathbf{z} \tag{25.4}$$

Proof. Using 25.2 one has

$$\begin{aligned}
\{\mu_A, \mu_{A'}\} &= -\sum_{j,k,l,m} \{z_j A_{jk} \overline{z}_k, z_l A'_{lm} \overline{z}_m\} \\
&= -\sum_{j,k,l,m} A_{jk} A'_{lm} \{z_j \overline{z}_k, z_l \overline{z}_m\} \\
&= i \sum_{j,k,l,m} A_{jk} A'_{lm} (z_j \overline{z}_m \delta_{kl} - z_l \overline{z}_k \delta_{jm}) \\
&= i \sum_{j,k} z_j [A, A']_{jk} \overline{z}_k = \mu_{[A,A']}
\end{aligned}$$

To show 25.4, compute

$$\begin{aligned}
\{\mu_A, \overline{z}_l\} &= \{i \sum_{j,k} z_j A_{jk} \overline{z}_k, \overline{z}_l\} = i \sum_{j,k} A_{jk} \{z_j, \overline{z}_l\} \overline{z}_k \\
&= -\sum_j A_{lj} \overline{z}_j
\end{aligned}$$

and

$$\begin{aligned}
\{\mu_A, z_l\} &= \{i \sum_{j,k} z_j A_{jk} \overline{z}_k, z_l\} = i \sum_{j,k} z_j A_{jk} \{\overline{z}_k, z_l\} \\
&= \sum_k z_k A_{kl}
\end{aligned}$$

$\square$

Note that here we have written formulas for $A \in \mathfrak{gl}(d, \mathbf{C})$, an arbitrary complex d by d matrix. It is only for $A \in \mathfrak{u}(d)$, the skew-adjoint $(\overline{A^T} = -A)$

matrices, that μ_A will be a real-valued moment map, lying in the real Lie algebra $sp(2d, \mathbf{R})$, and giving a unitary representation on the state space after quantization. For such A, we can write the relations 25.4 as a (complexified) example of 16.22

$$\left\{ \mu_A, \begin{pmatrix} \overline{\mathbf{z}} \\ \mathbf{z} \end{pmatrix} \right\} = \begin{pmatrix} \overline{A^T} & \mathbf{0} \\ \mathbf{0} & A^T \end{pmatrix} \begin{pmatrix} \overline{\mathbf{z}} \\ \mathbf{z} \end{pmatrix}$$

The standard harmonic oscillator Hamiltonian

$$h = \sum_{j=1}^{d} z_j \overline{z}_j \tag{25.5}$$

lies in this $\mathfrak{u}(d)$ subalgebra (it is the case $A = -i\mathbf{1}$), and its Poisson brackets with the rest of the subalgebra are zero. It gives a basis element of the one-dimensional $\mathfrak{u}(1)$ subalgebra that commutes with the rest of the $\mathfrak{u}(d)$ subalgebra.

While we are not entering here into the details of what happens for polynomials that are linear combinations of the $z_j z_k$ and $\overline{z}_j \overline{z}_k$, it may be worth noting one confusing point about these. Recall that in chapter 16, we found the moment map $\mu_L = -\mathbf{q} \cdot A\mathbf{p}$ for elements $L \in \mathfrak{sp}(2d, \mathbf{R})$ of the block-diagonal form

$$\begin{pmatrix} A & \mathbf{0} \\ \mathbf{0} & -A^T \end{pmatrix}$$

where A is a real d by d matrix and so in $\mathfrak{gl}(d, \mathbf{R})$. That block decomposition corresponded to the decomposition of basis vectors of $\mathcal{M}$ into the two sets q_j and p_j. Here we have complexified, and are working with respect to a different decomposition, that of basis vectors $\mathcal{M} \otimes \mathbf{C}$ into the two sets z_j and $\overline{z}_j$. The matrices A in this case are complex, skew-adjoint, and in a different non-isomorphic Lie subalgebra, $\mathfrak{u}(d)$ rather than $\mathfrak{gl}(d, \mathbf{R})$. For the simplest example of this, $d = 1$, the distinction is between the $\mathbf{R}$ Lie subgroup of $SL(2, \mathbf{R})$ (see section 20.3.4), for which the moment map is

$$-qp = \text{Im}(z^2) = \frac{1}{2i}(z^2 - \overline{z}^2)$$

and the $U(1)$ subgroup (see section 20.3.2), for which the moment map is

$$\frac{1}{2}(q^2 + p^2) = z\overline{z}$$

25.3   The metaplectic representation and $U(d) \subset Sp(2d, \mathbf{R})$

Turning to the quantization problem, we would like to extend the discussion of quantization of quadratic combinations of complex coordinates on phase space from the $d = 1$ case of chapter 24 to the general case. For any j, k one can take

$$\overline{z}_j \overline{z}_k \to -i a_j a_k, \quad z_j z_k \to -i a_j^\dagger a_k^\dagger$$

There is no ambiguity in the quantization of the two subalgebras given by pairs of the z_j coordinates or pairs of the $\overline{z}_j$ coordinates since creation operators commute with each other, and annihilation operators commute with each other.

If $j \neq k$ one can quantize by taking

$$z_j \overline{z}_k \to -i a_j^\dagger a_k = -i a_k a_j^\dagger$$

and there is again no ordering ambiguity. If $j = k$, as in the $d = 1$ case there is a choice to be made. One possibility is to take

$$z_j \overline{z}_j \to -i \frac{1}{2}(a_j a_j^\dagger + a_j^\dagger a_j) = -i \left(a_j^\dagger a_j + \frac{1}{2} \right)$$

which will have the proper $\mathfrak{sp}(2d, \mathbf{R})$ commutation relations (in particular for commutators of a_j^2 with $(a_j^\dagger)^2$), but require going to a double cover to get a true representation of the group. The Bargmann–Fock construction thus gives us a unitary representation of $\mathfrak{u}(d)$ on Fock space $\mathcal{F}_d$, but after exponentiation this is a representation not of the group $U(d)$, but of a double cover we call $\widetilde{U(d)}$.

One could instead quantize using normal-ordered operators, taking

$$z_j \overline{z}_j \to -i a_j^\dagger a_j$$

The definition of normal-ordering in section 24.3 generalizes simply, since the order of annihilation and creation operators with different values of j is immaterial. Using this normal-ordered choice, the usual quantized operators of the Bargmann–Fock representation are shifted by a scalar $\frac{1}{2}$ for each j, and after exponentiation the state space $\mathcal{H} = \mathcal{F}_d$ provides a representation of $U(d)$, with no need for a double cover. As a $\mathfrak{u}(d)$ representation however, this does not extend to a representation of $\mathfrak{sp}(2d, \mathbf{R})$, since commutation of a_j^2 with $(a_j^\dagger)^2$ can land one on the unshifted operators.

Since the normal-ordering does not change the commutation relations obeyed by products of the form $a_j^\dagger a_k$, the quadratic expression for μ_A can be quantized using normal-ordering and get quadratic combinations of the $a_j, a_k^\dagger$ with the same commutation relations as in theorem 25.1. Letting

$$U_A' = \sum_{j,k} a_j^\dagger A_{jk} a_k \qquad (25.6)$$

we have

Theorem 25.2. *For $A \in \mathfrak{gl}(d, \mathbf{C})$ a d by d complex matrix*

$$[U_A', U_{A'}'] = U_{[A,A']}'$$

As a result

$$A \in \mathfrak{gl}(d, \mathbf{C}) \rightarrow U_A'$$

is a Lie algebra representation of $\mathfrak{gl}(d, \mathbf{C})$ on $\mathcal{H} = \mathbf{C}[z_1, \ldots, z_d]$, the harmonic oscillator state space in d degrees of freedom.

In addition (for column vectors $\mathbf{a}$ with components $a_1, \ldots, a_d$)

$$[U_A', \mathbf{a}^\dagger] = A^T \mathbf{a}^\dagger, \quad [U_A', \mathbf{a}] = -A\mathbf{a} \qquad (25.7)$$

Proof. Essentially the same proof as 25.1. □

For $A \in \mathfrak{u}(d)$ the Lie algebra representation U_A' of $\mathfrak{u}(d)$ exponentiates to give a representation of $U(d)$ on $\mathcal{H} = \mathbf{C}[z_1, \ldots, z_d]$ by operators

$$U_{e^A} = e^{U_A'}$$

These satisfy

$$U_{e^A} \mathbf{a}^\dagger (U_{e^A})^{-1} = e^{A^T} \mathbf{a}^\dagger, \quad U_{e^A} \mathbf{a} (U_{e^A})^{-1} = e^{\overline{A^T}} \mathbf{a} \qquad (25.8)$$

(the relations 25.7 are the derivative of these). This shows that the U_{e^A} are intertwining operators for a $U(d)$ action on annihilation and creation operators that preserves the canonical commutation relations. Here the use of normal-ordered operators means that U_A' is a representation of $\mathfrak{u}(d)$ that differs by a constant from the metaplectic representation, and U_{e^A} differs by a phase-factor. This does not affect the commutation relations with U_A' or the conjugation action of U_{e^A}. The representation constructed this way differs in two ways from the metaplectic representation. It acts on the same space $\mathcal{H} = \mathcal{F}_d$, but it is a true representation of $U(d)$, no double cover is needed. It also does not extend to a representation of the larger group $Sp(2d, \mathbf{R})$.

The operators U_A' and U_{e^A} commute with the Hamiltonian operator for the harmonic oscillator (the quantization of equation 25.5). For physicists this is quite useful, as it provides a decomposition of energy eigenstates into irreducible representations of $U(d)$. For mathematicians, the quantum harmonic oscillator state space provides a construction of a large class of irreducible representations of $U(d)$, by considering the energy eigenstates of a given energy.

25.4   Examples in $d = 2$ and 3

25.4.1   Two degrees of freedom and $SU(2)$

In the case $d = 2$, the action of the group $U(2) \subset Sp(4, \mathbf{R})$ discussed in section 25.3 commutes with the standard harmonic oscillator Hamiltonian and thus acts as symmetries on the quantum harmonic oscillator state space, preserving energy eigenspaces. Restricting to the subgroup $SU(2) \subset U(2)$, we will see that we can recover our earlier (see section 8.2) construction of $SU(2)$ representations in terms of homogeneous polynomials, in a new context. This use of the energy eigenstates of a two-dimensional harmonic oscillator appears in the physics literature as the "Schwinger boson method" for studying representations of $SU(2)$.

The state space for the $d = 2$ Bargmann–Fock representation, restricting to finite linear combinations of energy eigenstates, is

$$\mathcal{H} = \mathcal{F}_2^{fin} = \mathbf{C}[z_1, z_2]$$

the polynomials in two complex variables z_1, z_2. Recall from our $SU(2)$ discussion that it was useful to organize these polynomials into finite dimensional sets of homogeneous polynomials of degree n for $n = 0, 1, 2, \ldots$

$$\mathcal{H} = \mathcal{H}^0 \oplus \mathcal{H}^1 \oplus \mathcal{H}^2 \oplus \cdots$$

There are four annihilation or creation operators

$$a_1^\dagger = z_1, \ a_2^\dagger = z_2, \ a_1 = \frac{\partial}{\partial z_1}, \ a_2 = \frac{\partial}{\partial z_2}$$

acting on $\mathcal{H}$. These are the quantizations of complexified phase space coordinates $z_1, z_2, \overline{z}_1, \overline{z}_2$, with quantization the Bargmann–Fock construction of the representation Γ'_{BF} of $\mathfrak{h}_{2d+1} = \mathfrak{h}_5$

$$\Gamma'_{BF}(1) = -i\mathbf{1}, \ \Gamma'_{BF}(z_j) = -ia_j^\dagger, \ \Gamma'_{BF}(\overline{z}_j) = -ia_j$$

Quadratic combinations of the creation and annihilation operators give representations on $\mathcal{H}$ of three subalgebras of the complexification $\mathfrak{sp}(4, \mathbf{C})$ of $\mathfrak{sp}(4, \mathbf{R})$:

- A three-dimensional commutative Lie subalgebra spanned by $\bar{z}_1 \bar{z}_2, \bar{z}_1^2, \bar{z}_2^2$, with quantization

$$\Gamma'_{BF}(\bar{z}_1 \bar{z}_2) = -i a_1 a_2, \quad \Gamma'_{BF}(\bar{z}_1^2) = -i a_1^2, \quad \Gamma'_{BF}(\bar{z}_2^2) = -i a_2^2$$

- A three-dimensional commutative Lie subalgebra spanned by $z_1 z_2, z_1^2, z_2^2$, with quantization

$$\Gamma'_{BF}(z_1 z_2) = -i a_1^\dagger a_2^\dagger, \quad \Gamma'_{BF}(z_1^2) = -i(a_1^\dagger)^2, \quad \Gamma'_{BF}(z_2^2) = -i(a_2^\dagger)^2$$

- A four-dimensional Lie subalgebra isomorphic to $\mathfrak{gl}(2, \mathbf{C})$ with basis

$$z_1 \bar{z}_1, z_2 \bar{z}_2, z_2 \bar{z}_1, z_1 \bar{z}_2$$

and quantization

$$\Gamma'_{BF}(z_1 \bar{z}_1) = -\frac{i}{2}(a_1^\dagger a_1 + a_1 a_1^\dagger), \quad \Gamma'_{BF}(z_2 \bar{z}_2) = -\frac{i}{2}(a_2^\dagger a_2 + a_2 a_2^\dagger)$$

$$\Gamma'_{BF}(z_2 \bar{z}_1) = -i a_2^\dagger a_1, \quad \Gamma'_{BF}(z_1 \bar{z}_2) = -i a_1^\dagger a_2$$

Real linear combinations of

$$z_1 \bar{z}_1, \quad z_2 \bar{z}_2, \quad z_1 \bar{z}_2 + z_2 \bar{z}_1, \quad i(z_1 \bar{z}_2 - z_2 \bar{z}_1)$$

span the Lie algebra $\mathfrak{u}(2) \subset \mathfrak{sp}(4, \mathbf{R})$, and Γ'_{BF} applied to these gives a unitary Lie algebra representation by skew-adjoint operators.

Inside this last subalgebra, there is a distinguished element $h = z_1 \bar{z}_1 + z_2 \bar{z}_2$ that Poisson-commutes with the rest of the subalgebra (but not with elements in the first two subalgebras). Quantization of h gives the Hamiltonian operator

$$H = \frac{1}{2}(a_1 a_1^\dagger + a_1^\dagger a_1 + a_2 a_2^\dagger + a_2^\dagger a_2) = N_1 + \frac{1}{2} + N_2 + \frac{1}{2} = z_1 \frac{\partial}{\partial z_1} + z_2 \frac{\partial}{\partial z_2} + 1$$

This operator will multiply a homogeneous polynomial by its degree plus one, so it acts by multiplication by $n + 1$ on $\mathcal{H}^n$. Exponentiating this operator (multiplied by $-i$) one gets a representation of a $U(1)$ subgroup of the metaplectic cover $Mp(4, \mathbf{R})$. Taking instead the normal-ordered version

$$:H: = a_1^\dagger a_1 + a_2^\dagger a_2 = N_1 + N_2 = z_1 \frac{\partial}{\partial z_1} + z_2 \frac{\partial}{\partial z_2}$$

one gets a representation of a $U(1)$ subgroup of $Sp(4, \mathbf{R})$. Neither H nor $:H:$ commutes with operators coming from quantization of the first two subalgebras. These will be linear combinations of pairs of either creation or annihilation operators, so will change the eigenvalue of H or $:H:$ by ± 2, mapping

$$\mathcal{H}^n \to \mathcal{H}^{n \pm 2}$$

and in particular taking $|0\rangle$ to either 0 or a state in $\mathcal{H}^2$.

h is a basis element for the $\mathfrak{u}(1)$ in $\mathfrak{u}(2) = \mathfrak{u}(1) \oplus \mathfrak{su}(2)$. For the $\mathfrak{su}(2)$ part, on basis elements $X_j = -i\frac{\sigma_j}{2}$ the moment map 25.3 gives the following quadratic polynomials

$$\mu_{X_1} = \frac{1}{2}(z_1\bar{z}_2 + z_2\bar{z}_1), \quad \mu_{X_2} = \frac{i}{2}(z_2\bar{z}_1 - z_1\bar{z}_2), \quad \mu_{X_3} = \frac{1}{2}(z_1\bar{z}_1 - z_2\bar{z}_2)$$

This relates two different but isomorphic ways of describing $\mathfrak{su}(2)$: as 2 by 2 matrices with Lie bracket the commutator, or as quadratic polynomials, with Lie bracket the Poisson bracket.

Quantizing using the Bargmann–Fock representation give a representation of $\mathfrak{su}(2)$ on $\mathcal{H}$

$$\Gamma'_{BF}(X_1) = -\frac{i}{2}(a_1^\dagger a_2 + a_2^\dagger a_1), \quad \Gamma'_{BF}(X_2) = \frac{1}{2}(a_2^\dagger a_1 - a_1^\dagger a_2)$$

$$\Gamma'_{BF}(X_3) = -\frac{i}{2}(a_1^\dagger a_1 - a_2^\dagger a_2)$$

Comparing this to the representation π' of $\mathfrak{su}(2)$ on homogeneous polynomials discussed in chapter 8, one finds that Γ'_{BF} and π' are the same representation. The inner product that makes the representation unitary is the one of equation 8.2. The Bargmann–Fock representation extends this $SU(2)$ representation as a unitary representation to a much larger group $(H_5 \rtimes Mp(4, \mathbf{R}))$, with all polynomials in z_1, z_2 now making up a single irreducible representation of H_5.

The fact that we have an $SU(2)$ group acting on the state space of the $d = 2$ harmonic oscillator and commuting with the action of the Hamiltonian H means that energy eigenstates can be organized as irreducible representations of $SU(2)$. In particular, one sees that the space $\mathcal{H}^n$ of energy eigenstates of energy $n + 1$ will be a single irreducible $SU(2)$ representation, the spin $\frac{n}{2}$ representation of dimension $n + 1$ (so $n + 1$ will be the multiplicity of energy eigenstates of that energy).

Another physically interesting subgroup here is the $SO(2) \subset SU(2) \subset Sp(4, \mathbf{R})$ consisting of simultaneous rotations in the position and momentum planes, which was studied in detail using the coordinates q_1, q_2, p_1, p_2 in section 20.3.1. There we found that the moment map was given by

$$\mu_L = l = q_1 p_2 - q_2 p_1$$

and quantization by the Schrödinger representation gave a representation of the Lie algebra $\mathfrak{so}(2)$ with

$$U'_L = -i(Q_1 P_2 - Q_2 P_1)$$

Note that this is a different $SO(2)$ action than the one with moment map the Hamiltonian, it acts separately on positions and momenta rather than mixing them.

To see what happens if one instead uses the Bargmann–Fock representation, using

$$q_j = \frac{1}{\sqrt{2}}(z_j + \overline{z}_j), \quad p_j = i\frac{1}{\sqrt{2}}(z_j - \overline{z}_j)$$

the moment map is

$$
\begin{aligned}
\mu_L =& \frac{i}{2}((z_1 + \overline{z}_1)(z_2 - \overline{z}_2) - (z_2 + \overline{z}_2)(z_1 - \overline{z}_1)) \\
=& i(z_2 \overline{z}_1 - z_1 \overline{z}_2)
\end{aligned}
$$

Quantizing, the operator

$$U'_L = a_2^\dagger a_1 - a_1^\dagger a_2 = \Gamma'(2X_2)$$

gives a unitary representation of $\mathfrak{so}(2)$. The factor of two here reflects the fact that exponentiation gives a representation of $SO(2) \subset Sp(4, \mathbf{R})$, with no need for a double cover.

25.4.2 Three degrees of freedom and $SO(3)$

The case $d = 3$ corresponds physically to the so-called "isotropic quantum harmonic oscillator system," and it is an example of the sort of central potential problem we studied in chapter 21 (since the potential just depends on $r^2 = q_1^2 + q_2^2 + q_3^2$). For such problems, we saw that since the classical Hamiltonian is rotationally invariant, the quantum Hamiltonian will commute with the action of $SO(3)$ on wavefunctions, and energy eigenstates can be decomposed into irreducible representations of $SO(3)$.

Here the Bargmann–Fock representation gives an action of $H_7 \rtimes Mp(6, \mathbf{R})$ on the state space, with a $U(3)$ subgroup commuting with the Hamiltonian (more precisely one has a double cover of $U(3)$, but by normal-ordering one can get an actual $U(3)$). The eigenvalue of the $U(1)$ corresponding to the Hamiltonian gives the energy of a state, and states of a given energy will be sums of irreducible representations of $SU(3)$. This works much like in the $d = 2$ case, although here our irreducible representations are on the spaces $\mathcal{H}^n$ of homogeneous polynomials of degree n in three variables rather than two. These spaces have dimension $\frac{1}{2}(n + 1)(n + 2)$. A difference with the $SU(2)$ case is that one does not get all irreducible representations of $SU(3)$ this way.

The rotation group $SO(3)$ will be a subgroup of this $U(3)$ and one can ask how the $SU(3)$ irreducible $\mathcal{H}^n$ decomposes into a sum of irreducibles of the subgroup (which will be characterized by an integral spin $l = 0, 1, 2, \cdots$). One can show that for even n one gets all even values of l from 0 to n, and for odd n one gets all odd values of l from 1 to n. A derivation can be found in some quantum mechanics textbooks, see, for example, pages 456–460 of [60].

To construct the angular momentum operators in the Bargmann–Fock representation, recall that in the Schrödinger representation these were

$$L_1 = Q_2 P_3 - Q_3 P_2, \quad L_2 = Q_3 P_1 - Q_1 P_3, \quad L_3 = Q_1 P_2 - Q_2 P_1$$

and these operators can be rewritten in terms of annihilation and creation operators. Alternatively, theorem 25.2 can be used, for Lie algebra basis elements $l_j \in \mathfrak{so}(3) \subset \mathfrak{u}(3) \subset \mathfrak{gl}(3, \mathbf{C})$ which are (see chapter 6)

$$l_1 = \begin{pmatrix} 0 & 0 & 0 \\ 0 & 0 & -1 \\ 0 & 1 & 0 \end{pmatrix}, \quad l_2 = \begin{pmatrix} 0 & 0 & 1 \\ 0 & 0 & 0 \\ -1 & 0 & 0 \end{pmatrix}, \quad l_3 = \begin{pmatrix} 0 & -1 & 0 \\ 1 & 0 & 0 \\ 0 & 0 & 0 \end{pmatrix}$$

to calculate

$$-iL_j = U'_{l_j} = \sum_{m,n=1}^{3} a_m^\dagger (l_j)_{mn} a_n$$

This gives

$$U'_{l_1} = a_3^\dagger a_2 - a_2^\dagger a_3, \quad U'_{l_2} = a_1^\dagger a_3 - a_3^\dagger a_1, \quad U'_{l_3} = a_2^\dagger a_1 - a_1^\dagger a_2$$

Exponentiating these operators gives a representation of the rotation group $SO(3)$ on the state space $\mathcal{F}_3$, commuting with the Hamiltonian, so acting on energy eigenspaces (which will be the homogeneous polynomials of fixed degree).

25.5 Normal-ordering and the anomaly in finite dimensions

For $A \in \mathfrak{u}(d) \subset \mathfrak{sp}(2d, \mathbf{R})$ we have seen that we can construct the Lie algebra version of the metaplectic representation as

$$\widetilde{U}'_A = \frac{1}{2} \sum_{j,k} A_{jk}(a_j^\dagger a_k + a_k a_j^\dagger)$$

which gives a representation that extends to $\mathfrak{sp}(2d, \mathbf{R})$, or we can normal-order, getting

$$U'_A = :\widetilde{U}'_A: = \sum_{j,k} a_j^\dagger A_{jk} a_k = \widetilde{U}'_A - \frac{1}{2}\sum_{j=1}^{d} A_{jj}\mathbf{1}$$

To see that this normal-ordered version does not extend to $\mathfrak{sp}(2d, \mathbf{R})$, observe that basis elements of $\mathfrak{sp}(2d, \mathbf{R})$ that are not in $\mathfrak{u}(d)$ are the linear combinations of $z_j z_k$ and $\overline{z}_j \overline{z}_k$ that correspond to real-valued functions. These are given by

$$\frac{1}{2}\sum_{jk}(B_{jk}z_j z_k + \overline{B}_{jk}\overline{z}_j \overline{z}_k)$$

for a complex symmetric matrix B with matrix entries B_{jk}. There is no normal-ordering ambiguity here, and quantization will give the unitary Lie algebra representation operators

$$-\frac{i}{2}\sum_{jk}(B_{jk}a_j^\dagger a_k^\dagger + \overline{B}_{jk}a_j a_k)$$

Exponentiating such operators will give operators which take the state $|0\rangle$ to a distinct state (one not proportional to $|0\rangle$).

Using the canonical commutation relations one can show

$$[a_j^\dagger a_k^\dagger, a_l a_m] = -a_l a_j^\dagger \delta_{km} - a_l a_k^\dagger \delta_{jm} - a_j^\dagger a_m \delta_{kl} - a_k^\dagger a_m \delta_{jl}$$

and these relations can in turn be used to compute the commutator of two such Lie algebra representation operators, with the result

$$[-\frac{i}{2}\sum_{jk}(B_{jk}a_j^\dagger a_k^\dagger + \overline{B}_{jk}a_j a_k), -\frac{i}{2}\sum_{lm}(C_{lm}a_l^\dagger a_m^\dagger + \overline{C}_{lm}a_l a_m)]$$

$$= \frac{1}{2} \sum_{jk} (B\overline{C} - C\overline{B})_{jk} (a_j^\dagger a_k + a_k a_j^\dagger) = \widetilde{U}'_{B\overline{C}-C\overline{B}}$$

Note that normal-ordering of these operators just shifts them by a constant, in particular

$$U'_{B\overline{C}-C\overline{B}} = :\widetilde{U}'_{B\overline{C}-C\overline{B}}: = \widetilde{U}'_{B\overline{C}-C\overline{B}} - \frac{1}{2} tr(B\overline{C} - C\overline{B})\mathbf{1} \qquad (25.9)$$

The normal-ordered operators fail to give a Lie algebra homomorphism when extended to $\mathfrak{sp}(2d, \mathbf{R})$, but this failure is just by a constant term. Recall from section 15.3 that even at the classical level, there was an ambiguity of a constant in the choice of a moment map which in principle could lead to an "anomaly," a situation where the moment map failed to be a Lie algebra homomorphism by a constant term. The situation here is that this potential anomaly is removable, by the shift

$$U'_A \to \widetilde{U}'_A = U'_A + \frac{1}{2} tr(A)\mathbf{1}$$

which gives representation operators that satisfy the Lie algebra homomorphism property. We will see in chapter 39 that for an infinite number of degrees of freedom, the anomaly may not be removable, since the trace of the operator A in that case may be divergent.

25.6 For further reading

The references from chapter 26 [26, 94] also contain the general case discussed here. Given a $U(d) \subset Sp(2d, \mathbf{R})$ action of phase space, the construction of corresponding metaplectic representation operators using quadratic expressions in annihilation and creation operators is of fundamental importance in quantum field theory, where d is infinite. This topic is, however, usually not discussed in physics textbooks for the finite dimensional case. We will encounter the quantum field theory version in later chapters where it will be examined in detail.

Chapter 26
Complex Structures
and Quantization

The Schrödinger representation Γ_S of H_{2d+1} uses a specific choice of extra structure on classical phase space: a decomposition of its coordinates into positions q_j and momenta p_j. For the unitarily equivalent Bargmann–Fock representation, a different sort of extra structure is needed, a decomposition of coordinates on phase space into complex coordinates z_j and their complex conjugates $\overline{z}_j$. Such a decomposition is called a "complex structure" J and will correspond after quantization to a choice that distinguishes annihilation and creation operators. In previous chapters, we used one particular standard choice $J = J_0$, but in this chapter will describe other possible choices. For each such choice we will get a different version Γ_J of the Bargmann–Fock construction of a Heisenberg group representation. In later chapters on relativistic quantum field theory, we will see that the phenomenon of antiparticles is best understood in terms of a new possibility for the choice of J that appears in that case.

26.1 Complex structures and phase space

Quantization of phase space $M = \mathbf{R}^{2d}$ using the Schrödinger representation gives a unitary Lie algebra representation Γ'_S of the Heisenberg Lie algebra $\mathfrak{h}_{2d+1}$ which takes the q_j and p_j coordinate functions on phase space to operators $-iQ_j$ and $-iP_j$ on $\mathcal{H}_S = L^2(\mathbf{R}^d)$. This involves a choice, that of taking states to be functions of the q_j, or (using the Fourier transform) of the p_j. It turns out to be a general phenomenon that quantization requires choosing some extra structure on phase space, beyond the Poisson bracket.

For the case of the harmonic oscillator, we found in chapter 22 that quantization was most conveniently performed using annihilation and creation

© Peter Woit 2017
P. Woit, *Quantum Theory, Groups and Representations*,
DOI 10.1007/978-3-319-64612-1_26

operators, which involve a different sort of choice of extra structure on phase space. There we introduced complex coordinates on phase space, making the choice

$$z_j = \frac{1}{\sqrt{2}}(q_j - ip_j), \quad \bar{z}_j = \frac{1}{\sqrt{2}}(q_j + ip_j)$$

The z_j were then quantized using creation operators $a_j^\dagger$, the $\bar{z}_j$ using annihilation operators a_j. In the Bargmann–Fock representation, where the state space is a space of functions of complex variables z_j, we have

$$a_j = \frac{\partial}{\partial z_j}, \quad a_j^\dagger = z_j$$

and there is a distinguished state, the constant function, which is annihilated by all the a_j.

In this section, we will introduce the notion of a complex structure on a real vector space, with such structures characterizing the possible ways of introducing complex coordinates $z_j, \bar{z}_j$ and thus annihilation and creation operators. The abstract notion of a complex structure can be formalized as follows. Given any real vector space $V = \mathbf{R}^n$, we have seen that taking complex linear combinations of vectors in V gives a complex vector space $V \otimes \mathbf{C}$, the complexification of V, and this can be identified with $\mathbf{C}^n$, a real vector space of twice the dimension. When $n = 2d$ is even there is another way to turn $V = \mathbf{R}^{2d}$ into a complex vector space, by using the following additional piece of information:

Definition (Complex structure). *A complex structure on a real vector space V is a linear operator*

$$J : V \to V$$

such that

$$J^2 = -\mathbf{1}$$

Given such a pair $(V = \mathbf{R}^{2d}, J)$ complex linear combinations of vectors in V can be decomposed into those on which J acts as i and those on which it acts as $-i$ (since $J^2 = -\mathbf{1}$, its eigenvalues must be $\pm i$), so we have

$$V \otimes \mathbf{C} = V_J^+ \oplus V_J^-$$

where V_J^+ is the $+i$ eigenspace of the operator J on $V \otimes \mathbf{C}$ and V_J^- is the $-i$ eigenspace. Note that we have extended the action of J on V to an action on $V \otimes \mathbf{C}$ using complex linearity. Complex conjugation takes elements of V_J^+ to V_J^- and vice versa. The choice of J has thus given us two complex vector spaces of complex dimension d, V_J^+ and V_J^-, related by this complex conjugation.

Since
$$J(v - iJv) = i(v - iJv)$$

for any $v \in V$, the real vector space V can by identified with the complex vector space V_J^+ by the map

$$v \in V \to \frac{1}{\sqrt{2}}(v - iJv) \in V_J^+ \qquad (26.1)$$

The pair (V, J) can be thought of as giving V the structure of a complex vector space, with J providing multiplication by i. Similarly, taking

$$v \in V \to \frac{1}{\sqrt{2}}(v + iJv) \in V_J^- \qquad (26.2)$$

identifies V with V_J^-, with J now providing multiplication by $-i$. V_J^+ and V_J^- are interchanged by changing the complex structure J to $-J$.

For the study of quantization, the real vector space we want to choose a complex structure on is the dual phase space $\mathcal{M} = M^*$, since it is elements of this space that are in a Heisenberg algebra, and taken to operators by quantization. There will be a decomposition

$$\mathcal{M} \otimes \mathbf{C} = \mathcal{M}_J^+ \oplus \mathcal{M}_J^-$$

and quantization will take elements of $\mathcal{M}_J^+$ to linear combinations of creation operators, elements of $\mathcal{M}_J^-$ to linear combinations of annihilation operators.

The standard choice of complex structure is to take $J = J_0$, where J_0 is the linear operator that acts on coordinate basis vectors q_j, p_j of $\mathcal{M}$ by

$$J_0 q_j = p_j, \quad J_0 p_j = -q_j$$

Making the choice

$$z_j = \frac{1}{\sqrt{2}}(q_j - ip_j)$$

implies

$$J_0 z_j = \frac{1}{\sqrt{2}}(p_j + iq_j) = i z_j$$

and the z_j are basis elements (over the complex numbers) of $\mathcal{M}_{J_0}^+$. The complex conjugates

$$\overline{z}_j = \frac{1}{\sqrt{2}}(q_j + ip_j)$$

provide basis elements of $\mathcal{M}_{J_0}^-$

With respect to the chosen basis q_j, p_j, the complex structure can be written as a matrix. For the case of J_0 and for $d = 1$, on an arbitrary element of $\mathcal{M}$ the action of J_0 is

$$J_0(c_q q + c_p p) = c_q p - c_p q$$

so J_0 in matrix form with respect to the basis (q, p) is

$$J_0 \begin{pmatrix} c_q \\ c_p \end{pmatrix} = \begin{pmatrix} 0 & -1 \\ 1 & 0 \end{pmatrix} \begin{pmatrix} c_q \\ c_p \end{pmatrix} = \begin{pmatrix} -c_p \\ c_q \end{pmatrix} \tag{26.3}$$

or, the action on basis vectors is the transpose

$$J_0 \begin{pmatrix} q \\ p \end{pmatrix} = \begin{pmatrix} 0 & 1 \\ -1 & 0 \end{pmatrix} \begin{pmatrix} q \\ p \end{pmatrix} = \begin{pmatrix} p \\ -q \end{pmatrix} \tag{26.4}$$

Note that, after complexifying, three different ways to identify the original $\mathcal{M}$ with a subspace of $\mathcal{M} \otimes \mathbf{C}$ are:

- $\mathcal{M}$ is identified with $\mathcal{M}^+_{J_0}$ by equation 26.1, with basis element q_j going to z_j, and p_j to iz_j.

- $\mathcal{M}$ is identified with $\mathcal{M}^-_{J_0}$ by equation 26.2, with basis element q_j going to $\overline{z}_j$, and p_j to $-i\overline{z}_j$.

- $\mathcal{M}$ is identified with elements of $\mathcal{M}^+_{J_0} \oplus \mathcal{M}^-_{J_0}$ that are invariant under conjugation, with basis element q_j going to $\frac{1}{\sqrt{2}}(z_j + \overline{z}_j)$ and p_j to $\frac{i}{\sqrt{2}}(z_j - \overline{z}_j)$.

26.2   Compatible complex structures and positivity

Our interest is in vector spaces $\mathcal{M}$ that come with a symplectic structure Ω, a non-degenerate antisymmetric bilinear form. To successfully use a complex structure J for quantization, it will turn out that it must be compatible with Ω in the following sense:

Definition (Compatible complex structure). *A complex structure on $\mathcal{M}$ is said to be compatible with Ω if*

$$\Omega(Jv_1, Jv_2) = \Omega(v_1, v_2) \tag{26.5}$$

Equivalently, $J \in Sp(2d, \mathbf{R})$, the group of linear transformations of $\mathcal{M}$ preserving Ω.

The standard complex structure $J = J_0$ is compatible with Ω, since (treating the $d = 1$ case, which generalizes easily, and using equations 16.2 and 26.3)

$$\Omega(J_0(c_q q + c_p p), J_0(c'_q q + c'_p p))$$

$$= \left(\begin{pmatrix} 0 & -1 \\ 1 & 0 \end{pmatrix} \begin{pmatrix} c_q \\ c_p \end{pmatrix} \right)^T \begin{pmatrix} 0 & 1 \\ -1 & 0 \end{pmatrix} \left(\begin{pmatrix} 0 & -1 \\ 1 & 0 \end{pmatrix} \begin{pmatrix} c'_q \\ c'_p \end{pmatrix} \right)$$

$$= (c_q \ c_p) \begin{pmatrix} 0 & 1 \\ -1 & 0 \end{pmatrix} \begin{pmatrix} 0 & 1 \\ -1 & 0 \end{pmatrix} \begin{pmatrix} 0 & -1 \\ 1 & 0 \end{pmatrix} \begin{pmatrix} c'_q \\ c'_p \end{pmatrix}$$

$$= (c_q \ c_p) \begin{pmatrix} 0 & 1 \\ -1 & 0 \end{pmatrix} \begin{pmatrix} c'_q \\ c'_p \end{pmatrix}$$

$$= \Omega(c_q q + c_p p, c'_q q + c'_p p)$$

More simply, the matrix for J_0 is obviously in $SL(2, \mathbf{R}) = Sp(2, \mathbf{R})$.

Note that elements g of the group $Sp(2d, \mathbf{R})$ act on the set of compatible complex structures by

$$J \to gJg^{-1} \tag{26.6}$$

This takes complex structures to complex structures since

$$(gJg^{-1})(gJg^{-1}) = gJ^2 g^{-1} = -\mathbf{1}$$

and preserves the compatibility condition since, if $J \in Sp(2d, \mathbf{R})$, so is gJg^{-1}.

A complex structure J can be characterized by the subgroup of $Sp(2d, \mathbf{R})$ that leaves it invariant, with the condition $gJg^{-1} = J$ equivalent to the commutativity condition $gJ = Jg$. For the case $d = 1$ and $J = J_0$, this becomes

$$\begin{pmatrix} a & b \\ c & d \end{pmatrix} \begin{pmatrix} 0 & -1 \\ 1 & 0 \end{pmatrix} = \begin{pmatrix} 0 & -1 \\ 1 & 0 \end{pmatrix} \begin{pmatrix} a & b \\ c & d \end{pmatrix}$$

so

$$\begin{pmatrix} b & -a \\ d & -c \end{pmatrix} = \begin{pmatrix} -c & -d \\ a & b \end{pmatrix}$$

which implies $b = -c$ and $a = d$. The elements of $SL(2, \mathbf{R})$ that preserve J_0 will be of the form

$$\begin{pmatrix} a & b \\ -b & a \end{pmatrix}$$

with unit determinant, so $a^2 + b^2 = 1$. This is the $U(1) = SO(2)$ subgroup of $SL(2, \mathbf{R})$ of matrices of the form

$$\begin{pmatrix} \cos\theta & \sin\theta \\ -\sin\theta & \cos\theta \end{pmatrix} = e^{\theta Z}$$

Other choices of J will correspond to other $U(1)$ subgroups of $SL(2, \mathbf{R})$, and the space of compatible complex structures conjugate to J_0 can be identified with the coset space $SL(2, \mathbf{R})/U(1)$. In higher dimensions, it turns out that the subgroup of $Sp(2d, \mathbf{R})$ that commutes with J_0 is isomorphic to the unitary group $U(d)$, and the space of compatible complex structures conjugate to J_0 is $Sp(2d, \mathbf{R})/U(d)$.

Even before we choose a complex structure J, we can use Ω to define an indefinite Hermitian form on $\mathcal{M} \otimes \mathbf{C}$ by:

Definition (Indefinite Hermitian form on $\mathcal{M} \otimes \mathbf{C}$). *For $u_1, u_2 \in \mathcal{M} \otimes \mathbf{C}$,*

$$\langle u_1, u_2 \rangle = i\Omega(\overline{u}_1, u_2) \tag{26.7}$$

is an indefinite Hermitian form on $\mathcal{M} \otimes \mathbf{C}$.

This is clearly antilinear in the first variable, linear in the second, and satisfies the Hermitian property, since

$$\langle u_2, u_1 \rangle = i\Omega(\overline{u}_2, u_1) = -i\Omega(u_1, \overline{u}_2) = \overline{i\Omega(\overline{u}_1, u_2)} = \overline{\langle u_1, u_2 \rangle}$$

Restricting $\langle \cdot, \cdot \rangle$ to $\mathcal{M}_J^+$ and using the identification 26.1 of $\mathcal{M}$ and $\mathcal{M}_J^+$ $\langle \cdot, \cdot \rangle$ gives a complex-valued bilinear form on $\mathcal{M}$. Any $u \in \mathcal{M}_J^+$ can be written as

$$u = \frac{1}{\sqrt{2}}(v - iJv) \tag{26.8}$$

for some nonzero $v \in \mathcal{M}$, so

$$
\begin{aligned}
\langle u_1, u_2 \rangle &= i\Omega(\overline{u}_1, u_2) \\
&= i\frac{1}{2}\Omega(v_1 + iJv_1, v_2 - iJv_2) \\
&= \frac{1}{2}(-\Omega(Jv_1, v_2) + \Omega(v_1, Jv_2)) + \frac{i}{2}(\Omega(v_1, v_2) + \Omega(Jv_1, Jv_2)) \\
&= \Omega(v_1, Jv_2) + i\Omega(v_1, v_2)
\end{aligned}
\tag{26.9}
$$

where we have used compatibility of J and $J^2 = -1$ to get

$$\Omega(Jv_1, v_2) = \Omega(J^2v_1, Jv_2) = -\Omega(v_1, Jv_2)$$

We thus can recover Ω on $\mathcal{M}$ as the imaginary part of the form $\langle \cdot, \cdot \rangle$.

This form $\langle \cdot, \cdot \rangle$ is not positive or negative-definite on $\mathcal{M} \otimes \mathbf{C}$. One can, however, restrict attention to those J that give a positive-definite form on $\mathcal{M}_J^+$:

Definition (Positive compatible complex structures). *A complex structure J on $\mathcal{M}$ is said to be positive and compatible with Ω if it satisfies the* *compatibility condition 26.5 (i.e., is in $Sp(2d, \mathbf{R})$) and one of the equivalent (by equation 26.9) positivity conditions*

$$\langle u, u \rangle = i\Omega(\overline{u}, u) > 0 \tag{26.10}$$

for nonzero $u \in \mathcal{M}_J^+$. or

$$\Omega(v, Jv) > 0 \tag{26.11}$$

for nonzero $v \in \mathcal{M}$.

For such a J, $\langle \cdot, \cdot \rangle$ restricted to $\mathcal{M}_J^-$ will be negative-definite since

$$\Omega(u, \overline{u}) = -\Omega(\overline{u}, u)$$

and complex conjugation interchanges $\mathcal{M}_J^+$ and $\mathcal{M}_J^-$. The standard complex structure J_0 is positive since

$$\langle z_j, z_k \rangle = i\Omega(\overline{z}_j, z_k) = i\{\overline{z}_j, z_k\} = \delta_{jk}$$

and $\langle \cdot, \cdot \rangle$ is thus the standard Hermitian form on $\mathcal{M}_{J_0}^+$ for which the z_j are orthonormal.

26.3 Complex structures and quantization

Recall that the Heisenberg Lie algebra is the Lie algebra of linear and constant functions on M, so can be thought of as

$$\mathfrak{h}_{2d+1} = \mathcal{M} \oplus \mathbf{R}$$

where the $\mathbf{R}$ component is the constant functions. The Lie bracket is the Poisson bracket. Complexifying gives

$$\mathfrak{h}_{2d+1} \otimes \mathbf{C} = (\mathcal{M} \oplus \mathbf{R}) \otimes \mathbf{C} = (\mathcal{M} \otimes \mathbf{C}) \oplus \mathbf{C} = \mathcal{M}_J^+ \oplus \mathcal{M}_J^- \oplus \mathbf{C}$$

so elements of $\mathfrak{h}_{2d+1} \otimes \mathbf{C}$ can be written as pairs $(u, c) = (u^+ + u^-, c)$ where

$$u \in \mathcal{M} \otimes \mathbf{C}, \quad u^+ \in \mathcal{M}_J^+, \quad u^- \in \mathcal{M}_J^-, \quad c \in \mathbf{C}$$

This complexified Lie algebra is still a Lie algebra, with the Lie bracket relations

$$[(u_1, c_1), (u_2, c_2)] = (0, \Omega(u_1, u_2)) \tag{26.12}$$

348 26 Complex Structures and Quantization

and antisymmetric bilinear form Ω on $\mathcal{M}$ extended from the real Lie algebra by complex linearity.

For each J, we would like to find a quantization that takes elements of $\mathcal{M}_J^+$ to linear combinations of creation operators, elements of $\mathcal{M}_J^-$ to linear combinations of annihilation operators. This will give a representation of the complexified Lie algebra

$$\Gamma_J' : (u,c) \in \mathfrak{h}_{2d+1} \otimes \mathbf{C} \to \Gamma_J'(u,c)$$

if it satisfies the Lie algebra homomorphism property

$$[\Gamma_J'(u_1,c_1), \Gamma_J'(u_2,c_2)] = \Gamma_J'([(u_1,c_1),(u_2,c_2)]) = \Gamma_J'(0, \Omega(u_1,u_2)) \quad (26.13)$$

Since we can write

$$(u,c) = (u^+,0) + (u^-,0) + (0,c)$$

where $u^+ \in \mathcal{M}_J^+$ and $u^- \in \mathcal{M}_J^-$, we have

$$\Gamma_J'(u,c) = \Gamma_J'(u^+,0) + \Gamma_J'(u^-,0) + \Gamma_J'(0,c)$$

Note that we only expect Γ_J' to be a unitary representation (with $\Gamma_J'(u,c)$ skew-adjoint operators) for (u,c) in the real Lie subalgebra $\mathfrak{h}_{2d+1}$ (meaning $u^+ = \overline{u^-}$, $c \in \mathbf{R}$).

For the case of $J = J_0$, the Lie algebra representation is given on basis elements

$$z_j \in \mathcal{M}_{J_0}^+, \quad \bar{z}_j \in \mathcal{M}_{J_0}^-$$

by

$$\Gamma_{J_0}'(0,1) = -i\mathbf{1}, \quad \Gamma_{J_0}'(z_j,0) = -ia_j^\dagger = -iz_j, \quad \Gamma_{J_0}'(\bar{z}_j,0) = -ia_j = -i\frac{\partial}{\partial z_j}$$

and is precisely the Bargmann–Fock representation Γ_{BF}' (see equation 22.5), Note that the operators a_j and $a_j^\dagger$ are not skew-adjoint, so Γ_{J_0}' is not unitary on the full Lie algebra $\mathfrak{h}_{2d+1} \otimes \mathbf{C}$, but only on the real subspace $\mathfrak{h}_{2d+1}$ of real linear combinations of $q_j, p_j, 1$.

For more general choices of J, we start by taking

$$\Gamma_J'(0,c) = -ic\mathbf{1} \quad (26.14)$$

which is chosen so that it commutes with all other operators of the representation, and for c real gives a skew-adjoint transformation and thus a unitary representation. We would like to construct $\Gamma_J'(u^+,0)$ as a linear combination of creation operators and $\Gamma'_J(u^-,0)$ as a linear combination of annihilation operators. The compatibility condition of equation 26.5 will ensure that the $\Gamma'_J(u^+,0)$ will commute, since if $u_1^+, u_2^+ \in \mathcal{M}_J^+$, by 26.13 we have

$$[\Gamma'_J(u_1^+,0), \Gamma'_J(u_2^+,0)] = \Gamma'_J(0, \Omega(u_1^+, u_2^+)) = -i\Omega(u_1^+, u_2^+)\mathbf{1}$$

and

$$\Omega(u_1^+, u_2^+) = \Omega(Ju_1^+, Ju_2^+) = \Omega(iu_1^+, iu_2^+) = -\Omega(u_1^+, u_2^+) = 0$$

The $\Gamma'_J(u^-,0)$ will commute with each other by essentially the same argument.

To see the necessity of the positivity condition 26.10 on J, recall that the annihilation and creation operators satisfy (for $d = 1$)

$$[a, a^\dagger] = \mathbf{1}$$

a condition which corresponds to

$$[-ia, -ia^\dagger] = [\Gamma'_{J_0}(\overline{z},0), \Gamma'_{J_0}(z,0)] = \Gamma'_{J_0}(0, \{\overline{z}, z\}) = \Gamma'_{J_0}(0, -i) = -\mathbf{1}$$

Use of the opposite sign for the commutator would correspond to interchanging the role of a and $a^\dagger$, with the state $|0\rangle$ now satisfying $a^\dagger|0\rangle = 0$ and no state $|\psi\rangle$ in the state space satisfying $a|\psi\rangle = 0$. In order to have a state $|0\rangle$ that is annihilated by all annihilation operators and a total number operator with nonnegative eigenvalues (and thus a Hamiltonian with a positive energy spectrum), we need all the commutators $[a_j, a_j^\dagger]$ to have the positive sign.

For any choice of a potential basis element u of $\mathcal{M}_J^+$, by 26.13 and 26.14 we have,

$$[\Gamma'_J(\overline{u},0), \Gamma'_J(u,0)] = \Gamma'_J(0, \Omega(\overline{u}, u)) = -i\Omega(\overline{u}, u)\mathbf{1} \qquad (26.15)$$

and the positivity condition 26.10 on J will ensure that quantizing such an element by a creation operator will give a representation with nonnegative number operator eigenvalues. We have the following general result about the Bargmann–Fock construction for suitable J:

Theorem. *Given a positive compatible complex structure J on $\mathcal{M}$, there is a basis z_j^J of $\mathcal{M}_J^+$ such that a representation of $\mathfrak{h}_{2d+1} \otimes \mathbf{C}$, unitary for the real subalgebra $\mathfrak{h}_{2d+1}$, is given by*

$$\Gamma'_J(z_j^J, 0) = -ia_j^\dagger, \quad \Gamma'_J(\overline{z}_j^J, 0) = -ia_j, \quad \Gamma'_J(0, c) = -ic\mathbf{1}$$

where $a_j, a_j^\dagger$ satisfy the conventional commutation relations, and $\overline{z}_j^J$ is the complex conjugate of z_j^J.

Proof. An outline of the construction goes as follows:

1. Define a positive inner product on $\mathcal{M}$ by $(u, v)_J = \Omega(u, Jv)$ on $\mathcal{M}$. By Gram–Schmidt orthonormalization, there is a basis of span$\{q_j\} \subset \mathcal{M}$ consisting of d vectors q_j^J satisfying

$$(q_j^J, q_k^J)_J = \delta_{jk}$$

2. The vectors Jq_j^J will also be orthonormal since

$$(Jq_j^J, Jq_k^J)_J = \Omega(Jq_j^J, J^2 q_j^J) = \Omega(q_j^J, Jq_j^J) = (q_j^J, q_k^J)_J$$

They will be orthogonal to the q_j^J since

$$(Jq_j^J, q_k^J)_J = \Omega(J^2 q_j^J, q_k^J) = -\Omega(q_j^J, q_k^J)$$

and $\Omega(q_j^J, q_k^J) = 0$ since any Poisson brackets of linear combinations of the q_j vanish.

3. Define
$$z_j^J = \frac{1}{\sqrt{2}}(q_j^J - iJq_j^J)$$

The z_j^J give a complex basis of $\mathcal{M}_J^+$, their complex conjugates $\overline{z}_j^J$ a complex basis of $\mathcal{M}_J^-$.

4. The operators

$$\Gamma'_J(z_j^J, 0) = -ia_j^\dagger = -iz_j, \quad \Gamma'_J(\overline{z}_j^J, 0) = -ia_j = -i\frac{\partial}{\partial z_j}$$

satisfy the desired commutation relations and give a unitary representation on linear combinations of the $(z_j^J, 0)$ and $(\overline{z}_j^J, 0)$ in the real subalgebra $\mathfrak{h}_{2d+1}$.

$\square$

26.4 Complex vector spaces with Hermitian inner product as phase spaces

In many cases of physical interest, the dual phase space $\mathcal{M}$ will be a complex vector space with a Hermitian inner product. This will occur for instance when $\mathcal{M}$ is a space of complex solutions to a field equation, with examples non-relativistic quantum field theory (see chapter 37) and the theory of a relativistic complex scalar field (see chapter 44.1.2). In such cases, the Bargmann–Fock quantization can be confusing, since it involves complexi-fying $\mathcal{M}$, which is already a complex vector space. One way to treat this situation is as follows, taking $\mathcal{M}_J^+ = \mathcal{M}$. In the non-relativistic quantum field theory, this gives a consistent Bargmann–Fock quantization of the the-ory, while in the relativistic case it does not, and in that case a different sort of complex structure J is needed, one not related to the complex nature of the field values.

Instead of trying to complexify $\mathcal{M}$, we introduce a conjugate complex vector space $\overline{\mathcal{M}}$ and an antilinear conjugation operation interchanging $\mathcal{M}$ and $\overline{\mathcal{M}}$, with square the identity. In the case of $\mathcal{M}$ complex solutions to a field equation, $\overline{\mathcal{M}}$ will be solutions to the complex conjugate equation. Then, Bargmann–Fock quantization proceeds with the decomposition

$$\mathcal{M} \oplus \overline{\mathcal{M}}$$

playing the role of the decomposition

$$\mathcal{M}_J^+ \oplus \mathcal{M}_J^-$$

in our previous discussion.

This determines J: It is the operator that is $+i$ on $\mathcal{M}$, and $-i$ on $\overline{\mathcal{M}}$. Given a Hermitian inner product $\langle \cdot, \cdot \rangle_\mathcal{M}$ on $\mathcal{M}$, a symplectic structure Ω and indefinite Hermitian product $\langle \cdot, \cdot \rangle$ on $\mathcal{M} \oplus \overline{\mathcal{M}}$ can be determined as follows, using the relation 26.7

$$\langle u_1, u_2 \rangle = i\Omega(\overline{u}_1, u_2)$$

Writing elements $u_1, u_2 \in \mathcal{M} \oplus \overline{\mathcal{M}}$ as

$$u_1 = u_1^+ + u_1^-, \quad u_2 = u_2^+ + u_2^-$$

where $u_1^+, u_2^+ \in \mathcal{M}$ and $u_1^-, u_2^- \in \overline{\mathcal{M}}$, Ω is defined to be the bilinear form such that

-
$$\Omega(u_1^+, u_2^+) = \Omega(u_1^-, u_2^-) = 0$$

- the Hermitian inner product is recovered on $\mathcal{M}$

$$i\Omega(u_1^-, u_2^+) = \langle \overline{u_1^-}, u_2^+ \rangle_\mathcal{M}$$

- Ω is antisymmetric, so

$$i\Omega(u_1^+, u_2^-) = -i\Omega(u_2^-, u_1^+) = -\langle \overline{u_2^-}, u_1^+ \rangle_\mathcal{M}$$

Basis vectors z_j of $\mathcal{M}$ orthonormal with respect to $\langle \cdot, \cdot \rangle_{\mathcal{M}}$ satisfy

$$\langle z_j, z_k \rangle = \delta_{jk}$$

$$\langle \overline{z}_j, \overline{z}_k \rangle = -\delta_{jk}$$

$$\langle z_j, \overline{z}_k \rangle = \langle \overline{z}_j, z_k \rangle = 0$$

The symplectic form Ω satisfies the usual Poisson bracket relation

$$\Omega(z_j, \overline{z}_k) = \{z_j, \overline{z}_k\} = i\delta_{jk}$$

and one has all the elements needed for the standard Bargmann–Fock quantization. Note that the Hermitian inner product here is indefinite, positive on $\mathcal{M}$, negative on $\overline{\mathcal{M}}$.

To understand better in a basis-independent way how quantization works in this case of a complex dual phase space $\mathcal{M}$, one can use the identification (see section 9.6) of polynomials with symmetric tensor products. In this case, polynomials in the z_j get identified with $S^*(\mathcal{M})$ (since $z_j \in \mathcal{M}$), while polynomials in the $\overline{z}_j$ get identified with $S^*(\overline{\mathcal{M}})$ (since $\overline{z}_j \in \overline{\mathcal{M}}$).

We see that the Fock space $\mathcal{F}_d$ gets identified with $S^*(\mathcal{M})$, and using this identification (instead of the one with polynomials), one can ask what operator gives the quantization of an element

$$u = u^+ + u^-$$

where $u^+ \in \mathcal{M}$ and $u^- \in \overline{\mathcal{M}}$. For basis elements $z_j \in \mathcal{M}$, the operator will be $-ia_j^\dagger$, while for $\overline{z}_j \in \overline{\mathcal{M}}$ it will be $-ia_j$. We will not enter here into details, which would require more discussion of how to manipulate symmetric tensor products (see, for instance, chapter 5.4 of [17]). One can show, however, that the operators $\Gamma'(u^+, 0), \Gamma'(u^-, 0)$ defined on symmetrized tensor products by (where P^+ is the symmetrization operator of section 9.6)

$$\Gamma'(u^+, 0)P^+(u_1 \otimes \cdots \otimes u_N) = -i\sqrt{N+1}P^+(u^+ \otimes u_1 \otimes \cdots \otimes u_N)$$

$$\Gamma'(u^-, 0)P^+(u_1 \otimes \cdots \otimes u_N) = \frac{-i}{\sqrt{N}}\sum_{j=1}^{N}\langle u^-, u_j \rangle P^+(u_1 \otimes \cdots \otimes \widehat{u}_j \otimes \cdots \otimes u_N)$$

satisfy the Heisenberg Lie algebra homomorphism relations

$$\begin{aligned}[\Gamma'(u^-, 0), \Gamma'(u^+, 0)] &= \Gamma'(0, \Omega(u^-, u^+)) \\ &= -i\Omega(u^-, u^+)\mathbf{1} \\ &= -\langle \overline{u^-}, u^+ \rangle \mathbf{1} \end{aligned} \qquad (26.16)$$

when acting on elements of $S^N(\mathcal{M})$ (which are given by applying the symmetrization operator P_+ to elements of the N-fold tensor product of $\mathcal{M}$).

26.5 Complex structures for $d = 1$ and squeezed states

To get a better understanding of what happens for other complex structures than J_0, in this section we will examine the case $d = 1$. We can generalize the choice $J = J_0$, where a basis of $\mathcal{M}_{J_0}^+$ is given by

$$z = \frac{1}{\sqrt{2}}(q - ip)$$

by replacing the i by an arbitrary complex number τ. Then, the condition that $q - \tau p$ be in $\mathcal{M}_J^+$ and its conjugate in $\mathcal{M}_J^-$ is

$$J(q - \tau p) = J(q) - \tau J(p) = i(q - \tau p)$$

$$J(q - \bar{\tau} p) = J(q) - \bar{\tau} J(p) = -i(q - \bar{\tau} p)$$

Subtracting and adding the two equations gives

$$J(p) = -\frac{1}{\mathrm{Im}(\tau)} q + \frac{\mathrm{Re}(\tau)}{\mathrm{Im}(\tau)} p$$

and

$$J(q) = -\frac{\mathrm{Re}(\tau)}{\mathrm{Im}(\tau)} q + \left(\mathrm{Im}(\tau) + \frac{(\mathrm{Re}(\tau))^2}{\mathrm{Im}(\tau)} \right) p$$

respectively. Generalizing 26.4, the matrix for J is

$$J = \begin{pmatrix} -\frac{\mathrm{Re}(\tau)}{\mathrm{Im}(\tau)} & \mathrm{Im}(\tau) \frac{(\mathrm{Re}(\tau))^2}{\mathrm{Im}(\tau)} \\ -\frac{1}{\mathrm{Im}(\tau)} & \frac{\mathrm{Re}(\tau)}{\mathrm{Im}(\tau)} \end{pmatrix} = \frac{1}{\mathrm{Im}(\tau)} \begin{pmatrix} -\mathrm{Re}(\tau) & |\tau|^2 \\ -1 & \mathrm{Re}(\tau) \end{pmatrix} \quad (26.17)$$

and it can easily be checked that $\det J = 1$, so $J \in SL(2, \mathbf{R})$ and is compatible with Ω.

The positivity condition here is that $\Omega(\cdot, J \cdot)$ is positive on $\mathcal{M}$, which in terms of matrices (see 16.2) becomes the condition that the matrix

$$\begin{pmatrix} 0 & 1 \\ -1 & 0 \end{pmatrix} J^T = \frac{1}{\mathrm{Im}(\tau)} \begin{pmatrix} |\tau|^2 & \mathrm{Re}(\tau) \\ \mathrm{Re}(\tau) & 1 \end{pmatrix}$$

gives a positive quadratic form. This will be the case when $\text{Im}(\tau) > 0$. We have thus constructed a set of J that are positive, compatible with Ω, and parametrized by an element τ of the upper half-plane, with J_0 corresponding to $\tau = i$.

To construct annihilation and creation operators satisfying the standard commutation relations

$$[a_\tau, a_\tau] = [a_\tau^\dagger, a_\tau^\dagger] = 0, \quad [a_\tau, a_\tau^\dagger] = 1$$

set

$$a_\tau = \frac{1}{\sqrt{2\,\text{Im}(\tau)}}(Q - \bar\tau P), \quad a_\tau^\dagger = \frac{1}{\sqrt{2\,\text{Im}(\tau)}}(Q - \tau P)$$

The Hamiltonian

$$H_\tau = \frac{1}{2}(a_\tau a_\tau^\dagger + a_\tau^\dagger a_\tau) = \frac{1}{2\,\text{Im}(\tau)}(Q^2 + |\tau|^2 P^2 - \text{Re}(\tau)(QP + PQ)) \quad (26.18)$$

will have eigenvalues $n + \frac{1}{2}$ for $n = 0, 1, 2, \cdots$. Its lowest energy state will satisfy

$$a_\tau |0\rangle_\tau = 0 \tag{26.19}$$

which in the Schrödinger representation is the differential equation

$$(Q - \bar\tau P)\psi(q) = \left(q + i\bar\tau \frac{d}{dq}\right)\psi(q) = 0$$

which has solutions

$$\psi(q) \propto e^{i\frac{\tau}{2|\tau|^2}q^2} \tag{26.20}$$

This will be a normalizable state for $\text{Im}(\tau) > 0$, again showing the necessity of the positivity condition.

Eigenstates of H_τ for $\tau = ic$, $c > 0$ real are known as "squeezed states" in physics. By equation 26.20, the lowest energy state $|0\rangle$ will have spatial dependence proportional to

$$e^{-\frac{1}{2c}q^2}$$

and higher energy eigenstates $|n\rangle$ will also have such a Gaussian factor in their position dependence. For $c < 1$, such states will have narrower spatial width than conventional quanta (thus the name "squeezed"), but wider width in momentum space. For $c > 1$ the opposite will be true. In some sense that we will not try to make precise, the limits as $c \to \infty$ and $c \to 0$ correspond to the Schrödinger representations in position and momentum space, respectively (with the distinguished Bargmann–Fock state $|0\rangle$ approaching the constant function in position or momentum space).

The subgroup $\mathbf{R} \subset SL(2, \mathbf{R})$ of equation 24.7 acts non-trivially on J_0, by

$$J_0 = \begin{pmatrix} 0 & 1 \\ -1 & 0 \end{pmatrix} \rightarrow \begin{pmatrix} e^r & 0 \\ 0 & e^{-r} \end{pmatrix} \begin{pmatrix} 0 & 1 \\ -1 & 0 \end{pmatrix} \begin{pmatrix} e^{-r} & 0 \\ 0 & e^r \end{pmatrix} = \begin{pmatrix} 0 & e^{2r} \\ -e^{-2r} & 0 \end{pmatrix}$$

taking J_0 to the complex structure with parameter $\tau = ie^{2r}$.

Recall from section 24.4 that changing from q, p coordinates to $z, \bar{z}$ coordinates on the complexified phase space, the group $SL(2, \mathbf{R})$ becomes the isomorphic group $SU(1, 1)$, the group of matrices

$$\begin{pmatrix} \alpha & \beta \\ \bar{\beta} & \bar{\alpha} \end{pmatrix}$$

satisfying

$$|\alpha|^2 - |\beta|^2 = 1$$

Looking at equation 24.11 that gives the conjugation relating the two groups, we see that

$$\begin{pmatrix} -i & 0 \\ 0 & i \end{pmatrix} \in SU(1, 1) \leftrightarrow \begin{pmatrix} 0 & 1 \\ -1 & 0 \end{pmatrix} \in SL(2, \mathbf{R})$$

and as expected, in these coordinates J_0 acts on z by multiplication by i, on $\bar{z}$ by multiplication by $-i$.

$SU(1, 1)$ matrices can be parametrized in terms of t, θ, θ' by taking

$$\alpha = e^{i\theta'} \cosh t, \quad \beta = e^{i\theta} \sinh t$$

Such matrices will have square -1 and give a positive complex structure when $\theta' = \frac{\pi}{2}$, so of the form

$$\begin{pmatrix} i \cosh t & -e^{i\theta} \sinh t \\ -e^{-i\theta} \sinh t & -i \cosh t \end{pmatrix}$$

The $U(1)$ subgroup of $SU(1, 1)$ preserving J_0 will be matrices of the form

$$\begin{pmatrix} e^{i\theta} & 0 \\ 0 & e^{-i\theta} \end{pmatrix}$$

Using the matrix from equation 24.9, the subgroup $\mathbf{R} \subset SL(2, \mathbf{R})$ of equation 24.7 takes J_0 to

$$\begin{pmatrix} \cosh r & \sinh r \\ \sinh r & \cosh r \end{pmatrix} \begin{pmatrix} -i & 0 \\ 0 & i \end{pmatrix} \begin{pmatrix} \cosh r & -\sinh r \\ -\sinh r & \cosh r \end{pmatrix} = i \begin{pmatrix} -\cosh 2r & \sinh 2r \\ -\sinh 2r & \cosh 2r \end{pmatrix}$$

26.6 Complex structures and Bargmann–Fock quantization for arbitrary d

Generalizing from $d = 1$ to arbitrary d, the additional piece of structure introduced by the method of annihilation and creation operators appears in the following ways:

- As a choice of positive compatible complex structure J, or equivalently a decomposition

$$\mathcal{M} \otimes \mathbf{C} = \mathcal{M}_J^+ \oplus \mathcal{M}_J^-$$

- As a division of the quantizations of elements of $\mathcal{M} \otimes \mathbf{C}$ into creation (coming from $\mathcal{M}_J^+$) and annihilation (coming from $\mathcal{M}_J^-$) operators. There will be a corresponding J-dependent definition of normal-ordering of an operator $\mathcal{O}$ that is a product of such operators, symbolized by $:\mathcal{O}:_J$.

- As a distinguished vector $|0\rangle_J$, the vector in $\mathcal{H}$ annihilated by all annihilation operators. Writing such a vector in the Schrödinger representation as a position space wavefunction in $L^2(\mathbf{R}^d)$, it will be a Gaussian function generalizing the $d = 1$ case of equation 26.20, with τ now a symmetric matrix with positive-definite imaginary part. Such matrices parametrize the space of positive compatible complex structures, a space called the "Siegel upper half-space."

- As a distinguished subgroup of $Sp(2n, \mathbf{R})$, the subgroup that commutes with J. This subgroup will be isomorphic to the group $U(n)$. The Siegel upper half-space of positive compatible complex structures can also be described as the coset space $Sp(2n, \mathbf{R})/U(n)$.

26.7 For further reading

For more detail on the space of positive compatible complex structures on a symplectic vector space, see chapter 1.4 of [8]. Few quantum mechanics textbooks discuss squeezed states, for one that does, see chapter 12 of [108].

Chapter 27
The Fermionic Oscillator

In this chapter, we will introduce a new quantum system by using a simple variation on techniques we used to study the harmonic oscillator, that of replacing commutators by anticommutators. This variant of the harmonic oscillator will be called a "fermionic oscillator," with the original sometimes called a "bosonic oscillator." The terminology of "boson" and "fermion" refers to the principle enunciated in chapter 9 that multiple identical particles are described by tensor product states that are either symmetric (bosons) or antisymmetric (fermions).

The bosonic and fermionic oscillator systems are single-particle systems, describing the energy states of a single particle, so the usage of the bosonic/fermionic terminology is not obviously relevant. In later chapters, we will study quantum field theories, which can be treated as infinite-dimensional oscillator systems. In that context, multiple particle states will automatically be symmetric or antisymmetric, depending on whether the field theory is treated as a bosonic or fermionic oscillator system, thus justifying the terminology.

27.1 Canonical anticommutation relations and the fermionic oscillator

Recall that the Hamiltonian for the quantum harmonic oscillator system in d degrees of freedom (setting $\hbar = m = \omega = 1$) is

$$H = \sum_{j=1}^{d} \frac{1}{2}(Q_j^2 + P_j^2)$$

and that it can be diagonalized by introducing number operators $N_j = a_j^\dagger a_j$ defined in terms of operators

$$a_j = \frac{1}{\sqrt{2}}(Q_j + iP_j), \quad a_j^\dagger = \frac{1}{\sqrt{2}}(Q_j - iP_j)$$

that satisfy the so-called "canonical commutation relation" (CCR)

$$[a_j, a_k^\dagger] = \delta_{jk}\mathbf{1}, \quad [a_j, a_k] = [a_j^\dagger, a_k^\dagger] = 0$$

The simple change in the harmonic oscillator problem that takes one from bosons to fermions is the replacement of the bosonic annihilation and creation operators (which we will now denote a_B and $a_B{}^\dagger$) by fermionic annihilation and creation operators called a_F and $a_F{}^\dagger$, and replacement of the commutator

$$[A, B] \equiv AB - BA$$

of operators by the anticommutator

$$[A, B]_+ \equiv AB + BA$$

The commutation relations are now (for $d = 1$, a single degree of freedom)

$$[a_F, a_F^\dagger]_+ = 1, \quad [a_F, a_F]_+ = 0, \quad [a_F^\dagger, a_F^\dagger]_+ = 0$$

with the last two relations implying that $a_F^2 = 0$ and $(a_F^\dagger)^2 = 0$
    The fermionic number operator

$$N_F = a_F^\dagger a_F$$

now satisfies

$$N_F^2 = a_F^\dagger a_F a_F^\dagger a_F = a_F^\dagger(1 - a_F^\dagger a_F)a_F = N_F - a_F^{\dagger^2} a_F^2 = N_F$$

(using the fact that $a_F^2 = a_F^{\dagger^2} = 0$). So one has

$$N_F^2 - N_F = N_F(N_F - 1) = 0$$

which implies that the eigenvalues of N_F are just 0 and 1. We will denote eigenvectors with such eigenvalues by $|0\rangle$ and $|1\rangle$. The simplest representation of the operators a_F and $a_F^\dagger$ on a complex vector space $\mathcal{H}_F$ will be on $\mathbf{C}^2$, and choosing the basis

$$|0\rangle = \begin{pmatrix} 0 \\ 1 \end{pmatrix}, \quad |1\rangle = \begin{pmatrix} 1 \\ 0 \end{pmatrix}$$

the operators are represented as

$$a_F = \begin{pmatrix} 0 & 0 \\ 1 & 0 \end{pmatrix}, \quad a_F^\dagger = \begin{pmatrix} 0 & 1 \\ 0 & 0 \end{pmatrix}, \quad N_F = \begin{pmatrix} 1 & 0 \\ 0 & 0 \end{pmatrix}$$

Since

$$H = \frac{1}{2}(a_F^\dagger a_F + a_F a_F^\dagger)$$

is just $\frac{1}{2}$ the identity operator, to get a non-trivial quantum system, instead we make a sign change and set

$$H = \frac{1}{2}(a_F^\dagger a_F - a_F a_F^\dagger) = N_F - \frac{1}{2}\mathbf{1} = \begin{pmatrix} \frac{1}{2} & 0 \\ 0 & -\frac{1}{2} \end{pmatrix}$$

The energies of the energy eigenstates $|0\rangle$ and $|1\rangle$ will then be $\pm\frac{1}{2}$ since

$$H|0\rangle = -\frac{1}{2}|0\rangle, \quad H|1\rangle = \frac{1}{2}|1\rangle$$

Note that the quantum system we have constructed here is nothing but our old friend the two-state system of chapter 3. Taking complex linear combinations of the operators

$$a_F, a_F^\dagger, N_F, \mathbf{1}$$

we get all linear transformations of $\mathcal{H}_F = \mathbf{C}^2$ (so this is an irreducible representation of the algebra of these operators). The relation to the Pauli matrices is

$$a_F^\dagger = \frac{1}{2}(\sigma_1 + i\sigma_2), \quad a_F = \frac{1}{2}(\sigma_1 - i\sigma_2), \quad H = \frac{1}{2}\sigma_3$$

27.2 Multiple degrees of freedom

For the case of d degrees of freedom, one has this variant of the canonical commutation relations (CCR) among the bosonic annihilation and creation operators a_{Bj} and $a_{Bj}^\dagger$:

Definition (Canonical anticommutation relations). *A set of $2d$ operators*

$$a_{Fj}, \ a_{Fj}^\dagger, \ \ j = 1, \ldots, d$$

is said to satisfy the canonical anticommutation relations (CAR) when one has

$$[a_{Fj}, a_{Fk}^\dagger]_+ = \delta_{jk}\mathbf{1}, \quad [a_{Fj}, a_{Fk}]_+ = 0, \quad [a_{Fj}^\dagger, a_{Fk}^\dagger]_+ = 0$$

In this case, one may choose as the state space the tensor product of N copies of the single fermionic oscillator state space

$$\mathcal{H}_F = (\mathbf{C}^2)^{\otimes d} = \underbrace{\mathbf{C}^2 \otimes \mathbf{C}^2 \otimes \cdots \otimes \mathbf{C}^2}_{d \text{ times}}$$

The dimension of $\mathcal{H}_F$ will be 2^d. On this space, an explicit construction of the operators a_{Fj} and $a_{Fj}^\dagger$ in terms of Pauli matrices is

$$a_{Fj} = \underbrace{\sigma_3 \otimes \sigma_3 \otimes \cdots \otimes \sigma_3}_{j-1 \text{ times}} \otimes \begin{pmatrix} 0 & 0 \\ 1 & 0 \end{pmatrix} \otimes \mathbf{1} \otimes \cdots \otimes \mathbf{1}$$

$$a_{Fj}^\dagger = \underbrace{\sigma_3 \otimes \sigma_3 \otimes \cdots \otimes \sigma_3}_{j-1 \text{ times}} \otimes \begin{pmatrix} 0 & 1 \\ 0 & 0 \end{pmatrix} \otimes \mathbf{1} \otimes \cdots \otimes \mathbf{1}$$

The factors of σ_3 are there as one possible way to ensure that

$$[a_{Fj}, a_{Fk}]_+ = [a_{Fj}^\dagger, a_{Fk}^\dagger]_+ = [a_{Fj}, a_{Fk}^\dagger]_+ = 0$$

are satisfied for $j \neq k$ since then one will get in the tensor product factors

$$[\sigma_3, \begin{pmatrix} 0 & 0 \\ 1 & 0 \end{pmatrix}]_+ = 0 \quad \text{or} \quad [\sigma_3, \begin{pmatrix} 0 & 1 \\ 0 & 0 \end{pmatrix}]_+ = 0$$

While this sort of tensor product construction is useful for discussing the physics of multiple qubits, in general it is easier to not work with large tensor products, and the Clifford algebra formalism we will describe in chapter 28 avoids this.

The number operators will be

$$N_{Fj} = a_{Fj}^\dagger a_{Fj}$$

These will commute with each other, so can be simultaneously diagonalized, with eigenvalues $n_j = 0, 1$. One can take as a basis of $\mathcal{H}_F$ the 2^d states

$$|n_1, n_2, \cdots, n_d\rangle$$

which are the natural basis states for $(\mathbf{C}^2)^{\otimes d}$ given by d choices of either $|0\rangle$ or $|1\rangle$.

As an example, for the case $d = 3$, the picture shows the pattern of states and their energy levels for the bosonic and fermionic cases. In the bosonic case, the lowest energy state is at positive energy and there are an infinite number of states of ever increasing energy. In the fermionic case, the lowest

Energy

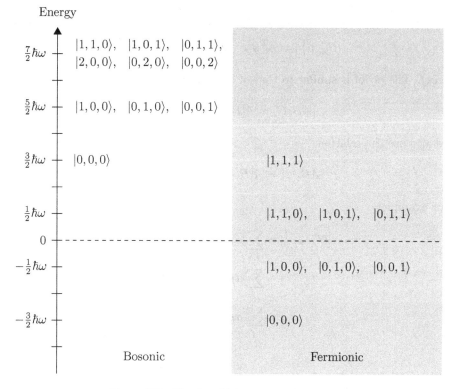

Figure 27.1: $N = 3$ oscillator energy eigenstates.

energy state is at negative energy, with the pattern of energy eigenvalues of the finite number of states symmetric about the zero energy level.

Just as in the bosonic case, we can consider quadratic combinations of creation and annihilation operators of the form

$$U'_A = \sum_{j,k} a^\dagger_{Fj} A_{jk} a_{Fk}$$

and we have

Theorem 27.1. *For* $A \in \mathfrak{gl}(d, \mathbf{C})$ *a* d *by* d *complex matrix one has*

$$[U'_A, U'_{A'}] = U_{[A,A']}$$

So

$$A \in \mathfrak{gl}(d, \mathbf{C}) \to U'_A$$

is a Lie algebra representation of $\mathfrak{gl}(d, \mathbf{C})$ *on* $\mathcal{H}_F$

One also has (for column vectors $\mathbf{a}_F$ with components $a_{F1}, \ldots, a_{Fd}$)

$$[U'_A, \mathbf{a}_F^\dagger] = A^T \mathbf{a}_F^\dagger, \quad [U'_A, \mathbf{a}_F] = -A\mathbf{a}_F \qquad (27.1)$$

Proof. The proof is similar to that of 25.1, except besides the relation

$$[AB, C] = A[B, C] + [A, B]C$$

we also use the relation

$$[AB, C] = A[B, C]_+ - [A, B]_+C$$

For example,

$$\begin{aligned}
[U'_A, a_{Fl}^\dagger] &= \sum_{j,k} [a_{Fj}^\dagger A_{jk} a_{Fk}, a_{Fl}^\dagger] \\
&= \sum_{j,k} a_{Fj}^\dagger A_{jk} [a_{Fk}, a_{Fl}^\dagger]_+ \\
&= \sum_j a_{Fj}^\dagger A_{jl}
\end{aligned}$$

$\square$

The Hamiltonian is

$$H = \sum_j \left(N_{Fj} - \frac{1}{2}\mathbf{1} \right)$$

which (up to the constant $\frac{1}{2}$ that doesn't contribute to commutation relations) is just U'_B for the case $B = \mathbf{1}$. Since this commutes with all other d by d matrices, we have

$$[H, U'_A] = 0$$

for all $A \in \mathfrak{gl}(d, \mathbf{C})$, so these are symmetries, and we have a representation of the Lie algebra $\mathfrak{gl}(d, \mathbf{C})$ on each energy eigenspace. Only for $A \in \mathfrak{u}(d)$ (A a skew-adjoint matrix) will the representation turn out to be unitary.

27.3 For further reading

Most quantum field theory books and a few quantum mechanics books contain some sort of discussion of the fermionic oscillator, see, for example, chapter 21.3 of [81] or chapter 5 of [16]. The standard discussion often starts with considering a form of classical analog using anticommuting "fermionic" variables and then quantizing to get the fermionic oscillator. Here we are doing things in the opposite order, starting in this chapter with the quantized oscillator, then considering the classical analog in chapter 30.

Chapter 28
Weyl and Clifford Algebras

We have seen that just changing commutators to anticommutators takes the harmonic oscillator quantum system to a very different one (the fermionic oscillator), with this new system having in many ways a parallel structure. It turns out that this parallelism goes much deeper, with every aspect of the harmonic oscillator story having a fermionic analog. We will begin in this chapter by studying the operators of the corresponding quantum systems.

28.1 The Complex Weyl and Clifford algebras

In mathematics, a "ring" is a set with addition and multiplication laws that are associative and distributive (but not necessarily commutative), and an "algebra" is a ring that is also a vector space over some field of scalars. The canonical commutation and anticommutation relations define interesting algebras, called the Weyl and Clifford algebras, respectively. The case of complex numbers as scalars is simplest, so we will start with that before moving on to the real number case.

28.1.1 One degree of freedom, bosonic case

Starting with the one degree of freedom case (corresponding to two operators Q, P, which is why the notation will have a 2), we can define:

Definition (Complex Weyl algebra, one degree of freedom). *The complex Weyl algebra in the one degree of freedom case is the algebra* $\mathrm{Weyl}(2, \mathbf{C})$ *generated by the elements* $1, a_B, a_B^\dagger$, *satisfying the canonical commutation relations:*

© Peter Woit 2017
P. Woit, *Quantum Theory, Groups and Representations*,
DOI 10.1007/978-3-319-64612-1_28

$$[a_B, a_B^\dagger] = 1, \quad [a_B, a_B] = [a_B^\dagger, a_B^\dagger] = 0$$

In other words, Weyl(2, **C**) is the algebra one gets by taking arbitrary products and complex linear combinations of the generators. By repeated use of the commutation relation

$$a_B a_B^\dagger = 1 + a_B^\dagger a_B$$

any element of this algebra can be written as a sum of elements in normal-order, of the form

$$c_{l,m}(a_B^\dagger)^l a_B^m$$

with all annihilation operators a_B on the right, for some complex constants $c_{l,m}$. As a vector space over **C**, Weyl(2, **C**) is infinite dimensional, with a basis

$$1, \ a_B, \ a_B^\dagger, \ a_B^2, \ a_B^\dagger a_B, \ (a_B^\dagger)^2, \ a_B^3, \ a_B^\dagger a_B^2, \ (a_B^\dagger)^2 a_B, \ (a_B^\dagger)^3, \dots$$

This algebra is isomorphic to a more familiar one. Setting

$$a_B^\dagger = z, \quad a_B = \frac{d}{dz}$$

one sees that Weyl(2, **C**) can be identified with the algebra of polynomial coefficient differential operators on functions of a complex variable z. As a complex vector space, the algebra is infinite dimensional, with a basis of elements

$$z^l \frac{d^m}{dz^m}$$

In our study of quantization by the Bargmann–Fock method, we saw that the subset of such operators consisting of complex linear combinations of

$$1, \ z, \ \frac{d}{dz}, \ z^2, \ \frac{d^2}{dz^2}, \ z\frac{d}{dz}$$

is closed under commutators, and is a representation of a Lie algebra of complex dimension *six*. This Lie algebra includes as subalgebras the Heisenberg Lie algebra $\mathfrak{h}_3 \otimes \mathbf{C}$ (first three elements) and the Lie algebra $\mathfrak{sl}(2, \mathbf{C}) = \mathfrak{sl}(2, \mathbf{R}) \otimes \mathbf{C}$ (last three elements). Note that here we are allowing complex linear combinations, so we are getting the complexification of the real six-dimensional Lie algebra that appeared in our study of quantization.

Since the a_B and $a_B^\dagger$ are defined in terms of P and Q, one could of course also define the Weyl algebra as the one generated by $1, P, Q$, with the Heisenberg commutation relations, taking complex linear combinations of all products of these operators.

28.1.2 One degree of freedom, fermionic case

Changing commutators to anticommutators, one gets a different algebra, the Clifford algebra:

Definition (Complex Clifford algebra, one degree of freedom). *The complex Clifford algebra in the one degree of freedom case is the algebra* $\mathrm{Cliff}(2, \mathbf{C})$ *generated by the elements* $1, a_F, a_F^\dagger$, *subject to the canonical anticommutation relations (CAR)*

$$[a_F, a_F^\dagger]_+ = 1, \quad [a_F, a_F]_+ = [a_F^\dagger, a_F^\dagger]_+ = 0$$

This algebra is a four-dimensional algebra over $\mathbf{C}$, with basis

$$1, \quad a_F, \quad a_F^\dagger, \quad a_F^\dagger a_F$$

since higher powers of the operators vanish, and the anticommutation relation between a_F and $a_F^\dagger$ can be used to normal-order and put factors of a_F on the right. We saw in chapter 27 that this algebra is isomorphic with the algebra $M(2, \mathbf{C})$ of 2 by 2 complex matrices, using

$$1 \leftrightarrow \begin{pmatrix} 1 & 0 \\ 0 & 1 \end{pmatrix}, \quad a_F \leftrightarrow \begin{pmatrix} 0 & 0 \\ 1 & 0 \end{pmatrix}, \quad a_F^\dagger \leftrightarrow \begin{pmatrix} 0 & 1 \\ 0 & 0 \end{pmatrix}, \quad a_F^\dagger a_F \leftrightarrow \begin{pmatrix} 1 & 0 \\ 0 & 0 \end{pmatrix}$$
$$(28.1)$$

We will see in chapter 30 that there is also a way of identifying this algebra with "differential operators in fermionic variables," analogous to what happens in the bosonic (Weyl algebra) case.

Recall that the bosonic annihilation and creation operators were originally defined in terms of the P and Q operators by

$$a_B = \frac{1}{\sqrt{2}}(Q + iP), \quad a_B^\dagger = \frac{1}{\sqrt{2}}(Q - iP)$$

Looking for the fermionic analogs of the operators Q and P, we use a slightly different normalization, and set

$$a_F = \frac{1}{2}(\gamma_1 + i\gamma_2), \quad a_F^\dagger = \frac{1}{2}(\gamma_1 - i\gamma_2)$$

so

$$\gamma_1 = a_F + a_F^\dagger, \quad \gamma_2 = \frac{1}{i}(a_F - a_F^\dagger)$$

and the CAR imply that the operators γ_j satisfy the anticommutation relations

$$[\gamma_1, \gamma_1]_+ = [a_F + a_F^\dagger, a_F + a_F^\dagger]_+ = 2$$

$$[\gamma_2, \gamma_2]_+ = -[a_F - a_F^\dagger, a_F - a_F^\dagger]_+ = 2$$

$$[\gamma_1, \gamma_2]_+ = \frac{1}{i}[a_F + a_F^\dagger, a_F - a_F^\dagger]_+ = 0$$

From this, we see that

- One could alternatively have defined Cliff$(2, \mathbf{C})$ as the algebra generated by $1, \gamma_1, \gamma_2$, subject to the relations

$$[\gamma_j, \gamma_k]_+ = 2\delta_{jk}$$

- Using just the generators 1 and γ_1, one gets an algebra Cliff$(1, \mathbf{C})$, generated by $1, \gamma_1$, with the relation

$$\gamma_1^2 = 1$$

This is a two-dimensional complex algebra, isomorphic to $\mathbf{C} \oplus \mathbf{C}$.

28.1.3 Multiple degrees of freedom

For a larger number of degrees of freedom, one can generalize the above and define Weyl and Clifford algebras as follows:

Definition (Complex Weyl algebras). *The complex Weyl algebra for d degrees of freedom is the algebra* Weyl$(2d, \mathbf{C})$ *generated by the elements* $1, a_{Bj}, a_{Bj}^\dagger$, $j = 1, \ldots, d$ *satisfying the CCR*

$$[a_{Bj}, a_{Bk}^\dagger] = \delta_{jk}1, \quad [a_{Bj}, a_{Bk}] = [a_{Bj}^\dagger, a_{Bk}^\dagger] = 0$$

Weyl$(2d, \mathbf{C})$ can be identified with the algebra of polynomial coefficient differential operators in d complex variables $z_1, z_2, \ldots, z_d$. The subspace of complex linear combinations of the elements

$$1, z_j, \frac{\partial}{\partial z_j}, z_j z_k, \frac{\partial^2}{\partial z_j \partial z_k}, z_j \frac{\partial}{\partial z_k}$$

is closed under commutators and provides a representation of the complexification of the Lie algebra $\mathfrak{h}_{2d+1} \rtimes \mathfrak{sp}(2d, \mathbf{R})$ built out of the Heisenberg Lie algebra for d degrees of freedom and the Lie algebra of the symplectic group $Sp(2d, \mathbf{R})$. Recall that this is the Lie algebra of polynomials of degree at most 2 on the phase space $\mathbf{R}^{2d}$, with the Poisson bracket as Lie bracket.

The complex Weyl algebra could also be defined by taking complex linear combinations of products of generators $1, P_j, Q_j$, subject to the Heisenberg commutation relations.

For Clifford algebras, one has:

Definition (Complex Clifford algebras, using annihilation and creation operators). *The complex Clifford algebra for d degrees of freedom is the algebra* $\text{Cliff}(2d, \mathbf{C})$ *generated by* $1, a_{Fj}, a_{Fj}^\dagger$ *for* $j = 1, 2, \ldots, d$ *satisfying the CAR*

$$[a_{Fj}, a_{Fk}^\dagger]_+ = \delta_{jk} 1, \quad [a_{Fj}, a_{Fk}]_+ = [a_{Fj}^\dagger, a_{Fk}^\dagger]_+ = 0$$

or, alternatively, one has the following more general definition that also works in the odd-dimensional case:

Definition (Complex Clifford algebras). *The complex Clifford algebra in n variables is the algebra* $\text{Cliff}(n, \mathbf{C})$ *generated by* $1, \gamma_j$ *for* $j = 1, 2, \ldots, n$ *satisfying the relations*

$$[\gamma_j, \gamma_k]_+ = 2\delta_{jk}$$

We will not try and prove this here, but one can show that, abstractly as algebras, the complex Clifford algebras are something well known. Generalizing the case $d = 1$ where we saw that $\text{Cliff}(2, \mathbf{C})$ was isomorphic to the algebra of 2 by 2 complex matrices, one has isomorphisms

$$\text{Cliff}(2d, \mathbf{C}) \leftrightarrow M(2^d, \mathbf{C})$$

in the even-dimensional case, and in the odd-dimensional case

$$\text{Cliff}(2d + 1, \mathbf{C}) \leftrightarrow M(2^d, \mathbf{C}) \oplus M(2^d, \mathbf{C})$$

Two properties of $\text{Cliff}(n, \mathbf{C})$ are

- As a vector space over $\mathbf{C}$, a basis of $\text{Cliff}(n, \mathbf{C})$ is the set of elements

$$1, \quad \gamma_j, \quad \gamma_j \gamma_k, \quad \gamma_j \gamma_k \gamma_l, \quad \ldots, \quad \gamma_1 \gamma_2 \gamma_3 \cdots \gamma_{n-1} \gamma_n$$

  for indices $j, k, l, \cdots \in 1, 2, \ldots, n$, with $j < k < l < \cdots$. To show this, consider all products of the generators, and use the commutation relations for the γ_j to identify any such product with an element of this basis. The relation $\gamma_j^2 = 1$ shows that repeated occurrences of a γ_j can be removed. The relation $\gamma_j \gamma_k = -\gamma_k \gamma_j$ can then be used to put elements of the product in the order of a basis element as above.

- As a vector space over $\mathbf{C}$, $\text{Cliff}(n, \mathbf{C})$ has dimension 2^n. One way to see this is to consider the product

$$(1 + \gamma_1)(1 + \gamma_2) \cdots (1 + \gamma_n)$$

which will have 2^n terms that are exactly those of the basis listed above.

28.2 Real Clifford algebras

We can define real Clifford algebras $\mathrm{Cliff}(n, \mathbf{R})$ just as for the complex case, by taking only real linear combinations:

Definition (Real Clifford algebras). *The real Clifford algebra in n variables is the algebra $\mathrm{Cliff}(n, \mathbf{R})$ generated over the real numbers by $1, \gamma_j$ for $j = 1, 2, \ldots, n$ satisfying the relations*

$$[\gamma_j, \gamma_k]_+ = 2\delta_{jk}$$

For reasons that will be explained in the next chapter, it turns out that a more general definition is useful. We write the number of variables as $n = r + s$, for r, s nonnegative integers, and now vary not just $r + s$, but also $r - s$, the so-called "signature":

Definition (Real Clifford algebras, arbitrary signature). *The real Clifford algebra in $n = r + s$ variables is the algebra $\mathrm{Cliff}(r, s, \mathbf{R})$ over the real numbers generated by $1, \gamma_j$ for $j = 1, 2, \ldots, n$ satisfying the relations*

$$[\gamma_j, \gamma_k]_+ = \pm 2\delta_{jk}1$$

where we choose the $+$ sign when $j = k = 1, \ldots, r$ and the $-$ sign when $j = k = r + 1, \ldots, n$.

In other words, as in the complex case, different γ_j anticommute, but only the first r of them satisfy $\gamma_j^2 = 1$, with the other s of them satisfying $\gamma_j^2 = -1$. Working out some of the low-dimensional examples, one finds:

- $\mathrm{Cliff}(0, 1, \mathbf{R})$. This has generators 1 and γ_1, satisfying

$$\gamma_1^2 = -1$$

  Taking real linear combinations of these two generators, the algebra one gets is just the algebra $\mathbf{C}$ of complex numbers, with γ_1 playing the role of $i = \sqrt{-1}$.

- $\mathrm{Cliff}(0, 2, \mathbf{R})$. This has generators $1, \gamma_1, \gamma_2$ and a basis

$$1, \quad \gamma_1, \quad \gamma_2, \quad \gamma_1\gamma_2$$

with

$$\gamma_1^2 = -1, \quad \gamma_2^2 = -1, \quad (\gamma_1\gamma_2)^2 = \gamma_1\gamma_2\gamma_1\gamma_2 = -\gamma_1^2\gamma_2^2 = -1$$

This four-dimensional algebra over the real numbers can be identified with the algebra $\mathbf{H}$ of quaternions by taking

$$\gamma_1 \leftrightarrow \mathbf{i}, \quad \gamma_2 \leftrightarrow \mathbf{j}, \quad \gamma_1\gamma_2 \leftrightarrow \mathbf{k}$$

• Cliff$(1,1,\mathbf{R})$. This is the algebra $M(2,\mathbf{R})$ of real 2 by 2 matrices, with one possible identification as follows:

$$1 \leftrightarrow \begin{pmatrix} 1 & 0 \\ 0 & 1 \end{pmatrix}, \quad \gamma_1 \leftrightarrow \begin{pmatrix} 0 & 1 \\ 1 & 0 \end{pmatrix}, \quad \gamma_2 \leftrightarrow \begin{pmatrix} 0 & -1 \\ 1 & 0 \end{pmatrix}, \quad \gamma_1\gamma_2 \leftrightarrow \begin{pmatrix} 1 & 0 \\ 0 & -1 \end{pmatrix}$$

Note that one can construct this using the $a_F, a_F^\dagger$ for the complex case Cliff$(2,\mathbf{C})$ (see 28.1) as

$$\gamma_1 = a_F + a_F^\dagger, \quad \gamma_2 = a_F - a_F^\dagger$$

since these are represented as real matrices.

• Cliff$(3,0,\mathbf{R})$. This is the algebra $M(2,\mathbf{C})$ of complex 2 by 2 matrices, with one possible identification using Pauli matrices given by

$$1 \leftrightarrow \begin{pmatrix} 1 & 0 \\ 0 & 1 \end{pmatrix}$$

$$\gamma_1 \leftrightarrow \sigma_1 = \begin{pmatrix} 0 & 1 \\ 1 & 0 \end{pmatrix}, \quad \gamma_2 \leftrightarrow \sigma_2 = \begin{pmatrix} 0 & -i \\ i & 0 \end{pmatrix}, \quad \gamma_3 \leftrightarrow \sigma_3 = \begin{pmatrix} 1 & 0 \\ 0 & -1 \end{pmatrix}$$

$$\gamma_1\gamma_2 \leftrightarrow i\sigma_3 = \begin{pmatrix} i & 0 \\ 0 & -i \end{pmatrix}, \quad \gamma_2\gamma_3 \leftrightarrow i\sigma_1 = \begin{pmatrix} 0 & i \\ i & 0 \end{pmatrix}, \quad \gamma_1\gamma_3 \leftrightarrow -i\sigma_2 = \begin{pmatrix} 0 & -1 \\ 1 & 0 \end{pmatrix}$$

$$\gamma_1\gamma_2\gamma_3 \leftrightarrow \begin{pmatrix} i & 0 \\ 0 & i \end{pmatrix}$$

It turns out that Cliff$(r, s, \mathbf{R})$ is always one or two copies of matrices of real, complex, or quaternionic elements, of dimension a power of 2, but this requires a rather intricate algebraic argument that we will not enter into here. For the details of this and the resulting pattern of algebras one gets, see, for instance, [55]. One special case where the pattern is relatively simple is when one has $r = s$. Then, $n = 2r$ is even-dimensional and one finds

$$\text{Cliff}(r, r, \mathbf{R}) = M(2^r, \mathbf{R})$$

28.3 For further reading

A good source for more details about Clifford algebras and spinors is chapter 12 of the representation theory textbook [94]. For the details of what happens for all $\text{Cliff}(r, s, \mathbf{R})$, another good source is chapter 1 of [55].

Chapter 29
Clifford Algebras and Geometry

The definitions given in chapter 28 of Weyl and Clifford algebras were purely algebraic, based on a choice of generators and relations. These definitions do though have a more geometrical formulation, with the definition in terms of generators corresponding to a specific choice of coordinates. For the Weyl algebra, the geometry involved is symplectic geometry, based on a non-degenerate antisymmetric bilinear form. We have already seen that in the bosonic case quantization of a phase space $\mathbf{R}^{2d}$ depends on the choice of a non-degenerate antisymmetric bilinear form Ω which determines the Poisson brackets and thus the Heisenberg commutation relations. Such a Ω also determines a group $Sp(2d, \mathbf{R})$, which is the group of linear transformations of $\mathbf{R}^{2d}$ preserving Ω.

The Clifford algebra also has a coordinate invariant definition, based on a more well-known structure on a vector space $\mathbf{R}^n$, that of a non-degenerate symmetric bilinear form, i.e., an inner product. In this case, the group that preserves the inner product is an orthogonal group. In the symplectic case, antisymmetric forms require an even number of dimensions, but this is not true for symmetric forms, which also exist in odd dimensions.

29.1 Non-degenerate bilinear forms

In the case of $\mathcal{M} = \mathbf{R}^{2d}$, the dual phase space, the Poisson bracket determines an antisymmetric bilinear form on $\mathcal{M}$, which, for a basis q_j, p_j and two vectors $u, u' \in \mathcal{M}$

$$u = c_{q_1}q_1 + c_{p_1}p_1 + \cdots + c_{q_d}q_d + c_{p_d}p_d \in \mathcal{M}$$

$$u' = c'_{q_1}q_1 + c'_{p_1}p_1 + \cdots + c'_{q_d}q_d + c'_{p_d}p_d \in \mathcal{M}$$

© Peter Woit 2017

P. Woit, *Quantum Theory, Groups and Representations*,
DOI 10.1007/978-3-319-64612-1_29

is given explicitly by

$$\Omega(u, u') = c_{q_1} c'_{p_1} - c_{p_1} c'_{q_1} + \cdots + c_{q_d} c'_{p_d} - c_{p_d} c'_{q_d}$$

$$= \begin{pmatrix} c_{q_1} & c_{p_1} & \cdots c_{q_d} & c_{p_d} \end{pmatrix} \begin{pmatrix} 0 & 1 & \cdots & 0 & 0 \\ -1 & 0 & \cdots & 0 & 0 \\ \vdots & \vdots & & \vdots & \vdots \\ 0 & 0 & \cdots & 0 & 1 \\ 0 & 0 & \cdots & -1 & 0 \end{pmatrix} \begin{pmatrix} c'_{q_1} \\ c'_{p_1} \\ \vdots \\ c'_{q_d} \\ c'_{p_d} \end{pmatrix}$$

Matrices $g \in M(2d, \mathbf{R})$ such that

$$g^T \begin{pmatrix} 0 & 1 & \cdots & 0 & 0 \\ -1 & 0 & \cdots & 0 & 0 \\ \vdots & \vdots & & \vdots & \vdots \\ 0 & 0 & \cdots & 0 & 1 \\ 0 & 0 & \cdots & -1 & 0 \end{pmatrix} g = \begin{pmatrix} 0 & 1 & \cdots & 0 & 0 \\ -1 & 0 & \cdots & 0 & 0 \\ \vdots & \vdots & & \vdots & \vdots \\ 0 & 0 & \cdots & 0 & 1 \\ 0 & 0 & \cdots & -1 & 0 \end{pmatrix}$$

make up the group $Sp(2d, \mathbf{R})$ and preserve Ω, satisfying

$$\Omega(gu, gu') = \Omega(u, u')$$

This choice of Ω is much less arbitrary than it looks. One can show that given any non-degenerate antisymmetric bilinear form on $\mathbf{R}^{2d}$ a basis can be found with respect to which it will be the Ω given here (for a proof, see [8]). This is also true if one complexifies $\mathbf{R}^{2d}$, using the same formula for Ω, which is now a bilinear form on $\mathbf{C}^{2d}$. In the real case, the group that preserves Ω is called $Sp(2d, \mathbf{R})$, in the complex case $Sp(2d, \mathbf{C})$.

To get a fermionic analog of this, all one needs to do is replace "non-degenerate antisymmetric bilinear form $\Omega(\cdot, \cdot)$" with "non-degenerate symmetric bilinear form $(\cdot, \cdot)$." Such a symmetric bilinear form is actually something much more familiar from geometry than the antisymmetric case analog: It is just a notion of inner product. Two things are different in the symmetric case:

- The underlying vector space does not have to be even dimensional, one can take $M = \mathbf{R}^n$ for any n, including n odd. To get a detailed analog of the bosonic case though, we will need to consider the even case $n = 2d$.

- For a given dimension n, there is not just one possible choice of $(\cdot, \cdot)$ up to change of basis, but one possible choice for each pair of nonnegative integers r, s such that $r + s = n$. Given r, s, any choice of $(\cdot, \cdot)$ can be put in the form

$$(\mathbf{u}, \mathbf{u}') = u_1 u_1' + u_2 u_2' + \cdots u_r u_r' - u_{r+1} u_{r+1}' - \cdots - u_n u_n'$$

$$= \begin{pmatrix} u_1 & \cdots & u_n \end{pmatrix} \underbrace{\begin{pmatrix} 1 & 0 & \cdots & 0 & 0 \\ 0 & 1 & \cdots & 0 & 0 \\ \vdots & \vdots & & \vdots & \vdots \\ 0 & 0 & \cdots & -1 & 0 \\ 0 & 0 & \cdots & 0 & -1 \end{pmatrix}}_{r \ + \ \text{signs,} \ s \ \text{-} \ \text{signs}} \begin{pmatrix} u_1' \\ u_2' \\ \vdots \\ u_{n-1}' \\ u_n' \end{pmatrix}$$

For a proof by Gram–Schmidt orthogonalization, see [8].

We can thus extend our definition of the orthogonal group as the group of transformations g preserving an inner product

$$(gu, gu') = (u, u')$$

to the case r, s arbitrary by:

Definition (Orthogonal group $O(r, s, \mathbf{R})$). *The group $O(r, s, \mathbf{R})$ is the group of real $r + s$ by $r + s$ matrices g that satisfy*

$$g^T \underbrace{\begin{pmatrix} 1 & 0 & \cdots & 0 & 0 \\ 0 & 1 & \cdots & 0 & 0 \\ \vdots & \vdots & & \vdots & \vdots \\ 0 & 0 & \cdots & -1 & 0 \\ 0 & 0 & \cdots & 0 & -1 \end{pmatrix}}_{r \ + \ \text{signs,} \ s \ \text{-} \ \text{signs}} g = \underbrace{\begin{pmatrix} 1 & 0 & \cdots & 0 & 0 \\ 0 & 1 & \cdots & 0 & 0 \\ \vdots & \vdots & & \vdots & \vdots \\ 0 & 0 & \cdots & -1 & 0 \\ 0 & 0 & \cdots & 0 & -1 \end{pmatrix}}_{r \ + \ \text{signs,} \ s \ \text{-} \ \text{signs}}$$

$SO(r, s, \mathbf{R}) \subset O(r, s, \mathbf{R})$ is the subgroup of matrices of determinant $+1$.

If one complexifies, taking components of vectors to be in $\mathbf{C}^n$, using the same formula for $(\cdot, \cdot)$, one can change basis by multiplying the s basis elements by a factor of i, and in this new basis all basis vectors e_j satisfy $(e_j, e_j) = 1$. One thus sees that on $\mathbf{C}^n$, as in the symplectic case, up to change of basis there is only one non-degenerate symmetric bilinear form. The group preserving this is called $O(n, \mathbf{C})$. Note that on $\mathbf{C}^n$ $(\cdot, \cdot)$ is not the Hermitian inner product (which is antilinear on the first variable), and it is not positive definite.

29.2 Clifford algebras and geometry

As defined by generators in the last chapter, Clifford algebras have no obvious geometrical significance. It turns out, however, that they are powerful tools in the study of the geometry of linear spaces with an inner product, including especially the study of linear transformations that preserve the inner product, i.e., rotations. To see the relation between Clifford algebras and geometry, consider first the positive-definite case $\mathrm{Cliff}(n, \mathbf{R})$. To an arbitrary vector

$$\mathbf{v} = (v_1, v_2, \ldots, v_n) \in \mathbf{R}^n$$

we associate the Clifford algebra element $\not{v} = \gamma(\mathbf{v})$ where γ is the map

$$\mathbf{v} \in \mathbf{R}^n \to \gamma(\mathbf{v}) = v_1\gamma_1 + v_2\gamma_2 + \cdots + v_n\gamma_n \in \mathrm{Cliff}(n, \mathbf{R}) \qquad (29.1)$$

Using the Clifford algebra relations for the γ_j, given two vectors $\mathbf{v}$, $\mathbf{w}$ the product of their associated Clifford algebra elements satisfies

$$\begin{aligned}
\not{v}\not{w} + \not{w}\not{v} &= [v_1\gamma_1 + v_2\gamma_2 + \cdots + v_n\gamma_n,\ w_1\gamma_1 + w_2\gamma_2 + \cdots + w_n\gamma_n]_+ \\
&= 2(v_1 w_1 + v_2 w_2 + \cdots + v_n w_n) \\
&= 2(\mathbf{v}, \mathbf{w}) \qquad (29.2)
\end{aligned}$$

where $(\cdot, \cdot)$ is the symmetric bilinear form on $\mathbf{R}^n$ corresponding to the standard inner product of vectors. Note that taking $\mathbf{v} = \mathbf{w}$ one has

$$\not{v}^2 = (\mathbf{v}, \mathbf{v}) = ||\mathbf{v}||^2$$

The Clifford algebra $\mathrm{Cliff}(n, \mathbf{R})$ thus contains $\mathbf{R}^n$ as the subspace of linear combinations of the generators γ_j. It can be thought of as a sort of enhancement of the vector space $\mathbf{R}^n$ that encodes information about the inner product, and it will sometimes be written $\mathrm{Cliff}(\mathbf{R}^n, (\cdot, \cdot))$. In this larger structure vectors can be multiplied as well as added, with the multiplication determined by the inner product and given by equation 29.2. Note that different people use different conventions, with

$$\not{v}\not{w} + \not{w}\not{v} = -2(\mathbf{v}, \mathbf{w})$$

another common choice. One also sees variants without the factor of 2.

For n-dimensional vector spaces over $\mathbf{C}$, we have seen that for any non-degenerate symmetric bilinear form a basis can be found such that $(\cdot, \cdot)$ has the standard form

$$(\mathbf{z}, \mathbf{w}) = z_1 w_1 + z_2 w_2 + \cdots + z_n w_n$$

As a result, up to isomorphism, there is just one complex Clifford algebra in dimension n, the one we defined as $\mathrm{Cliff}(n, \mathbf{C})$. For n-dimensional vector spaces over $\mathbf{R}$ with a non-degenerate symmetric bilinear form of type r, s such that $r + s = n$, the corresponding Clifford algebras $\mathrm{Cliff}(r, s, \mathbf{R})$ are the ones defined in terms of generators in section 28.2.

In special relativity, space–time is a real four-dimensional vector space with an indefinite inner product corresponding to (depending on one's choice of convention) either the case $r = 1, s = 3$ or the case $r = 3, s = 1$. The group of linear transformations preserving this inner product is called the Lorentz group, and its orientation preserving component is written as $SO(3, 1)$ or $SO(1, 3)$ depending on the choice of convention. In later chapters, we will consider what happens to quantum mechanics in the relativistic case, and there encounter the corresponding Clifford algebras $\mathrm{Cliff}(3, 1, \mathbf{R})$ or $\mathrm{Cliff}(1, 3, \mathbf{R})$. The generators γ_j of such a Clifford algebra are well known in the subject as the "Dirac γ- matrices."

For now though, we will restrict attention to the positive-definite case, so just will be considering $\mathrm{Cliff}(n, \mathbf{R})$ and seeing how it is used to study the group $O(n)$ of n-dimensional rotations in $\mathbf{R}^n$.

29.2.1 Rotations as iterated orthogonal reflections

We will consider two different ways of seeing the relationship between the Clifford algebra $\mathrm{Cliff}(n, \mathbf{R})$ and the group $O(n)$ of rotations in $\mathbf{R}^n$. The first is based upon the geometrical fact (known as the Cartan–Dieudonné theorem) that one can get any rotation by doing at most n orthogonal reflections in different hyperplanes. Orthogonal reflection in the hyperplane perpendicular to a vector $\mathbf{w}$ takes a vector $\mathbf{v}$ to the vector

$$\mathbf{v}' = \mathbf{v} - 2\frac{(\mathbf{v}, \mathbf{w})}{(\mathbf{w}, \mathbf{w})}\mathbf{w}$$

something that can easily be seen from the following picture

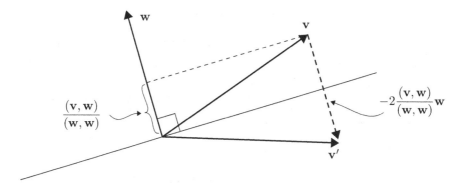

Figure 29.1: Orthogonal reflection in the hyperplane perpendicular to $\mathbf{w}$.

From now on, we identify vectors $\mathbf{v}, \mathbf{v}', \mathbf{w}$ with the corresponding Clifford algebra elements by the map γ of equation 29.1. The linear transformation given by reflection in $\mathbf{w}$ is

$$\slashed{v} \rightarrow \slashed{v}' = \slashed{v} - 2\frac{(\mathbf{v}, \mathbf{w})}{(\mathbf{w}, \mathbf{w})}\slashed{w}$$

$$= \slashed{v} - (\slashed{v}\slashed{w} + \slashed{w}\slashed{v})\frac{\slashed{w}}{(\mathbf{w}, \mathbf{w})}$$

Since

$$\slashed{w}\frac{\slashed{w}}{(\mathbf{w}, \mathbf{w})} = \frac{(\mathbf{w}, \mathbf{w})}{(\mathbf{w}, \mathbf{w})} = 1$$

we have (for nonzero vectors $\mathbf{w}$)

$$\slashed{w}^{-1} = \frac{\slashed{w}}{(\mathbf{w}, \mathbf{w})}$$

and the reflection transformation is just conjugation by $\slashed{w}$ times a minus sign

$$\slashed{v} \rightarrow \slashed{v}' = \slashed{v} - \slashed{v} - \slashed{w}\slashed{v}\slashed{w}^{-1} = -\slashed{w}\slashed{v}\slashed{w}^{-1}$$

Identifying vectors with Clifford algebra elements, the orthogonal transformation that is the result of one reflection is given by a conjugation (with a minus sign). These reflections lie in the group $O(n)$, but not in the subgroup $SO(n)$, since they change orientation. The result of two reflections in hyperplanes orthogonal to $\mathbf{w}_1, \mathbf{w}_2$ will be a conjugation by $\slashed{w}_2\slashed{w}_1$

$$\slashed{v} \rightarrow \slashed{v}' = -\slashed{w}_2(-\slashed{w}_1\slashed{v}\slashed{w}_1^{-1})\slashed{w}_2^{-1} = (\slashed{w}_2\slashed{w}_1)\slashed{v}(\slashed{w}_2\slashed{w}_1)^{-1}$$

This will be a rotation preserving the orientation, so of determinant one and in the group $SO(n)$.

This construction not only gives an efficient way of representing rotations (as conjugations in the Clifford algebra), but it also provides a construction of the group $Spin(n)$ in arbitrary dimension n. One can define:

Definition (Spin(n)). *The group $Spin(n, \mathbf{R})$ is the group of invertible elements of the Clifford algebra $Cliff(n)$ of the form*

$$\slashed{w}_1\slashed{w}_2 \cdots \slashed{w}_k$$

where the vectors $\mathbf{w}_j$ for $j = 1, \cdots, k$ $(k \leq n)$ are vectors in $\mathbf{R}^n$ satisfying $|\mathbf{w}_j|^2 = 1$ and k is even. Group multiplication is Clifford algebra multiplication.

The action of $Spin(n)$ on vectors $\mathbf{v} \in \mathbf{R}^n$ will be given by conjugation

$$\not{v} \to (\not{w}_1 \not{w}_2 \cdots \not{w}_k)\, \not{v}\, (\not{w}_1 \not{w}_2 \cdots \not{w}_k)^{-1} \tag{29.3}$$

and this will correspond to a rotation of the vector $\mathbf{v}$. This construction generalizes to arbitrary n the one we gave in chapter 6 of $Spin(3)$ in terms of unit-length elements of the quaternion algebra $\mathbf{H}$. One can see here the characteristic fact that there are two elements of the $Spin(n)$ group giving the same rotation in $SO(n)$ by noticing that changing the sign of the Clifford algebra element $\not{w}_1 \not{w}_2 \cdots \not{w}_k$ does not change the conjugation action, where signs cancel.

29.2.2 The Lie algebra of the rotation group and quadratic elements of the Clifford algebra

For a second approach to understanding rotations in arbitrary dimension, one can use the fact that these are generated by taking products of rotations in the coordinate planes. A rotation by an angle θ in the jk coordinate plane $(j < k)$ will be given by

$$\mathbf{v} \to e^{\theta \epsilon_{jk}} \mathbf{v}$$

where ϵ_{jk} is an n by n matrix with only two nonzero entries: jk entry -1 and kj entry $+1$ (see equation 5.2.1). Restricting attention to the jk plane, $e^{\theta \epsilon_{jk}}$ acts as the standard rotation matrix in the plane

$$\begin{pmatrix} v_j \\ v_k \end{pmatrix} \to \begin{pmatrix} \cos\theta & -\sin\theta \\ \sin\theta & \cos\theta \end{pmatrix} \begin{pmatrix} v_j \\ v_k \end{pmatrix}$$

In the $SO(3)$ case, we saw that there were three of these matrices

$$\epsilon_{23} = l_1, \quad \epsilon_{13} = -l_2, \quad \epsilon_{12} = l_3$$

providing a basis of the Lie algebra $\mathfrak{so}(3)$. In n dimensions there will be $\frac{1}{2}(n^2 - n)$ of them, providing a basis of the Lie algebra $\mathfrak{so}(n)$.

Just as in the case of $SO(3)$ where unit-length quaternions were used, in dimension n we can use elements of the Clifford algebra to get these same rotation transformations, but as conjugations in the Clifford algebra. To see how this works, consider the quadratic Clifford algebra element $\gamma_j \gamma_k$ for $j \neq k$ and notice that

$$(\gamma_j \gamma_k)^2 = \gamma_j \gamma_k \gamma_j \gamma_k = -\gamma_j \gamma_j \gamma_k \gamma_k = -1$$

so one has

$$e^{\frac{\theta}{2}\gamma_j\gamma_k} = \left(1 - \frac{(\theta/2)^2}{2!} + \cdots\right) + \gamma_j\gamma_k\left(\theta/2 - \frac{(\theta/2)^3}{3!} + \cdots\right)$$

$$= \cos\left(\frac{\theta}{2}\right) + \gamma_j\gamma_k\sin\left(\frac{\theta}{2}\right)$$

Conjugating a vector $v_j\gamma_j + v_k\gamma_k$ in the jk plane by this, one can show that

$$e^{-\frac{\theta}{2}\gamma_j\gamma_k}(v_j\gamma_j + v_k\gamma_k)e^{\frac{\theta}{2}\gamma_j\gamma_k} = (v_j\cos\theta - v_k\sin\theta)\gamma_j + (v_j\sin\theta + v_k\cos\theta)\gamma_k$$

which is a rotation by θ in the jk plane. Such a conjugation will also leave invariant the γ_l for $l \neq j, k$. Thus one has

$$e^{-\frac{\theta}{2}\gamma_j\gamma_k}\gamma(\mathbf{v})e^{\frac{\theta}{2}\gamma_j\gamma_k} = \gamma(e^{\theta\epsilon_{jk}}\mathbf{v}) \tag{29.4}$$

and, taking the derivative at $\theta = 0$, the infinitesimal version

$$\left[-\frac{1}{2}\gamma_j\gamma_k, \gamma(\mathbf{v})\right] = \gamma(\epsilon_{jk}\mathbf{v}) \tag{29.5}$$

Note that these relations are closely analogous to what happens in the symplectic case, where the symplectic group $Sp(2d, \mathbf{R})$ acts on linear combinations of the Q_j, P_j by conjugation by the exponential of an operator quadratic in the Q_j, P_j. We will examine this analogy in greater detail in chapter 31.

One can also see that, just as in our earlier calculations in three dimensions, one gets a double cover of the group of rotations, with here the elements $e^{\frac{\theta}{2}\gamma_j\gamma_k}$ of the Clifford algebra giving a double cover of the group of rotations in the jk plane (as θ goes from 0 to 2π). General elements of the spin group can be constructed by multiplying these for different angles in different coordinate planes. The Lie algebra $\mathfrak{spin}(n)$ can be identified with the Lie algebra $\mathfrak{so}(n)$ by

$$\epsilon_{jk} \leftrightarrow -\frac{1}{2}\gamma_j\gamma_k$$

Yet another way to see this would be to compute the commutators of the $-\frac{1}{2}\gamma_j\gamma_k$ for different values of j, k and show that they satisfy the same commutation relations as the corresponding matrices ϵ_{jk}.

Recall that in the bosonic case we found that quadratic combinations of the Q_j, P_k (or of the $a_{Bj}, a_{Bj}^\dagger$) gave operators satisfying the commutation relations of the Lie algebra $\mathfrak{sp}(2n, \mathbf{R})$. This is the Lie algebra of the group $Sp(2n, \mathbf{R})$, the group preserving the non-degenerate antisymmetric bilinear form $\Omega(\cdot, \cdot)$ on the phase space $\mathbf{R}^{2n}$. The fermionic case is precisely analogous, with the role of the antisymmetric bilinear form $\Omega(\cdot, \cdot)$ replaced by the symmetric bilinear form $(\cdot, \cdot)$ and the Lie algebra $\mathfrak{sp}(2n, \mathbf{R})$ replaced by $\mathfrak{so}(n) = \mathfrak{spin}(n)$.

In the bosonic case the linear functions of the Q_j, P_j satisfied the commutation relations of another Lie algebra, the Heisenberg algebra, but in the fermionic case this is not true for the γ_j. In chapter 30 we will see that a notion of a "Lie superalgebra" can be defined that restores the parallelism.

29.3 For further reading

Some more detail about spin groups and the relationship between geometry and Clifford algebras can be found in [55], and an exhaustive reference is [68].

Chapter 30
Anticommuting Variables
and Pseudo-classical Mechanics

The analogy between the algebras of operators in the bosonic (Weyl algebra) and fermionic (Clifford algebra) cases can be extended by introducing a fermionic analog of phase space and the Poisson bracket. This gives a fermionic analog of classical mechanics, sometimes called "pseudo-classical mechanics," the quantization of which gives the Clifford algebra as operators, and spinors as state spaces. In this chapter, we will introduce "anticommuting variables" ξ_j that will be the fermionic analogs of the variables q_j, p_j. These objects will become generators of the Clifford algebra under quantization, and will later be used in the construction of fermionic state spaces, by analogy with the Schrödinger and Bargmann–Fock constructions in the bosonic case.

30.1 The Grassmann algebra of polynomials on anticommuting generators

Given a phase space $M = \mathbf{R}^{2d}$, one gets classical observables by taking polynomial functions on M. These are generated by the linear functions $q_j, p_j, j = 1, \ldots, d$, which lie in the dual space $\mathcal{M} = M^*$. One can instead start with a real vector space $V = \mathbf{R}^n$ with n not necessarily even, and again consider the space V^* of linear functions on V, but with a different notion of multiplication, one that is anticommutative on elements of V^*. Using such a multiplication, an anticommuting analog of the algebra of polynomials on V can be generated in the following manner, beginning with a choice of basis elements ξ_j of V^*:

Definition (Grassmann algebra). *The algebra over the real numbers generated by $\xi_j, j = 1, \ldots, n$, satisfying the relations*

© Peter Woit 2017
P. Woit, *Quantum Theory, Groups and Representations*,
DOI 10.1007/978-3-319-64612-1_30

$$\xi_j \xi_k + \xi_k \xi_j = 0$$

is called the Grassmann algebra.

Note that these relations imply that generators satisfy $\xi_j^2 = 0$. Also, note that sometimes the Grassmann algebra product of ξ_j and ξ_k is denoted $\xi_j \wedge \xi_k$. We will not use a different symbol for the product in the Grassmann algebra, relying on the notation for generators to keep straight what is a generator of a conventional polynomial algebra (e.g., q_j or p_j) and what is a generator of a Grassmann algebra (e.g., ξ_j).

The Grassmann algebra is the algebra $\Lambda^*(V^*)$ of antisymmetric multilinear forms on V discussed in section 9.6, except that we have chosen a basis of V and have written out the definition in terms of the dual basis ξ_j of V^*. It is sometimes also called the "exterior algebra." This algebra behaves in many ways like the polynomial algebra on $\mathbf{R}^n$, but it is finite dimensional as a real vector space, with basis

$$1, \quad \xi_j, \quad \xi_j \xi_k, \quad \xi_j \xi_k \xi_l, \quad \cdots, \quad \xi_1 \xi_2 \cdots \xi_n$$

for indices $j < k < l < \cdots$ taking values $1, 2, \ldots, n$. As with polynomials, monomials are characterized by a degree (number of generators in the product), which in this case takes values from 0 only up to n. $\Lambda^k(\mathbf{R}^n)$ is the subspace of $\Lambda^*(\mathbf{R}^n)$ of linear combinations of monomials of degree k.

Digression (Differential forms). *Readers may have already seen the Grassmann algebra in the context of differential forms on $\mathbf{R}^n$. These are known to physicists as "antisymmetric tensor fields," and given by taking elements of the exterior algebra $\Lambda^*(\mathbf{R}^n)$ with coefficients not constants, but functions on $\mathbf{R}^n$. This construction is important in the theory of manifolds, where at a point x in a manifold M, one has a tangent space $T_x M$ and its dual space $(T_x M)^*$. A set of local coordinates x_j on M gives basis elements of $(T_x M)^*$ denoted by dx_j, and differential forms locally can be written as sums of terms of the form*

$$f(x_1, x_2, \cdots, x_n) dx_j \wedge \cdots \wedge dx_k \wedge \cdots \wedge dx_l$$

where the indices j, k, l satisfy $1 \leq j < k < l \leq n$.

A fundamental principle of mathematics is that a good way to understand a space is in terms of the functions on it. What we have done here can be thought of as creating a new kind of space out of $\mathbf{R}^n$, where the algebra of functions on the space is $\Lambda^*(\mathbf{R}^n)$, generated by coordinate functions ξ_j with respect to a basis of $\mathbf{R}^n$. The enlargement of conventional geometry to include new kinds of spaces such that this makes sense is known as "supergeometry," but we will not attempt to pursue this subject here. Spaces with this new kind of geometry have functions on them but do not have conventional points since we have seen that one cannot ask what the value of an anticommuting function at a point is.

Remarkably, an analog of calculus can be defined on such unconventional spaces, introducing analogs of the derivative and integral for anticommuting functions (i.e., elements of the Grassmann algebra). For the case $n = 1$, an arbitrary function is

$$F(\xi) = c_0 + c_1 \xi$$

and one can take

$$\frac{\partial}{\partial \xi} F = c_1$$

For larger values of n, an arbitrary function can be written as follows:

$$F(\xi_1, \xi_2, \ldots, \xi_n) = F_A + \xi_j F_B$$

where F_A, F_B are functions that do not depend on the chosen ξ_j (one gets F_B by using the anticommutation relations to move ξ_j all the way to the left). Then, one can define

$$\frac{\partial}{\partial \xi_j} F = F_B$$

This derivative operator has many of the same properties as the conventional derivative, although there are unconventional signs one must keep track of. An unusual property of this derivative that is easy to see is that one has

$$\frac{\partial}{\partial \xi_j} \frac{\partial}{\partial \xi_j} = 0$$

Taking the derivative of a product, one finds this version of the Leibniz rule for monomials F and G

$$\frac{\partial}{\partial \xi_j}(FG) = \left(\frac{\partial}{\partial \xi_j} F\right) G + (-1)^{|F|} F \left(\frac{\partial}{\partial \xi_j} G\right)$$

where $|F|$ is the degree of the monomial F.

A notion of integration (often called the "Berezin integral") with many of the usual properties of an integral can also be defined. It has the peculiar feature of being the same operation as differentiation, defined in the $n = 1$ case by

$$\int (c_0 + c_1 \xi) d\xi = c_1$$

and for larger n by

$$\int F(\xi_1,\xi_2,\cdots,\xi_n)d\xi_1 d\xi_2\cdots d\xi_n = \frac{\partial}{\partial\xi_n}\frac{\partial}{\partial\xi_{n-1}}\cdots\frac{\partial}{\partial\xi_1}F = c_n$$

where c_n is the coefficient of the basis element $\xi_1\xi_2\cdots\xi_n$ in the expression of F in terms of basis elements.

This notion of integration is a linear operator on functions, and it satisfies an analog of integration by parts, since if

$$F = \frac{\partial}{\partial\xi_j}G$$

then

$$\int F d\xi_j = \frac{\partial}{\partial\xi_j}F = \frac{\partial}{\partial\xi_j}\frac{\partial}{\partial\xi_j}G = 0$$

using the fact that repeated derivatives give zero.

30.2 Pseudo-classical mechanics and the fermionic Poisson bracket

The basic structure of Hamiltonian classical mechanics depends on an even-dimensional phase space $M = \mathbf{R}^{2d}$ with a Poisson bracket $\{\cdot,\cdot\}$ on functions on this space. Time evolution of a function f on phase space is determined by

$$\frac{d}{dt}f = \{f,h\}$$

for some Hamiltonian function h. This says that taking the derivative of any function in the direction of the velocity vector of a classical trajectory is the linear map

$$f \to \{f,h\}$$

on functions. As we saw in chapter 14, since this linear map is a derivative, the Poisson bracket will have the derivation property, satisfying the Leibniz rule

$$\{f_1,f_2 f_3\} = f_2\{f_1,f_3\} + \{f_1,f_2\}f_3$$

for arbitrary functions f_1,f_2,f_3 on phase space. Using the Leibniz rule and antisymmetry, Poisson brackets can be calculated for any polynomials, just from knowing the Poisson bracket on generators q_j,p_j (or, equivalently, the antisymmetric bilinear form $\Omega(\cdot,\cdot)$), which we chose to be

$$\{q_j, q_k\} = \{p_j, p_k\} = 0, \quad \{q_j, p_k\} = -\{p_k, q_j\} = \delta_{jk}$$

Notice that we have a symmetric multiplication on generators, while the Poisson bracket is antisymmetric.

To get pseudo-classical mechanics, we think of the Grassmann algebra $\Lambda^*(\mathbf{R}^n)$ as our algebra of classical observables, an algebra we can think of as functions on a "fermionic" phase space $V = \mathbf{R}^n$ (note that in the fermionic case, the phase space does not need to be even-dimensional). We want to find an appropriate notion of fermionic Poisson bracket operation on this algebra, and it turns out that this can be done. While the standard Poisson bracket is an antisymmetric bilinear form $\Omega(\cdot, \cdot)$ on linear functions, the fermionic Poisson bracket will be based on a choice of symmetric bilinear form on linear functions, equivalently, a notion of inner product $(\cdot, \cdot)$.

Denoting the fermionic Poisson bracket by $\{\cdot, \cdot\}_+$, for a multiplication anticommutative on generators, one has to adjust signs in the Leibniz rule, and the derivation property analogous to the derivation property of the usual Poisson bracket is, for monomials F_1, F_2, F_3,

$$\{F_1 F_2, F_3\}_+ = F_1 \{F_2, F_3\}_+ + (-1)^{|F_2||F_3|} \{F_1, F_3\}_+ F_2$$

where $|F_2|$ and $|F_3|$ are the degrees of F_2 and F_3. It will also have the symmetry property

$$\{F_1, F_2\}_+ = -(-1)^{|F_1||F_2|} \{F_2, F_1\}_+$$

and these properties can be used to compute the fermionic Poisson bracket for arbitrary functions in terms of the relations for generators.

The ξ_j can be thought of as the "anticommuting coordinate functions" with respect to a basis $\mathbf{e}_j$ of $V = \mathbf{R}^n$. We have seen that the symmetric bilinear forms on $\mathbf{R}^n$ are classified by a choice of positive signs for some basis vectors, negative signs for the others. So, on generators ξ_j, one can choose

$$\{\xi_j, \xi_k\}_+ = \pm \delta_{jk}$$

with a plus sign for $j = k = 1, \cdots, r$ and a minus sign for $j = k = r + 1, \cdots, n$, corresponding to the possible inequivalent choices of non-degenerate symmetric bilinear forms.

Taking the case of a positive-definite inner product for simplicity, one can calculate explicitly the fermionic Poisson brackets for linear and quadratic combinations of the generators. One finds

$$\{\xi_j \xi_k, \xi_l\}_+ = \xi_j \{\xi_k, \xi_l\}_+ - \{\xi_j, \xi_l\}_+ \xi_k = \delta_{kl} \xi_j - \delta_{jl} \xi_k \tag{30.1}$$

and

$$\{\xi_j\xi_k, \xi_l\xi_m\}_+ = \{\xi_j\xi_k, \xi_l\}_+\xi_m + \xi_l\{\xi_j\xi_k, \xi_m\}_+$$
$$= \delta_{kl}\xi_j\xi_m - \delta_{jl}\xi_k\xi_m + \delta_{km}\xi_l\xi_j - \delta_{jm}\xi_l\xi_k \qquad (30.2)$$

The second of these equations shows that the quadratic combinations of the generators ξ_j satisfy the relations of the Lie algebra of the group of rotations in n dimensions ($\mathfrak{so}(n) = \mathfrak{spin}(n)$). The first shows that the $\xi_k\xi_l$ acts on the ξ_j as infinitesimal rotations in the kl plane.

In the case of the conventional Poisson bracket, the antisymmetry of the bracket and the fact that it satisfies the Jacobi identity implies that it is a Lie bracket determining a Lie algebra (the infinite dimensional Lie algebra of functions on a phase space $\mathbf{R}^{2d}$). The fermionic Poisson bracket provides an example of something called a Lie superalgebra. These can be defined for vector spaces with some usual and some fermionic coordinates:

Definition (Lie superalgebra). *A Lie superalgebra structure on a real or complex vector space V is given by a Lie superbracket $[\cdot, \cdot]_\pm$. This is a bilinear map on V which on generators X, Y, Z (which may be usual or fermionic ones) satisfies*

$$[X, Y]_\pm = -(-1)^{|X||Y|}[Y, X]_\pm$$

and a super-Jacobi identity

$$[X, [Y, Z]_\pm]_\pm = [[X, Y]_\pm, Z]_\pm + (-1)^{|X||Y|}[Y, [X, Z]_\pm]_\pm$$

where $|X|$ takes value 0 for a usual generator, 1 for a fermionic generator.

Analogously to the bosonic case, on polynomials in generators with order of the polynomial less than or equal to two, the fermionic Poisson bracket $\{\cdot, \cdot\}_+$ is a Lie superbracket, giving a Lie superalgebra of dimension $1 + n + \frac{1}{2}(n^2 - n)$ (since there is one constant, n linear terms ξ_j and $\frac{1}{2}(n^2 - n)$ quadratic terms $\xi_j\xi_k$). On functions of order two, this Lie superalgebra is a Lie algebra, $\mathfrak{so}(n)$. We will see in chapter 31 that the definition of a representation can be generalized to Lie superalgebras, and quantization will give a distinguished representation of this Lie superalgebra, in a manner quite parallel to that of the Schrödinger or Bargmann–Fock constructions of a representation in the bosonic case.

The relation between the quadratic and linear polynomials in the generators is parallel to what happens in the bosonic case. Here, we have the fermionic analog of the bosonic theorem 16.2:

Theorem 30.1. *The Lie algebra $\mathfrak{so}(n, \mathbf{R})$ is isomorphic to the Lie algebra $\Lambda^2(V^*)$ (with Lie bracket $\{\cdot, \cdot\}_+$) of order two anticommuting polynomials on $V = \mathbf{R}^n$, by the isomorphism*

$$L \leftrightarrow \mu_L$$

where $L \in \mathfrak{so}(n, \mathbf{R})$ is an antisymmetric n by n real matrix, and

$$\mu_L = \frac{1}{2}\boldsymbol{\xi} \cdot L\boldsymbol{\xi} = \frac{1}{2}\sum_{j,k} L_{jk}\xi_j\xi_k$$

The $\mathfrak{so}(n, \mathbf{R})$ action on anticommuting coordinate functions is

$$\{\mu_L, \xi_k\}_+ = \sum_j L_{jk}\xi_j$$

or

$$\{\mu_L, \boldsymbol{\xi}\}_+ = L^T\boldsymbol{\xi}$$

Proof. The theorem follows from equations 30.1 and 30.2, or one can proceed by analogy with the proof of theorem 16.2 as follows. First, prove the second part of the theorem by computing

$$\left\{\frac{1}{2}\sum_{j,k}\xi_j L_{jk}\xi_k, \xi_l\right\}_+ = \frac{1}{2}\sum_{j,k} L_{jk}(\xi_j\{\xi_k, \xi_l\}_+ - \{\xi_j, \xi_l\}_+\xi_k)$$

$$= \frac{1}{2}(\sum_j L_{jl}\xi_j - \sum_k L_{lk}\xi_k)$$

$$= \sum_j L_{jl}\xi_j \quad (\text{since } L = -L^T)$$

For the first part of the theorem, the map

$$L \to \mu_L$$

is a vector space isomorphism of the space of antisymmetric matrices and $\Lambda^2(\mathbf{R}^n)$. To show that it is a Lie algebra isomorphism, one can use an analogous argument to that of the proof of 16.2. Here, one considers the action

$$\xi \to \{\mu_L, \xi\}_+$$

of $\mu_L \in \mathfrak{so}(n, \mathbf{R})$ on an arbitrary

$$\xi = \sum_j c_j\xi_j$$

and uses the super-Jacobi identity relating the fermionic Poisson brackets of $\mu_L, \mu_{L'}, \xi$. $\qquad\square$

30.3 Examples of pseudo-classical mechanics

In pseudo-classical mechanics, the dynamics will be determined by choosing a Hamiltonian h in $\Lambda^*(\mathbf{R}^n)$. Observables will be other functions $F \in \Lambda^*(\mathbf{R}^n)$, and they will satisfy the analog of Hamilton's equations

$$\frac{d}{dt}F = \{F, h\}_+$$

We will consider two of the simplest possible examples.

30.3.1 The pseudo-classical spin degree of freedom

Using pseudo-classical mechanics, a "classical" analog can be found for something that is quintessentially quantum: the degree of freedom that appears in the qubit or spin $\frac{1}{2}$ system that first appeared in chapter 3. Taking $V = \mathbf{R}^3$ with the standard inner product as fermionic phase space, we have three generators $\xi_1, \xi_2, \xi_3 \in V^*$ satisfying the relations

$$\{\xi_j, \xi_k\}_+ = \delta_{jk}$$

and an eight-dimensional space of functions with basis

$$1, \ \xi_1, \ \xi_2, \ \xi_3, \ \xi_1\xi_2, \ \xi_1\xi_3, \ \xi_2\xi_3, \ \xi_1\xi_2\xi_3$$

If we want the Hamiltonian function to be non-trivial and of even degree, it will have to be a linear combination

$$h = B_{12}\xi_1\xi_2 + B_{13}\xi_1\xi_3 + B_{23}\xi_2\xi_3$$

for some constants B_{12}, B_{13}, B_{23}. This can be written

$$h = \frac{1}{2}\sum_{j,k=1}^{3} L_{jk}\xi_j\xi_k$$

where L_{jk} are the entries of the matrix

$$L = \begin{pmatrix} 0 & B_{12} & B_{13} \\ -B_{12} & 0 & B_{23} \\ -B_{13} & -B_{23} & 0 \end{pmatrix}$$

The equations of motion on generators will be

$$\frac{d}{dt}\xi_j(t) = \{\xi_j, h\}_+ = -\{h, \xi_j\}_+$$

which, since $L = -L^T$, by theorem 30.1 can be written

$$\frac{d}{dt}\xi_j(t) = L\xi_j(t)$$

with solution

$$\xi_j(t) = e^{tL}\xi_j(0)$$

This will be a time-dependent rotation of the ξ_j in the plane perpendicular to

$$\mathbf{B} = (B_{23}, -B_{13}, B_{12})$$

at a constant speed proportional to $|\mathbf{B}|$.

30.3.2 The pseudo-classical fermionic oscillator

We have already studied the fermionic oscillator as a quantum system (in section 27.2), and one can ask whether there is a corresponding pseudo-classical system. For the case of d oscillators, such a system is given by taking an even-dimensional fermionic phase space $V = \mathbf{R}^{2d}$, with a basis of coordinate functions $\xi_1, \cdots, \xi_{2d}$ that generate $\Lambda^*(\mathbf{R}^{2d})$. On generators, the fermionic Poisson bracket relations come from the standard choice of positive-definite symmetric bilinear form

$$\{\xi_j, \xi_k\}_+ = \delta_{jk}$$

As shown in theorem 30.1, quadratic products $\xi_j\xi_k$ act on the generators by infinitesimal rotations in the jk plane and satisfy the commutation relations of $\mathfrak{so}(2d)$.

To get a pseudo-classical system corresponding to the fermionic oscillator, one makes the choice

$$h = \frac{1}{2}\sum_{j=1}^{d}(\xi_{2j}\xi_{2j-1} - \xi_{2j-1}\xi_{2j}) = \sum_{j=1}^{d}\xi_{2j}\xi_{2j-1}$$

This makes h the moment map for a simultaneous rotation in the $2j - 1, 2j$ planes, corresponding to a matrix in $\mathfrak{so}(2d)$ given by

$$L = \sum_{j=1}^{d} \epsilon_{2j-1,2j}$$

As in the bosonic case, we can make the standard choice of complex structure $J = J_0$ on $\mathbf{R}^{2d}$ and get a decomposition

$$V^* \otimes \mathbf{C} = \mathbf{R}^{2d} \otimes \mathbf{C} = \mathbf{C}^d \oplus \mathbf{C}^d$$

into eigenspaces of J of eigenvalue $\pm i$. This is done by defining

$$\theta_j = \frac{1}{\sqrt{2}}(\xi_{2j-1} - i\xi_{2j}), \quad \overline{\theta}_j = \frac{1}{\sqrt{2}}(\xi_{2j-1} + i\xi_{2j})$$

for $j = 1, \ldots, d$. These satisfy the fermionic Poisson bracket relations

$$\{\theta_j, \theta_k\}_+ = \{\overline{\theta}_j, \overline{\theta}_k\}_+ = 0, \quad \{\overline{\theta}_j, \theta_k\}_+ = \delta_{jk}$$

(where we have extended the inner product $\{\cdot, \cdot\}_+$ to $V^* \otimes \mathbf{C}$ by complex linearity).

In terms of the θ_j, the Hamiltonian is

$$h = -\frac{i}{2}\sum_{j=1}^{d}(\theta_j\overline{\theta}_j - \overline{\theta}_j\theta_j) = -i\sum_{j=1}^{d}\theta_j\overline{\theta}_j$$

Using the derivation property of $\{\cdot, \cdot\}_+$, one finds

$$\{h, \theta_j\}_+ = -i\sum_{k=1}^{d}(\theta_k\{\overline{\theta}_k, \theta_j\}_+ - \{\theta_k, \theta_j\}_+\overline{\theta}_k) = -i\theta_j$$

and, similarly,

$$\{h, \overline{\theta}_j\}_+ = i\overline{\theta}_j$$

so one sees that h is the generator of $U(1) \subset U(d)$ phase rotations on the variables θ_j. The equations of motion are

$$\frac{d}{dt}\theta_j = \{\theta_j, h\}_+ = i\theta_j, \quad \frac{d}{dt}\overline{\theta}_j = \{\overline{\theta}_j, h\}_+ = -i\overline{\theta}_j$$

with solutions

$$\theta_j(t) = e^{it}\theta_j(0), \quad \overline{\theta}_j(t) = e^{-it}\overline{\theta}_j(0)$$

30.4   For further reading

For more details on pseudo-classical mechanics, a very readable original reference is [7]. There is a detailed discussion in the textbook [90], chapter 7.

Chapter 31
Fermionic Quantization and Spinors

In this chapter, we will begin by investigating the fermionic analog of the notion of quantization, which takes functions of anticommuting variables on a phase space with symmetric bilinear form $(\cdot, \cdot)$ and gives an algebra of operators with generators satisfying the relations of the corresponding Clifford algebra. We will then consider analogs of the constructions used in the bosonic case which there gave us the Schrödinger and Bargmann–Fock representations of the Weyl algebra on a space of states.

We know that for a fermionic oscillator with d degrees of freedom, the algebra of operators will be $\mathrm{Cliff}(2d, \mathbf{C})$, the algebra generated by annihilation and creation operators $a_{Fj}, a_{Fj}^{\dagger}$. These operators will act on $\mathcal{H}_F = \mathcal{F}_d^+$, a complex vector space of dimension 2^d, and this will provide a fermionic analog of the bosonic Γ'_{BF} acting on $\mathcal{F}_d$. Since the spin group consists of invertible elements of the Clifford algebra, it has a representation on $\mathcal{F}_d^+$. This is known as the "spinor representation," and it can be constructed by analogy with the construction of the metaplectic representation in the bosonic case. We will also consider the analog in the fermionic case of the Schrödinger representation, which turns out to have a problem with unitarity, but finds a use in physics as "ghost" degrees of freedom.

31.1 Quantization of pseudo-classical systems

In the bosonic case, quantization was based on finding a representation of the Heisenberg Lie algebra of linear functions on phase space, or more explicitly, for basis elements q_j, p_j of this Lie algebra finding operators Q_j, P_j satisfying the Heisenberg commutation relations. In the fermionic case, the analog of

© Peter Woit 2017

P. Woit, *Quantum Theory, Groups and Representations*,

DOI 10.1007/978-3-319-64612-1_31

the Heisenberg Lie algebra is not a Lie algebra, but a Lie superalgebra, with basis elements $1, \xi_j, j = 1, \ldots, n$ and a Lie superbracket given by the fermionic Poisson bracket, which on basis elements is

$$\{\xi_j, \xi_k\}_+ = \pm \delta_{jk}, \quad \{\xi_j, 1\}_+ = 0, \quad \{1, 1\}_+ = 0$$

Quantization is given by finding a representation of this Lie superalgebra. The definition of a Lie algebra representation can be generalized to that of a Lie superalgebra representation by:

Definition (Representation of a Lie superalgebra). *A representation of a Lie superalgebra is a homomorphism Φ preserving the superbracket*

$$[\Phi(X), \Phi(Y)]_\pm = \Phi([X, Y]_\pm)$$

This takes values in a Lie superalgebra of linear operators, with $|\Phi(X)| = |X|$ and

$$[\Phi(X), \Phi(Y)]_\pm = \Phi(X)\Phi(Y) - (-)^{|X||Y|}\Phi(Y)\Phi(X)$$

A representation of the pseudo-classical Lie superalgebra (and thus a quantization of the pseudo-classical system) will be given by finding a linear map Γ^+ that takes basis elements ξ_j to operators $\Gamma^+(\xi_j)$ satisfying the relations

$$[\Gamma^+(\xi_j), \Gamma^+(\xi_k)]_+ = \pm \delta_{jk}\Gamma^+(1), \quad [\Gamma^+(\xi_j), \Gamma^+(1)] = [\Gamma^+(1), \Gamma^+(1)] = 0$$

These relations can be satisfied by taking

$$\Gamma^+(\xi_j) = \frac{1}{\sqrt{2}}\gamma_j, \quad \Gamma^+(1) = 1$$

since then

$$[\Gamma^+(\xi_j), \Gamma^+(\xi_k)]_+ = \frac{1}{2}[\gamma_j, \gamma_k]_+ = \pm \delta_{jk}$$

are exactly the Clifford algebra relations. This can be extended to a representation of the functions of the ξ_j of order two or less by

Theorem. *A representation of the Lie superalgebra of anticommuting functions of coordinates ξ_j on $\mathbf{R}^n$ of order two or less is given by*

$$\Gamma^+(1) = 1, \quad \Gamma^+(\xi_j) = \frac{1}{\sqrt{2}}\gamma_j, \quad \Gamma^+(\xi_j\xi_k) = \frac{1}{2}\gamma_j\gamma_k$$

Proof. We have already seen that this is a representation for polynomials in ξ_j of degree zero and one. For simplicity just considering the case $s = 0$ (positive-definite inner product), in degree two the fermionic Poisson bracket relations are given by equations 30.1 and 30.2. For 30.1, one can show that the products of Clifford algebra generators

$$\Gamma^+(\xi_j\xi_k) = \frac{1}{2}\gamma_j\gamma_k$$

satisfy

$$\left[\frac{1}{2}\gamma_j\gamma_k, \gamma_l\right] = \delta_{kl}\gamma_j - \delta_{jl}\gamma_k$$

by using the Clifford algebra relations, or by noting that this is the special case of equation 29.5 for $\mathbf{v} = \mathbf{e}_l$. That equation shows that commuting by $-\frac{1}{2}\gamma_j\gamma_k$ acts by the infinitesimal rotation ϵ_{jk} in the jk coordinate plane.

For 30.2, the Clifford algebra relations can again be used to show

$$\left[\frac{1}{2}\gamma_j\gamma_k, \frac{1}{2}\gamma_l\gamma_m\right] = \delta_{kl}\frac{1}{2}\gamma_j\gamma_m - \delta_{jl}\frac{1}{2}\gamma_k\gamma_m + \delta_{km}\frac{1}{2}\gamma_l\gamma_j - \delta_{jm}\frac{1}{2}\gamma_l\gamma_k$$

One could instead use the commutation relations for the $\mathfrak{so}(n)$ Lie algebra satisfied by the basis elements ϵ_{jk} corresponding to infinitesimal rotations. One must get identical commutation relations for the $-\frac{1}{2}\gamma_j\gamma_k$ and can show that these are the relations needed for commutators of $\Gamma^+(\xi_j\xi_k)$ and $\Gamma^+(\xi_l\xi_m)$. □

Note that here we are not introducing the factors of i into the definition of quantization that in the bosonic case were necessary to get a unitary representation of the Lie group corresponding to the real Heisenberg Lie algebra $\mathfrak{h}_{2d+1}$. In the bosonic, case we worked with all complex linear combinations of powers of the Q_j, P_j (the complex Weyl algebra Weyl$(2d, \mathbf{C})$), and thus had to identify the specific complex linear combinations of these that gave unitary representations of the Lie algebra $\mathfrak{h}_{2d+1} \rtimes \mathfrak{sp}(2d, \mathbf{R})$. Here we are not complexifying for now, but working with the real Clifford algebra Cliff$(r, s, \mathbf{R})$, and it is the irreducible representations of this algebra that provide an analog of the unique interesting irreducible representation of $\mathfrak{h}_{2d+1}$. In the Clifford algebra case, the representations of interest are not just Lie algebra representations and may be on real vector spaces. There is no analog of the unitarity property of the $\mathfrak{h}_{2d+1}$ representation.

In the bosonic case, we found that $Sp(2d, \mathbf{R})$ acted on the bosonic dual phase space, preserving the antisymmetric bilinear form Ω that determined the Lie algebra $\mathfrak{h}_{2d+1}$, so it acted on this Lie algebra by automorphisms. We saw (see chapter 20) that intertwining operators there gave us a representation of the double cover of $Sp(2d, \mathbf{R})$ (the metaplectic representation), with the Lie algebra representation given by the quantization of quadratic functions of the q_j, p_j phase space coordinates. There is a closely analogous story in the fermionic case, where $SO(r, s, \mathbf{R})$ acts on the fermionic phase space V, preserving the symmetric bilinear form $(\cdot, \cdot)$ that determines the Clifford algebra relations. Here, a representation of the spin group $Spin(r, s, \mathbf{R})$ double covering $SO(r, s, \mathbf{R})$ is constructed using intertwining operators, with the Lie algebra representation given by quadratic combinations of the quantizations of the fermionic coordinates ξ_j. The case of $r = 3, s = 1$ will be of importance later in our discussion of special relativity (see chapter 41), giving the spinor representation of the Lorentz group.

The fermionic analog of 20.1 is

$$U_k \Gamma^+(\xi) U_k^{-1} = \Gamma^+(\phi_{k_0}(\xi)) \tag{31.1}$$

Here $k_0 \in SO(r, s, \mathbf{R})$, $\xi \in V^* = \mathbf{R}^n$ $(n = r + s)$, ϕ_{k_0} is the action of k_0 on V^*. The U_k for $k = \Phi^{-1}(k_0) \in Spin(r, s)$ (Φ is the 2-fold covering map) are the intertwining operators we are looking for. The fermionic analog of 20.2 is

$$[U_L', \Gamma^+(\xi)] = \Gamma^+(L \cdot \xi)$$

where $L \in \mathfrak{so}(r, s, \mathbf{R})$ and L act on V^* as an infinitesimal orthogonal transformation. In terms of basis vectors of V^*

$$\boldsymbol{\xi} = \begin{pmatrix} \xi_1 \\ \vdots \\ \xi_n \end{pmatrix}$$

this says

$$[U_L', \Gamma^+(\boldsymbol{\xi})] = \Gamma^+(L^T \boldsymbol{\xi})$$

Just as in the bosonic case, the U_L' can be found by looking first at the pseudo-classical case, where one has theorem 30.1 which says

$$\{\mu_L, \boldsymbol{\xi}\}_+ = L^T \boldsymbol{\xi}$$

where

$$\mu_L = \frac{1}{2} \boldsymbol{\xi} \cdot L \boldsymbol{\xi} = \frac{1}{2} \sum_{j,k} L_{jk} \xi_j \xi_k$$

One then takes

$$U_L' = \Gamma^+(\mu_L) = \frac{1}{4} \sum_{j,k} L_{jk} \gamma_j \gamma_k$$

For the positive-definite case $s = 0$ and a rotation in the jk plane, with $L = \epsilon_{jk}$ one recovers formulas 29.4 and 29.5 from chapter 29, with

$$\left[-\frac{1}{2} \gamma_j \gamma_k, \gamma(\mathbf{v}) \right] = \gamma(\epsilon_{jk} \mathbf{v})$$

the infinitesimal action of a rotation on the γ matrices, and

$$\gamma(\mathbf{v}) \to e^{-\frac{\theta}{2}\gamma_j\gamma_k}\gamma(\mathbf{v})e^{\frac{\theta}{2}\gamma_j\gamma_k} = \gamma(e^{\theta\epsilon_{jk}}\mathbf{v})$$

the group version. Just as in the symplectic case, exponentiating the U'_L only gives a representation up to sign, and one needs to go to the double cover of $SO(n)$ to get a true representation. As in that case, the necessity of the double cover is best seen by use of a complex structure and an analog of the Bargmann–Fock construction, an example will be given in section 31.4.

In order to have a full construction of a quantization of a pseudo-classical system, we need to construct the $\Gamma^+(\xi_j)$ as linear operators on a state space. As mentioned in section 28.2, it can be shown that the real Clifford algebras $\mathrm{Cliff}(r, s, \mathbf{R})$ are isomorphic to either one or two copies of the matrix algebras $M(2^l, \mathbf{R}), M(2^l, \mathbf{C})$, or $M(2^l, \mathbf{H})$, with the power l depending on r, s. The irreducible representations of such a matrix algebra are just the column vectors of dimension 2^l, and there will be either one or two such irreducible representations for $\mathrm{Cliff}(r, s, \mathbf{R})$ depending on the number of copies of the matrix algebra. This is the fermionic analog of the Stone–von Neumann uniqueness result in the bosonic case.

31.1.1 Quantization of the pseudo-classical spin

As an example, one can consider the quantization of the pseudo-classical spin degree of freedom of section 30.3.1. In that case, Γ^+ takes values in $\mathrm{Cliff}(3, 0, \mathbf{R})$, for which an explicit identification with the algebra $M(2, \mathbf{C})$ of two-by-two complex matrices was given in section 28.2. One has

$$\Gamma^+(\xi_j) = \frac{1}{\sqrt{2}}\gamma_j = \frac{1}{\sqrt{2}}\sigma_j$$

and the Hamiltonian operator is

$$\begin{aligned} -iH = \Gamma^+(h) =& \Gamma^+(B_{12}\xi_1\xi_2 + B_{13}\xi_1\xi_3 + B_{23}\xi_2\xi_3) \\ =& \frac{1}{2}(B_{12}\sigma_1\sigma_2 + B_{13}\sigma_1\sigma_3 + B_{23}\sigma_2\sigma_3) \\ =& i\frac{1}{2}(B_1\sigma_1 + B_2\sigma_2 + B_3\sigma_3) \end{aligned}$$

This is nothing but our old example from chapter 7 of a fixed spin particle in a magnetic field.

The pseudo-classical equation of motion

$$\frac{d}{dt}\xi_j(t) = -\{h, \xi_j\}_+$$

after quantization becomes the Heisenberg picture equation of motion for the spin operators (see equation 7.3)

$$\frac{d}{dt}\mathbf{S}_H(t) = -i[\mathbf{S}_H \cdot \mathbf{B}, \mathbf{S}_H]$$

for the case of Hamiltonian
$$H = -\boldsymbol{\mu} \cdot \mathbf{B}$$

(see equation 7.2) and magnetic moment operator

$$\boldsymbol{\mu} = \mathbf{S}$$

Here, the state space is $\mathcal{H} = \mathbf{C}^2$, with an explicit choice of basis given by our chosen identification of $\text{Cliff}(3, 0, \mathbf{R})$ with two-by-two complex matrices. In the next sections, we will consider the case of an even-dimensional fermionic phase space, but there provide a basis-independent construction of the state space and the action of the Clifford algebra on it.

31.2 The Schrödinger representation for fermions: ghosts

We would like to construct representations of $\text{Cliff}(r, s, \mathbf{R})$ and thus fermionic state spaces by using analogous constructions to the Schrödinger and Bargmann–Fock ones in the bosonic case. The Schrödinger construction took the state space $\mathcal{H}$ to be a space of functions on a subspace of the classical phase space which had the property that the basis coordinate functions Poisson-commuted. Two examples of this are the position coordinates q_j, since $\{q_j, q_k\} = 0$, or the momentum coordinates p_j, since $\{p_j, p_k\} = 0$. Unfortunately, for symmetric bilinear forms $(\cdot, \cdot)$ of definite sign, such as the positive-definite case $\text{Cliff}(n, \mathbf{R})$, the only subspace the bilinear form is zero on is the zero subspace.

To get an analog of the bosonic situation, one needs to take the case of signature (d, d). The fermionic phase space will then be $2d$ dimensional, with d-dimensional subspaces on which $(\cdot, \cdot)$ and thus the fermionic Poisson bracket is zero. Quantization will give the Clifford algebra

$$\text{Cliff}(d, d, \mathbf{R}) = M(2^d, \mathbf{R})$$

which has just one irreducible representation, $\mathbf{R}^{2^d}$. This can be complexified to get a complex state space

$$\mathcal{H}_F = \mathbf{C}^{2^d}$$

This state space will come with a representation of $Spin(d, d, \mathbf{R})$ from exponentiating quadratic combinations of the generators of $\mathrm{Cliff}(d, d, \mathbf{R})$. However, this is a non-compact group, and one can show that on general grounds, it cannot have faithful unitary finite dimensional representations, so there must be a problem with unitarity.

To see what happens explicitly, consider the simplest case $d = 1$ of one degree of freedom. In the bosonic case, the classical phase space is $\mathbf{R}^2$, and quantization gives operators Q, P which in the Schrödinger representation act on functions of q, with $Q = q$ and $P = -i\frac{\partial}{\partial q}$. In the fermionic case with signature $(1, 1)$, basis coordinate functions on phase space are ξ_1, ξ_2, with

$$\{\xi_1, \xi_1\}_+ = 1, \quad \{\xi_2, \xi_2\}_+ = -1, \quad \{\xi_1, \xi_2\}_+ = 0$$

Defining

$$\eta = \frac{1}{\sqrt{2}}(\xi_1 + \xi_2), \quad \pi = \frac{1}{\sqrt{2}}(\xi_1 - \xi_2)$$

one gets objects with fermionic Poisson bracket analogous to those of q and p

$$\{\eta, \eta\}_+ = \{\pi, \pi\}_+ = 0, \quad \{\eta, \pi\}_+ = 1$$

Quantizing, we get analogs of the Q, P operators

$$\widehat{\eta} = \Gamma^+(\eta) = \frac{1}{\sqrt{2}}(\Gamma^+(\xi_1) + \Gamma^+(\xi_2)), \quad \widehat{\pi} = \Gamma^+(\pi) = \frac{1}{\sqrt{2}}(\Gamma^+(\xi_1) - \Gamma^+(\xi_2))$$

which satisfy anticommutation relations

$$\widehat{\eta}^2 = \widehat{\pi}^2 = 0, \quad \widehat{\eta}\widehat{\pi} + \widehat{\pi}\widehat{\eta} = 1$$

and can be realized as operators on the space of functions of one fermionic variable η as

$$\widehat{\eta} = \text{multiplication by } \eta, \quad \widehat{\pi} = \frac{\partial}{\partial \eta}$$

This state space is two-complex dimensional, with an arbitrary state

$$f(\eta) = c_1 1 + c_2 \eta$$

with c_j complex numbers. The inner product on this space is given by the fermionic integral

$$(f_1(\eta), f_2(\eta)) = \int f_1^*(\eta) f_2(\eta) d\eta$$

with

$$f^*(\xi) = \bar{c}_1 1 + \bar{c}_2 \eta$$

With respect to this inner product, one has

$$(1,1) = (\eta, \eta) = 0, \quad (1, \eta) = (\eta, 1) = 1$$

This inner product is indefinite and can take on negative values, since

$$(1 - \eta, 1 - \eta) = -2$$

Having such negative-norm states ruins any standard interpretation of this as a physical system, since this negative number is supposed to the probability of finding the system in this state. Such quantum systems are called "ghosts," and do have applications in the description of various quantum systems, but only when a mechanism exists for the negative-norm states to cancel or otherwise be removed from the physical state space of the theory.

31.3 Spinors and the Bargmann–Fock construction

While the fermionic analog of the Schrödinger construction does not give a unitary representation of the spin group, it turns out that the fermionic analog of the Bargmann–Fock construction does, on the fermionic oscillator state space discussed in chapter 27. This will work for the case of a positive-definite symmetric bilinear form $(\cdot, \cdot)$. Note though that this will only work for fermionic phase spaces $\mathbf{R}^n$ with n even, since a complex structure on the phase space is needed.

The corresponding pseudo-classical system will be the classical fermionic oscillator studied in section 30.3.2. Recall that this uses a choice of complex structure J on the fermionic phase space $\mathbf{R}^{2d}$, with the standard choice $J = J_0$ coming from the relations

$$\theta_j = \frac{1}{\sqrt{2}}(\xi_{2j-1} - i\xi_{2j}), \quad \bar{\theta}_j = \frac{1}{\sqrt{2}}(\xi_{2j-1} + i\xi_{2j}) \tag{31.2}$$

for $j = 1, \ldots, d$ between real and complex coordinates. Here $(\cdot, \cdot)$ is positive definite, and the ξ_j are coordinates with respect to an orthonormal basis, so we have the standard relation $\{\xi_j, \xi_k\}_+ = \delta_{jk}$ and the $\theta_j, \bar{\theta}_j$ satisfy

$$\{\theta_j, \theta_k\}_+ = \{\bar{\theta}_j, \bar{\theta}_k\}_+ = 0, \quad \{\bar{\theta}_j, \theta_k\}_+ = \delta_{jk}$$

In the bosonic case (see equation 26.7), extending the Poisson bracket from $\mathcal{M}$ to $\mathcal{M} \otimes \mathbf{C}$ by complex linearity gave an indefinite Hermitian form on $\mathcal{M} \otimes \mathbf{C}$

$$\langle \cdot, \cdot \rangle = i\{\overline{\cdot}, \cdot\} = i\Omega(\overline{\cdot}, \cdot)$$

positive definite on $\mathcal{M}_J^+$ for positive J. In the fermionic case, we can extend the fermionic Poisson bracket from $\mathcal{V}$ to $\mathcal{V} \otimes \mathbf{C}$ by complex linearity, getting a Hermitian form on $\mathcal{V} \otimes \mathbf{C}$

$$\langle \cdot, \cdot \rangle = \{\overline{\cdot}, \cdot\}_+ = (\overline{\cdot}, \cdot)$$

This is positive definite on $\mathcal{V}_J^+$ (and also on $\mathcal{V}_J^-$) if the initial symmetric bilinear form was positive.

To quantize this system, we need to find operators $\Gamma^+(\theta_j)$ and $\Gamma^+(\overline{\theta}_j)$ that satisfy

$$[\Gamma^+(\theta_j), \Gamma^+(\theta_k)]_+ = [\Gamma^+(\overline{\theta}_j), \Gamma^+(\overline{\theta}_k)]_+ = 0$$

$$[\Gamma^+(\overline{\theta}_j), \Gamma^+(\theta_k)]_+ = \delta_{jk}\mathbf{1}$$

but these are just the CAR satisfied by fermionic annihilation and creation operators. We can choose

$$\Gamma^+(\theta_j) = a_{F\,j}^\dagger, \quad \Gamma^+(\overline{\theta}_j) = a_{F\,j}$$

and realize these operators as

$$a_{F\,j} = \frac{\partial}{\partial\theta_j}, \quad a_{F\,j}^\dagger = \text{multiplication by } \theta_j$$

on the state space $\Lambda^* \mathbf{C}^d$ of polynomials in the anticommuting variables θ_j. This is a complex vector space of dimension 2^d, isomorphic with the state space $\mathcal{H}_F$ of the fermionic oscillator in d degrees of freedom, with the isomorphism given by

$$1 \leftrightarrow |0\rangle_F$$
$$\theta_j \leftrightarrow a_{F\,j}^\dagger |0\rangle_F$$
$$\theta_j \theta_k \leftrightarrow a_{F\,j}^\dagger a_{F\,k}^\dagger |0\rangle$$
$$\cdots$$
$$\theta_1 \dots \theta_d \leftrightarrow a_{F\,1}^\dagger a_{F\,2}^\dagger \cdots a_{F\,d}^\dagger |0\rangle_F$$

where the indices $j, k, \dots$ take values $1, 2, \dots, d$ and satisfy $j < k < \cdots$.

If one defines a Hermitian inner product $\langle \cdot, \cdot \rangle$ on $\mathcal{H}_F$ by taking these basis elements to be orthonormal, the operators $a_{F\,j}$ and $a_{F\,j}^\dagger$ will be adjoints with

respect to this inner product. This same inner product can also be defined
using fermionic integration by analogy with the Bargmann–Fock definition
in the bosonic case as

$$\langle f_1(\theta_1,\cdots,\theta_d), f_2(\theta_1,\cdots,\theta_d)\rangle = \int e^{-\sum_{j=1}^d \theta_j \bar{\theta}_j} \overline{f_1} f_2 d\bar{\theta} d\theta_1 \cdots d\bar{\theta}_d d\theta_d$$

(31.3)

where f_1 and f_2 are complex linear combinations of the powers of the anticommuting variables θ_j. For the details of the construction of this inner product,
see chapter 7.2 of [90] or chapters 7.5 and 7.6 of [109]. We will denote this
state space as $\mathcal{F}_d^+$ and refer to it as the fermionic Fock space. Since it is finite
dimensional, there is no need for a completion as in the bosonic case.

The quantization using fermionic annihilation and creation operators given
here provides an explicit realization of a representation of the Clifford algebra
$\text{Cliff}(2d, \mathbf{R})$ on the complex vector space $\mathcal{F}_d^+$. The generators of the Clifford
algebra are identified as operators on $\mathcal{F}_d^+$ by

$$\gamma_{2j-1} = \sqrt{2}\Gamma^+(\xi_{2j-1}) = \sqrt{2}\Gamma^+\left(\frac{1}{\sqrt{2}}(\theta_j + \bar{\theta}_j)\right) = a_{Fj} + a_{Fj}^\dagger$$

$$\gamma_{2j} = \sqrt{2}\Gamma^+(\xi_{2j}) = \sqrt{2}\Gamma^+\left(\frac{i}{\sqrt{2}}(\theta_j - \bar{\theta}_j)\right) = i(a_{Fj}^\dagger - a_{Fj})$$

Quantization of the pseudo-classical fermionic oscillator Hamiltonian h of
section 30.3.2 gives

$$\Gamma^+(h) = \Gamma^+\left(-\frac{i}{2}\sum_{j=1}^d(\theta_j\bar{\theta}_j - \bar{\theta}_j\theta_j)\right) = -\frac{i}{2}\sum_{j=1}^d(a_{Fj}^\dagger a_{Fj} - a_{Fj}a_{Fj}^\dagger) = -iH$$

(31.4)

where H is the Hamiltonian operator for the fermionic oscillator used in
chapter 27.

Taking quadratic combinations of the operators γ_j provides a representation of the Lie algebra $\mathfrak{so}(2d) = \mathfrak{spin}(2d)$. This representation exponentiates
to a representation up to sign of the group $SO(2d)$, and a true representation
of its double cover $Spin(2d)$. The representation that we have constructed
here on the fermionic oscillator state space $\mathcal{F}_d^+$ is called the spinor representation of $Spin(2d)$, and we will sometimes denote $\mathcal{F}_d^+$ with this group
action as S.

In the bosonic case, $\mathcal{H} = \mathcal{F}_d$ is an irreducible representation of the Heisenberg group, but as a representation of $Mp(2d, \mathbf{R})$, it has two irreducible components, corresponding to even and odd polynomials. The fermionic analog
is that $\mathcal{F}_d^+$ is irreducible under the action of the Clifford algebra $\text{Cliff}(2d, \mathbf{C})$.
One way to show this is to show that $\text{Cliff}(2d, \mathbf{C})$ is isomorphic to the matrix

algebra $M(2^d, \mathbf{C})$ and its action on $\mathcal{H}_F = \mathbf{C}^{2^d}$ is isomorphic to the action of matrices on column vectors.

While $\mathcal{F}_d^+$ is irreducible as a representation of the Clifford algebra, it is the sum of two irreducible representations of $Spin(2d)$, the so-called "half-spinor" representations. $Spin(2d)$ is generated by quadratic combinations of the Clifford algebra generators, so these will preserve the subspaces

$$S_+ = \text{span}\{|0\rangle_F, \quad a_{Fj}^\dagger a_{Fk}^\dagger |0\rangle_F, \cdots\} \subset S = \mathcal{F}_d^+$$

and

$$S_- = \text{span}\{a_{Fj}^\dagger |0\rangle_F, \quad a_{Fj}^\dagger a_{Fk}^\dagger a_{Fl}^\dagger |0\rangle_F, \cdots\} \subset S = \mathcal{F}_d^+$$

corresponding to the action of an even or odd number of creation operators on $|0\rangle_F$. This is because quadratic combinations of the $a_{Fj}, a_{Fj}^\dagger$ preserve the parity of the number of creation operators used to get an element of S by action on $|0\rangle_F$.

31.4 Complex structures, $U(d) \subset SO(2d)$ and the spinor representation

The construction of the spinor representation given here has used a specific choice of the $\theta_j, \overline{\theta}_j$ (see equations 31.2) and the fermionic annihilation and creation operators. This corresponds to a standard choice of complex structure J_0, which appears in a manner closely parallel to that of the Bargmann–Fock case of section 26.1. The difference here is that, for the analogous construction of spinors, the complex structure J must be chosen so as to preserve not an antisymmetric bilinear form Ω, but the inner product, and one has

$$(J(\cdot), J(\cdot)) = (\cdot, \cdot)$$

We will here restrict to the case of $(\cdot, \cdot)$ positive definite, and unlike in the bosonic case, no additional positivity condition on J will then be required.

J splits the complexification of the real dual phase space $V^* = \mathcal{V} = \mathbf{R}^{2d}$ with its coordinates ξ_j into a d-dimensional complex vector space on which $J = +i$ and a conjugate complex vector space on which $J = -i$. As in the bosonic case, one has

$$\mathcal{V} \otimes \mathbf{C} = \mathcal{V}_J^+ \oplus \mathcal{V}_J^-$$

and quantization of vectors in $\mathcal{V}_J^+$ gives linear combinations of creation operators, while vectors in $\mathcal{V}_J^-$ are taken to linear combinations of annihilation operators. The choice of J is reflected in the existence of a distinguished direction $|0\rangle_F$ in the spinor space $S = \mathcal{F}_d^+$ which is determined (up to phase) by the condition that it is annihilated by all linear combinations of annihilation operators.

The choice of J also picks out a subgroup $U(d) \subset SO(2d)$ of those orthogonal transformations that commute with J. Just as in the bosonic case, two different representations of the Lie algebra $\mathfrak{u}(d)$ of $U(d)$ are used:

- The restriction to $\mathfrak{u}(d) \subset \mathfrak{so}(2d)$ of the spinor representation is described above. This exponentiates to give a representation not of $U(d)$, but of a double cover of $U(d)$ that is a subgroup of $Spin(2d)$.

- By normal-ordering operators, one shifts the spinor representation of $\mathfrak{u}(d)$ by a constant and gets a representation that exponentiates to a true representation of $U(d)$. This representation is reducible, with irreducible components the $\Lambda^k(\mathbf{C}^d)$ for $k = 0, 1, \ldots, d$.

In both cases, the representation of $\mathfrak{u}(d)$ is constructed using quadratic combinations of annihilation and creation operators involving one annihilation operator and one creation operator, operators which annihilate $|0\rangle_F$. Nonzero pairs of two creation operators act non-trivially on $|0\rangle_F$, corresponding to the fact that elements of $SO(2d)$ not in the $U(d)$ subgroup take $|0\rangle_F$ to a different state in the spinor representation.

Given any group element

$$g_0 = e^A \subset U(d)$$

acting on the fermionic dual phase space preserving J and the inner product, we can use exactly the same method as in theorems 25.1 and 25.2 to construct its action on the fermionic state space by the second of the above representations. For A a skew-adjoint matrix, we have a fermionic moment map

$$A \in \mathfrak{u}(d) \to \mu_A = \sum_{j,k} \theta_j A_{jk} \overline{\theta}_k$$

satisfying

$$\{\mu_A, \mu_{A'}\}_+ = \mu_{[A,A']}$$

and

$$\{\mu_A, \boldsymbol{\theta}\}_+ = A^T \boldsymbol{\theta}, \quad \{\mu_A, \overline{\boldsymbol{\theta}}\}_+ = \overline{A^T \boldsymbol{\theta}} = -A\overline{\boldsymbol{\theta}} \qquad (31.5)$$

The Lie algebra representation operators are the

$$U'_A = \sum_{j,k} a^\dagger_{Fj} A_{jk} a_{Fk}$$

which satisfy (see theorem 27.1)

$$[U'_A, U'_{A'}] = U_{[A,A']}$$

and

$$[U'_A, \mathbf{a}^\dagger_F] = A^T \mathbf{a}^\dagger_F, \quad [U'_A, \mathbf{a}_F] = \overline{A^T} \mathbf{a}_F$$

Exponentiating these gives the intertwining operators, which act on the annihilation and creation operators as

$$U_{e^A} \mathbf{a}^\dagger_F (U_{e^A})^{-1} = e^{A^T} \mathbf{a}^\dagger_F, \quad U_{e^A} \mathbf{a}_F (U_{e^A})^{-1} = e^{\overline{A^T}} \mathbf{a}_F$$

For the simplest example, consider the $U(1) \subset U(d) \subset SO(2d)$ that acts by

$$\theta_j \to e^{i\phi} \theta_j, \quad \overline{\theta}_j \to e^{-i\phi} \overline{\theta}_j$$

corresponding to $A = i\phi \mathbf{1}$. The moment map will be

$$\mu_A = -\phi h$$

where

$$h = -i \sum_{j=1}^{d} \theta_j \overline{\theta}_j$$

is the Hamiltonian for the classical fermionic oscillator. Quantizing h (see equation 31.4) will give $(-i)$ times the Hamiltonian operator

$$-iH = -\frac{i}{2} \sum_{j=1}^{d} (a_{Fj}^\dagger a_{Fj} - a_{Fj} a_{Fj}^\dagger) = -i \sum_{j=1}^{d} \left(a_{Fj}^\dagger a_{Fj} - \frac{1}{2} \right)$$

and a Lie algebra representation of $\mathfrak{u}(1)$ with half-integral eigenvalues $(\pm i \frac{1}{2})$. Exponentiation will give a representation of a double cover of $U(1) \subset U(d)$.

Quantizing h instead using normal-ordering gives

$$: -iH: = -i \sum_{j=1}^{d} a_{Fj}^\dagger a_{Fj}$$

and a true representation of $U(1) \subset U(d)$, with

$$U'_A = i\phi \sum_{j=1}^{d} a_{Fj}^\dagger a_{Fj}$$

satisfying

$$[U'_A, \mathbf{a}^\dagger_F] = i\phi \mathbf{a}^\dagger_F, \quad [U'_A, \mathbf{a}_F] = -i\phi \mathbf{a}_F$$

Exponentiating, the action on annihilation and creation operators is

$$e^{-i\phi \sum_{j=1}^{d} a_{Fj}^{\dagger} a_{Fj}} \mathbf{a}_F^{\dagger} e^{i\phi \sum_{j=1}^{d} a_{Fj}^{\dagger} a_{Fj}} = e^{i\phi} \mathbf{a}_F^{\dagger}$$

$$e^{-i\phi \sum_{j=1}^{d} a_{Fj}^{\dagger} a_{Fj}} \mathbf{a}_F e^{i\phi \sum_{j=1}^{d} a_{Fj}^{\dagger} a_{Fj}} = e^{-i\phi} \mathbf{a}_F$$

31.5 An example: spinors for $SO(4)$

We saw in chapter 6 that the spin group $Spin(4)$ was isomorphic to $Sp(1) \times Sp(1) = SU(2) \times SU(2)$. Its action on $\mathbf{R}^4$ was then given by identifying $\mathbf{R}^4 = \mathbf{H}$ and acting by unit quaternions on the left and the right (thus the two copies of $Sp(1)$). While this constructs the representation of $Spin(4)$ on $\mathbf{R}^4$, it does not provide the spin representation of $Spin(4)$.

A conventional way of defining the spin representation is to choose an explicit matrix representation of the Clifford algebra (in this case Cliff $(4, 0, \mathbf{R})$), for instance

$$\gamma_0 = \begin{pmatrix} 0 & 1 \\ 1 & 0 \end{pmatrix}, \gamma_1 = -i \begin{pmatrix} 0 & \sigma_1 \\ -\sigma_1 & 0 \end{pmatrix}, \gamma_2 = -i \begin{pmatrix} 0 & \sigma_2 \\ -\sigma_2 & 0 \end{pmatrix}, \gamma_3 = -i \begin{pmatrix} 0 & \sigma_3 \\ -\sigma_3 & 0 \end{pmatrix}$$

where we have written the matrices in 2 by 2 block form and are indexing the four dimensions from 0 to 3. One can easily check that these satisfy the Clifford algebra relations: They anticommute with each other and

$$\gamma_0^2 = \gamma_1^2 = \gamma_2^2 = \gamma_3^2 = 1$$

The quadratic Clifford algebra elements $-\frac{1}{2}\gamma_j\gamma_k$ for $j < k$ satisfy the commutation relations of $\mathfrak{so}(4) = \mathfrak{spin}(4)$. These are explicitly

$$-\frac{1}{2}\gamma_0\gamma_1 = -\frac{i}{2} \begin{pmatrix} \sigma_1 & 0 \\ 0 & -\sigma_1 \end{pmatrix}, \quad -\frac{1}{2}\gamma_2\gamma_3 = -\frac{i}{2} \begin{pmatrix} \sigma_1 & 0 \\ 0 & \sigma_1 \end{pmatrix}$$

$$-\frac{1}{2}\gamma_0\gamma_2 = -\frac{i}{2} \begin{pmatrix} \sigma_2 & 0 \\ 0 & -\sigma_2 \end{pmatrix}, \quad -\frac{1}{2}\gamma_1\gamma_3 = -\frac{i}{2} \begin{pmatrix} \sigma_2 & 0 \\ 0 & \sigma_2 \end{pmatrix}$$

$$-\frac{1}{2}\gamma_0\gamma_3 = -\frac{i}{2} \begin{pmatrix} \sigma_3 & 0 \\ 0 & -\sigma_3 \end{pmatrix}, \quad -\frac{1}{2}\gamma_1\gamma_2 = -\frac{i}{2} \begin{pmatrix} \sigma_3 & 0 \\ 0 & \sigma_3 \end{pmatrix}$$

The Lie algebra spin representation is just matrix multiplication on $S = \mathbf{C}^4$, and it is obviously a reducible representation on two copies of $\mathbf{C}^2$ (the upper and lower two components). One can also see that the Lie algebra $\mathfrak{spin}(4) = \mathfrak{su}(2) \oplus \mathfrak{su}(2)$, with the two $\mathfrak{su}(2)$ Lie algebras having bases

$$-\frac{1}{4}(\gamma_0\gamma_1 + \gamma_2\gamma_3), \quad -\frac{1}{4}(\gamma_0\gamma_2 + \gamma_1\gamma_3), \quad -\frac{1}{4}(\gamma_0\gamma_3 + \gamma_1\gamma_2)$$

and

$$-\frac{1}{4}(\gamma_0\gamma_1 - \gamma_2\gamma_3), \quad -\frac{1}{4}(\gamma_0\gamma_2 - \gamma_1\gamma_3), \quad -\frac{1}{4}(\gamma_0\gamma_3 - \gamma_1\gamma_2)$$

The irreducible spin representations of $Spin(4)$ are just the tensor product of spin $\frac{1}{2}$ representations of the two copies of $SU(2)$ (with each copy acting on a different factor of the tensor product).

In the fermionic oscillator construction, we have

$$S = S^+ + S^-, \quad S^+ = \mathrm{span}\{1, \theta_1\theta_2\}, \quad S^- = \mathrm{span}\{\theta_1, \theta_2\}$$

and the Clifford algebra action on S is given for the generators as (now indexing dimensions from 1 to 4)

$$\gamma_1 = \frac{\partial}{\partial\theta_1} + \theta_1, \quad \gamma_2 = i\left(\frac{\partial}{\partial\theta_1} - \theta_1\right)$$

$$\gamma_3 = \frac{\partial}{\partial\theta_2} + \theta_2, \quad \gamma_4 = i\left(\frac{\partial}{\partial\theta_2} - \theta_2\right)$$

Note that in this construction, there is a choice of complex structure $J = J_0$. This gives a distinguished vector $|0\rangle = 1 \in S^+$, as well as a distinguished sub-Lie algebra $\mathfrak{u}(2) \subset \mathfrak{so}(4)$ of transformations that act trivially on $|0\rangle$, given by linear combinations of

$$\theta_1\frac{\partial}{\partial\theta_1}, \quad \theta_2\frac{\partial}{\partial\theta_2}, \quad \theta_1\frac{\partial}{\partial\theta_2}, \quad \theta_2\frac{\partial}{\partial\theta_1},$$

There is also a distinguished sub-Lie algebra $\mathfrak{u}(1) \subset \mathfrak{u}(2)$ that has zero Lie bracket with the rest, with basis element

$$\theta_1\frac{\partial}{\partial\theta_1} + \theta_2\frac{\partial}{\partial\theta_2}$$

$Spin(4)$ elements that act by unitary (for the Hermitian inner product 31.3) transformations on the spinor state space, but change $|0\rangle$ and correspond to a change in complex structure, are given by exponentiating the Lie algebra representation operators

$$i(a_{F1}^\dagger a_{F2}^\dagger + a_{F2}a_{F1}), \quad a_{F1}^\dagger a_{F2}^\dagger - a_{F2}a_{F1}$$

The possible choices of complex structure are parametrized by $SO(4)/U(2)$, which can be identified with the complex projective sphere $\mathbf{CP}^1 = S^2$.

The construction in terms of matrices is well suited to calculations, but it is inherently dependent on a choice of coordinates. The fermionic version of Bargmann–Fock is given here in terms of a choice of basis, but, like the closely analogous bosonic construction, only actually depends on a choice of inner product and a choice of compatible complex structure J, producing a representation on the coordinate-independent object $\mathcal{F}_d^+ = \Lambda^* V_J^+$.

In chapter 41, we will consider explicit matrix representations of the Clifford algebra for the case of $Spin(3,1)$. The fermionic oscillator construction could also be used, complexifying to get a representation of

$$\mathfrak{so}(4) \otimes \mathbf{C} = \mathfrak{sl}(2, \mathbf{C}) \oplus \mathfrak{sl}(2, \mathbf{C})$$

and then restricting to the subalgebra

$$\mathfrak{so}(3,1) \subset \mathfrak{so}(3,1) \otimes \mathbf{C} = \mathfrak{so}(4) \otimes \mathbf{C}$$

This will give a representation of $Spin(3,1)$ in terms of quadratic combinations of Clifford algebra generators, but unlike the case of $Spin(4)$, it will not be unitary. The lack of positivity for the inner product causes the same sort of wrong-sign problems with the CAR that were found in the bosonic case for the CCR when J and Ω gave a non-positive symmetric bilinear form. In the fermion case, the wrong-sign problem does not stop one from constructing a representation, but it will not be a unitary representation.

31.6 For further reading

For more about pseudo-classical mechanics and quantization, see [90] chapter 7. The Clifford algebra and fermionic quantization are discussed in chapter 20.3 of [46]. The fermionic quantization map, Clifford algebras, and the spinor representation are discussed in detail in [59]. For another discussion of the spinor representation from a similar point of view to the one here, see chapter 12 of [94]. Chapter 12 of [70] contains an extensive discussion of the role of different complex structures in the construction of the spinor representation.

Chapter 32
A Summary: Parallels Between Bosonic and Fermionic Quantization

To summarize much of the material we have covered, it may be useful to consider the following table, which explicitly gives the correspondence between the parallel constructions we have studied in the bosonic and fermionic cases.

Bosonic	Fermionic
Dual phase space $\mathcal{M} = \mathbf{R}^{2d}$	Dual phase space $\mathcal{V} = \mathbf{R}^n$
Non-degenerate antisymmetric bilinear form $\Omega(\cdot, \cdot)$ on $\mathcal{M}$	Non-degenerate symmetric bilinear form $(\cdot, \cdot)$ on $\mathcal{V}$
Poisson bracket $\{\cdot, \cdot\}$ on functions on $M = \mathbf{R}^{2d}$	Poisson bracket $\{\cdot, \cdot\}_+$ on anticommuting functions on $V = \mathbf{R}^n$
Lie algebra of polynomials of degree $0, 1, 2$	Lie superalgebra of anticommuting polynomials of degree $0, 1, 2$
Coordinates q_j, p_j, basis of $\mathcal{M}$	Coordinates ξ_j, basis of $\mathcal{V}$
Quadratics in q_j, p_j, basis for $\mathfrak{sp}(2d, \mathbf{R})$	Quadratics in ξ_j, basis for $\mathfrak{so}(n)$
$Sp(2d, \mathbf{R})$ preserves $\Omega(\cdot, \cdot)$	$SO(n, \mathbf{R})$ preserves $(\cdot, \cdot)$
Weyl algebra Weyl$(2d, \mathbf{C})$	Clifford algebra Cliff$(n, \mathbf{C})$
Momentum, position operators P_j, Q_j	Clifford algebra generators γ_j
Quadratics in P_j, Q_j provide representation of $\mathfrak{sp}(2d, \mathbf{R})$	Quadratics in γ_j provide representation of $\mathfrak{so}(2d)$
Metaplectic representation	Spinor representation
Stone–von Neumann, uniqueness of $\mathfrak{h}_{2d+1}$ representation	Uniqueness of Cliff$(2d, \mathbf{C})$ representation on spinors
$Mp(2d, \mathbf{R})$ double cover of $Sp(2d, \mathbf{R})$	$Spin(n)$ double cover of $SO(n)$
$J : J^2 = -1$, $\Omega(Ju, Jv) = \Omega(u, v)$	$J : J^2 = -1$, $(Ju, Jv) = (u, v)$
$\mathcal{M} \otimes \mathbf{C} = \mathcal{M}_J^+ \oplus \mathcal{M}_J^-$	$\mathcal{V} \otimes \mathbf{C} = \mathcal{V}_J^+ \oplus \mathcal{V}_J^-$

© Peter Woit 2017

P. Woit, *Quantum Theory, Groups and Representations*,
DOI 10.1007/978-3-319-64612-1_32

Coordinates $z_j \in \mathcal{M}_J^+$, $\bar{z}_j \in \mathcal{M}_J^-$

$U(d) \subset Sp(2d, \mathbf{R})$ commutes with J

Compatible $J \in Sp(2d, \mathbf{R})/U(d)$

$a_j, a_j^\dagger$ satisfying CCR

$a_j |0\rangle = 0$, $|0\rangle$ depends on J

$\mathcal{H} = \mathcal{F}_d^{fin} = \mathbf{C}[z_1, \ldots, z_d] = S^*(\mathbf{C}^d)$

$a_j^\dagger = z_j$, $a_j = \frac{\partial}{\partial z_j}$

Coordinates $\theta_j \in \mathcal{V}_J^+$, $\bar{\theta}_j \in \mathcal{V}_J^-$

$U(d) \subset SO(2d, \mathbf{R})$ commutes with J

Compatible $J \in O(2d)/U(d)$

$a_{Fj}, a_{Fj}^\dagger$ satisfying CAR

$a_{Fj} |0\rangle = 0$, $|0\rangle$ depends on J

$\mathcal{H} = \mathcal{F}_d^+ = \Lambda^*(\mathbf{C}^d)$

$a_j^\dagger = \theta_j$, $a_j = \frac{\partial}{\partial \theta_j}$

Positivity conditions, leading to unitary state space:

$\Omega(v, Jv) > 0$ for nonzero $v \in \mathcal{M}$

$\langle u, u \rangle = i\Omega(\bar{u}, u) > 0$ for nonzero $u \in \mathcal{M}_J^+$

$(v, v) > 0$ for nonzero $v \in V$

$\langle u, u \rangle = (\bar{u}, u) > 0$ for nonzero $u \in \mathcal{V}_J^-$ or $\mathcal{V}_J^+$

Chapter 33
Supersymmetry, Some Simple Examples

If one considers fermionic and bosonic quantum systems that each separately has operators coming from Lie algebra or superalgebra representations on their state spaces, when one combines the systems by taking the tensor product, these operators will continue to act on the combined system. In certain special cases, new operators with remarkable properties will appear that mix the fermionic and bosonic systems and commute with the Hamiltonian (these operators are often given by some sort of "square root" of the Hamiltonian). These are generically known as "supersymmetries" and provide new information about energy eigenspaces. In this chapter, we will examine in detail some of the simplest such quantum systems, examples of "supersymmetric quantum mechanics."

33.1 The supersymmetric oscillator

In the previous chapters, we discussed in detail

- The bosonic harmonic oscillator in d degrees of freedom, with state space $\mathcal{F}_d$ generated by applying d creation operators $a_{Bj}^\dagger$ an arbitrary number of times to a lowest energy state $|0\rangle_B$. The Hamiltonian is

$$H = \frac{1}{2}\hbar\omega \sum_{j=1}^{d}(a_{Bj}^\dagger a_{Bj} + a_{Bj}a_{Bj}^\dagger) = \sum_{j=1}^{d}\left(N_{Bj} + \frac{1}{2}\right)\hbar\omega$$

where N_{Bj} is the number operator for the jth degree of freedom, with eigenvalues $n_{Bj} = 0, 1, 2, \cdots$.

© Peter Woit 2017
P. Woit, *Quantum Theory, Groups and Representations*,
DOI 10.1007/978-3-319-64612-1_33

- The fermionic oscillator in d degrees of freedom, with state space $\mathcal{F}_d^+$ generated by applying d creation operators a_{Fj} to a lowest energy state $|0\rangle_F$. The Hamiltonian is

$$H = \frac{1}{2}\hbar\omega \sum_{j=1}^{d}(a_{Fj}^{\dagger}a_{Fj} - a_{Fj}a_{Fj}^{\dagger}) = \sum_{j=1}^{d}\left(N_{Fj} - \frac{1}{2}\right)\hbar\omega$$

  where N_{Fj} is the number operator for the jth degree of freedom, with eigenvalues $n_{Fj} = 0, 1$.

Putting these two systems together, we get a new quantum system with state space

$$\mathcal{H} = \mathcal{F}_d \otimes \mathcal{F}_d^+$$

and Hamiltonian

$$H = \sum_{j=1}^{d}(N_{Bj} + N_{Fj})\hbar\omega$$

Notice that the lowest energy state $|0\rangle$ for the combined system has energy 0, due to cancellation between the bosonic and fermionic degrees of freedom.

For now, taking for simplicity the case $d = 1$ of one degree of freedom, the Hamiltonian is

$$H = (N_B + N_F)\hbar\omega$$

with eigenvectors $|n_B, n_F\rangle$ satisfying

$$H|n_B, n_F\rangle = (n_B + n_F)\hbar\omega$$

While there is a unique lowest energy state $|0, 0\rangle$ of zero energy, all nonzero energy states come in pairs, with two states

$$|n, 0\rangle \quad \text{and} \quad |n - 1, 1\rangle$$

both having energy $n\hbar\omega$.

This kind of degeneracy of energy eigenvalues usually indicates the existence of some new symmetry operators commuting with the Hamiltonian operator. We are looking for operators that will take $|n, 0\rangle$ to $|n - 1, 1\rangle$ and vice versa, and the obvious choice is the two operators

$$Q_+ = a_B a_F^{\dagger}, \quad Q_- = a_B^{\dagger}a_F$$

which are not self-adjoint, but are each other's adjoints $((Q_-)^{\dagger} = Q_+)$.

The pattern of energy eigenstates looks like this

Energy

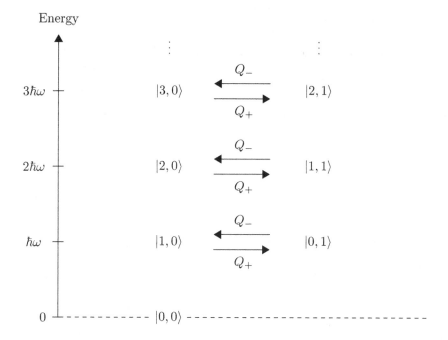

Figure 33.1: Energy eigenstates in the supersymmetric oscillator.

Computing anticommutators using the CCR and CAR for the bosonic and fermionic operators (and the fact that the bosonic operators commute with the fermionic ones since they act on different factors of the tensor product), one finds that

$$Q_+^2 = Q_-^2 = 0$$

and

$$(Q_+ + Q_-)^2 = [Q_+, Q_-]_+ = H$$

One could instead work with self-adjoint combinations

$$Q_1 = Q_+ + Q_-, \quad Q_2 = \frac{1}{i}(Q_+ - Q_-)$$

which satisfy

$$[Q_1, Q_2]_+ = 0, \quad Q_1^2 = Q_2^2 = H \tag{33.1}$$

The Hamiltonian H is a square of the self-adjoint operator $Q_+ + Q_-$, and this fact alone tells us that the energy eigenvalues will be nonnegative. It also tells us that energy eigenstates of nonzero energy will come in pairs

$$|\psi\rangle, \quad (Q_+ + Q_-)|\psi\rangle$$

with the same energy. To find states of zero energy (there will just be one, $|0,0\rangle$), instead of trying to solve the equation $H|0\rangle = 0$ for $|0\rangle$, one can look for solutions to

$$Q_1|0\rangle = 0 \quad \text{or} \quad Q_2|0\rangle = 0$$

The simplification here is much like what happens with the usual bosonic harmonic oscillator, where the lowest energy state in various representations can be found by looking for solutions to $a|0\rangle = 0$.

There is an example of a physical quantum mechanical system that has exactly the behavior of this supersymmetric oscillator. A charged particle confined to a plane, coupled to a magnetic field perpendicular to the plane, can be described by a Hamiltonian that can be put in the bosonic oscillator form (to show this, we need to know how to couple quantum systems to electromagnetic fields, which we will come to in chapter 45). The equally spaced energy levels are known as "Landau levels." If the particle has spin, there will be an additional term in the Hamiltonian coupling the spin and the magnetic field, exactly the one we have seen in our study of the two-state system. This additional term is precisely the Hamiltonian of a fermionic oscillator. For the case of gyromagnetic ratio $g = 2$, the coefficients match up so that we have exactly the supersymmetric oscillator described above, with exactly the pattern of energy levels seen there.

33.2 Supersymmetric quantum mechanics with a superpotential

The supersymmetric oscillator system can be generalized to a much wider class of potentials, while still preserving the supersymmetry of the system. For simplicity, we will here choose constants $\hbar = \omega = 1$. Recall that our bosonic annihilation and creation operators were defined by

$$a_B = \frac{1}{\sqrt{2}}(Q + iP), \quad a_B^\dagger = \frac{1}{\sqrt{2}}(Q - iP)$$

Introducing an arbitrary function $W(q)$ (called the "superpotential") with derivative $W'(q)$, we can define new annihilation and creation operators:

$$a_B = \frac{1}{\sqrt{2}}(W'(Q) + iP), \quad a_B^\dagger = \frac{1}{\sqrt{2}}(W'(Q) - iP)$$

Here, $W'(Q)$ is the multiplication operator $W'(q)$ in the Schrödinger position space representation on functions of q. The harmonic oscillator is the special case

$$W(q) = \frac{q^2}{2}$$

We keep our definition of the operators

$$Q_+ = a_B a_F^\dagger, \quad Q_- = a_B^\dagger a_F$$

These satisfy

$$Q_+^2 = Q_-^2 = 0$$

for the same reason as in the oscillator case: Repeated factors of a_F or $a_F^\dagger$ vanish. Taking as the Hamiltonian the same square as before, we find

$$
\begin{aligned}
H =& (Q_+ + Q_-)^2 \\
=& \frac{1}{2}(W'(Q) + iP)(W'(Q) - iP)a_F^\dagger a_F + \frac{1}{2}(W'(Q) - iP)(W'(Q) + iP)a_F a_F^\dagger \\
=& \frac{1}{2}(W'(Q)^2 + P^2)(a_F^\dagger a_F + a_F a_F^\dagger) + \frac{1}{2}(i[P, W'(Q)])(a_F^\dagger a_F - a_F a_F^\dagger) \\
=& \frac{1}{2}(W'(Q)^2 + P^2) + \frac{1}{2}(i[P, W'(Q)])\sigma_3
\end{aligned}
$$

But iP is the operator corresponding to infinitesimal translations in Q, so we have

$$i[P, W'(Q)] = W''(Q)$$

and

$$H = \frac{1}{2}(W'(Q)^2 + P^2) + \frac{1}{2}W''(Q)\sigma_3$$

For different choices of W, this gives a large class of quantum systems that can be used as toy models to investigate properties of ground states. All have the same state space

$$\mathcal{H} = \mathcal{H}_B \otimes \mathcal{F}_d^+ = L^2(\mathbf{R}) \otimes \mathbf{C}^2$$

(using the Schrödinger representation for the bosonic factor). The energy eigenvalues will be nonnegative, and energy eigenvectors with positive energy will occur in pairs

$$|\psi\rangle, \quad (Q_+ + Q_-)|\psi\rangle$$

For any quantum system, an important question is that of whether it has a unique lowest energy state. If the lowest energy state is not unique, and a symmetry group acts non-trivially on the space of lowest energy states, the symmetry is said to be "spontaneously broken," a situation that will be discussed in section 39.4. In supersymmetric quantum mechanics systems, thinking in terms of Lie superalgebras, one calls Q_1 the generator of the action of a supersymmetry, with H invariant under the supersymmetry in the sense that the commutator of Q_1 and H is zero. The question of how the supersymmetry acts on the lowest energy state depends on whether or not solutions can be found to the equation

$$(Q_+ + Q_-)|0\rangle = Q_1|0\rangle = 0$$

which will be a lowest energy state with zero energy. If such a solution does exist, one describes the ground state $|0\rangle$ as "invariant under the supersymmetry." If no such solution exists, Q_1 will take a lowest energy state to another, different, lowest energy state, in which case one says that one has "spontaneously broken supersymmetry." The question of whether a given supersymmetric theory has its supersymmetry spontaneously broken or not is one that has become of great interest in the case of much more sophisticated supersymmetric quantum field theories. There, hopes (so far unrealized) of making contact with the real world rely on finding theories where the supersymmetry is spontaneously broken.

In this simple quantum mechanical system, one can try and explicitly solve the equation $Q_1|\psi\rangle = 0$. States can be written as two-component complex functions

$$|\psi\rangle = \begin{pmatrix} \psi_+(q) \\ \psi_-(q) \end{pmatrix}$$

and the equation to be solved is

$$
\begin{aligned}
(Q_+ + Q_-)&|\psi\rangle \\
&= \frac{1}{\sqrt{2}}((W'(Q) + iP)a_F^\dagger + (W'(Q) - iP)a_F) \begin{pmatrix} \psi_+(q) \\ \psi_-(q) \end{pmatrix} \\
&= \frac{1}{\sqrt{2}} \left(\left(W'(Q) + \frac{d}{dq}\right) \begin{pmatrix} 0 & 1 \\ 0 & 0 \end{pmatrix} + \left(W'(Q) - \frac{d}{dq}\right) \begin{pmatrix} 0 & 0 \\ 1 & 0 \end{pmatrix} \right) \begin{pmatrix} \psi_+(q) \\ \psi_-(q) \end{pmatrix} \\
&= \frac{1}{\sqrt{2}} \left(W'(Q) \begin{pmatrix} 0 & 1 \\ 1 & 0 \end{pmatrix} + \frac{d}{dq} \begin{pmatrix} 0 & 1 \\ -1 & 0 \end{pmatrix} \right) \begin{pmatrix} \psi_+(q) \\ \psi_-(q) \end{pmatrix} \\
&= \frac{1}{\sqrt{2}} \begin{pmatrix} 0 & 1 \\ -1 & 0 \end{pmatrix} \left(\frac{d}{dq} - W'(Q)\sigma_3 \right) \begin{pmatrix} \psi_+(q) \\ \psi_-(q) \end{pmatrix} = 0
\end{aligned}
$$

which has general solution

$$\begin{pmatrix} \psi_+(q) \\ \psi_-(q) \end{pmatrix} = e^{W(q)\sigma_3} \begin{pmatrix} c_+ \\ c_- \end{pmatrix} = \begin{pmatrix} c_+ e^{W(q)} \\ c_- e^{-W(q)} \end{pmatrix}$$

for complex constants c_+, c_-. Such solutions can only be normalizable if

$$c_+ = 0, \quad \lim_{q \to \pm \infty} W(q) = +\infty$$

or

$$c_- = 0, \quad \lim_{q \to \pm \infty} W(q) = -\infty$$

If, for example, $W(q)$ is an odd polynomial, one will not be able to satisfy either of these conditions, so there will be no solution, and the supersymmetry will be spontaneously broken.

33.3 Supersymmetric quantum mechanics and differential forms

If one considers supersymmetric quantum mechanics in the case of d degrees of freedom and in the Schrödinger representation, one has

$$\mathcal{H} = L^2(\mathbf{R}^d) \otimes \Lambda^*(\mathbf{C}^d)$$

the tensor product of complex-valued functions on $\mathbf{R}^d$ (acted on by the Weyl algebra $\text{Weyl}(2d, \mathbf{C})$) and anticommuting functions on $\mathbf{C}^d$ (acted on by the Clifford algebra $\text{Cliff}(2d, \mathbf{C})$). There are two operators Q_+ and Q_-, adjoints of each other and of square zero. If one has studied differential forms, this should look familiar. This space $\mathcal{H}$ is well known to mathematicians, as the complex-valued differential forms on $\mathbf{R}^d$, often written $\Omega^*(\mathbf{R}^d)$, where here the $*$ denotes an index taking values from 0 (the 0-forms, or functions) to d (the d-forms). In the theory of differential forms, it is well known that one has an operator d on $\Omega^*(\mathbf{R}^d)$ with square zero, called the de Rham differential. Using the inner product on $\mathbf{R}^d$, a Hermitian inner product can be put on $\Omega^*(\mathbf{R}^d)$ by integration, and then d has an adjoint δ, also of square zero. The Laplacian operator on differential forms is

$$\square = (d + \delta)^2$$

The supersymmetric quantum system we have been considering corresponds precisely to this, once one conjugates d, δ as follows

$$Q_+ = e^{-W(q)} d e^{W(q)}, \quad Q_- = e^{W(q)} \delta e^{-W(q)}$$

In mathematics, the interest in differential forms mainly comes from the fact that they can be constructed not just on $\mathbf{R}^d$, but on a general differentiable manifold M, with a corresponding construction of $d, \delta, \square$ operators. In Hodge theory, one studies solutions of

$$\square \psi = 0$$

(these are called "harmonic forms") and finds that the dimension of the space of solutions $\psi \in \Omega^k(M)$ gives a topological invariant called the kth Betti number of the manifold M.

33.4 For further reading

For a reference at the level of these notes, see [31]. For more details about supersymmetric quantum mechanics, see the quantum mechanics textbook of Tahktajan [90], and lectures by Orlando Alvarez [1]. These references also describe the relation of these systems to the calculation of topological invariants, a topic pioneered in Witten's 1982 paper on supersymmetry and Morse theory [102].

Chapter 34
The Pauli Equation and the Dirac Operator

In Chapter 33, we considered supersymmetric quantum mechanical systems where both the bosonic and fermionic variables that get quantized take values in an even-dimensional space $\mathbf{R}^{2d} = \mathbf{C}^d$. There are then two different operators Q_1 and Q_2 that are square roots of the Hamiltonian operator. It turns out that there are much more interesting quantum mechanics systems that can be defined by quantizing bosonic variables in phase space $\mathbf{R}^{2d}$, and fermionic variables in $\mathbf{R}^d$. The operators appearing in such a theory will be given by the tensor product of the Weyl algebra in $2d$ variables and the Clifford algebra in d variables, and there will be a distinguished operator that provides a square root of the Hamiltonian.

This is equivalent to the fact that introduction of fermionic variables and the Clifford algebra provides the Casimir operator $-|\mathbf{P}|^2$ for the Euclidean group $E(3)$ with a square root: the Dirac operator $\not{\partial}$. This leads to a new way to construct irreducible representations of the group of spatial symmetries, using a new sort of quantum free particle, one carrying an internal "spin" degree of freedom due to the use of the Clifford algebra. Remarkably, fundamental matter particles are well described in exactly this way, both in the non-relativistic theory we study in this chapter and in the relativistic theory to be studied later.

34.1 The Pauli-Schrödinger equation and free spin $\frac{1}{2}$ particles in $d = 3$

We have so far seen two quite different quantum systems based on three-dimensional space:

© Peter Woit 2017
P. Woit, *Quantum Theory, Groups and Representations*,
DOI 10.1007/978-3-319-64612-1_34

- The free particle of chapter 19. This had classical phase space $\mathbf{R}^6$ with coordinates $q_1, q_2, q_3, p_1, p_2, p_3$ and Hamiltonian $\frac{1}{2m}|\mathbf{p}|^2$. Quantization using the Schrödinger representation gave operators Q_1, Q_2, Q_3, P_1, P_2, P_3 on the space $\mathcal{H}_B = L^2(\mathbf{R}^3)$ of square-integrable functions of the position coordinates. The Hamiltonian operator is

$$H = \frac{1}{2m}|\mathbf{P}|^2 = -\frac{1}{2m}\left(\frac{\partial^2}{\partial q_1^2} + \frac{\partial^2}{\partial q_2^2} + \frac{\partial^2}{\partial q_3^2}\right)$$

- The spin $\frac{1}{2}$ quantum system, discussed first in chapter 7 and later in section 31.1.1. This had a pseudo-classical fermionic phase space $\mathbf{R}^3$ with coordinates ξ_1, ξ_2, ξ_3 which after quantization became the operators

$$\frac{1}{\sqrt{2}}\sigma_1, \quad \frac{1}{\sqrt{2}}\sigma_2, \quad \frac{1}{\sqrt{2}}\sigma_3$$

on the state space $\mathcal{H}_F = \mathbf{C}^2$. For this system, we considered the Hamiltonian describing its interaction with a constant background magnetic field

$$H = -\frac{1}{2}(B_1\sigma_1 + B_2\sigma_2 + B_3\sigma_3) \tag{34.1}$$

It turns out to be an experimental fact that fundamental matter particles are described by a quantum system that is the tensor product of these two systems, with state space

$$\mathcal{H} = \mathcal{H}_B \otimes \mathcal{H}_F = L^2(\mathbf{R}^3) \otimes \mathbf{C}^2 \tag{34.2}$$

which can be thought of as two-component complex wavefunctions. This system has a pseudo-classical description using a phase space with six conventional coordinates q_j, p_j and three fermionic coordinates ξ_j. On functions of these coordinates, one has a generalized Poisson bracket $\{\cdot, \cdot\}_\pm$ which provides a Lie superalgebra structure on such functions. On generators, the *nonzero* bracket relations are

$$\{q_j, p_k\}_\pm = \delta_{jk}, \quad \{\xi_j, \xi_k\}_\pm = \delta_{jk}$$

For now, we will take the background magnetic field $\mathbf{B} = 0$. In chapter 45, we will see how to generalize the free particle to the case of a particle in a general background electromagnetic field, and then the Hamiltonian term 34.1 involving the $\mathbf{B}$ field will appear. In the absence of electromagnetic fields, the classical Hamiltonian function will still be

$$h = \frac{1}{2m}(p_1^2 + p_2^2 + p_3^2)$$

but now this can be written in the following form (using the Leibniz rule for a Lie superbracket)

$$h = \frac{1}{2m}\{\sum_{j=1}^{3} p_j \xi_j, \sum_{k=1}^{3} p_k \xi_k\}_\pm = \frac{1}{2m} \sum_{j,k=1}^{3} p_j \{\xi_j, \xi_k\}_\pm p_k = \frac{1}{2m} \sum_{j=1}^{3} p_j^2$$

Note the appearance of the function $p_1 \xi_1 + p_2 \xi_2 + p_3 \xi_3$ which now plays a role even more fundamental than that of the Hamiltonian (which can be expressed in terms of it). In this pseudo-classical theory, $p_1 \xi_1 + p_2 \xi_2 + p_3 \xi_3$ is the function generating a "supersymmetry," Poisson commuting with the Hamiltonian, while at the same time playing the role of a sort of "square root" of the Hamiltonian. It provides a new sort of symmetry that can be thought of as a "square root" of an infinitesimal time translation.

Quantization takes

$$p_1 \xi_1 + p_2 \xi_2 + p_3 \xi_3 \rightarrow \frac{1}{\sqrt{2}} \boldsymbol{\sigma} \cdot \mathbf{P}$$

and the Hamiltonian operator can now be written as an anticommutator or a square

$$H = \frac{1}{2m} [\frac{1}{\sqrt{2}} \boldsymbol{\sigma} \cdot \mathbf{P}, \frac{1}{\sqrt{2}} \boldsymbol{\sigma} \cdot \mathbf{P}]_+ = \frac{1}{2m}(\boldsymbol{\sigma} \cdot \mathbf{P})^2 = \frac{1}{2m}(P_1^2 + P_2^2 + P_3^2)$$

(using the fact that the σ_j satisfy the Clifford algebra relations for $\mathrm{Cliff}(3, 0, \mathbf{R})$).

We will define the three-dimensional Dirac operator as

$$\partial\!\!\!/ = \sigma_1 \frac{\partial}{\partial q_1} + \sigma_2 \frac{\partial}{\partial q_2} + \sigma_3 \frac{\partial}{\partial q_3} = \boldsymbol{\sigma} \cdot \boldsymbol{\nabla}$$

It operates on two-component wavefunctions

$$\psi(\mathbf{q}) = \begin{pmatrix} \psi_1(\mathbf{q}) \\ \psi_2(\mathbf{q}) \end{pmatrix}$$

Using this Dirac operator (often called in this context the "Pauli operator"), we can write a two-component version of the Schrödinger equation (often called the "Pauli equation" or "Pauli-Schrödinger equation")

$$i\frac{\partial}{\partial t} \begin{pmatrix} \psi_1(\mathbf{q}) \\ \psi_2(\mathbf{q}) \end{pmatrix} = -\frac{1}{2m} \left(\sigma_1 \frac{\partial}{\partial q_1} + \sigma_2 \frac{\partial}{\partial q_2} + \sigma_3 \frac{\partial}{\partial q_3} \right)^2 \begin{pmatrix} \psi_1(\mathbf{q}) \\ \psi_2(\mathbf{q}) \end{pmatrix} \qquad (34.3)$$

$$= -\frac{1}{2m} \left(\frac{\partial^2}{\partial q_1^2} + \frac{\partial^2}{\partial q_2^2} + \frac{\partial^2}{\partial q_3^2} \right) \begin{pmatrix} \psi_1(\mathbf{q}) \\ \psi_2(\mathbf{q}) \end{pmatrix}$$

This equation is two copies of the standard free particle Schrödinger equation, so physically corresponds to a quantum theory of two types of free particles of mass m. It becomes much more non-trivial when a coupling to an electromagnetic field is introduced, as will be seen in chapter 45.

The equation for the energy eigenfunctions of energy eigenvalue E will be

$$\frac{1}{2m}(\boldsymbol{\sigma}\cdot\mathbf{P})^2\begin{pmatrix}\psi_1(\mathbf{q})\\\psi_2(\mathbf{q})\end{pmatrix}=E\begin{pmatrix}\psi_1(\mathbf{q})\\\psi_2(\mathbf{q})\end{pmatrix}$$

In terms of the inverse Fourier transform

$$\psi_{1,2}(\mathbf{q})=\frac{1}{(2\pi)^{\frac{3}{2}}}\int_{\mathbf{R}^3}e^{i\mathbf{p}\cdot\mathbf{q}}\widetilde{\psi}_{1,2}(\mathbf{p})d^3\mathbf{p}$$

this equation becomes

$$((\boldsymbol{\sigma}\cdot\mathbf{p})^2-2mE)\begin{pmatrix}\widetilde{\psi}_1(\mathbf{p})\\\widetilde{\psi}_2(\mathbf{p})\end{pmatrix}=(|\mathbf{p}|^2-2mE)\begin{pmatrix}\widetilde{\psi}_1(\mathbf{p})\\\widetilde{\psi}_2(\mathbf{p})\end{pmatrix}=0\qquad(34.4)$$

and as in chapter 19, our solution space is given by distributions supported on the sphere of radius $\sqrt{2mE}=|\mathbf{p}|$ in momentum space which we will write as

$$\begin{pmatrix}\widetilde{\psi}_1(\mathbf{p})\\\widetilde{\psi}_2(\mathbf{p})\end{pmatrix}=\delta(|\mathbf{p}|^2-2mE)\begin{pmatrix}\widetilde{\psi}_{E,1}(\mathbf{p})\\\widetilde{\psi}_{E,2}(\mathbf{p})\end{pmatrix}\qquad(34.5)$$

where $\widetilde{\psi}_{E,1}(\mathbf{p})$ and $\widetilde{\psi}_{E,2}(\mathbf{p})$ are functions on the sphere $|\mathbf{p}|^2=2mE$.

34.2 Solutions of the Pauli equation and representations of $\widehat{E(3)}$

Since

$$\frac{\boldsymbol{\sigma}\cdot\mathbf{p}}{|\mathbf{p}|}$$

is an invertible operator with eigenvalues ±1, solutions to 34.4 will be given by solutions to

$$\frac{\boldsymbol{\sigma}\cdot\mathbf{p}}{|\mathbf{p}|}\begin{pmatrix}\widetilde{\psi}_{E,1}(\mathbf{p})\\\widetilde{\psi}_{E,2}(\mathbf{p})\end{pmatrix}=\pm\begin{pmatrix}\widetilde{\psi}_{E,1}(\mathbf{p})\\\widetilde{\psi}_{E,2}(\mathbf{p})\end{pmatrix}\qquad(34.6)$$

where $|\mathbf{p}|=\sqrt{2mE}$. We will write solutions to this equation with the $+$ sign as $\widetilde{\psi}_{E,+}(\mathbf{p})$, those for the $-$ sign as $\widetilde{\psi}_{E,-}(\mathbf{p})$. Note that $\widetilde{\psi}_{E,+}(\mathbf{p})$ and $\widetilde{\psi}_{E,-}(\mathbf{p})$

are each two-component complex functions on the sphere $\sqrt{2mE} = |\mathbf{p}|$ (or more generally distributions on the sphere). Our goal in the rest of this section will be to show

Theorem. *The spaces of solutions* $\widetilde{\psi}_{E,\pm}(\mathbf{p})$ *to equations 34.6 provide irreducible representations of* $\widetilde{E(3)}$, *the double cover of* $E(3)$, *with eigenvalue* $2mE$ *for the first Casimir operator*

$$|\mathbf{P}|^2 = (\boldsymbol{\sigma} \cdot \mathbf{P})^2$$

and eigenvalues $\pm\frac{1}{2}\sqrt{2mE}$ *for the second Casimir operator* $\mathbf{J} \cdot \mathbf{P}$.

We will not try to prove irreducibility, but just show that these solution spaces give representations with the claimed eigenvalues of the Casimir operators (see sections 19.2 and 19.3 for more about the Casimir operators and general theory of representations of $E(3)$). We will write the representation operators as $u(\mathbf{a}, \Omega)$ in position space and $\widetilde{u}(\mathbf{a}, \Omega)$ in momentum space, with $\mathbf{a}$ a translation, $\Omega \in SU(2)$ and $R = \Phi(\Omega) \in SO(3)$.

The translation part of the group acts as in the one-component case of chapter 19, by the multiplication operator

$$\widetilde{u}(\mathbf{a}, 1)\widetilde{\psi}_{E,\pm}(\mathbf{p}) = e^{-i\mathbf{a}\cdot\mathbf{P}}\widetilde{\psi}_{E,\pm}(\mathbf{p})$$

and

$$\widetilde{u}(\mathbf{a}, 1) = e^{-i\mathbf{a}\cdot\mathbf{P}}$$

so the Lie algebra representation is given by the usual $\mathbf{P}$ operator. This action of the translations is easily seen to commute with $\boldsymbol{\sigma} \cdot \mathbf{P}$ and thus act on the solutions to 34.6. It is the action of rotations that requires a more complicated discussion than in the single-component case.

In chapter 19, we saw that $R \in SO(3)$ acts on single-component momentum space solutions of the Schrödinger equation by

$$\widetilde{\psi}_E(\mathbf{p}) \rightarrow \widetilde{u}(0, R)\widetilde{\psi}_E(\mathbf{p}) = \widetilde{\psi}_E(R^{-1}\mathbf{p})$$

This takes solutions to solutions since the operator $\widetilde{u}(0, R)$ commutes with the Casimir operator $|\mathbf{P}|^2$

$$\widetilde{u}(0, R)|\mathbf{P}|^2 = |\mathbf{P}|^2\widetilde{u}(0, R) \iff \widetilde{u}(0, R)|\mathbf{P}|^2\widetilde{u}(0, R)^{-1} = |\mathbf{P}|^2$$

This is true since

$$\widetilde{u}(0, R)|\mathbf{P}|^2\widetilde{u}(0, R)^{-1}\widetilde{\psi}(\mathbf{p}) = \widetilde{u}(0, R)|\mathbf{P}|^2\widetilde{\psi}(R\mathbf{p})$$
$$= |R^{-1}\mathbf{P}|^2\widetilde{\psi}(R^{-1}R\mathbf{p}) = |\mathbf{P}|^2\widetilde{\psi}(\mathbf{p})$$

To get a representation on two-component wavefunctions that commutes with the operator $\boldsymbol{\sigma} \cdot \mathbf{P}$, we need to change the action of rotations to

$$\widetilde{\psi}_{E,\pm}(\mathbf{p}) \to \widetilde{u}(0,\Omega)\widetilde{\psi}_{E,\pm}(\mathbf{p}) = \Omega\widetilde{\psi}_{E,\pm}(R^{-1}\mathbf{p})$$

With this action on solutions, we have

$$
\begin{aligned}
\widetilde{u}(0,\Omega)(\boldsymbol{\sigma} \cdot \mathbf{P})\widetilde{u}(0,\Omega)^{-1}\widetilde{\psi}_{E,\pm}(\mathbf{p}) &= \widetilde{u}(0,\Omega)(\boldsymbol{\sigma} \cdot \mathbf{P})\Omega^{-1}\widetilde{\psi}_{E,\pm}(R\mathbf{p}) \\
&= \Omega(\boldsymbol{\sigma} \cdot R^{-1}\mathbf{P})\Omega^{-1}\widetilde{\psi}_{E,\pm}(R^{-1}R\mathbf{p}) \\
&= (\boldsymbol{\sigma} \cdot \mathbf{P})\widetilde{\psi}_{E,\pm}(\mathbf{p})
\end{aligned}
$$

where we have used equation 6.5 to show

$$\Omega(\boldsymbol{\sigma} \cdot R^{-1}\mathbf{P})\Omega^{-1} = \boldsymbol{\sigma} \cdot RR^{-1}\mathbf{P} = \boldsymbol{\sigma} \cdot \mathbf{P}$$

The $SU(2)$ part of the group acts by a product of two commuting different actions on the two factors of the tensor product 34.2. These are:

1. The same action on the momentum coordinates as in the one-component case, just using $R = \Phi(\Omega)$, the $SO(3)$ rotation corresponding to the $SU(2)$ group element Ω. For example, for a rotation about the x-axis by angle ϕ we have

$$\widetilde{\psi}_{E,\pm}(\mathbf{p}) \to \widetilde{\psi}_{E,\pm}(R(\phi,\mathbf{e}_1)^{-1}\mathbf{p})$$

    Recall that the operator that does this is $e^{-i\phi L_1}$ where

$$-iL_1 = -i(Q_2P_3 - Q_3P_2) = -\left(q_2\frac{\partial}{\partial q_3} - q_3\frac{\partial}{\partial q_2}\right)$$

    and in general we have operators

$$-i\mathbf{L} = -i\mathbf{Q} \times \mathbf{P}$$

    that provide the Lie algebra version of the representation (recall that at the Lie algebra level, $SO(3)$ and $Spin(3)$ are isomorphic).

2. The action of the matrix $\Omega \in SU(2)$ on the two-component wavefunction by

$$\widetilde{\psi}_{E,\pm}(\mathbf{p}) \to \Omega\widetilde{\psi}_{E,\pm}(\mathbf{p})$$

    For R a rotation by angle ϕ about the x-axis, one choice of Ω is

$$\Omega = e^{-i\phi\frac{\sigma_1}{2}}$$

and the operators that provide the Lie algebra version of the representation are the

$$-i\mathbf{S} = -i\frac{1}{2}\boldsymbol{\sigma}$$

The Lie algebra representation corresponding to the action of these transformations on the two factors of the tensor product is given as usual (see chapter 9) by a sum of operators that act on each factor

$$-i\mathbf{J} = -i(\mathbf{L} + \mathbf{S})$$

The standard terminology is to call $\mathbf{L}$ the "orbital" angular momentum, $\mathbf{S}$ the "spin" angular momentum, and $\mathbf{J}$ the "total" angular momentum.

The second Casimir operator for this case is

$$\mathbf{J} \cdot \mathbf{P}$$

and as in the one-component case (see section 19.3), a straightforward calculation shows that the $\mathbf{L} \cdot \mathbf{P}$ part of this acts trivially on our solutions $\widetilde{\psi}_{E,\pm}(\mathbf{p})$. The spin component acts non-trivially, and we have

$$(\mathbf{J} \cdot \mathbf{P})\widetilde{\psi}_{E,\pm}(\mathbf{p}) = (\frac{1}{2}\boldsymbol{\sigma} \cdot \mathbf{p})\widetilde{\psi}_{E,\pm}(\mathbf{p}) = \pm\frac{1}{2}|\mathbf{p}|\widetilde{\psi}_{E,\pm}(\mathbf{p})$$

so we see that our solutions have helicity (eigenvalue of $\mathbf{J} \cdot \mathbf{P}$ divided by the square root of the eigenvalue of $|\mathbf{P}|^2$) values $\pm\frac{1}{2}$, as opposed to the integral helicity values discussed in chapter 19, where $E(3)$ appeared and not its double cover. These two representations on the spaces of solutions $\widetilde{\psi}_{E,\pm}(\mathbf{p})$ are thus the $\widetilde{E(3)}$ representations described in section 19.3, the ones labeled by the helicity $\pm\frac{1}{2}$ representations of the stabilizer group $SO(2)$.

Solutions for either sign of equation 34.6 are given by a one-dimensional subspace of $\mathbf{C}^2$ for each $\mathbf{p}$, and it is sometimes convenient to represent them as follows. Note that for each $\mathbf{p}$ one can decompose

$$\mathbf{C}^2 = \mathbf{C} \oplus \mathbf{C}$$

into $\pm$-eigenspaces of the matrix $\frac{\boldsymbol{\sigma}\cdot\mathbf{p}}{|\mathbf{p}|}$. In our discussion of the Bloch sphere in section 7.5, we explicitly found that (see equation 7.6)

$$u_+(\mathbf{p}) = \frac{1}{\sqrt{2(1+p_3)}}\begin{pmatrix} 1+p_3 \\ p_1 + ip_2 \end{pmatrix} \tag{34.7}$$

provides a normalized element of the $+$ eigenspace of $\frac{\boldsymbol{\sigma}\cdot\mathbf{p}}{|\mathbf{p}|}$ that satisfies

$$\Omega u_+(\mathbf{p}) = u_+(R\mathbf{p})$$

Similarly, we saw that

$$u_-(\mathbf{p}) = \frac{1}{\sqrt{2(1+p_3)}} \begin{pmatrix} -(p_1 - ip_2) \\ 1 + p_3 \end{pmatrix} \tag{34.8}$$

provides such an element for the $-$ eigenspace.

Another way to construct such elements is to use projection operators. The operators

$$P_\pm(\mathbf{p}) = \frac{1}{2}(1 \pm \frac{\boldsymbol{\sigma} \cdot \mathbf{p}}{|\mathbf{p}|})$$

provide projection operators onto these two spaces, since one can easily check that

$$P_+^2 = P_+, \quad P_-^2 = P_-, \quad P_+ P_- = P_- P_+ = 0, \quad P_+ + P_- = 1$$

Solutions can now be written as

$$\widetilde{\psi}_{E,\pm}(\mathbf{p}) = \alpha_{E,\pm}(\mathbf{p}) u_\pm(\mathbf{p}) \tag{34.9}$$

for arbitrary functions $\alpha_{E,\pm}(\mathbf{p})$ on the sphere $|\mathbf{p}| = \sqrt{2mE}$, where the $u_\pm(\mathbf{p})$ in this context are called "spin polarization vectors." There is, however, a subtlety involved in representing solutions in this manner. Recall from section 7.5 that $u_+(\mathbf{p})$ is discontinuous at $p_3 = -1$ (the same will be true for $u_-(\mathbf{p})$) and any unit-length eigenvector of $\boldsymbol{\sigma} \cdot \mathbf{p}$ must have such a discontinuity somewhere. If $\alpha_{E,\pm}(\mathbf{p})$ has a zero at $p_3 = -1$, the product $\widetilde{\psi}_{E,\pm}(\mathbf{p})$ can be continuous. It remains a basic topological fact that the combination $\widetilde{\psi}_{E,\pm}(\mathbf{p})$ must have a zero, or it will have to be discontinuous. Our choice of $u_\pm(\mathbf{p})$ works well if this zero is at $p_3 = -1$, but if it is elsewhere one might want to make a different choice. In the end, one needs to check that computed physical quantities are independent of such choices.

Keeping in mind the above subtlety, the $u_\pm(\mathbf{p})$ can be used to write an arbitrary solution of the Pauli equation 34.3 of energy E as

$$\begin{pmatrix} \psi_1(\mathbf{q}, t) \\ \psi_2(\mathbf{q}, t) \end{pmatrix} = e^{-iEt} \begin{pmatrix} \psi_1(\mathbf{q}) \\ \psi_2(\mathbf{q}) \end{pmatrix}$$

where

$$\begin{pmatrix} \psi_1(\mathbf{q}) \\ \psi_2(\mathbf{q}) \end{pmatrix} = \frac{1}{(2\pi)^{\frac{3}{2}}} \int_{\mathbf{R}^3} \delta(|\mathbf{p}|^2 - 2mE)(\widetilde{\psi}_{E,+}(\mathbf{p}) + \widetilde{\psi}_{E,-}(\mathbf{p})) e^{i\mathbf{p} \cdot \mathbf{q}} d^3 \mathbf{p}$$

$$= \frac{1}{(2\pi)^{\frac{3}{2}}} \int_{\mathbf{R}^3} \delta(|\mathbf{p}|^2 - 2mE)(\alpha_{E,+}(\mathbf{p}) u_+(\mathbf{p}) + \alpha_{E,-}(\mathbf{p}) u_-(\mathbf{p})) e^{i\mathbf{p} \cdot \mathbf{q}} d^3 \mathbf{p}$$

$$\tag{34.10}$$

34.3 The $\widetilde{E(3)}$-invariant inner product

One can parametrize solutions to the Pauli equation and write an $\widetilde{E(3)}$-invariant inner product on the space of solutions in several different ways. Three different parametrizations of solutions that can be considered are:

- Using the initial data at a fixed time

$$\begin{pmatrix} \psi_1(\mathbf{q}) \\ \psi_2(\mathbf{q}) \end{pmatrix}$$

Here the $\widetilde{E(3)}$-invariant inner product is

$$\left\langle \begin{pmatrix} \psi_1(\mathbf{q}) \\ \psi_2(\mathbf{q}) \end{pmatrix}, \begin{pmatrix} \psi_1'(\mathbf{q}) \\ \psi_2'(\mathbf{q}) \end{pmatrix} \right\rangle = \int_{\mathbf{R}^3} \begin{pmatrix} \psi_1(\mathbf{q}) \\ \psi_2(\mathbf{q}) \end{pmatrix}^\dagger \begin{pmatrix} \psi_1'(\mathbf{q}) \\ \psi_2'(\mathbf{q}) \end{pmatrix} d^3\mathbf{q}$$

This parametrization does not make visible the decomposition into irreducible representations of $\widetilde{E(3)}$.

- Using the Fourier transforms

$$\begin{pmatrix} \widetilde{\psi}_1(\mathbf{p}) \\ \widetilde{\psi}_2(\mathbf{p}) \end{pmatrix}$$

to parametrize solutions, the invariant inner product is

$$\left\langle \begin{pmatrix} \widetilde{\psi}_1(\mathbf{p}) \\ \widetilde{\psi}_2(\mathbf{p}) \end{pmatrix}, \begin{pmatrix} \widetilde{\psi'}_1(\mathbf{p}) \\ \widetilde{\psi'}_2(\mathbf{p}) \end{pmatrix} \right\rangle = \int_{\mathbf{R}^3} \begin{pmatrix} \widetilde{\psi}_1(\mathbf{p}) \\ \widetilde{\psi}_2(\mathbf{p}) \end{pmatrix}^\dagger \begin{pmatrix} \widetilde{\psi'}_1(\mathbf{p}) \\ \widetilde{\psi'}_2(\mathbf{p}) \end{pmatrix} d^3\mathbf{p}$$

The decomposition of equation 34.5 can be used to express solutions of energy E in terms of two-component functions $\widetilde{\psi}_E(\mathbf{p})$, with an invariant inner product on the space of such solutions given by

$$\langle \widetilde{\psi}_E(\mathbf{p}), \widetilde{\psi'}_E(\mathbf{p}) \rangle = \frac{1}{4\pi} \int_{S^2} \widetilde{\psi}_E(\mathbf{p})^\dagger \widetilde{\psi}_E'(\mathbf{p}) \sin(\phi) d\phi d\theta$$

where (p, ϕ, θ) are spherical coordinates on momentum space and S^2 is the sphere of radius $\sqrt{2mE}$.

The $\widetilde{\psi}_E(\mathbf{p})$ parametrize not a single irreducible representation of $\widetilde{E(3)}$ but two of them, including both helicities.

- The spin polarization vectors and equation 34.9 can be used to parametrize solutions of fixed energy E in terms of two functions $\alpha_{E,+}(\mathbf{p})$, $\alpha_{E,+}(\mathbf{p})$ on the sphere $|\mathbf{p}|^2 = 2mE$. This gives an explicit decomposition into irreducible representations of $\widetilde{E(3)}$, with the representation space a space of complex functions on the sphere, with invariant inner product for each helicity choice given by

$$\langle \alpha_{E,\pm}(\mathbf{p}), \alpha'_{E,\pm}(\mathbf{p}) \rangle = \frac{1}{4\pi} \int_{S^2} \alpha_{E,\pm}(\mathbf{p})^\dagger \alpha'_{E,\pm}(\mathbf{p}) \sin(\phi) d\phi d\theta$$

In each case, the space of solutions is a complex vector space and one can imagine trying to take it as a phase space (with the imaginary part of the Hermitian inner product providing the symplectic structure) and quantizing. This will be an example of a quantum field theory, one discussed in more detail in section 38.3.3.

34.4 The Dirac operator

The above construction can be generalized to the case of any dimension d as follows. Recall from chapter 29 that associated to $\mathbf{R}^d$ with a standard inner product, but of a general signature (r, s) (where $r + s = d$, r is the number of $+$ signs, s the number of $-$ signs), we have a Clifford algebra $\text{Cliff}(r, s)$ with generators γ_j satisfying

$$\gamma_j \gamma_k = -\gamma_k \gamma_j, \quad j \neq k$$

$$\gamma_j^2 = +1 \text{ for } j = 1, \cdots, r \quad \gamma_j^2 = -1, \text{ for } j = r + 1, \cdots, d$$

To any vector $\mathbf{v} \in \mathbf{R}^d$ with components v_j, recall that we can associate a corresponding element $\not{v}$ in the Clifford algebra by

$$\mathbf{v} \in \mathbf{R}^d \rightarrow \not{v} = \sum_{j=1}^{d} \gamma_j v_j \in \text{Cliff}(r, s)$$

Multiplying this Clifford algebra element by itself and using the relations above, we get a scalar, the length-squared of the vector

$$\not{v}^2 = v_1^2 + v_2^2 \cdots + v_r^2 - v_{r+1}^2 - \cdots - v_d^2 = |\mathbf{v}|^2$$

This shows that by introducing a Clifford algebra, we can find an interesting new sort of square root for expressions like $|\mathbf{v}|^2$. We can define:

Definition (Dirac operator). *The Dirac operator is the operator*

$$\displaystyle{\not{\partial} = \sum_{j=1}^{d} \gamma_j \frac{\partial}{\partial q_j}}$$

This will be a first-order differential operator with the property that its square is the Laplacian

$$\not{\partial}^2 = \frac{\partial^2}{\partial q_1^2} + \cdots + \frac{\partial^2}{\partial q_r^2} - \frac{\partial^2}{\partial q_{r+1}^2} - \cdots - \frac{\partial^2}{\partial q_d^2}$$

The Dirac operator $\not{\partial}$ acts not on functions but on functions taking values in the spinor vector space S that the Clifford algebra acts on. Picking a matrix representation of the γ_j, the Dirac operator will be a constant coefficient first-order differential operator acting on wavefunctions with dim S components. In chapter 47, we will study in detail what happens for the case of $r = 3, s = 1$ and see how the Dirac operator there provides an appropriate wave equation with the symmetries of special relativistic space–time.

34.5 For further reading

The point of view here in terms of representations of $\widetilde{E(3)}$ is not very conventional, but the material here about spin and the Pauli equation can be found in any quantum mechanics book, see, for example, chapter 14 of [81]. For more details about supersymmetric quantum mechanics and the appearance of the Dirac operator as the generator of a supersymmetry in the quantization of a pseudo-classical system, see [90] and [1].

Chapter 35
Lagrangian Methods and the Path Integral

In this chapter, we will give a rapid survey of a different starting point for developing quantum mechanics, based on the Lagrangian rather than Hamiltonian classical formalism. The Lagrangian point of view is the one taken in most modern physics textbooks, and we will refer to these for more detail, concentrating here on explaining the relation to the Hamiltonian approach. Lagrangian methods have quite different strengths and weaknesses than those of the Hamiltonian formalism, and we will try and point these out.

The Lagrangian formalism leads naturally to an apparently very different notion of quantization, one based upon formulating quantum theory in terms of infinite dimensional integrals known as path integrals. A serious investigation of these would require another and very different volume, so we will have to restrict ourselves to a quick outline of how path integrals work, giving references to standard texts for the details. We will try and provide some indication of both the advantages of the path integral method, as well as the significant problems it entails.

35.1 Lagrangian mechanics

In the Lagrangian formalism, instead of a phase space $\mathbf{R}^{2d}$ of positions q_j and momenta p_j, one considers just the position (or configuration) space $\mathbf{R}^d$. Instead of a Hamiltonian function $h(\mathbf{q}, \mathbf{p})$, one has:

Definition (Lagrangian). *The Lagrangian L for a classical mechanical system with configuration space $\mathbf{R}^d$ is a function*

$$L : (\mathbf{q}, \mathbf{v}) \in \mathbf{R}^d \times \mathbf{R}^d \to L(\mathbf{q}, \mathbf{v}) \in \mathbf{R}$$

© Peter Woit 2017
P. Woit, *Quantum Theory, Groups and Representations*,
DOI 10.1007/978-3-319-64612-1_35

Given differentiable paths in the configuration space defined by functions

$$\gamma : t \in [t_1, t_2] \to \mathbf{R}^d$$

which we will write in terms of their position and velocity vectors as

$$\gamma(t) = (\mathbf{q}(t), \dot{\mathbf{q}}(t))$$

one can define a functional on the space of such paths:

Definition (Action). *The action S for a path γ is*

$$S[\gamma] = \int_{t_1}^{t_2} L(\mathbf{q}(t), \dot{\mathbf{q}}(t)) dt$$

The fundamental principle of classical mechanics in the Lagrangian formalism is that classical trajectories are given by critical points of the action functional. These may correspond to minima of the action (so this is sometimes called the "principle of least action"), but one gets classical trajectories also for critical points that are not minima of the action. One can define the appropriate notion of critical point as follows:

Definition (Critical point for S). *A path γ is a critical point of the functional $S[\gamma]$ if*

$$\delta S(\gamma) \equiv \frac{d}{ds} S(\gamma_s)_{|s=0} = 0$$

where

$$\gamma_s : [t_1, t_2] \to \mathbf{R}^d$$

is a smooth family of paths parametrized by an interval $s \in (-\epsilon, \epsilon)$, with $\gamma_0 = \gamma$.

We will now ignore analytical details and adopt the physicist's interpretation of δS as the first-order change in S due to an infinitesimal change $\delta \gamma = (\delta \mathbf{q}(t), \delta \dot{\mathbf{q}}(t))$ in the path.

When $(\mathbf{q}(t), \dot{\mathbf{q}}(t))$ satisfy a certain differential equation, the path γ will be a critical point and thus a classical trajectory:

Theorem (Euler–Lagrange equations). *One has*

$$\delta S[\gamma] = 0$$

for all variations of γ with endpoints $\gamma(t_1)$ and $\gamma(t_2)$ fixed if

$$\frac{\partial L}{\partial q_j}(\mathbf{q}(t), \dot{\mathbf{q}}(t)) - \frac{d}{dt}\left(\frac{\partial L}{\partial \dot{q}_j}(\mathbf{q}(t), \dot{\mathbf{q}}(t))\right) = 0$$

for $j = 1, \cdots, d$. These are called the Euler–Lagrange equations.

Proof. Ignoring analytical details, the Euler–Lagrange equations follow from the following calculations, which we will just do for $d = 1$, with the generalization to higher d straightforward. We are calculating the first-order change in S due to an infinitesimal change $\delta\gamma = (\delta q(t), \delta \dot{q}(t))$

$$\delta S[\gamma] = \int_{t_1}^{t_2} \delta L(q(t), \dot{q}(t))dt$$

$$= \int_{t_1}^{t_2} \left(\frac{\partial L}{\partial q}(q(t), \dot{q}(t))\delta q(t) + \frac{\partial L}{\partial \dot{q}}(q(t), \dot{q}(t))\delta \dot{q}(t)\right) dt$$

But

$$\delta \dot{q}(t) = \frac{d}{dt}\delta q(t)$$

and, using integration by parts

$$\frac{\partial L}{\partial \dot{q}}\delta \dot{q}(t) = \frac{d}{dt}\left(\frac{\partial L}{\partial \dot{q}}\delta q\right) - \left(\frac{d}{dt}\frac{\partial L}{\partial \dot{q}}\right)\delta q$$

so

$$\delta S[\gamma] = \int_{t_1}^{t_2} \left(\left(\frac{\partial L}{\partial q} - \frac{d}{dt}\frac{\partial L}{\partial \dot{q}}\right)\delta q - \frac{d}{dt}\left(\frac{\partial L}{\partial \dot{q}}\delta q\right)\right) dt$$

$$= \int_{t_1}^{t_2} \left(\frac{\partial L}{\partial q} - \frac{d}{dt}\frac{\partial L}{\partial \dot{q}}\right)\delta q\, dt - \left(\frac{\partial L}{\partial \dot{q}}\delta q\right)(t_2) + \left(\frac{\partial L}{\partial \dot{q}}\delta q\right)(t_1) \quad (35.1)$$

If we keep the endpoints fixed so $\delta q(t_1) = \delta q(t_2) = 0$, then for solutions to

$$\frac{\partial L}{\partial q}(q(t), \dot{q}(t)) - \frac{d}{dt}\left(\frac{\partial L}{\partial \dot{q}}(q(t), \dot{q}(t))\right) = 0$$

the integral will be zero for arbitrary variations δq. □

As an example, a particle moving in a potential $V(\mathbf{q})$ will be described by a Lagrangian

$$L(\mathbf{q}, \dot{\mathbf{q}}) = \frac{1}{2}m|\dot{\mathbf{q}}|^2 - V(\mathbf{q})$$

for which the Euler–Lagrange equations will be

$$-\frac{\partial V}{\partial q_j} = \frac{d}{dt}(m\dot{q}_j) = m\ddot{q}_j$$

This is just Newton's second law, which says that the force due to a potential is equal to the mass times the acceleration of the particle.

Given a Lagrangian classical mechanical system, one would like to be able to find a corresponding Hamiltonian system that will give the same equations of motion. To do this, we proceed by defining (for each configuration coordinate q_j) a corresponding momentum coordinate p_j by

$$p_j = \frac{\partial L}{\partial \dot{q}_j}$$

Then, instead of working with trajectories characterized at time t by

$$(\mathbf{q}(t), \dot{\mathbf{q}}(t)) \in \mathbf{R}^{2d}$$

we would like to instead use

$$(\mathbf{q}(t), \mathbf{p}(t)) \in \mathbf{R}^{2d}$$

where $p_j = \frac{\partial L}{\partial \dot{q}_j}$ and identify this $\mathbf{R}^{2d}$ (e.g., at $t = 0$) as the phase space of the conventional Hamiltonian formalism.

The transformation

$$(q_j, \dot{q}_k) \rightarrow \left(q_j, p_k = \frac{\partial L}{\partial \dot{q}_k} \right)$$

between position-velocity and phase space is known as the Legendre transform, and in good cases (for instance, when L is quadratic in all the velocities), it is an isomorphism. In general though, this is not an isomorphism, with the Legendre transform often taking position-velocity space to a lower dimensional subspace of phase space. Such cases are not unusual and require a much more complicated formalism, even as classical mechanical systems (this subject is known as "constrained Hamiltonian dynamics"). One important example we will study in Chapter 46 is that of the free electromagnetic field, with equations of motion the Maxwell equations. In that case, the configuration space coordinates are the components (A_0, A_1, A_2, A_3) of the vector potential, with the problem arising because the Lagrangian does not depend on $\dot{A}_0$.

Besides a phase space, for a Hamiltonian system one needs a Hamiltonian function. Choosing

$$h = \sum_{j=1}^{d} p_j \dot{q}_j - L(\mathbf{q}, \dot{\mathbf{q}})$$

will work, provided the relation

$$p_j = \frac{\partial L}{\partial \dot{q}_j}$$

can be used to solve for the velocities $\dot{q}_j$ and express them in terms of the momentum variables. In that case, computing the differential of h one finds (for $d = 1$, the generalization to higher d is straightforward)

$$dh = p d\dot{q} + \dot{q} dp - \frac{\partial L}{\partial q} dq - \frac{\partial L}{\partial \dot{q}} d\dot{q}$$
$$= \dot{q} dp - \frac{\partial L}{\partial q} dq$$

So one has

$$\frac{\partial h}{\partial p} = \dot{q}, \quad \frac{\partial h}{\partial q} = -\frac{\partial L}{\partial q}$$

but these are precisely Hamilton's equations since the Euler–Lagrange equations imply

$$\frac{\partial L}{\partial q} = \frac{d}{dt}\frac{\partial L}{\partial \dot{q}} = \dot{p}$$

While the Legendre transform method given above works in some situations, more generally and more abstractly, one can pass from the Lagrangian to the Hamiltonian formalism by taking as phase space the space of solutions of the Euler–Lagrange equations. This is sometimes called the "covariant phase space," and it can often concretely be realized by fixing a time $t = 0$ and parametrizing solutions by their initial conditions at such a $t = 0$. One can also go directly from the action to a sort of Poisson bracket on this covariant phase space (this is called the "Peierls bracket"). For a general Lagrangian, one can pass to a version of the Hamiltonian formalism either by this method or by the method of Hamiltonian mechanics with constraints. Only for a special class of Lagrangians though will one get a non-degenerate Poisson bracket on a linear phase space and recover the usual properties of the standard Hamiltonian formalism.

35.2 Noether's theorem and symmetries in the Lagrangian formalism

The derivation of the Euler–Lagrange equations given above can also be used to study the implications of Lie group symmetries of a Lagrangian system. When a Lie group G acts on the space of paths, preserving the action S, it will take classical trajectories to classical trajectories, so we have a Lie group action on the space of solutions to the equations of motion (the Euler–Lagrange equations). On this space of solutions, we have, from equation 35.1 (generalized to multiple coordinate variables),

$$\delta S[\gamma] = \left(\sum_{j=1}^{d} \frac{\partial L}{\partial \dot{q}_j} \delta q_j(X) \right)(t_1) - \left(\sum_{j=1}^{d} \frac{\partial L}{\partial \dot{q}_j} \delta q_j(X) \right)(t_2)$$

where now $\delta q_j(X)$ is the infinitesimal change in a classical trajectory coming from the infinitesimal group action by an element X in the Lie algebra of G. From invariance of the action S under G, we must have $\delta S = 0$, so

$$\left(\sum_{j=1}^{d} \frac{\partial L}{\partial \dot{q}_j} \delta q_j(X) \right)(t_2) = \left(\sum_{j=1}^{d} \frac{\partial L}{\partial \dot{q}_j} \delta q_j(X) \right)(t_1)$$

This is an example of a more general result known as "Noether's theorem." In this context, it says that given a Lie group action on a Lagrangian system that leaves the action invariant, for each element X of the Lie algebra we will have a conserved quantity

$$\sum_{j=1}^{d} \frac{\partial L}{\partial \dot{q}_j} \delta q_j(X)$$

which is independent of time along the trajectory.

A basic example occurs when the Lagrangian is independent of the position variables q_j, depending only on the velocities $\dot{q}_j$, for example, in the case of a free particle, when $V(q_j) = 0$. In such a case, one has invariance of the Lagrangian under the Lie group $\mathbf{R}^d$ of space translations. Taking X to be an infinitesimal translation in the j-direction, one has as conserved quantity

$$\frac{\partial L}{\partial \dot{q}_j} = p_j$$

For the case of the free particle, this will be

$$\frac{\partial L}{\partial \dot{q}_j} = m\dot{q}_j$$

and the conservation law is conservation of the jth component of momentum. Another example is given (in $d = 3$) by rotational invariance of the Lagrangian under the group $SO(3)$ acting by rotations of the q_j. One can show that, for X an infinitesimal rotation about the k axis, the kth component of the angular momentum vector

$$\mathbf{q} \times \mathbf{p}$$

will be a conserved quantity.

The Lagrangian formalism has the advantage that the dynamics depends only on the choice of action functional on the space of possible trajectories, and it can be straightforwardly generalized to theories where the configuration space is an infinite dimensional space of classical fields. Unlike the usual Hamiltonian formalism for such theories, the Lagrangian formalism allows one to treat space and time symmetrically. For relativistic field theories, this allows one to exploit the full set of space–time symmetries, which can mix space and time directions. In such theories, Noether's theorem provides a powerful tool for finding the conserved quantities corresponding to symmetries of the system that are due to invariance of the action under some group of transformations.

On the other hand, in the Lagrangian formalism, since Noether's theorem only considers group actions on configuration space, it does not cover the case of Hamiltonian group actions that mix position and momentum coordinates. Recall that in the Hamiltonian formalism the moment map provides functions corresponding to group actions preserving the Poisson bracket. These functions will give the same conserved quantities as the ones one gets from Noether's theorem for the case of symmetries (i.e., functions that Poisson-commute with the Hamiltonian function), when the group action is given by an action on configuration space.

As an important example not covered by Noether's theorem, our study of the harmonic oscillator exploited several techniques (use of a complex structure on phase space, and of the $U(1)$ symmetry of rotations in the qp plane) that are unavailable in the Lagrangian formalism, which just uses configuration space, not phase space.

35.3 Quantization and path integrals

After use of the Legendre transform to pass to a Hamiltonian system, one then faces the question of how to construct a corresponding quantum theory. The method of "canonical quantization" is the one we have studied, taking the position coordinates q_j to operators Q_j and momentum coordinates p_j to operators P_j, with Q_j and P_j satisfying the Heisenberg commutation relations. By the Stone von-Neumann theorem, up to unitary equivalence there is only one way to do this and realize these operators on a state space $\mathcal{H}$. Recall though that the Groenewold-van Hove no-go theorem says that there is an inherent operator ordering ambiguity for operators of higher order than quadratic, thus for such operators providing many different possible quantizations of the same classical system (different though only by terms proportional to $\hbar$). In cases where the Legendre transform is not an isomorphism, a new set of problems appear when one tries to pass to a quantum system since the standard method of canonical quantization will no longer apply, and new methods are needed.

There is, however, a very different approach to relating classical and quantum theories, which completely bypasses the Hamiltonian formalism, just using the Lagrangian. This is the path integral formalism, which is based upon a method for calculating matrix elements of the time evolution operator

$$\langle q_T | e^{-\frac{i}{\hbar} H T} | q_0 \rangle$$

in the position eigenstate basis in terms of an integral over the space of paths that go from position q_0 to position q_T in time T (we will here only treat the $d = 1$ case). Here, $|q_0\rangle$ is an eigenstate of Q with eigenvalue q_0 (a delta-function at q_0 in the position space representation), and $|q_T\rangle$ has Q eigenvalue q_T. This matrix element has a physical interpretation as the amplitude for a particle starting at q_0 at $t = 0$ to have position q_T at time T, with its norm-squared giving the probability density for observing the particle at position q_T. It is also the kernel function that allows one to determine the wavefunction $\psi(q,T)$ at any time $t = T$ in terms of its initial value at $t = 0$, by calculating

$$\psi(q_T, T) = \int_{-\infty}^{\infty} \langle q_T | e^{-\frac{i}{\hbar} H T} | q_0 \rangle \psi(q_0, 0) dq_0$$

Note that for the free particle case this is the propagator that we studied in section 12.5.

To try and derive a path integral expression for this, one breaks up the interval $[0, T]$ into N equal-sized subintervals and calculates

$$\langle q_T | (e^{-\frac{i}{N\hbar} H T})^N | q_0 \rangle$$

If the Hamiltonian is a sum $H = K + V$, the Trotter product formula shows that

$$\langle q_T | e^{-\frac{i}{\hbar} H T} | q_0 \rangle = \lim_{N \to \infty} \langle q_T | (e^{-\frac{i}{N\hbar} K T} e^{-\frac{i}{N\hbar} V T})^N | q_0 \rangle \qquad (35.2)$$

If $K(P)$ can be chosen to depend only on the momentum operator P and $V(Q)$ depends only on the operator Q, then one can insert alternate copies of the identity operator in the forms

$$\int_{-\infty}^{\infty} |q\rangle\langle q|dq = 1, \quad \int_{-\infty}^{\infty} |p\rangle\langle p|dp = 1$$

This gives a product of terms of the form

$$\langle q_{t_j}|e^{-\frac{i}{N\hbar}K(P)T}|p_{t_j}\rangle\langle p_{t_j}|e^{-\frac{i}{N\hbar}V(Q)T}|q_{t_{j-1}}\rangle$$

where the index j goes from 0 to N, $t_j = jT/N$ and the variables q_{t_j} and p_{t_j} will be integrated over.

Such a term can be evaluated as

$$\langle q_{t_j}|p_{t_j}\rangle\langle p_{t_j}|q_{t_{j-1}}\rangle e^{-\frac{i}{N\hbar}K(p_{t_j})T} e^{-\frac{i}{N\hbar}V(q_{t_{j-1}})T}$$

$$= \frac{1}{\sqrt{2\pi\hbar}}e^{\frac{i}{\hbar}q_{t_j}p_{t_j}} \frac{1}{\sqrt{2\pi\hbar}} e^{-\frac{i}{\hbar}q_{t_{j-1}}p_{t_j}} e^{-\frac{i}{N\hbar}K(p_{t_j})T} e^{-\frac{i}{N\hbar}V(q_{t_{j-1}})T}$$

$$= \frac{1}{2\pi\hbar}e^{\frac{i}{\hbar}p_{t_j}(q_{t_j}-q_{t_{j-1}})}e^{-\frac{i}{N\hbar}(K(p_{t_j})+V(q_{t_{j-1}}))T}$$

The N factors of this kind give an overall factor of $(\frac{1}{2\pi\hbar})^N$ times something which is a discretized approximation to

$$e^{\frac{i}{\hbar}\int_0^T (p\dot{q}-h(q(t),p(t)))dt}$$

where the phase in the exponential is just the action. Taking into account the integrations over q_{t_j} and p_{t_j}, one should have something like

$$\langle q_T|e^{-\frac{i}{\hbar}HT}|q_0\rangle = \lim_{N\to\infty}(\frac{1}{2\pi\hbar})^N \prod_{j=1}^{N}\int_{-\infty}^{\infty}\int_{-\infty}^{\infty} dp_{t_j}dq_{t_j} e^{\frac{i}{\hbar}\int_0^T(p\dot{q}-h(q(t),p(t)))dt}$$

although one should not do the first and last integrals over q but fix the first value of q to q_0 and the last one to q_T. One can try and interpret this sort of integration in the limit as an integral over the space of paths in phase space, thus a "phase space path integral."

This is an extremely simple and seductive expression, apparently saying that, once the action S is specified, a quantum system is defined just by considering integrals

$$\int D\gamma \, e^{\frac{i}{\hbar}S[\gamma]}$$

over paths γ in phase space, where $D\gamma$ is some sort of measure on this space of paths. Since the integration just involves factors of $dpdq$ and the exponential just pdq and $h(p,q)$, this formalism seems to share the same sort of behavior under the infinite dimensional group of canonical transformations (transformations of the phase space preserving the Poisson bracket) as the classical Hamiltonian formalism. It also appears to solve our problem with operator ordering ambiguities, since the effect of products of P and Q operators at various times can be computed by computing path integrals with various p and q factors in the integrand. These integrand factors commute, giving just one way of producing products at equal times of any number of P and Q operators.

Unfortunately, we know from the Groenewold-van Hove theorem that this is too good to be true. This expression cannot give a unitary representation of the full group of canonical transformations, at least not one that is irreducible and restricts to what we want on transformations generated by linear functions q and p. Another way to see the problem is that a simple argument shows that by canonical transformations any Hamiltonian can be transformed into a free particle Hamiltonian, so all quantum systems would just be free particles in some choice of variables. For the details of these arguments and a careful examination of what goes wrong, see chapter 31 of [77]. One aspect of the problem is that for successive values of j the coordinates q_{t_j} or p_{t_j} have no reason to be close together. This is an integral over "paths" that do not acquire any expected continuity property as $N \to \infty$, so the answer one gets can depend on the details of the discretization chosen, reintroducing the operator ordering ambiguity problem.

One can intuitively see that there is something disturbing about such paths, since one is alternately at each time interval switching back and forth between a q-space representation where q has a fixed value and nothing is known about p, and a p space representation where p has a fixed value but nothing is known about q. The "paths" of the limit are objects with little relation to continuous paths in phase space, so while one may be able to define the limit of equation 35.2, it will not necessarily have any of the properties one expects of an integral over continuous paths.

When the Hamiltonian h is quadratic in the momentum p, the p_{t_j} integrals will be Gaussian integrals that can be performed exactly. Equivalently, the kinetic energy part K of the Hamiltonian operator will have a kernel in position space that can be computed exactly (see equation 12.9). As a result, the p_{t_j} integrals can be eliminated, along with the problematic use of alternating q-space and p-space representations. The remaining integrals over the q_{t_j} are then interpreted as a path integral over paths not in phase space, but in position space. One finds, if $K = \frac{P^2}{2m}$

$$\langle q_T | e^{-\frac{i}{\hbar} H T} | q_0 \rangle =$$

$$\lim_{N\to\infty}\left(\frac{i2\pi\hbar T}{Nm}\right)^{\frac{N}{2}}\sqrt{\frac{m}{i2\pi\hbar T}}\prod_{j=1}^{N}\int_{-\infty}^{\infty}dq_{t_j}\,e^{\frac{i}{\hbar}\sum_{j=1}^{N}\left(\frac{m(q_{t_j}-q_{t_{j-1}})^2}{2T/N}-V(q_{t_j})\frac{T}{N}\right)}$$

In the limit $N \to \infty$ the phase of the exponential becomes

$$S(\gamma) = \int_0^T dt(\frac{1}{2}m(\dot{q}^2) - V(q(t)))$$

One can try and properly normalize things so that this limit becomes an integral

$$\int D\gamma\; e^{\frac{i}{\hbar}S[\gamma]} \tag{35.3}$$

where now the paths $\gamma(t)$ are paths in the position space.

An especially attractive aspect of this expression is that it provides a simple understanding of how classical behavior emerges in the classical limit as $\hbar \to 0$. The stationary phase approximation method for oscillatory integrals says that, for a function f with a single critical point at $x = x_c$ (i.e., $f'(x_c) = 0$) and for a small parameter ϵ, one has

$$\frac{1}{\sqrt{i2\pi\epsilon}}\int_{-\infty}^{+\infty} dx\; e^{if/\epsilon} = \frac{1}{\sqrt{f''(c)}}e^{if(x_c)/\epsilon}(1 + O(\epsilon))$$

Using the same principle for the infinite dimensional path integral, with $f = S$ the action functional on paths, and $\epsilon = \hbar$, one finds that for $\hbar \to 0$ the path integral will simplify to something that just depends on the classical trajectory, since by the principle of least action, this is the critical point of S.

Such position space path integrals do not have the problems of principle of phase space path integrals coming from the Groenewold-van Hove theorem, but they still have serious analytical problems since they involve an attempt to integrate a wildly oscillating phase over an infinite dimensional space. Away from the limit $\hbar \to 0$, it is not clear that whatever results one gets will be independent of the details of how one takes the limit to define the infinite dimensional integral or that one will naturally get a unitary result for the time evolution operator.

One method for making path integrals better defined is an analytic continuation in the time variable, as discussed in section 12.5 for the case of a free particle. In such a free particle case, replacing the use of equation 12.9 by equation 12.8 in the definition of the position space path integral, one finds that this leads to a well-defined measure on paths, Wiener measure. More generally, Wiener measure techniques can be used to define the path integral when the potential energy is nonzero, getting results that ultimately need to be analytically continued back to the physical time variable.

35.4 Advantages and disadvantages of the path integral

In summary, the path integral method has the following advantages:

- An intuitive picture of the classical limit and a calculational method for "semi-classical" effects (quantum effects at small $\hbar$).

- Calculations for free particles or potentials V at most quadratic in q can be done just using Gaussian integrals, and these are relatively easy to evaluate and make sense of, despite the infinite dimensionality of the space of paths. For higher-order terms in $V(q)$, one can get a series expansion by expanding out the exponential, giving terms that are moments of Gaussians so can be evaluated exactly.

- After analytical continuation, path integrals can be rigorously defined using Wiener measure techniques and often evaluated numerically even in cases where no exact solution is known.

On the other hand, there are disadvantages:

- Some path integrals such as phase space path integrals do not at all have the properties one might expect for an integral, so great care is required in any use of them.

- How to get unitary results can be quite unclear. The analytic continuation necessary to make path integrals well defined can make their physical interpretation obscure.

- Symmetries with their origin in symmetries of phase space that are not just symmetries of configuration space are difficult to see using the configuration space path integral, with the harmonic oscillator providing a good example. Such symmetries can be seen using the phase space path integral, but this is not reliable.

Path integrals for anticommuting variables can also be defined by analogy with the bosonic case, using the notion of fermionic integration discussed earlier. Such fermionic path integrals will usually be analogs of the phase space path integral, but in the fermionic case there are no points and no problem of continuity of paths. In this case, the "integral" is not really an integral, but rather an algebraic operation with some of the same properties, and it will not obviously suffer from the same problems as the phase space path integral.

35.5   For further reading

For much more about Lagrangian mechanics and its relation to the Hamiltonian formalism, see [2]. More details along the lines of the discussion here can be found in most quantum mechanics and quantum field theory textbooks. An extensive discussion at an introductory level of the Lagrangian formalism and the use of Noether's theorem to find conserved quantities when it is invariant under a group action can be found in [63]. For the formalism of constrained Hamiltonian dynamics, see [89], and for a review article about the covariant phase space and the Peierls bracket, see [50].

For the path integral, Feynman's original paper [21] or his book [24] is quite readable. A typical textbook discussion is the one in Chapter 8 of Shankar [81]. The book by Schulman [77] has quite a bit more detail, both about applications and about the problems of phase space path integrals. Yet another fairly comprehensive treatment, including the fermionic case, is the book by Zinn-Justin [109].

Chapter 36
Multiparticle Systems: Momentum Space Description

In chapter 9, we saw how to use symmetric or antisymmetric tensor products to describe a fixed number of identical quantum systems (for instance, free particles). From very early on in the history of quantum mechanics, it became clear that at least certain kinds of quantum particles, photons, required a formalism that could describe arbitrary numbers of particles, as well as phenomena involving their creation and annihilation. This could be accomplished by thinking of photons as quantized excitations of a classical electromagnetic field. In our modern understanding of fundamental physics, all elementary particles, not just photons, are best described in this way, by quantum theories of fields. For free particles, the necessary theory can be understood as the quantum theory of the harmonic oscillator, but with an infinite number of degrees of freedom, one for each possible value of the momentum (or, Fourier transforming, each possible value of the position). The symmetric (bosons) or antisymmetric (fermions) nature of multiparticle quantum states is automatic in such a description as quanta of oscillators.

Conventional textbooks on quantum field theory often begin with relativistic systems, but we will start instead with the non-relativistic case. This is significantly simpler, lacking the phenomenon of antiparticles that appears in the relativistic case. It is also the case of relevance to condensed matter physics and applies equally well to bosonic or fermionic particles.

Quantum field theory is a large and complicated subject, suitable for a full-year course at an advanced level. We will be giving only a very basic introduction, mostly just considering free fields, which correspond to systems of non-interacting particles. Much of the complexity of the subject only appears when one tries to construct quantum field theories of interacting particles.

For simplicity, we will start with the case of a single spatial dimension. We will also begin using x to denote a spatial variable instead of the q conventional when this is the coordinate variable in a finite dimensional phase

© Peter Woit 2017
P. Woit, *Quantum Theory, Groups and Representations*,
DOI 10.1007/978-3-319-64612-1_36

space. In quantum field theory, position or momentum variables parametrize the fundamental degrees of freedom, the field variables, rather than providing such degrees of freedom themselves. In this chapter, emphasis will be on the momentum parametrization and the description of collections of free particles in terms of quanta of degrees of freedom labeled by momenta.

36.1 Multiparticle quantum systems as quanta of a harmonic oscillator

It turns out that quantum systems of identical particles are best understood by thinking of such particles as quanta of a harmonic oscillator system. We will begin with the bosonic case and then later consider the fermionic case, which uses the fermionic oscillator system.

36.1.1 Bosons and the quantum harmonic oscillator

A fundamental postulate of quantum mechanics (see chapter 9) is that given a space of states $\mathcal{H}_1$ describing a bosonic single particle, a collection of N identical such particles has state space

$$S^N(\mathcal{H}_1) = \underbrace{(\mathcal{H}_1 \otimes \cdots \otimes \mathcal{H}_1)}_{N-times}{}^S$$

where the superscript S means we take elements of the tensor product invariant under the action of the group S_N by permutation of the N factors. To describe states that include superpositions of an arbitrary number of identical particles, one should take as state space the sum of these

$$S^*(\mathcal{H}_1) = \mathbf{C} \oplus \mathcal{H}_1 \oplus (\mathcal{H}_1 \otimes \mathcal{H}_1)^S \oplus \cdots \tag{36.1}$$

This same symmetric part of a tensor product occurs in the Bargmann–Fock construction of the quantum state space for a phase space $M = \mathbf{R}^{2d}$, where the Fock space $\mathcal{F}_d$ can be described in three different but isomorphic ways. Note that we generally will not take care to distinguish here between $\mathcal{F}_d^{fin}$ (superpositions of states with finite number of quanta) and its completion $\mathcal{F}_d$ (which includes states with an infinite number of quanta). The three different descriptions of $\mathcal{F}_d$ are:

- $\mathcal{F}_d$ has an orthonormal basis

$$|n_1, n_2, \cdots, n_d\rangle$$

labeled by the eigenvalues n_j of the number operators N_j for $j = 1, \cdots, d$. Here, $n_j \in \{0, 1, 2, \cdots\}$ and

$$n = \sum_{j=1}^{d} n_j$$

is finite. This is called the "occupation number" basis of $\mathcal{F}_d$. In this basis, the annihilation and creation operators are

$$a_j |n_1, n_2, \cdots, n_d\rangle = \sqrt{n_j} |n_1, n_2, \cdots, n_j - 1, \cdots, n_d\rangle$$

$$a_j^\dagger |n_1, n_2, \cdots, n_d\rangle = \sqrt{n_j + 1} |n_1, n_2, \cdots, n_j + 1, \cdots, n_d\rangle$$

- $\mathcal{F}_d$ is the space of polynomials $\mathbf{C}[z_1, z_2, \cdots, z_d]$ in d complex variables z_j, with inner product the d-dimensional version of equation 22.4 and orthonormal basis elements corresponding to the occupation number basis the monomials

$$\frac{1}{\sqrt{n_1! \cdots n_d!}} z_1^{n_1} z_2^{n_2} \cdots z_d^{n_d} \tag{36.2}$$

Here, the annihilation and creation operators are

$$a_j = \frac{\partial}{\partial z_j}, \quad a_j^\dagger = z_j$$

- $\mathcal{F}_d$ is the algebra

$$S^*(\mathcal{M}_J^+) = \mathbf{C} \oplus \mathcal{M}_J^+ \oplus (\mathcal{M}_J^+ \otimes \mathcal{M}_J^+)^S \oplus \cdots \tag{36.3}$$

with product the symmetrized tensor product given by equation 9.3. Here, $\mathcal{M}_J^+$ is the space of complex linear functions on M that are eigenvectors with eigenvalue $+i$ for the complex structure J. Using the isomorphism between monomials and symmetric tensor products (given on monomials in one variable by equation 9.4), the monomials 36.2 provide an orthonormal basis of this space. Expressions for the annihilation and creation operators acting on $S^*(\mathcal{M}_J^+)$ can be found using this isomorphism, using their action on monomials as derivative and multiplication operators.

For each of these descriptions of $\mathcal{F}_d$, the choice of orthonormal basis elements given above provides an inner product, with the annihilation and creation operators $a_j, a_j^\dagger$ each other's adjoints, satisfying the canonical commutation relations

$$[a_j, a_k^\dagger] = \delta_{jk}$$

We will describe the basis state $|n_1, n_2, \cdots, n_d\rangle$ as one containing n_1 quanta of type 1, n_2 quanta of type 2, etc., and a total number of quanta n.

Comparing 36.1 and 36.3, we see that these are the same state spaces if $\mathcal{H}_1 = \mathcal{M}_J^+$. This construction of multiparticle states by taking as dual classical phase space $\mathcal{M}$ a space of solutions to a wave equation, then quantizing by the Bargmann–Fock method, with $\mathcal{H}_1 = \mathcal{M}_J^+$ the quantum state space for a single particle, is sometimes known as "second quantization." Choosing a (complex) basis of $\mathcal{H}_1$, for each basis element one gets an independent quantum harmonic oscillator, with corresponding occupation number, the number of "quanta" labeled by that basis element. This formalism automatically implies indistinguishability of quanta and symmetry under interchange of quanta since only the numbers of quanta appear in the description of the state. The separate symmetry postulate needed in the conventional quantum mechanical description of multiple identical particles by tensor products is no longer needed.

In chapter 43, we will see that in the case of relativistic scalar quantum field theory, the dual phase space $\mathcal{M}$ will be the space of real solutions of an equation called the Klein–Gordon equation. The J needed for Bargmann–Fock quantization will be determined by the decomposition into positive and negative energy solutions, and $\mathcal{H}_1 = \mathcal{M}_J^+$ will be a space of states describing a single relativistic particle.

In this chapter, we will consider a non-relativistic theory, with wave equation the Schrödinger equation. Here, the dual phase space $\mathcal{M}$ of complex solutions will already be a complex vector space, and we can use the version of Bargmann–Fock quantization described in section 26.4, so $\mathcal{H}_1 = \mathcal{M}$. The "second quantization" terminology is appropriate, since we take as (dual) classical phase space a quantum state space, and quantize that.

36.1.2 Fermions and the fermionic oscillator

For the case of fermionic particles, if $\mathcal{H}_1$ is the state space for a single particle, an arbitrary number of particles will be described by the state space

$$\Lambda^*(\mathcal{H}_1) = \mathbf{C} \oplus \mathcal{H}_1 \oplus (\mathcal{H}_1 \otimes \mathcal{H}_1)^A \oplus \cdots$$

where (unlike the bosonic case) this is a finite sum if $\mathcal{H}_1$ is finite dimensional. One can proceed as in the bosonic case, using instead of $\mathcal{F}_d$ the fermionic oscillator state space $\mathcal{H}_F = \mathcal{F}_d^+$. This again has three isomorphic descriptions:

- $\mathcal{F}_d^+$ has an orthonormal basis

$$|n_1, n_2, \cdots, n_d\rangle$$

labeled by the eigenvalues of the number operators N_j for $j = 1, \cdots, d$ where $n_j \in \{0, 1\}$. This is called the "occupation number" basis of $\mathcal{F}_d^+$.

- $\mathcal{F}_d^+$ is the Grassmann algebra $\mathbf{C}[\theta_1, \theta_2, \cdots, \theta_d]$ (see section 30.1) of polynomials in d anticommuting complex variables, with orthonormal basis elements corresponding to the occupation number basis the monomials

$$\theta_1^{n_1} \theta_2^{n_2} \cdots \theta_d^{n_d} \tag{36.4}$$

- $\mathcal{F}_d^+$ is the algebra of antisymmetric multilinear forms

$$\Lambda^*(\mathcal{V}_J^+) = \mathbf{C} \oplus \mathcal{V}_J^+ \oplus (\mathcal{V}_J^+ \otimes \mathcal{V}_J^+)^A \oplus \cdots$$

discussed in section 9.6, with product the wedge product (see equation 9.6). Here, $\mathcal{V}_J^+$ is the space of complex linear functions on a vector space $V = \mathbf{R}^{2d}$ (the pseudo-classical phase space), eigenvectors with eigenvalue $+i$ for the complex structure J.

For each of these descriptions of $\mathcal{F}_d^+$, we have basis elements we can take to be orthonormal, providing an inner product on $\mathcal{F}_d^+$. We also have a set of d annihilation and creation operators $a_{Fj}, a_{F_j}^\dagger$ that are each other's adjoints, and satisfy the canonical anticommutation relations

$$[a_{Fj}, a_{F_k}^\dagger]_+ = \delta_{jk}$$

We will describe the basis state $|n_1, n_2, \cdots, n_d\rangle$ as one containing n_1 quanta of type 1, n_2 quanta of type 2, etc., and a total number of quanta

$$n = \sum_{j=1}^{d} n_j$$

Analogously to the bosonic case, a multiparticle fermionic theory can be constructed using $\mathcal{F}_d^+$, by taking $\mathcal{V}_J^+ = \mathcal{H}_1$. This is a fermionic version of second quantization, with the multiparticle state space given by quantization of a pseudo-classical dual phase space $\mathcal{V}$ of solutions to some wave equation. The formalism automatically implies the Pauli principle (no more than one quantum per state) as well as the antisymmetry property for states of multiple fermionic quanta that is a separate postulate in our earlier description of multiple particle states as tensor products.

36.2 Multiparticle quantum systems of free particles: finite cutoff formalism

To describe multiparticle quantum systems in terms of quanta of a harmonic oscillator system, we would like to proceed as described in section 36.1, taking solutions to the free particle Schrödinger equation (discussed in chapters 10 and 11) as the single-particle state space. Recall that for a free particle in one spatial dimension, such solutions are given by complex-valued functions on $\mathbf{R}$, with observables the self-adjoint operators for momentum

$$P = -i\frac{d}{dx}$$

and energy (the Hamiltonian)

$$H = \frac{P^2}{2m} = -\frac{1}{2m}\frac{d^2}{dx^2}$$

Eigenfunctions for both P and H are the functions of the form

$$\psi_p(x) \propto e^{ipx}$$

for $p \in \mathbf{R}$, with eigenvalues p for P and $\frac{p^2}{2m}$ for H. Recall that these eigenfunctions are not normalizable and thus not in the conventional choice of state space as $L^2(\mathbf{R})$.

As we saw in section 11.1, one way to deal with this issue is to do what physicists sometimes refer to as "putting the system in a box," by imposing periodic boundary conditions

$$\psi(x + L) = \psi(x)$$

for some number L, effectively restricting the relevant values of x to be considered to those on an interval of length L. For our eigenfunctions, this condition is

$$e^{ip(x+L)} = e^{ipx}$$

so we must have

$$e^{ipL} = 1$$

which implies that

$$p = \frac{2\pi}{L}j \equiv p_j$$

for j an integer. Then, the momentum will take on a countable number of discrete values corresponding to the $j \in \mathbf{Z}$, and

$$|j\rangle = \psi_j(x) = \frac{1}{\sqrt{L}} e^{ip_j x} = \frac{1}{\sqrt{L}} e^{i\frac{2\pi j}{L}x}$$

will be orthonormal eigenfunctions satisfying

$$\langle j'|j\rangle = \delta_{jj'}$$

This use of periodic boundary conditions is one form of what physicists call an "infrared cutoff," a way of removing degrees of freedom that correspond to arbitrarily large sizes, in order to make the quantum system well defined. One starts with a fixed value of L and only later studies the limit $L \to \infty$.

The number of degrees of freedom is now countable, but still infinite, and something more must be done in order to make the single-particle state space finite dimensional. This can be accomplished with an additional cutoff, an "ultraviolet cutoff," which means restricting attention to $|p| \leq \Lambda$ for some finite Λ, or equivalently $|j| < \frac{\Lambda L}{2\pi}$. This makes the space of solutions finite dimensional, allowing quantization by the use of the Bargmann–Fock method used for the finite dimensional harmonic oscillator. The $\Lambda \to \infty$ and $L \to \infty$ limits can then be taken at the end of a calculation.

The Schrödinger equation is a first-order differential equation in time, t, and solutions can be completely characterized by their initial value at $t = 0$

$$\psi(x,0) = \sum_{j=-\frac{\Lambda L}{2\pi}}^{+\frac{\Lambda L}{2\pi}} \alpha(p_j) e^{i\frac{2\pi j}{L}x}$$

determined by a choice of complex coefficients $\alpha = \{\alpha(p_j)\}$. At later times, the solution will be given by

$$\psi(x,t) = \sum_{j=-\frac{\Lambda L}{2\pi}}^{+\frac{\Lambda L}{2\pi}} \alpha(p_j) e^{ip_j x} e^{-i\frac{p_j^2}{2m}t} \tag{36.5}$$

Our space of solutions is the space of all sets of complex numbers $\alpha(p_j)$. In principle, we could take this space as our dual phase space and quantize using the Schrödinger representation, for instance taking the real parts of the $\alpha(p_j)$ as position-like coordinates. Especially since our dual phase space is already complex, it is much more convenient to use the Bargmann–Fock method of quantization. Recalling the discussion of section 26.4, we will need both a dual phase space $\mathcal{M}$ and its conjugate space $\overline{\mathcal{M}}$, which means that we will need to consider not just solutions of the Schrödinger equation, but of its conjugate

$$-i\frac{\partial}{\partial t}\overline{\psi} = -\frac{1}{2m}\frac{\partial^2}{\partial x^2}\overline{\psi} \tag{36.6}$$

which is satisfied by conjugates $\overline{\psi}$ of solutions ψ of the usual Schrödinger equation. We will take $\mathcal{M} = \mathcal{H}_1$ to be the space of Schrödinger equation solutions 36.5. $\overline{\mathcal{M}} = \overline{\mathcal{H}}_1$ will be the space of solutions of 36.6, which can be written

$$\sum_{j=-\frac{\Delta L}{2\pi}}^{+\frac{\Delta L}{2\pi}} \overline{\alpha}(p_j)e^{-ip_j x}e^{i\frac{p_j^2}{2m}t}$$

for some complex numbers $\overline{\alpha}(p_j)$.

A basis for $\mathcal{M}$ will be given by the

$$A(p_j) = \begin{cases} \alpha(p_k) = 0 & k \neq j \\ \alpha(p_k) = 1 & k = j \end{cases}$$

with conjugates $\overline{A}(p_j)$ a basis for $\overline{\mathcal{M}}$. The Poisson bracket on $\mathcal{M} \oplus \overline{\mathcal{M}}$ will be determined by the following Poisson bracket relations on basis elements

$$\{A(p_j), A(p_k)\} = \{\overline{A}(p_j), \overline{A}(p_k)\} = 0, \quad \{A(p_j), \overline{A}(p_k)\} = i\delta_{jk} \tag{36.7}$$

Bargmann–Fock quantization gives as state space a Fock space $\mathcal{F}_D$, where D is the number of values of p_j. This is (ignoring issues of completion) the space of polynomials in the D variables $A(p_j)$. One has a pair of annihilation and creation operators

$$a(p_j) = \frac{\partial}{\partial A(p_j)}, \quad a(p_j)^\dagger = A(p_j)$$

for each possible value of j, which indexes the possible values p_j. These operators satisfy the commutation relations

$$[a(p_j), a(p_j)^\dagger] = \delta_{jk}$$

In the occupation number representation of the Fock space, orthonormal basis elements are

$$|\cdots, n_{p_{j-1}}, n_{p_j}, n_{p_{j+1}}, \cdots\rangle$$

with annihilation and creation operators acting by

$$a_{p_j}|\cdots, n_{p_{j-1}}, n_{p_j}, n_{p_{j+1}}, \cdots\rangle = \sqrt{n_{p_j}}|\cdots, n_{p_{j-1}}, n_{p_j} - 1, n_{p_{j+1}}, \cdots\rangle$$

$$a_{p_j}^\dagger|\cdots, n_{p_{j-1}}, n_{p_j}, n_{p_{j+1}}, \cdots\rangle = \sqrt{n_{p_j} + 1}|\cdots, n_{p_{j-1}}, n_{p_j} + 1, n_{p_{j+1}}, \cdots\rangle$$

The occupation number n_{p_j} is the eigenvalue of the operator $a(p_j)^\dagger a(p_j)$ and takes values $0, 1, 2, \cdots, \infty$. It has a physical interpretation as the number of particles in the state with momentum p_j (recall that such momentum values are discretized in units of $\frac{2\pi}{L}$ and in the interval $[-\Lambda, \Lambda]$). The state with all occupation numbers equal to zero is denoted

$$|\cdots, 0, 0, 0, \cdots\rangle = |0\rangle$$

and called the "vacuum" state.

Observables that can be built out of the annihilation and creation operators include

- The total number operator

$$\widehat{N} = \sum_k a(p_k)^\dagger a(p_k) \tag{36.8}$$

which will have as eigenvalues the total number of particles

$$\widehat{N}|\cdots, n_{p_{j-1}}, n_{p_j}, n_{p_{j+1}}, \cdots\rangle = \left(\sum_k n_{p_k}\right)|\cdots, n_{p_{j-1}}, n_{p_j}, n_{p_{j+1}}, \cdots\rangle$$

- The momentum operator

$$\widehat{P} = \sum_k p_k a(p_k)^\dagger a(p_k) \tag{36.9}$$

with eigenvalues the total momentum of the multiparticle system

$$\widehat{P}|\cdots, n_{p_{j-1}}, n_{p_j}, n_{p_{j+1}}, \cdots\rangle = \left(\sum_k n_{p_k} p_k\right)|\cdots, n_{p_{j-1}}, n_{p_j}, n_{p_{j+1}}, \cdots\rangle$$

- The Hamiltonian

$$\widehat{H} = \sum_k \frac{p_k^2}{2m} a(p_k)^\dagger a(p_k) \tag{36.10}$$

which has eigenvalues the total energy

$$\widehat{H}|\cdots, n_{p_{j-1}}, n_{p_j}, n_{p_{j+1}}, \cdots\rangle = \left(\sum_k n_{p_k} \frac{p_k^2}{2m}\right)|\cdots, n_{p_{j-1}}, n_{p_j}, n_{p_{j+1}}, \cdots\rangle$$

With ultraviolet and infrared cutoffs in place, the possible values of p_j are of a finite number D which is also the complex dimension of $\mathcal{H}_1$. The Hamiltonian operator is the standard harmonic oscillator Hamiltonian, with different frequencies

$$\omega_j = \frac{p_j^2}{2m}$$

for different values of j. Note that we are using normal-ordered operators here, which is necessary since, in the limit as one or both cutoffs are removed, $\mathcal{H}_1$ becomes infinite dimensional, and only the normal-ordered version of the Hamiltonian used here is well defined (the non-normal-ordered version will differ by an infinite sum of $\frac{1}{2}$s).

Everything in this section has a straightforward analog describing a multi-particle system of fermionic particles with energy–momentum relation given by the free particle Schrödinger equation. The annihilation and creation operators will be the fermionic ones, satisfying the canonical anticommutation relations

$$[a_{F\,p_j}, a_{F\,p_k}^\dagger]_+ = \delta_{jk}$$

implying that states will have occupation numbers $n_{p_j} = 0, 1$, automatically implementing the Pauli principle.

36.3 Continuum formalism

The use of cutoffs allows for a finite dimensional phase space and makes it possible to straightforwardly use the Bargmann–Fock quantization method. Such cutoffs, however, introduce very significant problems, by making unavailable some of the continuum symmetries and mathematical structures that we would like to exploit. In particular, the use of an infrared cutoff (periodic boundary conditions) makes the momentum space a discrete set of points, and this set of points will not have the same symmetries as the usual continuous momentum space (for instance, in three dimensions, it will not carry an action of the rotation group $SO(3)$). In our study of quantum field theory, we would like to exploit the action of space–time symmetry groups on the state space of the theory, so need a formalism that preserves such symmetries. In this section, we will outline such a formalism, without attempting a detailed rigorous version. One reason for this choice is that for the case of physically interesting interacting quantum field theories, this continuum formalism is inadequate, since a rigorous definition will require first defining a finite, cut-off version, then using renormalization group methods to analyze the very non-trivial continuum limit.

If we try and work directly with the infinite dimensional space of solutions of the free Schrödinger equation, for the three forms of the Fock space construction discussed in section 36.1.1, we find:

- The occupation number construction of Fock space is not available (since it requires a discrete basis).

- For the Bargmann–Fock holomorphic function state space and inner product on it, one needs to make sense of holomorphic functions on an infinite dimensional space, as well as the Gaussian measure on this space. See section 36.6 for references that discuss this.

- For the symmetric tensor product representation, one needs to make sense of symmetric tensor products of infinite dimensional Hilbert spaces $\mathcal{H}_1$ and the induced Hilbert space structure on such tensor products. We will adopt this point of view here, with details available in the references of section 36.6.

In the continuum normalization, an arbitrary solution to the free particle Schrödinger equation is given by

$$\psi(x,t) = \frac{1}{\sqrt{2\pi}} \int_{-\infty}^{\infty} \alpha(p) e^{ipx} e^{-i\frac{p^2}{2m}t} dp \qquad (36.11)$$

At $t = 0$

$$\psi(x,0) = \frac{1}{\sqrt{2\pi}} \int_{-\infty}^{\infty} \alpha(p) e^{ipx} dp$$

which is the Fourier inversion formula, expressing a function $\psi(x,0)$ in terms of its Fourier transform, $\alpha(p) = \widetilde{\psi(x,0)}(p)$. We see that the functions $\alpha(p)$ parametrize initial data and $\mathcal{H}_1$, the solution space of the free particle Schrödinger equation, can be identified with the space of such α.

Using the notation $A(\alpha)$ to denote the element of $\mathcal{H}_1$ determined by initial data α, in the Fock space description of multiparticle states as symmetric tensor products of $\mathcal{H}_1$, we have the following annihilation and creation operators (these were discussed in the finite dimensional case in section 26.4)

$$a^\dagger(\alpha) P^+ (A(\alpha_1) \otimes \cdots \otimes A(\alpha_n)) = \sqrt{n+1} P^+ (A(\alpha) \otimes A(\alpha_1) \otimes \cdots \otimes A(\alpha_n)) \qquad (36.12)$$

$$a(\alpha) P^+ (A(\alpha_1) \otimes \cdots \otimes A(\alpha_n)) =$$
$$\frac{1}{\sqrt{n}} \sum_{j=1}^{n} \langle \alpha, \alpha_j \rangle P^+ (A(\alpha_1) \otimes \cdots \otimes \widehat{A(\alpha_j)} \otimes \cdots \otimes A(\alpha_n))$$

$$(36.13)$$

(the $\widehat{A(\alpha_j)}$ means omit that term in the tensor product, and P^+ is the symmetrization operator defined in section 9.6) satisfying the commutation relations

$$[a(\alpha_1), a(\alpha_2)] = [a^\dagger(\alpha_1), a^\dagger(\alpha_2)] = 0$$

$$[a(\alpha_1), a^\dagger(\alpha_2)] = \langle \alpha_1, \alpha_2 \rangle = \int \overline{\alpha_1(p)} \alpha_2(p) dp \qquad (36.14)$$

Different choices for which space of functions α to take as $\mathcal{H}_1$ lead to different problems. Three possibilities are:

- $\mathcal{H}_1 = L^2(\mathbf{R})$

  This choice allows for an isomorphism between $\mathcal{H}_1$ and its dual, using the inner product

$$\langle \alpha_1, \alpha_2 \rangle = \int \overline{\alpha_1(p)} \alpha_2(p) dp$$

  As in the single-particle case the problem here is that position

$$\alpha(p) = \frac{1}{\sqrt{2\pi}} e^{ipx'}$$

  and momentum

$$\alpha(p) = \delta(p - p')$$

  eigenstates are not in $\mathcal{H}_1 = L^2(\mathbf{R})$. In addition, as in the single-particle case, there are domain issues to consider, since differentiating by p or multiplying by p can take something in $L^2(\mathbf{R})$ to something not in $L^2(\mathbf{R})$.

- $\mathcal{H}_1 = \mathcal{S}(\mathbf{R})$

  This choice, taking α to be in the well-behaved space of Schwartz functions, avoids the domain issues of $L^2(\mathbf{R})$ and the Hermitian inner product is well defined. It, however, shares the problem with $L^2(\mathbf{R})$ of not including position or momentum eigenstates. In addition, the inner product no longer provides an isomorphism of $\mathcal{H}_1$ with its dual.

- $\mathcal{H}_1 = \mathcal{S}'(\mathbf{R})$

  This choice, allowing α to be distributional solutions, will solve domain issues and includes position and momentum eigenstates. It, however, introduces a serious problem: The Hermitian inner product on functions does not extend to distributions. With this choice, $\mathcal{H}_1$ is not an inner product space and neither are its symmetric tensor products.

To get a rigorous mathematical formalism, for some purposes it is possible to adopt the first choice, $\mathcal{H}_1 = L^2(\mathbf{R})$. With this choice, the symmetric tensor product version of the Fock space can be given a Hilbert space structure, with operators $a^\dagger(\alpha)$ and $a(\alpha)$ defined by equations 36.12 and 36.13 satisfying the Heisenberg commutation relations of equation 36.14. We will, however, want to consider operators quadratic in the $a(\alpha)$ and $a^\dagger(\alpha)$, and for these to be well defined, we may need to use $\mathcal{H}_1 = \mathcal{S}(\mathbf{R})$.

If we ignore the problem with the inner product, and take $\mathcal{H}_1 = \mathcal{S}'(\mathbf{R})$, then in particular we can take α to be a delta-function, and when doing this, we will use the notation

$$a(p) = a(\delta(p' - p)), \quad a^\dagger(p) = a^\dagger(\delta(p' - p))$$

and write

$$a(\alpha) = \int \overline{\alpha(p)} a(p) dp, \quad a^\dagger(\alpha) = \int \alpha(p) a^\dagger(p) dp \qquad (36.15)$$

The choice of the conjugations here reflects that fact that $a^\dagger(\alpha)$ is complex linear in α, $a(\alpha)$ complex antilinear.

While the operator $a(p)$ may be well defined, the problem with the operator $a^\dagger(p)$ is clear: It takes in particular the vacuum state $|0\rangle$ to the non-normalizable state $|p\rangle$. We will, like most other authors, often write equations in terms of operators $a(p)$ and $a^\dagger(p)$, acting as if $\mathcal{H}_1 = \mathcal{S}'(\mathbf{R})$. For a legitimate interpretation though, such equations will always require an interpretation either

- using cutoffs which make the values of p discrete and of finite number, p_j labeled by an index j with a finite number of values, as in section 36.2. In this case, the $a(p), a^\dagger(p)$ are the $a(p_j), a^\dagger(p_j)$ of that section.

- using equations 36.15 formally, with $a(\alpha)$ and $a^\dagger(\alpha)$ the objects that are well defined, for some specified class of functions α, generally $\mathcal{S}(\mathbf{R})$. In this case, the $a(p), a^\dagger(p)$ are often described as "operator-valued distributions."

The nonzero commutators of $a(p), a^\dagger(p)$ can be written as

$$[a(p), a^\dagger(p')] = \delta(p - p')$$

a formula that should be interpreted as meaning either a continuum limit of

$$[a(p_j), a^\dagger(p_k)] = \delta_{jk}$$

or

$$[a(\alpha_1), a^\dagger(\alpha_2)] = \left[\int \overline{\alpha_1(p)} a(p) dp, \int \alpha_2(p') a(p') dp' \right]$$

$$= \int \int \overline{\alpha_1(p)} \alpha_2(p') \delta(p - p') dp dp'$$

$$= \int \overline{\alpha_1(p)} \alpha_2(p) dp = \langle \alpha_1, \alpha_2 \rangle$$

for some class (e.g., $\mathcal{S}(\mathbf{R})$ or $L^2(\mathbf{R})$) of functions for which the inner product of α_1 and α_2 makes sense.

While we have defined here first the quantum theory in terms of a state space and operators, one could instead start by writing down a classical theory, with dual phase space $\mathcal{H}_1$. This is already a complex vector space with Hermitian inner product, so we are in the situation described for the finite dimensional case in section 26.4. We need to apply Bargmann–Fock quantization in the manner described there, introducing a complex conjugate space $\overline{\mathcal{H}}_1$, as well as a symplectic structure and indefinite Hermitian inner product on $\mathcal{H}_1 \oplus \overline{\mathcal{H}}_1$. Restricted to $\mathcal{H}_1$, the Hermitian inner product will be the given one, and the symplectic structure will be its imaginary part.

If we denote by $A(\alpha) \in \mathcal{H}_1$ the solution of the Schrödinger equation with Fourier transform of initial data given by α, and by $\overline{A}(\alpha) \in \overline{\mathcal{H}}_1$ the conjugate solution of the conjugate Schrödinger equation, the Poisson bracket relations are then

$$\{A(\alpha_1), A(\alpha_2)\} = \{\overline{A}(\alpha_1), \overline{A}(\alpha_2)\} = 0, \quad \{A(\alpha_1), \overline{A}(\alpha_2)\} = i\langle \alpha_1, \alpha_2\rangle$$
(36.16)

Quantization then takes

$$A(\alpha) \to -ia^\dagger(\alpha), \quad \overline{A}(\alpha) \to -ia(\alpha)$$

where $a^\dagger(\alpha)$ and $a(\alpha)$ are given by equations 36.12 and 36.13. This gives a representation of the Lie algebra relations 36.16 for an infinite dimensional Heisenberg Lie algebra.

As with annihilation and creation operators, adopting a notation that formally extends the state space to $\mathcal{H}_1 = \mathcal{S}'(\mathbf{R})$, we define

$$A(p) = A(\delta(p' - p)), \quad \overline{A}(p) = \overline{A}(\delta(p' - p))$$

$$A(\alpha) = \int \alpha(p) A(p) dp, \quad \overline{A}(\alpha) = \int \overline{\alpha(p)\overline{A}(p)} dp$$

with Poisson bracket relations written

$$\{A(p), A(p')\} = \{\overline{A}(p), \overline{A}(p'\} = 0, \quad \{A(p), \overline{A}(p')\} = i\delta(p - p')$$
(36.17)

To get observables, we would like to define quadratic products of operators such as

$$\widehat{N} = \int_{-\infty}^{+\infty} a(p)^\dagger a(p) dp$$

for the number operator,

$$\widehat{P} = \int_{-\infty}^{+\infty} p a(p)^\dagger a(p) dp$$

for the momentum operator, and

$$\widehat{H} = \int_{-\infty}^{+\infty} \frac{p^2}{2m} a(p)^\dagger a(p) dp$$

for the Hamiltonian operator. One way to make rigorous sense of these is as limits of the operators 36.8, 36.9, and 36.10. Another is as bilinear forms on $S^*(\mathcal{H}_1) \times S^*(\mathcal{H}_1)$ for $\mathcal{H}_1 = \mathcal{S}(\mathbf{R})$, sending pairs of states $|\phi_1\rangle, |\phi_2\rangle$ to, for instance, $\langle \phi_1 | \widehat{N} | \phi_2 \rangle$ (for details see [17], section 5.4.2).

36.4 Multiparticle wavefunctions

To recover the conventional formalism in which an N-particle state is described by a wavefunction

$$\psi_N(p_1, p_2, \cdots, p_N)$$

symmetric in the N arguments, one needs to recall (see chapter 9) that the tensor product of the vector space of functions on a set X_1 and the vector space of functions on a set X_2 is the vector space of functions on the product set $X_1 \times X_2$. The symmetric tensor product will be the symmetric functions. Applying this to whatever space $\mathcal{H}_1$ of functions on $\mathbf{R}$ we choose to use, the symmetric tensor product $S^N(\mathcal{H}_1)$ will be a space of symmetric functions on $\mathbf{R}^N$. For details of this construction, see, for instance, chapter 5 of [17].

From the point of view of distributional operators $a(p), a^\dagger(p)$, given an arbitrary state $|\psi\rangle$ in the multiparticle state space, the momentum space wavefunction component with particle number N can be expressed as

$$\psi_N(p_1, p_2, \cdots, p_N) = \langle 0 | a(p_1) a(p_2) \cdots a(p_N) | \psi \rangle$$

36.5 Dynamics

To describe the time evolution of a quantum field theory system, it is generally easier to work with the Heisenberg picture (in which the time-dependence is in the operators) than the Schrödinger picture (in which the time-dependence is in the states). This is especially true in relativistic systems where one wants to as much as possible treat space and time on the same footing. It is, however, also true in the case of non-relativistic multiparticle systems due to the complexity of the description of the states (inherent since one is trying to describe arbitrary numbers of particles) versus the description of the operators, which are built simply out of the annihilation and creation operators.

In the Heisenberg picture, the time evolution of an operator $\widehat{f}$ is given by

$$\widehat{f}(t) = e^{iHt}\widehat{f}(0)e^{-iHt}$$

and such operators satisfy the differential equation

$$\frac{d}{dt}\widehat{f} = [\widehat{f}, -iH]$$

For the operators that create and annihilate states with momentum p_j in the finite cutoff formalism of section 36.2, $H = \widehat{H}$ (given by equation 36.10) and we have

$$\frac{d}{dt}a^{\dagger}(p_j, t) = [a^{\dagger}(p_j, t), -i\widehat{H}] = i\frac{p_j^2}{2m}a^{\dagger}(p_j, t)$$

with solutions

$$a^{\dagger}(p_j, t) = e^{i\frac{p_j^2}{2m}t}a^{\dagger}(p_j, 0) \tag{36.18}$$

Recall that in classical Hamiltonian mechanics, the Hamiltonian function h determines how an observable f evolves in time by the differential equation

$$\frac{d}{dt}f = \{f, h\}$$

Quantization takes f to an operator $\widehat{f}$ and h to a self-adjoint operator H.

In our case, the "classical" dynamical equation is meant to be the Schrödinger equation. In the finite cutoff formalism, one can take as Hamiltonian

$$h = \sum_j \frac{p_j^2}{2m}A(p_j)\overline{A}(p_j)$$

Here, the $A(p_j)$ should be interpreted as linear functions that on a solution given by α take the value α_j, and h a quadratic function on solutions that takes the value

$$h(\alpha) = \sum_j \frac{p_j^2}{2m}\alpha(p_j)\overline{\alpha}(p_j)$$

Hamilton's equations are

$$\frac{d}{dt}A(p_j, t) = \{A(p_j, t), h\} = i\frac{p_j^2}{2m}A(p_j, t)$$

with solutions

$$A(p_j, t) = e^{i\frac{p_j^2}{2m}t}(p_j, 0)$$

In the continuum formalism, one can write

$$h = \int_{-\infty}^{+\infty} \frac{p^2}{2m} A(p)\overline{A}(p)dp$$

which should be interpreted as a limit of the finite cutoff version. We will not try and give a rigorous continuum interpretation of a quadratic product of distributions such as this one. As discussed at the end of section 36.3, the quantization of h can be given a rigorous interpretation as a bilinear form. We will, however, assume the $A(p_j, t)$ have as continuum limits distributions $A(p, t)$ that satisfy

$$\frac{d}{dt}A(p,t) = i\frac{p^2}{2m}A(p,t)$$

so have time-dependence

$$A(p,t) = e^{i\frac{p^2}{2m}t}A(p,0)$$

Note that this time-dependence is opposite to that of the Schrödinger solutions, since, as a distribution, $\frac{d}{dt}A(p,t)$ evaluated on a function f is $A(p,t)$ evaluated on $-\frac{d}{dt}f$.

36.6 For further reading

The material of this chapter is just the conventional multiparticle formalism described or implicit in most quantum field theory textbooks. Many do not explicitly discuss the non-relativistic case, two that do are [35] and [45]. Three books aimed at mathematicians that cover this subject much more comprehensively than done here are [28] and [17]. For example, section 4.5 of [28] gives a detailed description of the bosonic and fermionic Fock space constructions in terms of tensor products. For a rigorous version of the construction of annihilation and creation operators as operator-valued distributions, see, for instance, section 5.4 of [17], chapter X.7 of [73], or [64]. The construction of the Fock space in the infinite dimensional case using Bargmann–Fock methods of holomorphic functions (rather than tensor products) goes back to Berezin [6] and Segal (for whom it is the "complex-wave representation," see [4]) and is explained in chapter 6 of [61].

Some good sources for learning about quantum field theory from the point of view of non-relativistic many-body theory are Feynman's lecture notes on statistical mechanics [23], as well as [53] and [86].

Chapter 37
Multiparticle Systems and Field Quantization

The multiparticle formalism developed in chapter 36 is based on the idea of taking as dual phase space the space of solutions to the free particle Schrödinger equation, then quantizing using the Bargmann–Fock method. Continuous basis elements $A(p)$, $\overline{A}(p)$ are momentum operator eigenstates which after quantization become creation and annihilation operators $a^\dagger(p)$, $a(p)$. These act on the multiparticle state space by adding or subtracting a free particle of momentum p.

Instead of the solutions $A(p)$, which are momentum space delta-functions localized at p at $t = 0$, one could use solutions $\Psi(x)$ that are position space delta-functions localized at x at $t = 0$. Quantization using these basis elements of $\mathcal{H}_1$ (and conjugate basis elements of $\overline{\mathcal{H}}_1$) will give operators $\widehat{\Psi}^\dagger(x)$ and $\widehat{\Psi}(x)$, which are the quantum field operators. These are related to the operators $a^\dagger(p), a(p)$ by the Fourier transform.

The operators $\widehat{\Psi}(x)$ and $\widehat{\Psi}^\dagger(x)$ can be given a physical interpretation as acting on states by subtracting or adding a particle localized at x. Such states with localized position have the same problem of non-normalizability as momentum eigenstates. In addition, unlike states with fixed momentum, they are not stable energy eigenstates so will immediately evolve into something non-localized (just as in the single-particle case discussed in section 12.5). Quantum fields are, however, very useful in the study of theories of interacting particles, since the interactions in such theories are typically local, taking place at a point x and describable by adding terms to the Hamiltonian operator involving multiplying field operators at the same point x. The difficulties involved in properly defining products of these operators and calculating their dynamical effects will keep us from starting the study of such interacting theories.

© Peter Woit 2017

P. Woit, *Quantum Theory, Groups and Representations*,
DOI 10.1007/978-3-319-64612-1_37

37.1 Quantum field operators

The multiparticle formalism developed in chapter 36 works well to describe states of multiple free particles, but does so purely in terms of states with well-defined momenta, with no information at all about their position. Instead of starting with $t = 0$ momentum eigenstates $|p\rangle$ and corresponding Schrödinger solutions $A(p)$ as continuous basis elements of the single-particle space $\mathcal{H}_1$, the position operator Q eigenstates $|x\rangle$ could be used. The solution that is such an eigenstate at $t = 0$ will be denoted $\Psi(x)$; the conjugate solution will be written as $\overline{\Psi}(x)$. The $\Psi(x), \overline{\Psi}(x)$ can be thought of as the complex coordinates of an oscillator for each value of x.

The corresponding quantum state space would naively be a Fock space for an infinite number of degrees of freedom, with an occupation number for each value of x. This could be made well defined by introducing a spatial cutoff and discretizing space, so that x only takes on a finite number of values. However, such states in the occupation number basis would not be free particle energy eigenstates. While a state with a well-defined momentum evolves as a state with the same momentum, a state with well-defined position at some time does not evolve into states with well-defined positions (its wavefunction immediately spreads out).

One does, however, want to be able to discuss states with well-defined positions, in order to describe amplitudes for particle propagation, and to introduce local interactions between particles. One approach is to try and define operators corresponding to creation or annihilation of a particle at a fixed position, by taking a Fourier transform of the annihilation and creation operators for momentum eigenstates. Quantum fields could be defined as

$$\widehat{\Psi}(x) = \frac{1}{\sqrt{2\pi}} \int_{-\infty}^{\infty} e^{ipx} a(p) dp \qquad (37.1)$$

and its adjoint

$$\widehat{\Psi}^{\dagger}(x) = \frac{1}{\sqrt{2\pi}} \int_{-\infty}^{\infty} e^{-ipx} a^{\dagger}(p) dp$$

Note that, just like $a(p)$ and $a^{\dagger}(p)$, these are not self-adjoint operators, and thus not themselves observables, but physical observables can be constructed by taking simple (typically quadratic) combinations of them. As explained in section 36.5, for multiparticle systems, the state space $\mathcal{H}$ is complicated to describe and work with; it is the operators that behave simply. These will generally be built out of either the field operators $\widehat{\Psi}(x), \widehat{\Psi}^{\dagger}(x)$, or annihilation and creation operators $a(p), a^{\dagger}(p)$, with the Fourier transform relating the two possibilities.

In the continuum formalism, the annihilation and creation operators satisfy the distributional equation

$$[a(p), a^\dagger(p')] = \delta(p - p')$$

and one can formally compute the commutators

$$[\widehat{\Psi}(x), \widehat{\Psi}(x')] = [\widehat{\Psi}^\dagger(x), \widehat{\Psi}^\dagger(x')] = 0$$

$$
\begin{aligned}
[\widehat{\Psi}(x), \widehat{\Psi}^\dagger(x')] &= \frac{1}{2\pi} \int_{-\infty}^{\infty} \int_{-\infty}^{\infty} e^{ipx} e^{-ip'x'} [a(p), a^\dagger(p')] \, dp \, dp' \\
&= \frac{1}{2\pi} \int_{-\infty}^{\infty} \int_{-\infty}^{\infty} e^{ipx} e^{-ip'x'} \delta(p - p') \, dp \, dp' \\
&= \frac{1}{2\pi} \int_{-\infty}^{\infty} e^{ip(x-x')} \, dp \\
&= \delta(x - x')
\end{aligned}
$$

getting results consistent with the interpretation of the field operator and its adjoint as operators that annihilate and create particles at a point x.

   This sort of definition relies upon making sense of the operators $a(p)$ and $a^\dagger(p)$ as distributional operators, using the action of elements of $\mathcal{H}_1$ and $\overline{\mathcal{H}}_1$ on Fock space given by equations 36.12 and 36.13. We could instead more directly proceed exactly as in section 36.3 but with solutions characterized by initial data given by a function $\psi(x)$ rather than its Fourier transform $\alpha(p)$. One gets all the same objects and formulas, related by Fourier transform. Our notation for these transformed objects will be

$$\alpha(p) \to \psi(x), \quad \overline{\alpha}(p) \to \overline{\psi}(x)$$

$$A(\alpha) \to \Psi(\psi), \quad \overline{A}(\alpha) \to \overline{\Psi}(\psi)$$

$$A(p) \to \Psi(x), \quad \overline{A}(p) \to \overline{\Psi}(x)$$

$\Psi(\psi)$ will be the solution in $\mathcal{H}_1$ with initial data $\psi(x)$ and $\overline{\Psi}(\psi)$ the conjugate solution in $\overline{\mathcal{H}}_1$. $\Psi(x)$ can be interpreted as the distributional solution equal to $\delta(x - x')$ at $t = 0$, and we can write

$$\Psi(\psi) = \int \psi(x)\Psi(x)\,dx, \quad \overline{\Psi}(\psi) = \int \overline{\psi(x)\Psi}(x)\,dx$$

These satisfy the Poisson bracket relations

$$\{\Psi(\psi_1), \Psi(\psi_2)\} = \{\overline{\Psi}(\psi_1), \overline{\Psi}(\psi_2)\} = 0, \quad \{\Psi(\psi_1), \overline{\Psi}(\psi_2)\} = i\langle \psi_1, \psi_2 \rangle \quad (37.2)$$

$$\{\Psi(x), \Psi(x')\} = \{\overline{\Psi}(x), \overline{\Psi}(x'\} = 0, \quad \{\Psi(x), \overline{\Psi}(x')\} = i\delta(x - x') \quad (37.3)$$

Quantization then takes (note the perhaps confusing choice of notational convention due to following the physicist's convention that $\widehat{\Psi}$ is the annihilation operator)

$$\Psi(\psi) \to -i\widehat{\Psi}^\dagger(\psi), \quad \overline{\Psi}(\psi) \to -i\widehat{\Psi}(\psi)$$

The quantum field operators can be defined in terms of tensor products using the same generalization of the finite dimensional case of section 26.4 that we used to define $a(\alpha)$ and $a^\dagger(\alpha)$. Here

$$\widehat{\Psi}^\dagger(\psi)P^+(\Psi(\psi_1) \otimes \cdots \otimes \Psi(\psi_n)) = \sqrt{n+1}\,P^+(\Psi(\psi) \otimes \Psi(\psi_1) \otimes \cdots \otimes \Psi(\psi_n))$$
$$(37.4)$$

$$\widehat{\Psi}(\psi)P^+(\Psi(\psi_1) \otimes \cdots \otimes \Psi(\psi_n)) =$$
$$\frac{1}{\sqrt{n}}\sum_{j=1}^{n}\langle \psi, \psi_j \rangle P^+(\Psi(\psi_1) \otimes \cdots \otimes \widehat{\Psi(\psi_j)} \otimes \cdots \otimes \Psi(\psi_n))$$
$$(37.5)$$

(the $\widehat{\Psi(\psi_j)}$ means omit that term in the tensor product, and P^+ is the symmetrization operator defined in section 9.6). This gives a representation of the Lie algebra relations 37.2, satisfying

$$[\widehat{\Psi}(\psi_1), \widehat{\Psi}(\psi_2)] = [\widehat{\Psi}^\dagger(\psi_1), \widehat{\Psi}^\dagger(\psi_2)] = 0, \quad [\widehat{\Psi}(\psi_1), \widehat{\Psi}^\dagger(\psi_2)] = \langle \psi_1, \psi_2 \rangle$$

Conventional multiparticle wavefunctions in position space have the same relation to symmetric tensor products as in the momentum space case of section 36.4. Given an arbitrary state $|\psi\rangle$ in the multiparticle state space, the position space wavefunction component with particle number N can be expressed as

$$\psi_N(x_1, x_2, \cdots, x_N) = \langle 0|\widehat{\Psi}(x_1)\widehat{\Psi}(x_2)\cdots\widehat{\Psi}(x_N)|\psi\rangle$$

37.2 Quadratic operators and dynamics

Other observables can be defined simply in terms of the field operators. These include (note that in all cases, these formulas require interpretation as limits of finite sums in the finite cutoff theory):

- The number operator $\widehat{N}$. A number density operator can be defined by

$$\widehat{n}(x) = \widehat{\Psi}^\dagger(x)\widehat{\Psi}(x)$$

and integrated to get an operator with eigenvalues the total number of particles in a state

$$
\begin{aligned}
\widehat{N} &= \int_{-\infty}^{\infty} \widehat{n}(x)dx \\
&= \int_{-\infty}^{\infty}\int_{-\infty}^{\infty}\int_{-\infty}^{\infty} \frac{1}{\sqrt{2\pi}}e^{-ip'x}a^{\dagger}(p')\frac{1}{\sqrt{2\pi}}e^{ipx}a(p)dpdp'dx \\
&= \int_{-\infty}^{\infty}\int_{-\infty}^{\infty} \delta(p-p')a^{\dagger}(p')a(p)dpdp' \\
&= \int_{-\infty}^{\infty} a^{\dagger}(p)a(p)dp
\end{aligned}
$$

- The total momentum operator $\widehat{P}$. This can be defined in terms of field operators as

$$
\begin{aligned}
\widehat{P} &= \int_{-\infty}^{\infty} \widehat{\Psi}^{\dagger}(x)(-i\frac{d}{dx}\widehat{\Psi}(x))dx \\
&= \int_{-\infty}^{\infty}\int_{-\infty}^{\infty}\int_{-\infty}^{\infty} \frac{1}{\sqrt{2\pi}}e^{-ip'x}a^{\dagger}(p')(-i)(ip)\frac{1}{\sqrt{2\pi}}e^{ipx}a(p)dpdp'dx \\
&= \int_{-\infty}^{\infty}\int_{-\infty}^{\infty} \delta(p-p')pa^{\dagger}(p')a(p)dpdp' \\
&= \int_{-\infty}^{\infty} pa^{\dagger}(p)a(p)dp
\end{aligned}
$$

For more discussion of this operator and its relation to spatial translations, see section 38.3.1.

- The Hamiltonian $\widehat{H}$. As an operator quadratic in the field operators, this can be chosen to be

$$
\widehat{H} = \int_{-\infty}^{\infty} \widehat{\Psi}^{\dagger}(x)\left(-\frac{1}{2m}\frac{d^2}{dx^2}\right)\widehat{\Psi}(x)dx = \int_{-\infty}^{\infty} \frac{p^2}{2m}a^{\dagger}(p)a(p)dp
$$

The dynamics of a quantum field theory is usually described in the Heisenberg picture, with the evolution of the field operators given by Fourier transformed versions of the discussion in terms of $a(p), a^{\dagger}(p)$ of section 36.5. The quantum fields satisfy the general dynamical equation

$$
\frac{d}{dt}\widehat{\Psi}^{\dagger}(x,t) = -i[\widehat{\Psi}^{\dagger}(x,t),\widehat{H}]
$$

which in this case is

$$\frac{\partial}{\partial t}\widehat{\Psi}^\dagger(x,t) = -\frac{i}{2m}\frac{\partial^2}{\partial x^2}\widehat{\Psi}^\dagger(x,t)$$

Note that the field operator $\widehat{\Psi}^\dagger(x,t)$ satisfies the (conjugate) Schrödinger equation, which now appears as a differential equation for distributional operators rather than for wavefunctions. Such a differential equation can be solved just as for wavefunctions, by Fourier transforming and turning differentiation into multiplication, and we find

$$\widehat{\Psi}^\dagger(x,t) = \frac{1}{\sqrt{2\pi}}\int_{-\infty}^\infty e^{-ipx}e^{i\frac{p^2}{2m}t}a^\dagger(p)dp$$

Just as in the case of 36.5, this formal calculation involving the quantum field operators has an analog in terms of the $\Psi(x)$ and a quadratic function on the phase space. One can write

$$h = \int_{-\infty}^{+\infty}\Psi(x)\frac{-1}{2m}\frac{\partial^2}{\partial x^2}\overline{\Psi}(x)dx$$

and the dynamical equations as

$$\frac{d}{dt}\Psi(x,t) = \{\Psi(x,t),h\}$$

which can be evaluated to give

$$\frac{\partial}{\partial t}\Psi(x,t) = -\frac{i}{2m}\frac{\partial^2}{\partial x^2}\Psi(x,t)$$

Note that there are other possible forms of the Hamiltonian function that give the same dynamics, related to the one we chose by integration by parts, in particular

$$\Psi(x)\frac{d^2}{dx^2}\overline{\Psi}(x) = \frac{d}{dx}\left(\Psi(x)\frac{d}{dx}\overline{\Psi}(x)\right) - |\frac{d}{dx}\Psi(x)|^2$$

or

$$\Psi(x)\frac{d^2}{dx^2}\overline{\Psi}(x) = \frac{d}{dx}\left(\Psi(x)\frac{d}{dx}\overline{\Psi}(x)\right) - \left(\frac{d}{dx}\Psi(x)\right)\overline{\Psi}(x) + \left(\frac{d^2}{dx^2}\Psi(x)\right)\overline{\Psi}(x)$$

Neglecting integrals of derivatives (assuming boundary terms go to zero at infinity), one could have used

$$h = \frac{1}{2m}\int_{-\infty}^{+\infty}|\frac{d}{dx}\Psi(x)|^2dx \quad \text{or} \quad h = -\frac{1}{2m}\int_{-\infty}^{+\infty}\left(\frac{d^2}{dx^2}\Psi(x)\right)\overline{\Psi}(x)dx$$

37.3 The propagator in non-relativistic quantum field theory

In quantum field theory, the Heisenberg picture operators that provide observables will be products of the field operators, and the time-dependence of these for the free particle theory was determined in section 37.2. For the time-independent state, the natural choice is the vacuum state $|0\rangle$, although other possibilities such as coherent states may also be useful. States with a finite number of particles will be given by applying field operators to the vacuum, so such states just corresponds to a different product of field operators.

We will not enter here into details, but a standard topic in quantum field theory textbooks is "Wick's theorem," which says that the calculation of expectation values of products of field operators in the state $|0\rangle$ can be reduced to the problem of calculating the following special case:

Definition (Propagator for non-relativistic quantum field theory). *The propagator for a non-relativistic quantum field theory is the amplitude, for $t_2 > t_1$*

$$U(x_2, t_2, x_1, t_1) = \langle 0 | \widehat{\Psi}(x_2, t_2) \widehat{\Psi}^\dagger(x_1, t_1) | 0 \rangle$$

The physical interpretation of these functions is that they describe the amplitude for a process in which a one-particle state localized at x_1 is created at time t_1, propagates for a time $t_2 - t_1$, and is annihilated at position x_2. Using the solution for the time-dependent field operator given earlier, we find

$$
\begin{aligned}
U(x_2, t_2, x_1, t_1) &= \frac{1}{2\pi} \iint_{\mathbf{R}^2} \langle 0 | e^{i p_2 x_2} e^{-i \frac{p_2^2}{2m} t} a(p_2) e^{-i p_1 x_1} e^{i \frac{p_1^2}{2m} t_1} a^\dagger(p_1) | 0 \rangle dp_2 dp_1 \\
&= \frac{1}{2\pi} \iint_{\mathbf{R}^2} e^{i p_2 x_2} e^{-i \frac{p_2^2}{2m} t_2} e^{-i p_1 x_1} e^{i \frac{p_1^2}{2m} t_1} \delta(p_2 - p_1) dp_2 dp_1 \\
&= \frac{1}{2\pi} \int_{-\infty}^{+\infty} e^{-i p(x_1 - x_2)} e^{-i \frac{p^2}{2m}(t_2 - t_1)} dp
\end{aligned}
$$

This is exactly the same calculation (see equation 12.5) already discussed in detail in section 12.5. As described there, the result (equation 12.9) is

$$U(x_2, t_2, x_1, t_1) = U(t_2 - t_1, x_2 - x_1) = \left(\frac{m}{i 2\pi(t_2 - t_1)} \right)^{\frac{1}{2}} e^{\frac{m}{2i(t_2 - t_1)}(x_2 - x_1)^2}$$

which satisfies

$$\lim_{t \to 0^+} U(t, x_2 - x_1) = \delta(x_2 - x_1)$$

If we extend the definition of $U(t, x_2 - x_1)$ to $t < 0$ by taking it to be zero there, as in section we get the retarded propagator $U_+(t, x_2 - x_1)$ and its Fourier transformed version in frequency–momentum space of section 12.6 as well as the relation to Green's functions of section 12.7.

37.4 Interacting quantum fields

To describe an arbitrary number of particles moving in an external potential $V(x)$, the Hamiltonian can be taken to be

$$\widehat{H} = \int_{-\infty}^{\infty} \widehat{\Psi}^\dagger(x) \left(-\frac{1}{2m} \frac{d^2}{dx^2} + V(x) \right) \widehat{\Psi}(x) dx$$

If a complete set of orthonormal solutions $\psi_n(x)$ to the Schrödinger equation with potential can be found, they can be used to describe this quantum system using similar techniques to those for the free particle, taking as basis for $\mathcal{H}_1$ the $\psi_n(x)$ instead of plane waves of momentum p. A creation-annihilation operator pair $a_n, a_n^\dagger$ is associated to each eigenfunction, and quantum fields are defined by

$$\widehat{\Psi}(x) = \sum_n \overline{\psi}_n(x) a_n, \quad \widehat{\Psi}^\dagger(x) = \sum_n \psi_n(x) a_n^\dagger$$

For Hamiltonians quadratic in the quantum fields, quantum field theories are relatively tractable objects. They are in some sense decoupled quantum oscillator systems, although with an infinite number of degrees of freedom. Higher-order terms in the Hamiltonian are what makes quantum field theory a difficult and complicated subject; one that requires a year-long graduate level course to master basic computational techniques, and one that to this day resists mathematician's attempts to prove that many examples of such theories have even the basic expected properties. In the quantum theory of charged particles interacting with an electromagnetic field (see chapter 45), when the electromagnetic field is treated classically, one still has a Hamiltonian quadratic in the field operators for the particles. But if the electromagnetic field is treated as a quantum system, it acquires its own field operators, and the Hamiltonian is no longer quadratic in the fields but instead gives an interacting quantum field theory.

Even if one restricts attention to the quantum fields describing one kind of particle, there may be interactions between particles that add terms to the Hamiltonian that will be higher order than quadratic. For instance, if there is an interaction between such particles described by an interaction energy $v(y - x)$, this can be described by adding the following quartic term to the

Hamiltonian

$$\frac{1}{2}\int_{-\infty}^{\infty}\int_{-\infty}^{\infty}\widehat{\Psi}^{\dagger}(x)\widehat{\Psi}^{\dagger}(y)v(y-x)\widehat{\Psi}(y)\widehat{\Psi}(x)dxdy$$

The study of "many-body" quantum systems with interactions of this kind is a major topic in condensed matter physics.

Digression (The Lagrangian density and the path integral). *While we have worked purely in the Hamiltonian formalism, another approach would have been to start with an action for this system and use Lagrangian methods. An action that will give the Schrödinger equation as an Euler-Lagrange equation is*

$$
\begin{aligned}
S &= \int_{-\infty}^{\infty}\int_{-\infty}^{\infty}\left(i\psi\frac{\partial}{\partial t}\overline{\psi}-h\right)dxdt\\
&= \int_{-\infty}^{\infty}\int_{-\infty}^{\infty}\left(i\psi\frac{\partial}{\partial t}\overline{\psi}+\psi\frac{1}{2m}\frac{\partial^2}{\partial x^2}\overline{\psi}\right)dxdt\\
&= \int_{-\infty}^{\infty}\int_{-\infty}^{\infty}\left(i\psi\frac{\partial}{\partial t}\overline{\psi}-\frac{1}{2m}|\frac{\partial}{\partial x}\psi|^2\right)dxdt
\end{aligned}
$$

where the last form comes by using integration by parts to get an alternate form of h as mentioned in section 37.2. In the Lagrangian approach to field theory, the action is an integral over space and time of a Lagrangian density, which in this case is

$$L(x,t)=i\psi\frac{\partial}{\partial t}\overline{\psi}-\frac{1}{2m}|\frac{\partial}{\partial x}\psi|^2$$

Defining a canonical conjugate momentum for $\overline{\psi}$ as $\frac{\partial L}{\partial\dot{\overline{\psi}}}$ gives as momentum variable $i\psi$. This justifies the Poisson bracket relation

$$\{\overline{\Psi}(x),i\Psi(x')\}=\delta(x-x')$$

but, as expected for a case where the equation of motion is first order in time, the canonical momentum coordinate $i\Psi(x)$ is not independent of the coordinate $\overline{\Psi}(x)$. The space $\mathcal{H}_1$ of wavefunctions is already a phase space rather than just a configuration space, and one does not need to introduce new momentum variables. One could try and quantize this system by path integral methods, for instance computing the propagator by doing the integral

$$\int D\psi(x,t)e^{\frac{i}{\hbar}S[\psi]}$$

over paths in $\mathcal{H}_1$ parametrized by t, taking values from $t = 0$ to $t = T$. This is a highly infinite dimensional integral, over paths in an infinite dimensional space. In addition, recall the warnings given in chapter 35 about the problematic nature of path integrals over paths in a phase space, which is the case here.

37.5 Fermion fields

Most everything discussed in this chapter and in chapter 36 applies with little change to the case of fermionic quantum fields using fermionic instead of bosonic oscillators, and changing commutators to anticommutators for the annihilation and creation operator. This gives fermionic fields that satisfy anticommutation relations

$$[\widehat{\Psi}(x), \widehat{\Psi}^\dagger(x')]_+ = \delta(x - x')$$

and states that in the occupation number representation have $n_p = 0, 1$, while also having a description in terms of antisymmetric tensor products, or polynomials in anticommuting coordinates. Field operators will in this case generate an infinite dimensional Clifford algebra. Elements of this Clifford algebra act on states by an infinite dimensional version of the construction of spinors in terms of fermionic oscillators described in chapter 31.

For applications to physical systems in three-dimensional space, it is often the fermionic version that is relevant, with the systems of interest for instance describing arbitrary numbers of electrons, which are fermionic particles so need to be described by anticommuting fields. The quantum field theory of non-relativistic free electrons is the quantum theory one gets by taking as single-particle phase space $\mathcal{H}_1$ the space of solutions of the two-component Pauli–Schrödinger equation 34.3 described in section 34.2 and then quantizing using the fermionic version of Bargmann–Fock quantization. The fermionic Poisson bracket is determined by the inner product on this $\mathcal{H}_1$ discussed in section 34.3.

More explicitly, this is a theory of two quantum fields $\widehat{\Psi}_1(\mathbf{x}), \widehat{\Psi}_2(\mathbf{x})$ satisfying the anticommutation relations

$$[\widehat{\Psi}_j(\mathbf{x}), \widehat{\Psi}_k^\dagger(\mathbf{x}')]_+ = \delta_{jk}\delta^3(\mathbf{x} - \mathbf{x}')$$

These are related by Fourier transform

$$\widehat{\Psi}_j(\mathbf{x}) = \frac{1}{(2\pi)^{\frac{3}{2}}} \int_{\mathbf{R}^3} e^{i\mathbf{p}\cdot\mathbf{x}} a_j(\mathbf{p}) d^3\mathbf{p}$$

$$\widehat{\Psi}_j^\dagger(\mathbf{x}) = \frac{1}{(2\pi)^{\frac{3}{2}}} \int_{\mathbf{R}^3} e^{-i\mathbf{p}\cdot\mathbf{x}} a_j^\dagger(\mathbf{p}) d^3\mathbf{p}$$

to annihilation and creation operators satisfying

$$[a_j(\mathbf{p}), a_k^\dagger(\mathbf{p}')]_+ = \delta_{jk}\delta^3(\mathbf{p} - \mathbf{p}')$$

The theory of non-relativistic electrons is something different than simply two copies of a single fermionic field, since it describes spin $\frac{1}{2}$ particles, not pairs of spin 0 particles. In section 38.3.3, we will see how the group $\widetilde{E(3)}$ acts on the theory, giving angular momentum observables corresponding to spin $\frac{1}{2}$ rather than two copies of spin 0.

37.6 For further reading

See the same references at the end of chapter 36 for more details about the material of this one. A discussion of the physics described by the formalism of this chapter can be found in most quantum field theory textbooks and in textbooks dealing with the many-particle formalism and condensed matter theory. Two textbooks that explicitly discuss the non-relativistic case are [35] and [45].

Chapter 38
Symmetries and Non-relativistic Quantum Fields

In our study (chapters 25 and 26) of quantization using complex structures on phase space, we found that using the Poisson bracket, quadratic polynomials of the (complexified) phase space coordinates provided a symplectic Lie algebra $\mathfrak{sp}(2d, \mathbf{C})$, with a distinguished $\mathfrak{gl}(d, \mathbf{C})$ sub-Lie algebra determined by the complex structure (see section 25.2). In section 25.3, we saw that these quadratic polynomials could be quantized as quadratic combinations of the annihilation and creation operators, giving a representation on the harmonic oscillator state space, one that was unitary on the unitary sub-Lie algebra $\mathfrak{u}(d) \subset \mathfrak{gl}(d, \mathbf{C})$.

The non-relativistic quantum field theory of chapters 36 and 37 is an infinite dimensional version of this, with the dual phase space now the space $\mathcal{H}_1$ of solutions of the Schrödinger equation. When a group G acts on $\mathcal{H}_1$ preserving the Hermitian inner product (and thus the symplectic and complex structures), generalizing the formulas of sections 25.2 and 25.3 should give a unitary representation of such a group G on the multiparticle state space $S^*(\mathcal{H}_1)$.

In sections 36.5 and 37.2, we saw how this works for $G = \mathbf{R}$, the group of time translations, which determines the dynamics of the theory. For the case of a free particle, the field theory Hamiltonian is a quadratic polynomial of the fields, providing a basic example of how such polynomials provide a unitary representation on the states of the quantum theory by use of a quadratic combination of the quantum field operators. In this chapter, we will see some other examples of how group actions on the single-particle space $\mathcal{H}_1$ lead to quadratic operators and unitary transformations on the full quantum field theory. The moment map for these group actions gives the quadratic polynomials on $\mathcal{H}_1$, which after quantization become the quadratic operators of the Lie algebra representation.

© Peter Woit 2017   477
P. Woit, *Quantum Theory, Groups and Representations*,
DOI 10.1007/978-3-319-64612-1_38

38.1 Unitary transformations on $\mathcal{H}_1$

The single-particle state space $\mathcal{H}_1$ of non-relativistic quantum field theory can be parametrized by either wavefunctions $\psi(x)$ or their Fourier transforms $\widetilde{\psi}(p)$, and carries a Hermitian inner product

$$\langle \psi_1, \psi_2 \rangle = \int \overline{\psi_1(x)}\psi_2(x)dx = \int \overline{\widetilde{\psi}_1(p)}\widetilde{\psi}_2(p)dp$$

As a dual phase space, the symplectic structure is given by the imaginary part of this

$$\Omega(\psi_1, \psi_2) = \frac{1}{2i} \int (\overline{\psi_1(x)}\psi_2(x) - \overline{\psi_2(x)}\psi_1(x))dx$$

There is an infinite dimensional symplectic group that acts on $\mathcal{H}_1$ by linear transformations that preserve Ω. It has an infinite dimensional unitary subgroup, those transformations preserving the full inner product. In this chapter, we will consider various finite dimensional groups G that are subgroups of this unitary group, and see how they are represented on the quantum field theory state space. Note that there are also groups that act as symplectic but not unitary transformations of $\mathcal{H}_1$, after quantization acting by a unitary representation on the multiparticle state space. Such actions change particle number, and the vacuum state $|0\rangle$ in particular will not be invariant. For some indications of what happen in this more general situation, see sections 25.5 and 39.4.

The finite dimensional version of the case of unitary transformations of $\mathcal{H}_1$ was discussed in detail in sections 25.2 and 25.3 where we saw that the moment map for the $U(d)$ action was given by

$$\mu_A = i \sum_{j,k} z_j A_{jk}\overline{z}_k$$

for A a skew-adjoint matrix. Quantization took this quadratic function on phase space to the quadratic combination of annihilation and creation operators

$$\sum_{j,k} a_j^\dagger A_{jk} a_k$$

and exponentiation of these operators gave the unitary representation on state space.

In the quantum field theory case with dual phase space $\mathcal{H}_1$, the generalization of the finite dimensional case will be

$$\text{index } j \to x \text{ or } p$$

$$z_j \to \Psi(x) \text{ or } \alpha(p), \quad \overline{z}_j \to \overline{\Psi}(x) \text{ or } \overline{\alpha}(p)$$

$$a_j^\dagger \to \widehat{\Psi}^\dagger(x) \text{ or } a^\dagger(p), \quad a_j \to \widehat{\Psi}(x) \text{ or } a(p)$$

The quadratic functions we will consider will be "local," multiplying elements parametrized by the same points in position space. Often these will be differential operators. As a result, the generalization from the finite dimensional case will take

$$\sum_{j,k} \to \int dx \text{ or } \int dp$$

and

$$A_{jk} \to \mathcal{O}(x) \text{ or } \mathcal{O}(p)$$

38.2 Internal symmetries

Since the phase space $\mathcal{H}_1$ is a space of complex functions, there is an obvious group that acts unitarily on this space: the group $U(1)$ of phase transformations of the complex values of the function. Such a group action that acts trivially on the spatial coordinates but non-trivially on the values of $\psi(x)$ is called an "internal symmetry." If the fields ψ have multiple components, taking values in $\mathbf{C}^n$, there will be a unitary action of the larger group $U(n)$.

38.2.1 $U(1)$ symmetry

In chapter 2, we saw that the fact that irreducible representations of $U(1)$ are labeled by integers is responsible for the term "quantization": Since quantum states are representations of this group, they break up into states characterized by integers, with these integers counting the number of "quanta." In the non-relativistic quantum field theory, this integer will be the total particle number. Such a theory can be thought of as a harmonic oscillator with an infinite number of degrees of freedom, and the total particle number is the total occupation number, summed over all degrees of freedom.

Consider the $U(1)$ action on the fields $\Psi(x), \overline{\Psi}(x)$ given by

$$\Psi(x) \to e^{-i\theta}\Psi(x), \quad \overline{\Psi}(x) \to e^{i\theta}\overline{\Psi}(x) \tag{38.1}$$

This is an infinite dimensional generalization of the case worked out in section 24.2, where recall that the moment map was $\mu = \overline{z}z$ and

$$\{z\bar{z}, \bar{z}\} = i\bar{z} \quad \{z\bar{z}, z\} = -iz$$

There were two possible choices for the unitary operator that will be the quantization of $z\bar{z}$:

-

$$z\bar{z} \to -\frac{i}{2}(a^\dagger a + a a^\dagger)$$

This will have eigenvalues $-i(n + \frac{1}{2})$, $n = 0, 1, 2 \dots$.

-

$$z\bar{z} \to -ia^\dagger a$$

This is the normal-ordered form, with eigenvalues $-in$.

With either choice, we get a number operator

$$N = \frac{1}{2}(a^\dagger a + a a^\dagger), \quad \text{or} \quad N = \frac{1}{2}:(a^\dagger a + a a^\dagger): = a^\dagger a$$

In both cases, we have

$$[N, a] = -a, \quad [N, a^\dagger] = a^\dagger$$

so

$$e^{i\theta N} a e^{-i\theta N} = e^{-i\theta} a, \quad e^{i\theta N} a^\dagger e^{-i\theta N} = e^{i\theta} a^\dagger$$

Either choice of N will give the same action on operators. However, on states only the normal-ordered one will have the desirable feature that

$$N|0\rangle = 0, \quad e^{iN\theta}|0\rangle = |0\rangle$$

Since we now want to treat fields, adding together an infinite number of such oscillator degrees of freedom, we will need the normal-ordered version in order to not get $\infty \cdot \frac{1}{2}$ as the number eigenvalue for the vacuum state.

We now generalize as described in section 38.1 and get, in momentum space, the expression

$$\widehat{N} = \int_{-\infty}^{+\infty} a^\dagger(p)a(p)dp \tag{38.2}$$

which is just the number operator already discussed in chapter 36. Recall from section 36.3 that this sort of operator product requires some interpretation in order to give it a well-defined meaning, either as a limit of a finite dimensional definition, or by giving it a distributional interpretation.

Fourier transforming to position space, one can work with $\widehat{\Psi}(x), \widehat{\Psi}^\dagger(x)$ instead of $a(p), a^\dagger(p)$ and find that

$$\widehat{N} = \int_{-\infty}^{+\infty} \widehat{\Psi}^\dagger(x)\widehat{\Psi}(x)dx \tag{38.3}$$

$\widehat{\Psi}^\dagger(x)\widehat{\Psi}(x)$ can be interpreted as an operator-valued distribution, with the physical interpretation of measuring the number density at x. On field operators, $\widehat{N}$ satisfies

$$[\widehat{N},\widehat{\Psi}] = -\widehat{\Psi}, \quad [\widehat{N},\widehat{\Psi}^\dagger] = \widehat{\Psi}^\dagger$$

so $\widehat{\Psi}$ acts on states by reducing the eigenvalue of $\widehat{N}$ by one, while $\widehat{\Psi}^\dagger$ acts on states by increasing the eigenvalue of $\widehat{N}$ by one. Exponentiating gives

$$e^{i\theta\widehat{N}}\widehat{\Psi}e^{-i\theta\widehat{N}} = e^{-i\theta}\widehat{\Psi}, \quad e^{i\theta\widehat{N}}\widehat{\Psi}^\dagger e^{-i\theta\widehat{N}} = e^{i\theta}\widehat{\Psi}^\dagger$$

which are the quantized versions of the $U(1)$ action on the phase space coordinates (see equation 38.1) that we began our discussion with.

An important property of $\widehat{N}$ that can be straightforwardly checked is that

$$[\widehat{N},\widehat{H}] = \left[\widehat{N}, \int_{-\infty}^{+\infty} \widehat{\Psi}^\dagger(x)\frac{-1}{2m}\frac{\partial^2}{\partial x^2}\widehat{\Psi}(x)dx\right] = 0$$

This implies that particle number is a conserved quantity: If we start out with a state with a definite particle number, this will remain constant. Note that the origin of this conservation law comes from the fact that $\widehat{N}$ is the quantized generator of the $U(1)$ symmetry of phase transformations on complex-valued fields Ψ. If we start with any Hamiltonian function h on $\mathcal{H}_1$ that is invariant under the $U(1)$ (i.e., built out of terms with an equal number of Ψs and $\overline{\Psi}$s), then for such a theory $\widehat{N}$ will commute with $\widehat{H}$ and particle number will be conserved.

38.2.2 $U(n)$ symmetry

By taking fields with values in $\mathbf{C}^n$, or, equivalently, n different species of complex-valued field Ψ_j, $j = 1, 2, \ldots, n$, quantum field theories with larger internal symmetry groups than $U(1)$ can easily be constructed. Taking as Hamiltonian function

$$h = \int_{-\infty}^{+\infty} \sum_{j=1}^{n} \Psi_j(x)\frac{-1}{2m}\frac{\partial^2}{\partial x^2}\overline{\Psi}_j(x)dx \tag{38.4}$$

gives a Hamiltonian that will be invariant not just under $U(1)$ phase transformations, but also under transformations

$$\begin{pmatrix} \Psi_1 \\ \Psi_2 \\ \vdots \\ \Psi_n \end{pmatrix} \rightarrow U \begin{pmatrix} \Psi_1 \\ \Psi_2 \\ \vdots \\ \Psi_n \end{pmatrix}$$

where U is an n by n unitary matrix. The Poisson brackets will be

$$\{\Psi_j(x), \overline{\Psi}_k(x')\} = i\delta(x - x')\delta_{jk}$$

and are also invariant under such transformations by $U \in U(n)$.

As in the $U(1)$ case, we begin by considering the case of one particular value of p or of x, for which the phase space is $\mathbf{C}^n$, with coordinates $z_j, \overline{z}_j$. As we saw in section 25.2, the n^2 quadratic combinations $z_j\overline{z}_k$ for $j = 1, \ldots, n$, $k = 1, \ldots, n$ will generalize the role played by $z\overline{z}$ in the $n = 1$ case, with their Poisson bracket relations exactly the Lie bracket relations of the Lie algebra $\mathfrak{u}(n)$ (or, considering all complex linear combinations, $\mathfrak{gl}(n, \mathbf{C})$).

After quantization, these quadratic combinations become quadratic combinations of annihilation and creation operators $a_j, a_j^\dagger$ satisfying

$$[a_j, a_k^\dagger] = \delta_{jk}$$

Recall (theorem 25.2) that for n by n matrices X and Y

$$\left[\sum_{j,k=1}^n a_j^\dagger X_{jk} a_k, \sum_{j,k=1}^n a_j^\dagger Y_{jk} a_k \right] = \sum_{j,k=1}^n a_j^\dagger [X, Y]_{jk} a_k$$

So, for each X in the Lie algebra $\mathfrak{gl}(n, \mathbf{C})$, quantization will give us a representation of $\mathfrak{gl}(n, \mathbf{C})$ where X acts as the operator

$$\sum_{j,k=1}^n a_j^\dagger X_{jk} a_k$$

When the matrices X are chosen to be skew-adjoint ($X_{jk} = -\overline{X_{kj}}$), this construction will give us a unitary representation of $\mathfrak{u}(n)$.

As in the $U(1)$ case, one gets an operator in the quantum field theory by integrating over quadratic combinations of the $a(p), a^\dagger(p)$ in momentum space, or the field operators $\widehat{\Psi}(x), \widehat{\Psi}^\dagger(x)$ in configuration space, finding for each $X \in \mathfrak{u}(n)$ an operator

$$\widehat{X} = \int_{-\infty}^{+\infty} \sum_{j,k=1}^n \widehat{\Psi}_j^\dagger(x) X_{jk} \widehat{\Psi}_k(x) dx = \int_{-\infty}^{+\infty} \sum_{j,k=1}^n a_j^\dagger(p) X_{jk} a_k(p) dp \quad (38.5)$$

This satisfies

$$[\widehat{X}, \widehat{Y}] = \widehat{[X, Y]} \tag{38.6}$$

and, acting on operators

$$[\widehat{X}, \widehat{\Psi}_j(x)] = -\sum_{k=1}^{n} X_{jk}\widehat{\Psi}_k(x), \quad [\widehat{X}, \widehat{\Psi}_j^\dagger(x)] = \sum_{k=1}^{n} X_{kj}\widehat{\Psi}_k^\dagger(x) \tag{38.7}$$

$\widehat{X}$ provides a Lie algebra representation of $\mathfrak{u}(n)$ on the multiparticle state space. After exponentiation, this representation takes

$$e^X \in U(n) \to U(e^X) = e^{\widehat{X}} = e^{\int_{-\infty}^{+\infty} \sum_{b,c=1}^{n} \widehat{\Psi}_j^\dagger(x) X_{jk}\widehat{\Psi}_k(x)dx}$$

The construction of the operator $\widehat{X}$ above is an infinite dimensional example of our standard method of creating a Lie algebra representation by quantizing moment map functions. In this case, the quadratic moment map function on the space of solutions of the Schrödinger equation is

$$\mu_X = i \int_{-\infty}^{+\infty} \sum_{j,k=1}^{n} \Psi_j(x) X_{jk}\overline{\Psi}_k(x)dx$$

which (generalizing the finite dimensional case of theorem 25.1) satisfies the Poisson bracket relations

$$\{\mu_X, \mu_Y\} = \mu_{[X,Y]}$$

$$\{\mu_X, \overline{\Psi}_j(x)\} = -X_{jk}\overline{\Psi}_j(x), \quad \{\mu_X, \Psi_j(x)\} = X_{kj}\Psi_k(x)$$

After quantization, these become the operator relations 38.6 and 38.7. Note that the factor of i in the expression for μ_X is there to make it a real function for $X \in \mathfrak{u}(n)$. Quantization of this would give a self-adjoint operator, so multiplication by $-i$ makes the expression for $\widehat{X}$ skew-adjoint, and thus a unitary Lie algebra representation.

When, as for the free particle case of equation 38.4, the Hamiltonian is invariant under $U(n)$ transformations of the fields $\Psi_j, \overline{\Psi}_j$, then we will have

$$[\widehat{X}, \widehat{H}] = 0$$

Energy eigenstates in the multiparticle state space will break up into irreducible representations of $U(n)$ and can be labeled accordingly.

38.3   Spatial symmetries

We saw in chapter 19 that the action of the group $E(3)$ on physical space $\mathbf{R}^3$ induces a unitary action on the space $\mathcal{H}_1$ of solutions to the free particle Schrödinger equation. Quantization of this phase space with this group action produces a multiparticle state space carrying a unitary representation of the group $E(3)$. There are several different actions of the group $E(3)$ that one needs to keep track of here. Given an element $(\mathbf{a}, R) \in E(3)$, one has:

- An action on $\mathbf{R}^3$, by

$$\mathbf{x} \to R\mathbf{x} + \mathbf{a}$$

- A unitary action on $\mathcal{H}_1$ induced by the action on $\mathbf{R}^3$, given by

$$\psi(\mathbf{x}) \to u(\mathbf{a}, R)\psi(\mathbf{x}) = \psi(R^{-1}(\mathbf{x} - \mathbf{a}))$$

on wavefunctions, or, on Fourier transforms by

$$\widetilde{\psi}(\mathbf{p}) \to \widetilde{u}(\mathbf{a}, R)\widetilde{\psi}(\mathbf{p}) = e^{-i\mathbf{a}\cdot R^{-1}\mathbf{p}}\widetilde{\psi}(R^{-1}\mathbf{p})$$

Recall from chapter 19 that this is not an irreducible representation of $E(3)$, but an irreducible representation can be constructed by taking the space of solutions that are energy eigenfunctions with fixed eigenvalue $E = \frac{|\mathbf{p}|^2}{2m}$.

- $E(3)$ will act on distributional fields $\Psi(\mathbf{x})$ by

$$\Psi(\mathbf{x}) \to (\mathbf{a}, R) \cdot \Psi(\mathbf{x}) = \Psi(R\mathbf{x} + \mathbf{a}) \tag{38.8}$$

This is because elements of $\mathcal{H}_1$ can be written in terms of these distributional fields as

$$\Psi(\psi) = \int_{\mathbf{R}^3} \Psi(\mathbf{x})\psi(\mathbf{x})d^3\mathbf{x}$$

and $E(3)$ will act on $\Psi(\psi)$ by

$$\Psi(\psi) \to (\mathbf{a}, R) \cdot \Psi(\psi) = \int_{\mathbf{R}^3} \Psi(\mathbf{x})\psi(R^{-1}(\mathbf{x} - \mathbf{a}))d^3\mathbf{x}$$

$$= \int_{\mathbf{R}^3} \Psi(R\mathbf{x} + \mathbf{a})\psi(\mathbf{x})d^3\mathbf{x}$$

(using invariance of the integration measure under $E(3)$ transformations).

More generally, if elements of $\mathcal{H}_1$ are multicomponent functions ψ_j (for instance, in the case of spin $\frac{1}{2}$ wavefunctions), the (double cover of) the

$E(3)$ group may act by

$$\psi_j(\mathbf{x}) \to \sum_k \Omega_{jk}\psi_k(R^{-1}(\mathbf{x}-\mathbf{a}))$$

on wavefunctions, and

$$\Psi_j(\mathbf{x}) \to \sum_k (\Omega^{-1})_{jk}\Psi_k(R\mathbf{x}+\mathbf{a})$$

on distributional fields (see section 38.3.3).

• The action of $E(3)$ on $\mathcal{H}_1$ is a linear map preserving the symplectic structure. We thus expect by the general method of section 20.2 to be able to construct intertwining operators, by taking the quadratic functions given by the moment map, quantizing to get a Lie algebra representation, and exponentiating to get a unitary representation of $E(3)$. More specifically, we will use Bargmann–Fock quantization, and the method carried out for a finite dimensional phase space in section 25.3. We end up with a representation of $E(3)$ on the quantum field theory state space $\mathcal{H}$, given by unitary operators $U(\mathbf{a}, R)$.

It is the last of these that we want to examine here, and as usual for quantum field theory, we do not want to try and explicitly construct the multiparticle state space $\mathcal{H}$ and see the $E(3)$ action on that construction, but instead want to use the analog of the Heisenberg picture in the time-translation case, taking the group to act on operators. For each $(\mathbf{a}, R) \in E(3)$, we want to find operators $U(\mathbf{a}, R)$ that will be built out of the field operators and act on the field operators as

$$\widehat{\Psi}(\mathbf{x}) \to U(\mathbf{a}, R)\widehat{\Psi}(\mathbf{x})U(\mathbf{a}, R)^{-1} = \widehat{\Psi}(R\mathbf{x}+\mathbf{a}) \tag{38.9}$$

38.3.1 Spatial translations

For spatial translations, we want to construct momentum operators $\widehat{\mathbf{P}}$ such that the $-i\widehat{\mathbf{P}}$ give a unitary Lie algebra representation of the translation group. Exponentiation will then give the unitary representation

$$U(\mathbf{a}, \mathbf{1}) = e^{-i\mathbf{a}\cdot\widehat{\mathbf{P}}}$$

Note that these are not the momentum operators $\mathbf{P}$ that act on $\mathcal{H}_1$, but are operators in the quantum field theory that will be built out of quadratic combinations of the field operators. By equation 38.9, we want

$$e^{-i\mathbf{a}\cdot\widehat{\mathbf{P}}}\widehat{\Psi}(\mathbf{x})e^{i\mathbf{a}\cdot\widehat{\mathbf{P}}} = \widehat{\Psi}(\mathbf{x}+\mathbf{a})$$

or the derivative of this equation

$$[-i\widehat{\mathbf{P}}, \widehat{\Psi}(\mathbf{x})] = \boldsymbol{\nabla}\widehat{\Psi}(\mathbf{x}) \tag{38.10}$$

Such an operator $\widehat{\mathbf{P}}$ can be constructed in terms of quadratic combinations of the field operators by our moment map methods. We find (generalizing theorem 25.1) that the quadratic expression

$$\mu_{-\boldsymbol{\nabla}} = i\int_{\mathbf{R}^3}\Psi(\mathbf{x})(-\boldsymbol{\nabla})\overline{\Psi}(\mathbf{x})d^3\mathbf{x}$$

is real (since $\boldsymbol{\nabla}$ is skew-adjoint) and satisfies

$$\{\mu_{-\boldsymbol{\nabla}}, \Psi(\mathbf{x})\} = \boldsymbol{\nabla}\Psi(\mathbf{x}), \quad \{\mu_{-\boldsymbol{\nabla}}, \overline{\Psi}(\mathbf{x})\} = \boldsymbol{\nabla}\overline{\Psi}(\mathbf{x})$$

Using the Poisson bracket relations, this can be checked by computing for instance (we will do this just for $d=1$)

$$\{\mu_{-\frac{d}{dx}}, \overline{\Psi}(x)\} = i\left\{\int\Psi(y)(-\frac{d}{dy})\overline{\Psi}(y)dy, \overline{\Psi}(x)\right\}$$

$$= -i\int\{\Psi(y), \overline{\Psi}(x)\}\frac{d}{dy}\overline{\Psi}(y)dy$$

$$= \int\delta(x-y)\frac{d}{dy}\overline{\Psi}(y)dy = \frac{d}{dx}\overline{\Psi}(x)$$

Quantization replaces $\Psi, \overline{\Psi}$ by $\widehat{\Psi}^\dagger, \widehat{\Psi}$ and gives the self-adjoint expression

$$\widehat{\mathbf{P}} = \int_{\mathbf{R}^3}\widehat{\Psi}^\dagger(\mathbf{x})(-i\boldsymbol{\nabla})\widehat{\Psi}(\mathbf{x})d^3\mathbf{x} \tag{38.11}$$

for the momentum operator. In chapter 37, we saw that in terms of momentum space annihilation and creation operators, this operator is

$$\widehat{\mathbf{P}} = \int_{\mathbf{R}^3}\mathbf{p}\, a^\dagger(\mathbf{p})a(\mathbf{p})d^3\mathbf{p}$$

which is the integral over momentum space of the momentum times the number density operator in momentum space.

38.3.2 Spatial rotations

For spatial rotations, we found in chapter 19 that these had as generators
the angular momentum operators

$$\mathbf{L} = \mathbf{X} \times \mathbf{P} = \mathbf{X} \times (-i\boldsymbol{\nabla})$$

acting on $\mathcal{H}_1$. Just as for energy and momentum, we can construct angular
momentum operators in the quantum field theory as quadratic field operators,
in this case getting

$$\widehat{\mathbf{L}} = \int_{\mathbf{R}^3} \widehat{\Psi}^\dagger(\mathbf{x})(\mathbf{x} \times (-i\boldsymbol{\nabla}))\widehat{\Psi}(\mathbf{x})d^3\mathbf{x} \tag{38.12}$$

These will generate the action of rotations on the field operators. For instance,
if $R(\theta)$ is a rotation about the x_3 axis by angle θ, we will have

$$\widehat{\Psi}(R(\theta)\mathbf{x}) = e^{-i\theta\widehat{L}_3}\widehat{\Psi}(\mathbf{x})e^{i\theta\widehat{L}_3}$$

The operators $\widehat{\mathbf{P}}$ and $\widehat{\mathbf{L}}$ together give a representation of the Lie algebra of $E(3)$ on the multiparticle state space, satisfying the $E(3)$ Lie algebra
commutation relations

$$[-i\widehat{P}_j, -i\widehat{P}_k] = 0, \quad [-i\widehat{L}_j, -i\widehat{P}_k] = \epsilon_{jkl}(-i\widehat{P}_l), \quad [-i\widehat{L}_j, -i\widehat{L}_k] = \epsilon_{jkl}(-i\widehat{L}_l) \tag{38.13}$$

$\widehat{\mathbf{L}}$ could also have been found by the moment map method. Recall from
section 8.3 that for the $SO(3)$ representation on functions on $\mathbf{R}^3$ induced from
the $SO(3)$ action on $\mathbf{R}^3$, the Lie algebra representation is (for $\mathbf{l} \in \mathfrak{so}(3)$)

$$\rho'(\mathbf{l}) = -\mathbf{x} \times \boldsymbol{\nabla}$$

The action on distributions will differ by a minus sign, so we are looking for
a moment map μ such that

$$\{\mu, \Psi(\mathbf{x})\} = \mathbf{x} \times \boldsymbol{\nabla}\Psi(\mathbf{x})$$

and this will be given by

$$\mu_{-\mathbf{x}\times\boldsymbol{\nabla}} = i\int_{\mathbf{R}^3} \Psi(\mathbf{x})(-\mathbf{x} \times \boldsymbol{\nabla})\overline{\Psi}(\mathbf{x})d^3\mathbf{x}$$

After quantization, this gives equation 38.12 for the angular momentum operator $\widehat{\mathbf{L}}$.

38.3.3 Spin $\frac{1}{2}$ fields

For the case of two-component wavefunctions describing spin $\frac{1}{2}$ particles satisfying the Pauli–Schrödinger equation (see chapter 34 and section 37.5), the
groups $U(1)$, $U(n)$ (for multiple kinds of spin $\frac{1}{2}$ particles), and the $\mathbf{R}^3$ of
translations act independently on the two spinor components, and the for-
mulas for $\widehat{N}$, $\widehat{X}$, and $\widehat{\mathbf{P}}$ are just the sum of two copies of the single component
equations. As discussed in section 34.2, the action of the rotation group on
solutions in this case requires the use of the double cover $SU(2)$ of $SO(3)$,
with $SU(2)$ group elements Ω acting on two-component solutions ψ by

$$\psi(\mathbf{x}) \to \Omega\psi(R^{-1}\mathbf{x})$$

(here R is the $SO(3)$ rotation corresponding to Ω). This action can be thought
of as an action on a tensor product of $\mathbf{C}^2$ and a space of functions on $\mathbf{R}^3$,
with the matrix Ω acting on the $\mathbf{C}^2$ factor, and the action on functions the
induced action from rotations on $\mathbf{R}^3$. On distributional fields, the action will
be by the inverse

$$\Psi(\mathbf{x}) \to \Omega^{-1}\Psi(R\mathbf{x})$$

(where Ψ has two components).

The $SU(2)$ action on quantum fields will be given by a unitary operator
$U(\Omega)$ satisfying

$$U(\Omega)\widehat{\Psi}(\mathbf{x})U^{-1}(\Omega) = \Omega^{-1}\widehat{\Psi}(R\mathbf{x})$$

which will give a unitary representation on the multiparticle state space. The
Lie algebra representation on this state space will be given by the sum of two
terms

$$\widehat{\mathbf{J}} = \widehat{\mathbf{L}} + \widehat{\mathbf{S}}$$

corresponding to the fact that this comes from a representation on a tensor
product. Here, the operator $\widehat{\mathbf{L}}$ is just two copies of the single component
version (equation 38.12) and comes from the same source, the induced action
on solutions from rotations of $\mathbf{R}^3$. The "spin" operator $\widehat{\mathbf{S}}$ comes from the
$SU(2)$ action on the $\mathbf{C}^2$ factor in the tensor product description of solutions
and is given by

$$\widehat{\mathbf{S}} = \int_{\mathbf{R}^3} \widehat{\Psi}^\dagger(\mathbf{x}) \left(\frac{1}{2}\boldsymbol{\sigma}\right) \widehat{\Psi}(\mathbf{x})d^3\mathbf{x} \tag{38.14}$$

It mixes the two components of the spin $\frac{1}{2}$ field and is a new feature not seen
in the single component ("spin 0") theory.

It is a straightforward exercise using the commutation relations to show that these operators $\widehat{\mathbf{J}}$ satisfy the $\mathfrak{su}(2)$ commutation relations and have the expected commutation relations with the two-component field operators. They also commute with the Hamiltonian, providing an action of $SU(2)$ by symmetries on the multiparticle state space.

States of this quantum field theory can be produced by applying products of operators $a_j^\dagger(\mathbf{p})$ for various choices of $\mathbf{p}$ and $j = 1, 2$ to the vacuum state. Note that the $\widetilde{E(3)}$ Casimir operator $\widehat{\mathbf{J}} \cdot \widehat{\mathbf{P}}$ does not commute with the $a_j^\dagger(\mathbf{p})$. If one wants to work with states with a definite helicity (eigenvalue of $\widehat{\mathbf{J}} \cdot \widehat{\mathbf{P}}$ divided by the square root of the eigenvalue of the operator $|\widehat{\mathbf{P}}|^2$), one could instead write wavefunctions as in equation 34.10, and field operators as

$$\widehat{\Psi}_\pm(\mathbf{x}) = \frac{1}{(2\pi)^{\frac{3}{2}}} \int_{\mathbf{R}^3} e^{i\mathbf{P} \cdot \mathbf{x}} a_\pm(\mathbf{p}) u_\pm(\mathbf{p}) d^3\mathbf{p}$$

$$\widehat{\Psi}_\pm^\dagger(\mathbf{x}) = \frac{1}{(2\pi)^{\frac{3}{2}}} \int_{\mathbf{R}^3} e^{-i\mathbf{P} \cdot \mathbf{x}} a_\pm^\dagger(\mathbf{p}) u_\pm^\dagger(\mathbf{p}) d^3\mathbf{p}$$

Here, the operators $a_\pm(\mathbf{p}), a_\pm^\dagger(\mathbf{p})$ would be annihilation and creation operators for helicity eigenstates. Such a formalism is not particularly useful in the non-relativistic case, but we mention it here because its analog in the relativistic case will be more significant.

38.4 Fermionic fields

It is an experimentally observed fact that elementary particles with spin $\frac{1}{2}$ behave as fermions and are described by fermionic fields. In non-relativistic quantum field theory, such spin $\frac{1}{2}$ elementary particles could in principle be bosons, described by bosonic fields as in section 38.3.3. There is, however, a "spin-statistics theorem" in relativistic quantum field theory that says that spin $\frac{1}{2}$ fields must be quantized with anticommutators. This provides an explanation of the observed correlation of values of the spin and of the particle statistics, due to the fact that the non-relativistic theories describing fundamental particles should be low-energy limits of relativistic theories.

The discussion of the symplectic and unitary group actions on $\mathcal{H}_1$ of section 38.1 has a straightforward analog in the case of a single-particle state space $\mathcal{H}_1$ with Hermitian inner product describing fermions, rather than bosons. The analog of the infinite dimensional symplectic group action (preserving the imaginary part of the Hermitian inner product) of the bosonic case is an infinite dimensional orthogonal group action (preserving the real part of the Hermitian inner product) in the fermionic case. The multiparticle state space will be an infinite dimensional version of the spinor representation for this orthogonal group. As in the bosonic case, there will be an infinite dimensional unitary group preserving the full Hermitian inner product, and the groups of symmetries we will be interested in will be subgroups of this group.

In section 31.3, we saw in finite dimensions how unitary group actions on a fermionic phase space gave a unitary representation on the fermionic oscillator state space, by the same method of annihilation and creation operators as in the bosonic case (changing commutators to anticommutators). Applying this to the infinite dimensional case of the single-particle space $\mathcal{H}_1$ of solutions to the free particle Schrödinger equation is done by taking

$$\theta_j \to \Psi(x) \quad \text{or} \quad A(p)$$

$$\overline{\theta}_j \to \overline{\Psi}(x) \quad \text{or} \quad \overline{A}(p)$$

$$\{\theta_j, \overline{\theta}_k\}_+ = \delta_{jk} \to \{\Psi(x), \overline{\Psi}(x')\}_+ = \delta(x - x') \quad \text{or} \quad \{A(p), \overline{A}(p')\}_+ = \delta(p - p')$$

Quantization generalizes the construction of the spinor representation from sections 31.3 and 31.4 to the $\mathcal{H}_1$ case, taking

$$a_{Fj} \to \widehat{\Psi}(x) \quad \text{or} \quad a(p)$$

$$a_{Fj}^\dagger \to \widehat{\Psi}^\dagger(x) \quad \text{or} \quad a^\dagger(p)$$

$$[a_{Fj}, a_{Fk}^\dagger]_+ = \delta_{jk} \to [\widehat{\Psi}(x), \widehat{\Psi}^\dagger(x')]_+ = \delta(x - x') \quad \text{or} \quad [a(p), a^\dagger(p')]_+ = \delta(p - p')$$

Quadratic combinations of the $\theta_j, \overline{\theta}_j$ give the Lie algebra of orthogonal transformations of the phase space M. We will again be interested in the generalization to $M = \mathcal{H}_1$, but for very specific quadratic combinations, corresponding to certain finite dimensional Lie algebras of unitary transformations of $\mathcal{H}_1$. Quantization will take these to quadratic combinations of fermionic field operators, giving a Lie algebra representation on the fermionic state space. We get the same formulas for operators $\widehat{N}$ (equation 38.3), $\widehat{X}$ (equation 38.5), $\widehat{\mathbf{P}}$ (equation 38.11), $\widehat{\mathbf{L}}$ (equation 38.12), and $\widehat{\mathbf{S}}$ (equation 38.14), but with anticommuting field operators. These give unitary representations on the multiparticle state space of the Lie algebras of $U(1)$, $U(n)$, $\mathbf{R}^3$ translations and $SO(3)$ rotations, respectively. For the free particle, these operators commute with the Hamiltonian and act as symmetries on the state space.

38.5 For further reading

The material of this chapter is often developed in conventional quantum field theory texts in the context of relativistic rather than non-relativistic quantum field theory. Symmetry generators are also more often derived via Lagrangian methods (Noether's theorem) rather than the Hamiltonian methods used here. For an example of a detailed physics textbook discussion relatively close to this one, getting quadratic operators based on group actions on the space of solutions to field equations, see [35].

Chapter 39
Quantization of Infinite dimensional Phase Spaces

While finite dimensional Lie groups and their representations are rather well-understood mathematical objects, this is not at all true for infinite dimensional Lie groups, where only a fragmentary such understanding is available. In earlier chapters, we have studied in detail what happens when quantizing a finite dimensional phase space, bosonic or fermionic. In these cases, a finite dimensional symplectic or orthogonal group acts and quantization uses a representation of these groups. For the case of quantum field theories with their infinite dimensional phase spaces, the symplectic or orthogonal groups acting on these spaces will be infinite dimensional. In this chapter, we will consider some of the new phenomena that arise when one looks for infinite dimensional analogs of the role these groups and their representations play in quantum theory in the finite dimensional case.

The most important difference in the infinite dimensional case is that the Stone–von Neumann theorem and its analog for Clifford algebras no longer hold. One no longer has a unique (up to unitary equivalence) representation of the canonical commutation (or anticommutation) relations. It turns out that only for a restricted sort of infinite dimensional symplectic or orthogonal group does one recover the Stone–von Neumann uniqueness of the finite dimensional case, and even then new phenomena appear. The arbitrary constants found in the definition of the moment map now cannot be ignored but may appear in commutation relations, leading to something called an "anomaly."

Physically, new phenomena due to an infinite number of degrees of freedom can have their origin in the degrees of freedom occurring at arbitrarily short distances ("ultraviolet divergences"), but also can be due to degrees of freedom corresponding to large distances. In the application of quantum field theories to the study of condensed matter systems, it is the second of these that is relevant, since the atomic scale provides a cutoff distance scale below which there are no degrees of freedom.

© Peter Woit 2017
P. Woit, *Quantum Theory, Groups and Representations*,
DOI 10.1007/978-3-319-64612-1_39

For general interacting quantum field theories, one must choose among inequivalent possibilities for representations of the canonical commutation relations, finding one on which the operators of the interacting field theory are well defined. This makes interacting quantum field theory a much more complex subject than free field theory and is the source of well-known difficulties with infinities that appear when standard calculational methods are applied. A proper definition of an interacting quantum field theory generally requires introducing cutoffs that make the number of degrees of freedom finite so that standard properties used in the finite dimensional case still hold, then studying what happens as the cutoffs are removed, trying to find a physically sensible limit ("renormalization").

The reader is warned that this chapter is of a much sketchier nature than earlier ones, intended only to indicate some outlines of how certain foundational ideas about representation theory and quantization developed for the finite dimensional case apply to quantum field theory. This material will not play a significant role in later chapters.

39.1 Inequivalent irreducible representations

In our discussion of quantization, an important part of this story was the Stone–von Neumann theorem, which says that the Heisenberg group has only one interesting irreducible representation, up to unitary equivalence (the Schrödinger representation). In infinite dimensions, this is no longer true: there will be an infinite number of inequivalent irreducible representations, with no known complete classification of the possibilities. Before one can even begin to compute things like expectation values of observables, one needs to find an appropriate choice of representation, adding a new layer of difficulty to the problem that goes beyond that of just increasing the number of degrees of freedom.

To get some idea of how the Stone–von Neumann theorem can fail, one can consider the Bargmann–Fock quantization of the harmonic oscillator degrees of freedom and the coherent states (see section 23.2)

$$|\alpha\rangle = D(\alpha)|0\rangle = e^{\alpha a^\dagger - \overline{\alpha} a}|0\rangle$$

where $D(\alpha)$ is a unitary operator. These satisfy

$$|\langle\alpha|0\rangle|^2 = e^{-|\alpha|^2}$$

Each choice of α gives a different, unitarily equivalent using $D(\alpha)$, representation of the Heisenberg group. This is on the space spanned by

$$D(\alpha)(a^\dagger)^n|0\rangle = (a(\alpha)^\dagger)^n|\alpha\rangle$$

where

$$a(\alpha)^\dagger = D(\alpha)a^\dagger D^{-1}(\alpha)$$

This is for $d = 1$, for arbitrary d one gets states parametrized by a vector $\alpha \in \mathbf{C}^d$, and

$$|\langle\alpha|0\rangle|^2 = e^{-\sum_{j=1}^d |\alpha_j|^2}$$

In the infinite dimensional case, for any sequence of α_j with divergent $\sum_{j=1}^\infty |\alpha_j|^2$, one will have

$$\langle\alpha|0\rangle = 0$$

For each such sequence α, this leads to a different representation of the Heisenberg group, spanned by acting with various products of the

$$a_j(\alpha)^\dagger = D(\alpha)a_j^\dagger D^{-1}(\alpha)$$

on $|\alpha\rangle$.

These representations will all be unitarily inequivalent. To show that the representation built on $|\alpha\rangle$ is inequivalent to the one built on $|0\rangle$, one shows that $|\alpha\rangle$ is not only orthogonal to $|0\rangle$ but to all the other $(a_j^\dagger)^n|0\rangle$ also. This is true because one has (see equation 23.8)

$$a_j(\alpha) = D(\alpha)a_j D^{-1}(\alpha) = a_j - \alpha_j$$

so

$$
\begin{aligned}
\langle 0|a_j^n|\alpha\rangle &= \langle 0|a_j^{n-1}(a_j(\alpha) + \alpha_j)|\alpha\rangle \\
&= \alpha_j\langle 0|a_j^{n-1}|\alpha\rangle = \ldots \\
&= \alpha_j^n\langle 0|\alpha\rangle = 0
\end{aligned}
$$

Examples of this kind of phenomenon can occur in quantum field theories, in cases where it is energetically favorable for many quanta of the field to "condense" into the lowest energy state. This could be a state like $|\alpha\rangle$, with

$$\sum_{j=1}^d \langle\alpha|a_j^\dagger a_j|\alpha\rangle$$

having a physical interpretation in terms of a nonzero particle density in the condensate state $|\alpha\rangle$.

Other examples of this phenomenon can be constructed by considering changes in the complex structure J used to define the Bargmann–Fock construction of the representation. For finite d, representations defined using $|0\rangle_J$ for different complex structures are all unitarily equivalent, but this can fail in the limit as d goes to infinity.

In both the standard oscillator case with $Sp(2d, \mathbf{R})$ acting, and the fermionic oscillator case with $SO(2d, \mathbf{R})$ acting, we found that there were "Bogoliubov transformations": elements of the group not in the $U(d)$ subgroup distinguished by the choice of J, which acted non-trivially on $|0\rangle_J$, taking it to a different state. As in the case of the Heisenberg group action on coherent states above, such action by Bogoliubov transformations can, in the limit of $d \to \infty$, take $|0\rangle$ to an orthogonal state. This introduces the possibility of inequivalent representations of the commutation relations, built by applying operators to orthogonal ground states. The physical interpretation again is that such states correspond to condensates of quanta. For the usual bosonic oscillator case, this phenomenon occurs in the theory of superfluidity, for fermionic oscillators it occurs in the theory of superconductivity. It was in the study of such systems that Bogoliubov discovered the transformations that now bear his name.

39.2 The restricted symplectic group

If one restricts the class of complex structures J to ones not that different from the standard one J_0, then one can recover a version of the Stone–von Neumann theorem and have much the same behavior as in the finite dimensional case. Note that for each invertible linear map g on phase space, g acts on the complex structure (see equation 26.6), taking J_0 to a complex structure we will call J_g. One can define subgroups of the infinite dimensional symplectic or orthogonal groups as follows:

Definition (Restricted symplectic and orthogonal groups). *The group of linear transformations g of an infinite dimensional symplectic vector space preserving the symplectic structure and also satisfying the condition*

$$tr(A^\dagger A) < \infty$$

on the operator

$$A = [J_g, J_0]$$

is called the restricted symplectic group and denoted Sp_{res}. The group of linear transformations g of an infinite dimensional inner product space preserving the inner product and satisfying the same condition as above on $[J_g, J_0]$ is called the restricted orthogonal group and denoted SO_{res}.

An operator A satisfying $tr(A^\dagger A) < \infty$ is said to be a Hilbert–Schmidt operator.

One then has the following replacement for the Stone–von Neumann theorem:

Theorem. *Given two complex structures J_1, J_2 on a Hilbert space such that $[J_1, J_2]$ is Hilbert–Schmidt, acting on the states*

$$|0\rangle_{J_1}, \quad |0\rangle_{J_2}$$

by annihilation and creation operators will give unitarily equivalent representations of the Weyl algebra (in the bosonic case) or the Clifford algebra (in the fermionic case).

The standard reference for the proof of this statement is the original papers of Shale [79] and Shale–Stinespring [80]. A detailed discussion of the theorem can be found in [64].

For some motivation for this theorem, consider the finite dimensional case studied in section 25.5 (this is for the symplectic group case, a similar calculation holds in the orthogonal group case). Elements of $\mathfrak{sp}(2d, \mathbf{R})$ corresponding to Bogoliubov transformations (i.e., with nonzero commutator with J_0) were of the form

$$\frac{1}{2} \sum_{jk} (B_{jk} z_j z_k + \overline{B}_{jk} \overline{z}_j \overline{z}_k)$$

for symmetric complex matrices B. These acted on the metaplectic representation by

$$-\frac{i}{2} \sum_{jk} (B_{jk} a_j^\dagger a_k^\dagger + \overline{B}_{jk} a_j a_k) \tag{39.1}$$

and commuting two of them gave a result (equation 25.9) corresponding to quantization of an element of the $\mathfrak{u}(d)$ subgroup, differing from its normal-ordered version by a term

$$-\frac{1}{2} tr(B\overline{C} - C\overline{B})\mathbf{1} = -\frac{1}{2} tr(BC^\dagger - CB^\dagger)\mathbf{1}$$

For $d = \infty$, this trace, in general, will be infinite and undefined. An alternate characterization of Hilbert–Schmidt operators is that for B and C Hilbert–Schmidt operators, the traces

$$tr(BC^\dagger) \quad \text{and} \quad tr(CB^\dagger)$$

will be finite and well defined. So, at least to the extent normal-ordered operators quadratic in annihilation and creation operators are well defined, the Hilbert–Schmidt condition on operators not commuting with the complex structure implies that they will have well-defined commutation relations with each other.

39.3   The anomaly and the Schwinger term

The argument above gives some motivation for the existence as d goes to ∞ of well-defined commutators of operators of the form 39.1 and thus for the existence of an analog of the metaplectic representation for the infinite dimensional Lie algebra $\mathfrak{sp}_{res}$ of Sp_{res}. There is one obvious problem though with this argument, in that while it tells us that normal-ordered operators will have well-defined commutation relations, they are not quite the right commutation relations, due to the occurrence of the extra scalar term

$$-\frac{1}{2}tr(BC^\dagger - CB^\dagger)\mathbf{1}$$

This term is sometimes called the "Schwinger term."

The Schwinger term causes a problem with the standard expectation that given some group G acting on the phase space preserving the Poisson bracket, one should get a unitary representation of G on the quantum state space $\mathcal{H}$. This problem is sometimes called the "anomaly," meaning that the expected unitary Lie algebra representation does not exist (due to extra scalar terms in the commutation relations). Recall from section 15.3 that this potential problem was already visible at the classical level, in the fact that given $L \in \mathfrak{g}$, the corresponding moment map μ_L is only well defined up to a constant. While for the finite dimensional cases we studied, the constants could be chosen so as to make the map

$$L \to \mu_L$$

a Lie algebra homomorphism, that turns out to no longer be true for the case $\mathfrak{g} = \mathfrak{sp}_{res}$ (or $\mathfrak{so}_{res}$) acting on an infinite dimensional phase space. The potential problem of the anomaly is thus already visible classically, but it is only when one constructs the quantum theory, and thus a representation on the state space that one can see whether the problem cannot be removed by a constant shift in the representation operators. This situation, despite its classical origin, is sometimes characterized as a form of symmetry breaking due to the quantization procedure.

Note that this problem will not occur for G that commute with the complex structure, since for these the normal-ordered Lie algebra representation operators will be a true representation of $\mathfrak{u}(\infty) \subset \mathfrak{sp}_{res}$. We will call $U(\infty) \subset Sp_{res}$ the subgroup of elements that commute with J_0 exactly, not just up to a Hilbert–Schmidt operator. It turns out that $G \subset U(\infty)$ for most of the cases we are interested in, allowing construction of the Lie algebra representation by normal-ordered quadratic combinations of the annihilation and creation operators (as in 25.6). Also note that since normal-ordering just shifts operators by something proportional to a constant, when this constant is finite there will be no anomaly since one can get operators with correct commutators by such a finite shift of the normal-ordered ones. The anomaly is an inherently infinite dimensional problem since it is only then that infinite shifts are necessary. When the anomaly does appear, it will appear as a phase ambiguity in the group representation operators (not just a sign ambiguity as in finite dimensional case of $Sp(2d, \mathbf{R})$), and $\mathcal{H}$ will be a projective representation of the group (a representation up to phase).

Such an undetermined phase factor only creates a problem for the action on states, not for the action on operators. Recall that in the finite dimensional case, the action of $Sp(2d, \mathbf{R})$ on operators (see 20.3) is independent of any constant shift in the Lie algebra representation operators. Equivalently, if one has a unitary projective representation on states, the phase ambiguity cancels out in the action on operators, which is by conjugation.

39.4 Spontaneous symmetry breaking

In the standard Bargmann–Fock construction, there is a unique state $|0\rangle$, and for the Hamiltonian of the free particle quantum field theory, this will be the lowest energy state. In interacting quantum field theories, one may have state spaces unitarily inequivalent to the standard Bargmann–Fock one. These can have their own annihilation and creation operators, and thus a notion of particle number and a particle number operator $\widehat{N}$, but the lowest energy $|0\rangle$ may not have the properties

$$\widehat{N}|0\rangle = 0, \quad e^{-i\theta\widehat{N}}|0\rangle = |0\rangle$$

Instead, the state $|0\rangle$ gets taken by $e^{-i\theta\widehat{N}}$ to some other state, with

$$\widehat{N}|0\rangle \neq 0, \quad e^{-i\theta\widehat{N}}|0\rangle \equiv |\theta\rangle \neq |0\rangle \quad (\text{for } \theta \neq 0)$$

and the vacuum state not an eigenstate of $\widehat{N}$, so it does not have a well-defined particle number. If $[\widehat{N}, \widehat{H}] = 0$, the states $|\theta\rangle$ will all have the same energy as $|0\rangle$ and there will be a multiplicity of different vacuum states, labeled by θ. In such a case, the $U(1)$ symmetry is said to be "spontaneously broken." This phenomenon occurs when non-relativistic quantum field theory is used to describe a superconductor. There the lowest energy state will be a state without a definite particle number, with electrons pairing up in a way that allows them to lower their energy, "condensing" in the lowest energy state.

When, as for the multicomponent free particle (the Hamiltonian of equation 38.4), the Hamiltonian is invariant under $U(n)$ transformations of the fields ψ_j, then we will have

$$[\widehat{X}, \widehat{H}] = 0$$

for $\widehat{X}$ the operator giving the Lie algebra representation of $U(n)$ on the mul-
tiparticle state space (see section 38.2.2). In this case, if $|0\rangle$ is invariant under
the $U(n)$ symmetry, then energy eigenstates of the quantum field theory will
break up into irreducible representations of $U(n)$ and can be labeled accordingly. As in the $U(1)$ case, the $U(n)$ symmetry may be spontaneously broken,
with

$$\widehat{X}|0\rangle \neq 0$$

for some directions X in $\mathfrak{u}(n)$. When this happens, just as in the $U(1)$ case
states did not have well-defined particle number, now they will not carry
well-defined irreducible $U(n)$ representation labels.

39.5 Higher-order operators and renormalization

We have generally restricted ourselves to considering only products of basis
elements of the Heisenberg Lie algebra (position and momentum in the finite
dimensional case, fields in the infinite dimensional case) of degree less than or
equal to two, since it is these that after quantization have an interpretation
as the operators of a Lie algebra representation. In the finite dimensional
case, one can consider higher-order products of operators, for instance systems with Hamiltonian operators of higher order than quadratic. Unlike the
quadratic case, typically no exact solution for eigenvectors and eigenvalues
will exist, but various approximation methods may be available. In particular, for Hamiltonians that are quadratic plus a term with a small parameter,
perturbation theory methods can be used to compute a power series approximation in the small parameter. This is an important topic in physics, covered
in detail in the standard textbooks.

The standard approach to quantization of infinite dimensional systems is
to begin with "regularization," somehow modifying the system to only have
a finite dimensional phase space, for instance by introducing cutoffs that
make the possible momenta discrete and finite. One quantizes this theory by
taking the state space and canonical commutation relations to be the unique
ones for the Heisenberg Lie algebra, somehow dealing with the calculational
difficulties in the interacting case (non-quadratic Hamiltonian).

One then tries to take a limit that recovers the infinite dimensional system. Such a limit will generally be quite singular, leading to an infinite result,
and the process of manipulating these potential infinities is called "renormalization." Techniques for taking limits of this kind in a manner that leads to
a consistent and physically sensible result typically take up a large part of
standard quantum field theory textbooks. For many theories, no appropriate
such techniques are known, and conjecturally none are possible. For others

there is good evidence that such a limit can be successfully taken, but the details of how to do this remain unknown, with for instance a \$1 million Millenium Prize offered for showing rigorously this is possible in the case of Yang–Mills gauge theory (the Hamiltonian, in this case, will be discussed in chapter 46).

39.6   For further reading

Berezin's *The Method of Second Quantization* [6] develops in detail the infinite dimensional version of the Bargmann–Fock construction, both in the bosonic and fermionic cases. Infinite dimensional versions of the metaplectic and spinor representations are given there in terms of operators defined by integral kernels. For a discussion of the infinite dimensional Weyl and Clifford algebras, together with a realization of their automorphism groups Sp_{res} and O_{res} (and the corresponding Lie algebras) in terms of annihilation and creation operators acting on the infinite dimensional metaplectic and spinor representations, see [64]. The book [70] contains an extensive discussion of the groups Sp_{res} and O_{res} and the infinite dimensional version of their metaplectic and spinor representations. It emphasizes the origin of novel infinite dimensional phenomena in the geometry of the complex structures used in infinite dimensional examples.

The use of Bogoliubov transformations in the theories of superfluidity and superconductivity is a standard topic in quantum field theory textbooks that emphasize condensed matter applications, see, for example, [53]. The book [11] discusses in detail the occurrence of inequivalent representations of the commutation relations in various physical systems.

For a discussion of "Haag's theorem," which can be interpreted as showing that to describe an interacting quantum field theory, one must use a representation of the canonical commutation relations inequivalent to the one for free field theory, see [19].

Chapter 40
Minkowski Space and the Lorentz Group

For the case of non-relativistic quantum mechanics, we saw that systems with an arbitrary number of particles, bosons or fermions, could be described by taking as dual phase space the state space $\mathcal{H}_1$ of the single-particle quantum theory. This space is infinite dimensional, but it is linear and it can be quantized using the same techniques that work for the finite dimensional harmonic oscillator. This is an example of a quantum field theory since it is a space of functions that is being quantized.

We would like to find some similar way to proceed for the case of relativistic systems, finding relativistic quantum field theories capable of describing arbitrary numbers of particles, with the energy–momentum relationship $E^2 = |\mathbf{p}|^2 c^2 + m^2 c^4$ characteristic of special relativity, not the non-relativistic limit $|\mathbf{p}| \ll mc$ where $E = \frac{|\mathbf{p}|^2}{2m}$. In general, a phase space can be thought of as the space of initial conditions for an equation of motion, or equivalently, as the space of solutions of the equation of motion. In the non-relativistic field theory, the equation of motion is the first order in time Schrödinger equation, and the phase space is the space of fields (wavefunctions) at a specified initial time, say $t = 0$. This space carries a representation of the time-translation group $\mathbf{R}$ and the Euclidean group $E(3) = \mathbf{R}^3 \rtimes SO(3)$. To construct a relativistic quantum field theory, we want to find an analog of this space of wavefunctions. It will be some sort of linear space of functions satisfying an equation of motion, and we will then quantize by applying harmonic oscillator methods.

Just as in the non-relativistic case, the space of solutions to the equation of motion provides a representation of the group of space–time symmetries of the theory. This group will now be the Poincaré group, a ten-dimensional group which includes a four-dimensional subgroup of translations in space–time, and a six-dimensional subgroup (the Lorentz group), which combines spatial rotations and "boosts" (transformations mixing spatial and time coor-

© Peter Woit 2017

P. Woit, *Quantum Theory, Groups and Representations*,

DOI 10.1007/978-3-319-64612-1_40

dinates). The representation of the Poincaré group on the solutions to the relativistic wave equation will in general be reducible. Irreducible such representations will be the objects corresponding to elementary particles. This chapter will deal with the Lorentz group itself, chapter 41 with its representations, and chapter 42 will move on to the Poincaré group and its representations.

40.1 Minkowski space

Special relativity is based on the principle that one should consider space and time together, and take them to be a four-dimensional space $\mathbf{R}^4$ with an indefinite inner product:

Definition (Minkowski space). *Minkowski space M^4 is the vector space $\mathbf{R}^4$ with an indefinite inner product given by*

$$(x, y) \equiv x \cdot y = -x_0 y_0 + x_1 y_1 + x_2 y_2 + x_3 y_3$$

where (x_0, x_1, x_2, x_3) are the coordinates of $x \in \mathbf{R}^4$, (y_0, y_1, y_2, y_3) the coordinates of $y \in \mathbf{R}^4$.

Digression. *We have chosen to use the $-+++$ instead of the $+---$ sign convention for the following reasons:*

- *Analytically continuing the time variable x_0 to ix_0 gives a positive-definite inner product.*

- *Restricting to spatial components, there is no change from our previous formulas for the symmetries of Euclidean space $E(3)$.*

- *Only for this choice will we have a real (as opposed to complex) spinor representation (since $\mathrm{Cliff}(3,1) = M(4, \mathbf{R}) \neq \mathrm{Cliff}(1,3)$).*

- *Weinberg's quantum field theory textbook [99] uses this convention (although, unlike him, we will put the 0 component first).*

This inner product will also sometimes be written using the matrix

$$\eta_{\mu\nu} = \begin{pmatrix} -1 & 0 & 0 & 0 \\ 0 & 1 & 0 & 0 \\ 0 & 0 & 1 & 0 \\ 0 & 0 & 0 & 1 \end{pmatrix}$$

as

$$x \cdot y = \sum_{\mu,\nu=0}^{3} \eta_{\mu\nu} x_\mu y_\nu$$

Digression (Upper and lower indices). *In many physics texts, it is conventional in discussions of special relativity to write formulas using both upper and lower indices, related by*

$$x_\mu = \sum_{\nu=0}^{3} \eta_{\mu\nu} x^\nu = \eta_{\mu\nu} x^\nu$$

with the last form of this using the Einstein summation convention.

One motivation for introducing both upper and lower indices is that special relativity is a limiting case of general relativity, which is a fully geometrical theory based on taking space–time to be a manifold M with a metric g that varies from point to point. In such a theory, it is important to distinguish between elements of the tangent space $T_x(M)$ at a point $x \in M$ and elements of its dual, the cotangent space $T_x^*(M)$, while using the fact that the metric g provides an inner product on $T_x(M)$ and thus an isomorphism $T_x(M) \simeq T_x^*(M)$. In the special relativity case, this distinction between $T_x(M)$ and $T_x^*(M)$ just comes down to an issue of signs, but the upper and lower index notation is useful for keeping track of those.

A second motivation is that position and momenta naturally live in dual vector spaces, so one would like to distinguish between the vector space M^4 of positions and the dual vector space of momenta. In the case though of a vector space like M^4 which comes with a fixed inner product $\eta_{\mu\nu}$, this inner product gives a fixed identification of M^4 and its dual, an identification that is also an identification as representations of the Lorentz group. Given this fixed identification, we will not here try and distinguish by notation whether a vector is in M^4 or its dual, so will just use lower indices, not both upper and lower indices.

The coordinates x_1, x_2, x_3 are interpreted as spatial coordinates, and the coordinate x_0 is a time coordinate, related to the conventional time coordinate t with respect to chosen units of time and distance by $x_0 = ct$ where c is the speed of light. Mostly we will assume units of time and distance have been chosen so that $c = 1$.

Vectors $v \in M^4$ such that $|v|^2 = v \cdot v > 0$ are called "space-like," those with $|v|^2 < 0$ "time-like" and those with $|v|^2 = 0$ are said to lie on the "light cone." Suppressing one space dimension, the picture to keep in mind of Minkowski space looks like this:

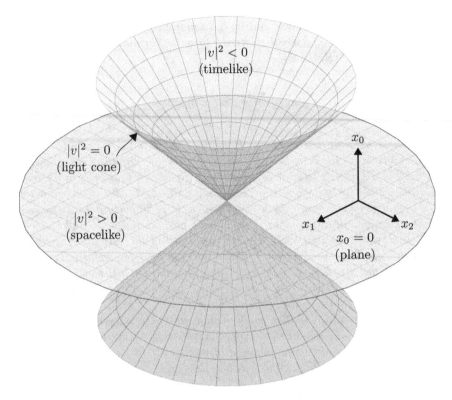

Figure 40.1: Light cone structure of Minkowski spacetime.

40.2   The Lorentz group and its Lie algebra

Recall that in 3 dimensions the group of linear transformations of $\mathbf{R}^3$ preserving the standard inner product was the group $O(3)$ of 3 by 3 orthogonal matrices. This group has two disconnected components: $SO(3)$, the subgroup of orientation preserving (determinant $+1$) transformations, and a component of orientation reversing (determinant -1) transformations. In Minkowski space, one has:

Definition (Lorentz group). *The Lorentz group $O(3,1)$ is the group of linear transformations preserving the Minkowski space inner product on $\mathbf{R}^4$.*

In terms of matrices, the condition for a 4 by 4 matrix Λ to be in $O(3,1)$ will be

$$\Lambda^T \begin{pmatrix} -1 & 0 & 0 & 0 \\ 0 & 1 & 0 & 0 \\ 0 & 0 & 1 & 0 \\ 0 & 0 & 0 & 1 \end{pmatrix} \Lambda = \begin{pmatrix} -1 & 0 & 0 & 0 \\ 0 & 1 & 0 & 0 \\ 0 & 0 & 1 & 0 \\ 0 & 0 & 0 & 1 \end{pmatrix}$$

The Lorentz group has four components, with the component of the identity a subgroup called $SO(3,1)$ (which some call $SO^+(3,1)$). The other three components arise by multiplication of elements in $SO(3,1)$ by P, T, PT where

$$
P = \begin{pmatrix} 1 & 0 & 0 & 0 \\ 0 & -1 & 0 & 0 \\ 0 & 0 & -1 & 0 \\ 0 & 0 & 0 & -1 \end{pmatrix}
$$

is called the "parity" transformation, reversing the orientation of the spatial variables, and

$$
T = \begin{pmatrix} -1 & 0 & 0 & 0 \\ 0 & 1 & 0 & 0 \\ 0 & 0 & 1 & 0 \\ 0 & 0 & 0 & 1 \end{pmatrix}
$$

reverses the time orientation.

The Lorentz group has a subgroup $SO(3)$ of transformations that just act on the spatial components, given by matrices of the form

$$
\Lambda = \begin{pmatrix} 1 & 0 & 0 & 0 \\ 0 & & & \\ 0 & & R & \\ 0 & & & \end{pmatrix}
$$

where R is in $SO(3)$. For each pair j, k of spatial directions one has the usual $SO(2)$ subgroup of rotations in the jk plane, but now in addition for each pair $0, j$ of the time direction with a spatial direction, one has $SO(1,1)$ subgroups of matrices of transformations called "boosts" in the j direction. For example, for $j = 1$, one has the subgroup of $SO(3,1)$ of matrices of the form

$$
\Lambda = \begin{pmatrix} \cosh\phi & \sinh\phi & 0 & 0 \\ \sinh\phi & \cosh\phi & 0 & 0 \\ 0 & 0 & 1 & 0 \\ 0 & 0 & 0 & 1 \end{pmatrix}
$$

for $\phi \in \mathbf{R}$.

The Lorentz group is six dimensional. For a basis of its Lie algebra, one can take six matrices $M_{\mu\nu}$ for $\mu, \nu \in 0, 1, 2, 3$ and $j < k$. For the spatial indices, these are

$$
M_{12} = \begin{pmatrix} 0 & 0 & 0 & 0 \\ 0 & 0 & -1 & 0 \\ 0 & 1 & 0 & 0 \\ 0 & 0 & 0 & 0 \end{pmatrix}, \quad
M_{13} = \begin{pmatrix} 0 & 0 & 0 & 0 \\ 0 & 0 & 0 & 1 \\ 0 & 0 & 0 & 0 \\ 0 & -1 & 0 & 0 \end{pmatrix}, \quad
M_{23} = \begin{pmatrix} 0 & 0 & 0 & 0 \\ 0 & 0 & 0 & 0 \\ 0 & 0 & 0 & -1 \\ 0 & 0 & 1 & 0 \end{pmatrix}
$$

which correspond to the basis elements of the Lie algebra of $SO(3)$ that we first saw in chapter 6. These can be renamed using the same names as earlier

$$l_1 = M_{23}, \quad l_2 = M_{13}, \quad l_3 = M_{12}$$

and recall that these satisfy the $\mathfrak{so}(3)$ commutation relations

$$[l_1, l_2] = l_3, \quad [l_2, l_3] = l_1, \quad [l_3, l_1] = l_2$$

and correspond to infinitesimal rotations about the three spatial axes.

Taking the first index 0, one gets three elements corresponding to infinitesimal boosts in the three spatial directions

$$M_{01} = \begin{pmatrix} 0 & 1 & 0 & 0 \\ 1 & 0 & 0 & 0 \\ 0 & 0 & 0 & 0 \\ 0 & 0 & 0 & 0 \end{pmatrix}, M_{02} = \begin{pmatrix} 0 & 0 & 1 & 0 \\ 0 & 0 & 0 & 0 \\ 1 & 0 & 0 & 0 \\ 0 & 0 & 0 & 0 \end{pmatrix}, M_{03} = \begin{pmatrix} 0 & 0 & 0 & 1 \\ 0 & 0 & 0 & 0 \\ 0 & 0 & 0 & 0 \\ 1 & 0 & 0 & 0 \end{pmatrix}$$

These can be renamed as

$$k_1 = M_{01}, \quad k_2 = M_{02}, \quad k_3 = M_{03}$$

One can easily calculate the commutation relations between the k_j and l_j, which show that the k_j transform as a vector under infinitesimal rotations. For instance, for infinitesimal rotations about the x_1 axis, one finds

$$[l_1, k_1] = 0, \quad [l_1, k_2] = k_3, \quad [l_1, k_3] = -k_2 \tag{40.1}$$

Commuting infinitesimal boosts, one gets infinitesimal spatial rotations

$$[k_1, k_2] = -l_3, \quad [k_3, k_1] = -l_2, \quad [k_2, k_3] = -l_1 \tag{40.2}$$

Digression. *A more conventional notation in physics is to use $J_j = il_j$ for infinitesimal rotations and $K_j = ik_j$ for infinitesimal boosts. The intention of the different notation used here is to start with basis elements of the real Lie algebra $\mathfrak{so}(3,1)$, (the l_j and k_j) which are purely real objects, before complexifying and considering representations of the Lie algebra.*

Taking the following complex linear combinations of the l_j and k_j

$$A_j = \frac{1}{2}(l_j + ik_j), \quad B_j = \frac{1}{2}(l_j - ik_j)$$

one finds

$$[A_1, A_2] = A_3, \quad [A_3, A_1] = A_2, \quad [A_2, A_3] = A_1$$

and
$$[B_1, B_2] = B_3, \quad [B_3, B_1] = B_2, \quad [B_2, B_3] = B_1$$

This construction of the A_j, B_j requires that we complexify (allow complex linear combinations of basis elements) the Lie algebra $\mathfrak{so}(3,1)$ of $SO(3,1)$ and work with the complex Lie algebra $\mathfrak{so}(3,1) \otimes \mathbf{C}$. It shows that this Lie algebra splits into a sum of two sub-Lie algebras, which are each copies of the (complexified) Lie algebra of $SO(3)$, $\mathfrak{so}(3) \otimes \mathbf{C}$. Since

$$\mathfrak{so}(3) \otimes \mathbf{C} = \mathfrak{su}(2) \otimes \mathbf{C} = \mathfrak{sl}(2, \mathbf{C})$$

we have

$$\mathfrak{so}(3,1) \otimes \mathbf{C} = \mathfrak{sl}(2, \mathbf{C}) \oplus \mathfrak{sl}(2, \mathbf{C})$$

In section 40.4, we will see the origin of this phenomenon at the group level.

40.3   The Fourier transform in Minkowski space

One can define a Fourier transform with respect to the four space–time variables, which will take functions of x_0, x_1, x_2, x_3 to functions of the Fourier transform variables p_0, p_1, p_2, p_3:

Definition (Minkowski space Fourier transform). *The Fourier transform of a function f on Minkowski space is given by*

$$\begin{aligned}
\widetilde{f}(p) &= \frac{1}{(2\pi)^2} \int_{M^4} e^{-ip \cdot x} f(x) d^4 x \\
&= \frac{1}{(2\pi)^2} \int_{M^4} e^{-i(-p_0 x_0 + p_1 x_1 + p_2 x_2 + p_3 x_3)} f(x) dx_0 d^3 \mathbf{x}
\end{aligned}$$

In this case, the Fourier inversion formula is

$$f(x) = \frac{1}{(2\pi)^2} \int_{M^4} e^{ip \cdot x} \widetilde{f}(p) d^4 p \tag{40.3}$$

Note that our definition puts one factor of $\frac{1}{\sqrt{2\pi}}$ with each Fourier (or inverse Fourier) transform with respect to a single variable. A common alternate convention among physicists is to put all factors of 2π with the p integrals (and thus in the inverse Fourier transform), none in the definition of $\widetilde{f}(p)$, the Fourier transform itself.

The sign change between the time and space variables that occurs in the exponent of this definition is there to ensure that this exponent is Lorentz invariant. Since Lorentz transformations have determinant 1, the measure d^4x will be Lorentz invariant and the Fourier transform of a function will behave under Lorentz transformations in the same ways as the function

$$\begin{aligned}
\widetilde{f}(\Lambda^{-1}p) &= \frac{1}{(2\pi)^2} \int_{M^4} e^{-i(\Lambda^{-1}p)\cdot x} f(x) d^4x \\
&= \frac{1}{(2\pi)^2} \int_{M^4} e^{-ip\cdot \Lambda x} f(x) d^4x \\
&= \frac{1}{(2\pi)^2} \int_{M^4} e^{-ip\cdot x} f(\Lambda^{-1}x) d^4x
\end{aligned}$$

The reason why one conventionally defines the Hamiltonian operator as $i\frac{\partial}{\partial t}$ (with eigenvalues $p_0 = E$) but the momentum operator P_j as $-i\frac{\partial}{\partial x_j}$ (with eigenvalues p_j) is due to this Lorentz invariant choice of the Fourier transform.

40.4 Spin and the Lorentz group

Just as the groups $SO(n)$ have double covers $Spin(n)$, the group $SO(3,1)$ has a double cover $Spin(3,1)$, which we will show can be identified with the group $SL(2,\mathbf{C})$ of 2 by 2 complex matrices with unit determinant. This group will have the same Lie algebra as $SO(3,1)$, and we will sometimes refer to either group as the "Lorentz group."

Recall from chapter 6 that for $SO(3)$, the spin double cover $Spin(3)$ can be identified with either $Sp(1)$ (the unit quaternions) or $SU(2)$, and then the action of $Spin(3)$ as $SO(3)$ rotations of $\mathbf{R}^3$ was given by conjugation of imaginary quaternions (using $Sp(1)$) or certain 2 by 2 complex matrices (using $SU(2)$). In the $SU(2)$ case, this was done explicitly by identifying

$$(x_1, x_2, x_3) \leftrightarrow \begin{pmatrix} x_3 & x_1 - ix_2 \\ x_1 + ix_2 & -x_3 \end{pmatrix}$$

and then showing that conjugating this matrix by an element of $SU(2)$ was a linear map leaving invariant

$$\det \begin{pmatrix} x_3 & x_1 - ix_2 \\ x_1 + ix_2 & -x_3 \end{pmatrix} = -(x_1^2 + x_2^2 + x_3^2)$$

and thus a rotation in $SO(3)$.

The same sort of thing works for the Lorentz group case. Now we identify $\mathbf{R}^4$ with the space of 2 by 2 complex self-adjoint matrices by

$$(x_0, x_1, x_2, x_3) \leftrightarrow \begin{pmatrix} x_0 + x_3 & x_1 - ix_2 \\ x_1 + ix_2 & x_0 - x_3 \end{pmatrix}$$

and observe that

$$\det \begin{pmatrix} x_0 + x_3 & x_1 - ix_2 \\ x_1 + ix_2 & x_0 - x_3 \end{pmatrix} = x_0^2 - x_1^2 - x_2^2 - x_3^2$$

This provides a very useful way to think of Minkowski space: as complex self-adjoint 2 by 2 matrices, with norm-squared minus the determinant of the matrix.

The linear transformation

$$\begin{pmatrix} x_0 + x_3 & x_1 - ix_2 \\ x_1 + ix_2 & x_0 - x_3 \end{pmatrix} \rightarrow \Omega \begin{pmatrix} x_0 + x_3 & x_1 - ix_2 \\ x_1 + ix_2 & x_0 - x_3 \end{pmatrix} \Omega^\dagger \qquad (40.4)$$

for $\Omega \in SL(2, \mathbf{C})$ preserves the determinant and thus the inner product, since

$$\det(\Omega \begin{pmatrix} x_0 + x_3 & x_1 - ix_2 \\ x_1 + ix_2 & x_0 - x_3 \end{pmatrix} \Omega^\dagger) = (\det \Omega) \det \begin{pmatrix} x_0 + x_3 & x_1 - ix_2 \\ x_1 + ix_2 & x_0 - x_3 \end{pmatrix} (\det \Omega^\dagger)$$

$$= x_0^2 - x_1^2 - x_2^2 - x_3^2$$

It also takes self-adjoint matrices to self-adjoints, and thus $\mathbf{R}^4$ to $\mathbf{R}^4$, since

$$(\Omega \begin{pmatrix} x_0 + x_3 & x_1 - ix_2 \\ x_1 + ix_2 & x_0 - x_3 \end{pmatrix} \Omega^\dagger)^\dagger = (\Omega^\dagger)^\dagger \begin{pmatrix} x_0 + x_3 & x_1 - ix_2 \\ x_1 + ix_2 & x_0 - x_3 \end{pmatrix}^\dagger \Omega^\dagger$$

$$= \Omega \begin{pmatrix} x_0 + x_3 & x_1 - ix_2 \\ x_1 + ix_2 & x_0 - x_3 \end{pmatrix} \Omega^\dagger$$

Note that both Ω and $-\Omega$ give the same linear transformation when they act by conjugation like this. One can show that all elements of $SO(3, 1)$ arise as such conjugation maps, by finding appropriate Ω that give rotations or boosts in the $\mu\nu$ planes, since these generate the group.

Recall that the double covering map

$$\Phi : SU(2) \rightarrow SO(3)$$

was given for $\Omega \in SU(2)$ by taking $\Phi(\Omega)$ to be the linear transformation in $SO(3)$

$$\begin{pmatrix} x_3 & x_1 - ix_2 \\ x_1 + ix_2 & -x_3 \end{pmatrix} \rightarrow \Omega \begin{pmatrix} x_3 & x_1 - ix_2 \\ x_1 + ix_2 & -x_3 \end{pmatrix} \Omega^{-1}$$

We have found an extension of this map to a double covering map from $SL(2, \mathbf{C})$ to $SO(3, 1)$. This restricts to Φ on the subgroup $SU(2)$ of $SL(2, \mathbf{C})$ matrices satisfying $\Omega^\dagger = \Omega^{-1}$.

Digression (The complex group $Spin(4, \mathbf{C})$ and its real forms). *Recall from chapter 6 that we found that $Spin(4) = Sp(1) \times Sp(1)$, with the corresponding $SO(4)$ transformation given by identifying $\mathbf{R}^4$ with the quaternions $\mathbf{H}$ and taking not just conjugations by unit quaternions, but both left and right multiplication by distinct unit quaternions. Rewriting this in terms of complex matrices instead of quaternions, we have $Spin(4) = SU(2) \times SU(2)$, and a pair Ω_1, Ω_2 of $SU(2)$ matrices acts as an $SO(4)$ rotation by*

$$\begin{pmatrix} x_0 - ix_3 & -x_2 - ix_1 \\ x_2 - ix_1 & x_0 + ix_3 \end{pmatrix} \to \Omega_1 \begin{pmatrix} x_0 - ix_3 & -x_2 - ix_1 \\ x_2 - ix_1 & x_0 + ix_3 \end{pmatrix} \Omega_2$$

preserving the determinant $x_0^2 + x_1^2 + x_2^2 + x_3^2$.

For another example, consider the identification of $\mathbf{R}^4$ with 2 by 2 real matrices given by

$$(x_0, x_1, x_2, x_3) \leftrightarrow \begin{pmatrix} x_0 + x_3 & x_2 + x_1 \\ x_2 - x_1 & x_0 - x_3 \end{pmatrix}$$

Given a pair of matrices Ω_1, Ω_2 in $SL(2, \mathbf{R})$, the linear transformation

$$\begin{pmatrix} x_0 + x_3 & x_2 + x_1 \\ x_2 - x_1 & x_0 - x_3 \end{pmatrix} \to \Omega_1 \begin{pmatrix} x_0 + x_3 & x_2 + x_1 \\ x_2 - x_1 & x_0 - x_3 \end{pmatrix} \Omega_2$$

preserves the reality condition on the matrix, and preserves

$$\det \begin{pmatrix} x_0 + x_3 & x_2 + x_1 \\ x_2 - x_1 & x_0 - x_3 \end{pmatrix} = x_0^2 + x_1^2 - x_2^2 - x_3^2$$

so gives an element of $SO(2, 2)$, and we see that $Spin(2, 2) = SL(2, \mathbf{R}) \times SL(2, \mathbf{R})$.

These three different constructions for the cases

$$Spin(4) = SU(2) \times SU(2), \quad Spin(3, 1) = SL(2, \mathbf{C})$$

and

$$Spin(2, 2) = SL(2, \mathbf{R}) \times SL(2, \mathbf{R})$$

correspond to different so-called "real forms" of a fact about complex groups that one can get by complexifying any of the examples (considering elements $(x_0, x_1, x_2, x_3) \in \mathbf{C}^4$, not just in $\mathbf{R}^4$). For instance, in the $Spin(4)$ case, taking the x_0, x_1, x_2, x_3 in the matrix

$$\begin{pmatrix} x_0 - ix_3 & -x_2 - ix_1 \\ x_2 - ix_1 & x_0 + ix_3 \end{pmatrix}$$

to have arbitrary complex values z_0, z_1, z_2, z_3 one gets arbitrary 2 by 2 complex matrices, and the transformation

$$\begin{pmatrix} z_0 - iz_3 & -z_2 - iz_1 \\ z_2 - iz_1 & z_0 + iz_3 \end{pmatrix} \rightarrow \Omega_1 \begin{pmatrix} z_0 - iz_3 & -z_2 - iz_1 \\ z_2 - iz_1 & z_0 + iz_3 \end{pmatrix} \Omega_2$$

preserves this space as well as the determinant $(z_0^2 + z_1^2 + z_2^2 + z_3^2)$ for Ω_1 and Ω_2 not just in $SU(2)$, but in the larger group $SL(2, \mathbf{C})$. So we find that the group $SO(4, \mathbf{C})$ of complex orthogonal transformations of $\mathbf{C}^4$ has spin double cover

$$Spin(4, \mathbf{C}) = SL(2, \mathbf{C}) \times SL(2, \mathbf{C})$$

Since $\mathfrak{spin}(4, \mathbf{C}) = \mathfrak{so}(3, 1) \otimes \mathbf{C}$, this relation between complex Lie groups corresponds to the Lie algebra relation

$$\mathfrak{so}(3, 1) \otimes \mathbf{C} = \mathfrak{sl}(2, \mathbf{C}) \oplus \mathfrak{sl}(2, \mathbf{C})$$

we found explicitly earlier when we showed that by taking complex coefficients of generators l_j and k_j of $\mathfrak{so}(3, 1)$ we could find generators A_j and B_j of two different $\mathfrak{sl}(2, \mathbf{C})$ subalgebras.

40.5   For further reading

Those not familiar with special relativity should consult a textbook on the subject for the physics background necessary to appreciate the significance of Minkowski space and its Lorentz group of invariances. An example of a suitable such book aimed at mathematics students is Woodhouse's *Special Relativity* [104].

Most quantum field theory textbooks have some sort of discussion of the Lorentz group and its Lie algebra, although the issue of how complexification works in this case is routinely ignored (recall the comments in section 5.5). Typical examples are Peskin–Schroeder [67], see the beginning of their chapter 3 or chapter II.3 [105] of Zee.

Chapter 41
Representations of the Lorentz Group

Having seen the importance in quantum mechanics of understanding the representations of the rotation group $SO(3)$ and its double cover $Spin(3) = SU(2)$, one would like to also understand the representations of the Lorentz group $SO(3,1)$ and its double cover $Spin(3,1) = SL(2,\mathbf{C})$. One difference from the $SO(3)$ case is that all non-trivial finite dimensional irreducible representations of the Lorentz group are non-unitary (there are infinite dimensional unitary irreducible representations, of no known physical significance, which we will not discuss). While these finite dimensional representations themselves only provide a unitary action of the subgroup $Spin(3) \subset Spin(3,1)$, they will later be used in the construction of quantum field theories whose state spaces will have a unitary action of the Lorentz group.

41.1 Representations of the Lorentz group

In the $SU(2)$ case, we found irreducible unitary representations (π_n, V^n) of dimension $n + 1$ for $n = 0, 1, 2, \ldots$. These could also be labeled by $s = \frac{n}{2}$, called the "spin" of the representation, and we will do that from now on. These representations can be realized explicitly as homogeneous polynomials of degree $n = 2s$ in two complex variables z_1, z_2. For the case of $Spin(4) = SU(2) \times SU(2)$, the irreducible representations will be tensor products

$$V^{s_1} \otimes V^{s_2}$$

of $SU(2)$ irreducibles, with the first $SU(2)$ acting on the first factor, the second on the second factor. The case $s_1 = s_2 = 0$ is the trivial representation, $s_1 = \frac{1}{2}, s_2 = 0$ is one of the half-spinor representations of $Spin(4)$ on $\mathbf{C}^2$,

© Peter Woit 2017   515
P. Woit, *Quantum Theory, Groups and Representations*,
DOI 10.1007/978-3-319-64612-1_41

$s_1 = 0, s_2 = \frac{1}{2}$ is the other, and $s_1 = s_2 = \frac{1}{2}$ is the representation on four-dimensional (complexified) vectors.

Turning now to $Spin(3,1) = SL(2,\mathbf{C})$, one can use the same construction using homogeneous polynomials as in the $SU(2)$ case to get irreducible representations of dimension $2s + 1$ for $s = 0, \frac{1}{2}, 1, \ldots$. Instead of acting by $SU(2)$ on z_1, z_2, one acts by $SL(2,\mathbf{C})$, and then as before uses the induced action on polynomials of z_1 and z_2. This gives representations (π_s, V^s) of $SL(2,\mathbf{C})$. Among the things that are different though about these representations:

- They are not unitary (except in the case of the trivial representation). For example, for the defining representation $V^{\frac{1}{2}}$ on $\mathbf{C}^2$, the Hermitian inner product

$$\left\langle \begin{pmatrix} \psi \\ \phi \end{pmatrix}, \begin{pmatrix} \psi' \\ \phi' \end{pmatrix} \right\rangle = \begin{pmatrix} \overline{\psi} & \overline{\phi} \end{pmatrix} \cdot \begin{pmatrix} \psi' \\ \phi' \end{pmatrix} = \overline{\psi}\psi' + \overline{\phi}\phi'$$

  is invariant under $SU(2)$ transformations Ω since

$$\left\langle \Omega \begin{pmatrix} \psi \\ \phi \end{pmatrix}, \Omega \begin{pmatrix} \psi' \\ \phi' \end{pmatrix} \right\rangle = \begin{pmatrix} \overline{\psi} & \overline{\phi} \end{pmatrix} \Omega^\dagger \cdot \Omega \begin{pmatrix} \psi' \\ \phi' \end{pmatrix}$$

  and $\Omega^\dagger\Omega = 1$ by unitarity. This is no longer true for $\Omega \in SL(2,\mathbf{C})$. The representation $V^{\frac{1}{2}}$ of $SL(2,\mathbf{C})$ does have a non-degenerate bilinear form, which we will denote by ϵ,

$$\epsilon\left(\begin{pmatrix} \psi \\ \phi \end{pmatrix}, \begin{pmatrix} \psi' \\ \phi' \end{pmatrix} \right) = \begin{pmatrix} \psi & \phi \end{pmatrix} \begin{pmatrix} 0 & 1 \\ -1 & 0 \end{pmatrix} \begin{pmatrix} \psi' \\ \phi' \end{pmatrix} = \psi\phi' - \phi\psi'$$

  that is invariant under the $SL(2,\mathbf{C})$ action on $V^{\frac{1}{2}}$ and can be used to identify the representation and its dual. This is the complexification of the symplectic form on $\mathbf{R}^2$ studied in section 16.1.1, and the same calculation there which showed that it was $SL(2,\mathbf{R})$ invariant here shows that the complex version is $SL(2,\mathbf{C})$ invariant.

- In the case of $SU(2)$ representations, the complex conjugate representation one gets by taking as representation matrices $\overline{\pi(g)}$ instead of $\pi(g)$ is equivalent to the original representation (the same representation, with a different basis choice, so matrices changed by a conjugation). To see this for the spin $\frac{1}{2}$ representation, note that $SU(2)$ matrices are of the form

$$\Omega = \begin{pmatrix} \alpha & \beta \\ -\overline{\beta} & \overline{\alpha} \end{pmatrix}$$

and one has

$$\begin{pmatrix} 0 & 1 \\ -1 & 0 \end{pmatrix} \begin{pmatrix} \alpha & \beta \\ -\overline{\beta} & \overline{\alpha} \end{pmatrix} \begin{pmatrix} 0 & 1 \\ -1 & 0 \end{pmatrix}^{-1} = \begin{pmatrix} \overline{\alpha} & \overline{\beta} \\ -\beta & \alpha \end{pmatrix}$$

so the matrix

$$\begin{pmatrix} 0 & 1 \\ -1 & 0 \end{pmatrix}$$

is the change of basis matrix relating the representation and its complex conjugate.

This is no longer true for $SL(2, \mathbf{C})$. Conjugation by a fixed matrix will not change the eigenvalues of the matrix, and these can be complex (unlike $SU(2)$ matrices, which have real eigenvalues). So such a (matrix) conjugation cannot change all $SL(2, \mathbf{C})$ matrices to their complex conjugates, since in general (complex) conjugation will change their eigenvalues.

The classification of irreducible finite dimensional $SU(2)$ representation was done in chapter 8 by considering its Lie algebra $\mathfrak{su}(2)$, complexified to give us raising and lowering operators, and this complexification is $\mathfrak{sl}(2, \mathbf{C})$. If one examines that argument, one finds that it mostly also applies to irreducible finite dimensional $\mathfrak{sl}(2, \mathbf{C})$ representations. There is a difference though: now, flipping positive to negative weights (which corresponds to change of sign of the Lie algebra representation matrices or conjugation of the Lie group representation matrices) no longer takes one to an equivalent representation. It turns out that to get all irreducibles, one must take both the representations we already know about and their complex conjugates. One can show (we will not prove this here) that the tensor product of one of each type of irreducible is still an irreducible, and that the complete list of finite dimensional irreducible representations of $\mathfrak{sl}(2, \mathbf{C})$ is given by:

Theorem (Classification of finite dimensional $\mathfrak{sl}(2, \mathbf{C})$ *representations*). *The irreducible representations of* $\mathfrak{sl}(2, \mathbf{C})$ *are labeled by* (s_1, s_2) *for* $s_j = 0, \frac{1}{2}, 1, \ldots$. *These representations are given by the tensor product representations*

$$(\pi_{s_1} \otimes \overline{\pi}_{s_2}, V^{s_1} \otimes V^{s_2})$$

where (π_s, V^s) *is the irreducible representation of dimension* $2s + 1$ *and* $(\overline{\pi}_s, V^s)$ *its complex conjugate. Such representations have dimension* $(2s_1 + 1)(2s_2 + 1)$.

All these representations are also representations of the group $SL(2, \mathbf{C})$, and one has the same classification theorem for the group, although we will not try and prove this. We will also not try and study these representations in general but will restrict attention to the cases of most physical interest, which are

- $(0,0)$: The trivial representation on $\mathbf{C}$, also called the "spin 0" or scalar representation.

- $(\frac{1}{2},0)$: These are called left-handed (for reasons we will see later on) "Weyl spinors." We will often denote the representation space $\mathbf{C}^2$ in this case as S_L and write an element of it as ψ_L.

- $(0,\frac{1}{2})$: These are called right-handed Weyl spinors. We will often denote the representation space $\mathbf{C}^2$ in this case as S_R and write an element of it as ψ_R.

- $(\frac{1}{2},\frac{1}{2})$: This is called the "vector" representation since it is the complexification of the action of $SL(2,\mathbf{C})$ as $SO(3,1)$ transformations of space–time vectors that we saw earlier. It is a representation of $SO(3,1)$ as well as $SL(2,\mathbf{C})$.

- $(\frac{1}{2},0) \oplus (0,\frac{1}{2})$: This reducible four-complex dimensional representation is known as the representation on "Dirac spinors."

One can manipulate these Weyl spinor representations $(\frac{1}{2},0)$ and $(0,\frac{1}{2})$ in a similar way to the treatment of tangent vectors and their duals in tensor analysis. Just like in that formalism, one can distinguish between a representation space and its dual by upper and lower indices, in this case using not the metric but the $SL(2,\mathbf{C})$ invariant bilinear form ϵ to raise and lower indices. With complex conjugates and duals, there are four kinds of irreducible $SL(2,\mathbf{C})$ representations on $\mathbf{C}^2$ to keep track of:

- S_L: This is the standard defining representation of $SL(2,\mathbf{C})$ on $\mathbf{C}^2$, with $\Omega \in SL(2,\mathbf{C})$ acting on $\psi_L \in S_L$ by

$$\psi_L \to \Omega \psi_L$$

A standard index notation for such things is called the "van der Waerden notation." It uses a lower index A taking values $1,2$ to label the components with respect to a basis of S_L as follows:

$$\psi_L = \begin{pmatrix} \psi_1 \\ \psi_2 \end{pmatrix} = \psi_A$$

and in this notation, Ω acts by

$$\psi_A \to \Omega_A^B \psi_B$$

For instance, the element

$$\Omega = e^{-i\frac{\theta}{2}\sigma_3}$$

corresponding to an $SO(3)$ rotation by an angle θ around the z-axis acts on S_L by

$$\begin{pmatrix} \psi_1 \\ \psi_2 \end{pmatrix} \rightarrow e^{-i\frac{\theta}{2}\sigma_3} \begin{pmatrix} \psi_1 \\ \psi_2 \end{pmatrix}$$

- S_L^*: This is the dual of the defining representation, with $\Omega \in SL(2,\mathbf{C})$ acting on $\psi_L^* \in S_L^*$ by

$$\psi_L^* \rightarrow (\Omega^{-1})^T \psi_L^*$$

This is a general property of representations: given any finite dimensional representation $(\pi(g), V)$, the pairing between V and its dual V^* is preserved by acting on V^* by matrices $(\pi(g)^{-1})^T$, and these provide a representation $((\pi(g)^{-1})^T, V^*)$. In van der Waerden notation, one uses upper indices and writes

$$\psi^A \rightarrow ((\Omega^{-1})^T)^A_B \psi^B$$

Writing elements of the dual as row vectors, our example above of a particular Ω acts by

$$\begin{pmatrix} \psi^1 & \psi^2 \end{pmatrix} \rightarrow \begin{pmatrix} \psi^1 & \psi^2 \end{pmatrix} e^{i\frac{\theta}{2}\sigma_3}$$

Note that the bilinear form ϵ gives an isomorphism of representations between S_L and S_L^*, written in index notation as follows:

$$\psi^A = \epsilon^{AB}\psi_B$$

where

$$\epsilon^{AB} = \begin{pmatrix} 0 & 1 \\ -1 & 0 \end{pmatrix}$$

- S_R: This is the complex conjugate representation to S_L, with $\Omega \in SL(2,\mathbf{C})$ acting on $\psi_R \in S_R$ by

$$\psi_R \rightarrow \overline{\Omega}\psi_R$$

The van der Waerden notation uses a separate set of dotted indices for these, writing this as follows:

$$\psi_{\dot{A}} \rightarrow \overline{\Omega}^{\dot{B}}_{\dot{A}}\psi_{\dot{B}}$$

Another common notation among physicists puts a bar over the ψ to denote that the vector is in this representation, but we will reserve that notation for complex conjugation. The Ω corresponding to a rotation about the z-axis acts as follows:

$$\begin{pmatrix} \psi_{\dot{1}} \\ \psi_{\dot{2}} \end{pmatrix} \rightarrow e^{i\frac{\theta}{2}\sigma_3} \begin{pmatrix} \psi_{\dot{1}} \\ \psi_{\dot{2}} \end{pmatrix}$$

- S_R^*: This is the dual representation to S_R, with $\Omega \in SL(2,\mathbf{C})$ acting on $\psi_R^* \in S_R^*$ by

$$\psi_R^* \rightarrow (\overline{\Omega}^{-1})^T \psi_R^*$$

and the index notation uses raised dotted indices

$$\psi^{\dot{A}} \rightarrow ((\overline{\Omega}^{-1})^T)_{\dot{B}}^{\dot{A}} \psi^{\dot{B}}$$

Our standard example of a Ω acts by

$$\begin{pmatrix} \psi^{\dot{1}} & \psi^{\dot{2}} \end{pmatrix} \rightarrow \begin{pmatrix} \psi^{\dot{1}} & \psi^{\dot{2}} \end{pmatrix} e^{-i\frac{\theta}{2}\sigma_3}$$

Another copy of ϵ

$$\epsilon^{\dot{A}\dot{B}} = \begin{pmatrix} 0 & 1 \\ -1 & 0 \end{pmatrix}$$

gives the isomorphism of S_R and S_R^* as representations, by

$$\psi^{\dot{A}} = \epsilon^{\dot{A}\dot{B}} \psi_{\dot{B}}$$

Restricting to the $SU(2)$ subgroup of $SL(2,\mathbf{C})$, all these representations are unitary and equivalent. As $SL(2,\mathbf{C})$ representations, they are not unitary, and while the representations are equivalent to their duals, S_L and S_R are inequivalent (since as we have seen, one cannot complex conjugate $SL(2,\mathbf{C})$ matrices by a matrix conjugation).

For the case of the $(\frac{1}{2}, \frac{1}{2})$ representation, to see explicitly the isomorphism between $S_L \otimes S_R$ and vectors, recall that we can identify Minkowski space with 2 by 2 self-adjoint matrices. $\Omega \in SL(2,\mathbf{C})$ acts by

$$\begin{pmatrix} x_0 + x_3 & x_1 - ix_2 \\ x_1 + ix_2 & x_0 - x_3 \end{pmatrix} \rightarrow \Omega \begin{pmatrix} x_0 + x_3 & x_1 - ix_2 \\ x_1 + ix_2 & x_0 - x_3 \end{pmatrix} \Omega^\dagger$$

We can identify such matrices as linear maps from S_R^* to S_L (and thus isomorphic to the tensor product $S_L \otimes (S_R^*)^* = S_L \otimes S_R$, see chapter 9).

41.2 Dirac γ-matrices and Cliff$(3, 1)$

In our discussion of the fermionic version of the harmonic oscillator, we defined the Clifford algebra Cliff(r, s) and found that elements quadratic in its generators gave a basis for the Lie algebra of $\mathfrak{so}(r, s) = \mathfrak{spin}(r, s)$. Exponentiating these gave an explicit construction of the group $Spin(r, s)$. We can apply that general theory to the case of Cliff$(3, 1)$ and this will give us the representations $(\frac{1}{2}, 0)$ and $(0, \frac{1}{2})$.

If we complexify our $\mathbf{R}^4$, then its Clifford algebra becomes the algebra of 4 by 4 complex matrices

$$\text{Cliff}(3, 1) \otimes \mathbf{C} = \text{Cliff}(4, \mathbf{C}) = M(4, \mathbf{C})$$

We will represent elements of Cliff$(3, 1)$ as such 4 by 4 matrices, but should keep in mind that we are working in the complexification of the Clifford algebra that corresponds to the Lorentz group, so there is some sort of condition on the matrices that needs to be kept track of to identify Cliff$(3, 1) \subset M(4, \mathbf{C})$. There are several different choices of how to explicitly represent these matrices, and for different purposes, different ones are most convenient. The one we will begin with and mostly use is sometimes called the chiral or Weyl representation and is the most convenient for discussing massless charged particles. We will try and follow the conventions used for this representation in [99]. Note that these 4 by 4 matrices act not on four-dimensional space–time, but on spinors. It is a special feature of 4 dimensions that these two different representations of the Lorentz group have the same dimension.

Writing 4 by 4 matrices in 2 by 2 block form and using the Pauli matrices σ_j, we assign the following matrices to Clifford algebra generators

$$\gamma_0 = -i\begin{pmatrix} 0 & 1 \\ 1 & 0 \end{pmatrix}, \gamma_1 = -i\begin{pmatrix} 0 & \sigma_1 \\ -\sigma_1 & 0 \end{pmatrix}, \gamma_2 = -i\begin{pmatrix} 0 & \sigma_2 \\ -\sigma_2 & 0 \end{pmatrix}, \gamma_3 = -i\begin{pmatrix} 0 & \sigma_3 \\ -\sigma_3 & 0 \end{pmatrix}$$

One can easily check that these satisfy the Clifford algebra relations for generators of Cliff$(3, 1)$: they anticommute with each other and

$$\gamma_0^2 = -1, \quad \gamma_1^2 = \gamma_2^2 = \gamma_3^2 = 1$$

The quadratic Clifford algebra elements $-\frac{1}{2}\gamma_j\gamma_k$ for $j < k$ satisfy the commutation relations of $\mathfrak{so}(3, 1)$. These are explicitly

$$-\frac{1}{2}\gamma_1\gamma_2 = -\frac{i}{2}\begin{pmatrix} \sigma_3 & 0 \\ 0 & \sigma_3 \end{pmatrix}, \quad -\frac{1}{2}\gamma_1\gamma_3 = -\frac{i}{2}\begin{pmatrix} \sigma_2 & 0 \\ 0 & \sigma_2 \end{pmatrix}, \quad -\frac{1}{2}\gamma_2\gamma_3 = -\frac{i}{2}\begin{pmatrix} \sigma_1 & 0 \\ 0 & \sigma_1 \end{pmatrix}$$

and

$$-\frac{1}{2}\gamma_0\gamma_1 = \frac{1}{2}\begin{pmatrix} -\sigma_1 & 0 \\ 0 & \sigma_1 \end{pmatrix}, \quad -\frac{1}{2}\gamma_0\gamma_2 = \frac{1}{2}\begin{pmatrix} -\sigma_2 & 0 \\ 0 & \sigma_2 \end{pmatrix}, \quad -\frac{1}{2}\gamma_0\gamma_3 = \frac{1}{2}\begin{pmatrix} -\sigma_3 & 0 \\ 0 & \sigma_3 \end{pmatrix}$$

They provide a representation $(\pi', \mathbf{C}^4)$ of the Lie algebra $\mathfrak{so}(3,1)$ with

$$\pi'(l_1) = -\frac{1}{2}\gamma_2\gamma_3, \ \pi'(l_2) = -\frac{1}{2}\gamma_1\gamma_3, \ \pi'(l_3) = -\frac{1}{2}\gamma_1\gamma_2$$

and

$$\pi'(k_1) = -\frac{1}{2}\gamma_0\gamma_1, \ \pi'(k_2) = -\frac{1}{2}\gamma_0\gamma_2, \ \pi'(k_3) = -\frac{1}{2}\gamma_0\gamma_3$$

Note that the $\pi'(l_j)$ are skew-adjoint, since this representation of the $\mathfrak{so}(3) \subset \mathfrak{so}(3,1)$ subalgebra is unitary. The $\pi'(k_j)$ are self-adjoint, and this representation π' of $\mathfrak{so}(3,1)$ is not unitary.

On the two commuting $\mathfrak{sl}(2,\mathbf{C})$ subalgebras of $\mathfrak{so}(3,1)\otimes\mathbf{C}$ with bases (see section 40.2)

$$A_j = \frac{1}{2}(l_j + ik_j), \ \ B_j = \frac{1}{2}(l_j - ik_j)$$

this representation is

$$\pi'(A_1) = -\frac{i}{2}\begin{pmatrix} \sigma_1 & 0 \\ 0 & 0 \end{pmatrix}, \ \pi'(A_2) = -\frac{i}{2}\begin{pmatrix} \sigma_2 & 0 \\ 0 & 0 \end{pmatrix}, \ \pi'(A_3) = -\frac{i}{2}\begin{pmatrix} \sigma_3 & 0 \\ 0 & 0 \end{pmatrix}$$

and

$$\pi'(B_1) = -\frac{i}{2}\begin{pmatrix} 0 & 0 \\ 0 & \sigma_1 \end{pmatrix}, \ \pi'(B_2) = -\frac{i}{2}\begin{pmatrix} 0 & 0 \\ 0 & \sigma_2 \end{pmatrix}, \ \pi'(B_3) = -\frac{i}{2}\begin{pmatrix} 0 & 0 \\ 0 & \sigma_3 \end{pmatrix}$$

We see explicitly that the action of the quadratic elements of the Clifford algebra on the spinor representation $\mathbf{C}^4$ is reducible, decomposing as the direct sum $S_L \oplus S_R^*$ of two inequivalent representations on $\mathbf{C}^2$

$$\Psi = \begin{pmatrix} \psi_L \\ \psi_R^* \end{pmatrix}$$

with complex conjugation (interchange of A_j and B_j) relating the $\mathfrak{sl}(2,\mathbf{C})$ actions on the components. The A_j act just on S_L, the B_j just on S_R^*. An alternative standard notation to the two-component van der Waerden notation is to use the four components of $\mathbf{C}^4$ with the action of the γ-matrices. The relation between the two notations is given by

$$\Psi_A \leftrightarrow \begin{pmatrix} \psi_B \\ \phi^{\dot{B}} \end{pmatrix}$$

where the index A on the left takes values $1, 2, 3, 4$, and the indices $B, \dot{B}$ on the right each take values $1, 2$.

Note that identifying Minkowski space with elements of the Clifford algebra by

$$(x_0, x_1, x_2, x_3) \to \rlap{$/$}x = x_0 \gamma_0 + x_1 \gamma_1 + x_2 \gamma_2 + x_3 \gamma_3$$

identifies Minkowski space with certain 4 by 4 matrices. This again gives the identification used earlier of Minkowski space with linear maps from S_R^* to S_L, since the upper right two by two block of the matrix will be given by

$$-i \begin{pmatrix} x_0 + x_3 & x_1 - ix_2 \\ x_1 + ix_2 & -x_0 - x_3 \end{pmatrix}$$

and takes S_R^* to S_L.

An important element of the Clifford algebra is constructed by multiplying all of the basis elements together. Physicists traditionally multiply this by i to make it self-adjoint and define

$$\gamma_5 = i\gamma_0 \gamma_1 \gamma_2 \gamma_3 = \begin{pmatrix} -1 & 0 \\ 0 & 1 \end{pmatrix}$$

This can be used to produce projection operators from the Dirac spinors onto the left- and right-handed Weyl spinors

$$\frac{1}{2}(1 - \gamma_5)\Psi = \psi_L, \quad \frac{1}{2}(1 + \gamma_5)\Psi = \phi_R^*$$

There are two other commonly used representations of the Clifford algebra relations, related to the one above by a change of basis. The Dirac representation is useful to describe massive charged particles, especially in the non-relativistic limit. Generators are given by

$$\gamma_0^D = -i \begin{pmatrix} 1 & 0 \\ 0 & -1 \end{pmatrix}, \gamma_1^D = -i \begin{pmatrix} 0 & \sigma_1 \\ -\sigma_1 & 0 \end{pmatrix}$$

$$\gamma_2^D = -i \begin{pmatrix} 0 & \sigma_2 \\ -\sigma_2 & 0 \end{pmatrix}, \gamma_3^D = -i \begin{pmatrix} 0 & \sigma_3 \\ -\sigma_3 & 0 \end{pmatrix}$$

and the projection operators for Weyl spinors are no longer diagonal, since

$$\gamma_5^D = \begin{pmatrix} 0 & 1 \\ 1 & 0 \end{pmatrix}$$

A third representation, the Majorana representation, is given by (now, no longer writing in 2 by 2 block form but as 4 by 4 matrices)

$$
\gamma_0^M = \begin{pmatrix} 0 & 0 & 0 & -1 \\ 0 & 0 & 1 & 0 \\ 0 & -1 & 0 & 0 \\ 1 & 0 & 0 & 0 \end{pmatrix}, \quad
\gamma_1^M = \begin{pmatrix} 1 & 0 & 0 & 0 \\ 0 & -1 & 0 & 0 \\ 0 & 0 & 1 & 0 \\ 0 & 0 & 0 & -1 \end{pmatrix}
$$

$$
\gamma_2^M = \begin{pmatrix} 0 & 0 & 0 & 1 \\ 0 & 0 & -1 & 0 \\ 0 & -1 & 0 & 0 \\ 1 & 0 & 0 & 0 \end{pmatrix}, \quad
\gamma_3^M = \begin{pmatrix} 0 & -1 & 0 & 0 \\ -1 & 0 & 0 & 0 \\ 0 & 0 & 0 & -1 \\ 0 & 0 & -1 & 0 \end{pmatrix}
$$

with

$$
\gamma_5^M = i \begin{pmatrix} 0 & -1 & 0 & 0 \\ 1 & 0 & 0 & 0 \\ 0 & 0 & 0 & 1 \\ 0 & 0 & -1 & 0 \end{pmatrix}
$$

The importance of the Majorana representation is that it shows the interesting possibility of having (in signature $(3,1)$) a spinor representation on a real vector space $\mathbf{R}^4$, since one sees that the Clifford algebra matrices can be chosen to be real. One has

$$
\gamma_0\gamma_1\gamma_2\gamma_3 = \begin{pmatrix} 0 & -1 & 0 & 0 \\ 1 & 0 & 0 & 0 \\ 0 & 0 & 0 & 1 \\ 0 & 0 & -1 & 0 \end{pmatrix}
$$

and

$$
(\gamma_0\gamma_1\gamma_2\gamma_3)^2 = -1
$$

The Majorana spinor representation is on $S_M = \mathbf{R}^4$, with $\gamma_0\gamma_1\gamma_2\gamma_3$ a real operator on this space with square -1, so it provides a complex structure on S_M. Recall that a complex structure on a real vector space gives a splitting of the complexification of the real vector space into a sum of two complex vector spaces, related by complex conjugation. In this case, this corresponds to

$$
S_M \otimes \mathbf{C} = S_L \oplus S_R^*
$$

the fact that complexifying Majorana spinors gives the two kinds of Weyl spinors.

41.3 For further reading

Most quantum field theory textbook have extensive discussions of spinor representations of the Lorentz group and gamma matrices, although most use the opposite convention for the signature of the Minkowski metric. Typical examples are Peskin–Schroeder [67] and chapter II.3 and Appendix E of Zee [105].

Chapter 42
The Poincaré Group
and its Representations

In chapter 19 we saw that the Euclidean group $E(3)$ has infinite dimensional irreducible unitary representations on the state space of a quantum free particle. The free particle Hamiltonian plays the role of a Casimir operator: to get irreducible representations one fixes the eigenvalue of the Hamiltonian (the energy), and then the representation is on the space of solutions to the Schrödinger equation with this energy. There is also a second Casimir operator, with integral eigenvalue the helicity, which further characterizes irreducible representations. The case of helicity $\pm\frac{1}{2}$ (which uses the double cover $\widetilde{E(3)}$) occurs for solutions of the Pauli equation, see section 34.2.

For a relativistic analog, treating space and time on the same footing, we will use instead the semi-direct product of space–time translations and Lorentz transformations, called the Poincaré group. Irreducible representations of this group will again be labeled by eigenvalues of two Casimir operators, giving in the cases relevant to physics one continuous parameter (the mass) and a discrete parameter (the spin or helicity). These representations can be realized as spaces of solutions for relativistic wave equations, with such representations corresponding to possible relativistic elementary particles.

For an element (a, Λ) of the Poincaré group, with a a space–time translation and Λ an element of the Lorentz group, there are three different sorts of actions of the group and Lie algebra to distinguish:

- The action

$$x \to \Lambda x + a$$

on a Minkowski space vector x. This is an action on a real vector space, it is not a unitary representation.

© Peter Woit 2017

P. Woit, *Quantum Theory, Groups and Representations*,
DOI 10.1007/978-3-319-64612-1_42

- The action

$$\psi \to u(a, \Lambda)\psi(x) = S(\Lambda)\psi(\Lambda^{-1}(x-a))$$

  on n-component wavefunctions, solutions to a wave equation (here S is an n-dimensional representation of the Lorentz group). These will be the unitary representations classified in this chapter.

- The space of single-particle wavefunctions can be used to construct a quantum field theory, describing arbitrary numbers of particles. This will come with an action on the state space by unitary operators $U(a, \Lambda)$. This will be a unitary representation, but very much not irreducible.

For the corresponding Lie algebra actions, we will use lower case letters (e.g., t_j, l_j) to denote the Lie algebra elements and their action on Minkowski space, upper case letters (e.g., P_j, L_j) to denote the Lie algebra representation on wavefunctions, and upper case hatted letters (e.g., $\widehat{P}_j, \widehat{L}_j$) to denote the Lie algebra representation on states of the quantum field theory.

42.1 The Poincaré group and its Lie algebra

Definition (Poincaré group). *The Poincaré group is the semi-direct product*

$$\mathcal{P} = \mathbf{R}^4 \rtimes SO(3,1)$$

with double cover

$$\tilde{\mathcal{P}} = \mathbf{R}^4 \rtimes SL(2, \mathbf{C})$$

The action of $SO(3,1)$ or $SL(2, \mathbf{C})$ on $\mathbf{R}^4$ is the action of the Lorentz group on Minkowski space.

We will refer to both of these groups as the "Poincaré group," meaning by this the double cover only when we need it because spinor representations of the Lorentz group are involved. The two groups have the same Lie algebra, so the distinction is not needed in discussions that only involve the Lie algebra. Elements of the group $\mathcal{P}$ will be written as pairs (a, Λ), with $a \in \mathbf{R}^4$ and $\Lambda \in SO(3,1)$. The group law is

$$(a_1, \Lambda_1)(a_2, \Lambda_2) = (a_1 + \Lambda_1 a_2, \Lambda_1 \Lambda_2)$$

The Lie algebra $Lie\ \mathcal{P} = Lie\ \tilde{\mathcal{P}}$ has dimension 10, with basis

$$t_0, t_1, t_2, t_3, l_1, l_2, l_3, k_1, k_2, k_3$$

where the first four elements are a basis of the Lie algebra of the translation group, and the next six are a basis of $\mathfrak{so}(3,1)$, with the l_j giving the subgroup of spatial rotations, the k_j the boosts. We already know the commutation relations for the translation subgroup, which is commutative so

$$[t_j, t_k] = 0$$

We have seen in chapter 40 that the commutation relations for $\mathfrak{so}(3,1)$ are

$$[l_1, l_2] = l_3, \quad [l_2, l_3] = l_1, \quad [l_3, l_1] = l_2$$

$$[k_1, k_2] = -l_3, \quad [k_3, k_1] = -l_2, \quad [k_2, k_3] = -l_1$$

or

$$[l_j, l_k] = \epsilon_{jkl} l_l, \quad [k_j, k_k] = -\epsilon_{jkl} l_l$$

and that the commutation relations between the l_j and k_j are

$$[l_j, k_k] = \epsilon_{jkl} k_l$$

corresponding to the fact that the k_j transform as a vector under spatial rotations.

The Poincaré group is a semi-direct product group of the sort discussed in chapter 18 and it can be represented as a group of 5 by 5 matrices in much the same way as elements of the Euclidean group $E(3)$ could be represented by 4 by 4 matrices (see chapter 19). Writing out this isomorphism explicitly for a basis of the Lie algebra, we have

$$l_1 \leftrightarrow \begin{pmatrix} 0 & 0 & 0 & 0 & 0 \\ 0 & 0 & 0 & 0 & 0 \\ 0 & 0 & 0 & -1 & 0 \\ 0 & 0 & 1 & 0 & 0 \\ 0 & 0 & 0 & 0 & 0 \end{pmatrix} \quad l_2 \leftrightarrow \begin{pmatrix} 0 & 0 & 0 & 0 & 0 \\ 0 & 0 & 0 & 1 & 0 \\ 0 & 0 & 0 & 0 & 0 \\ 0 & -1 & 0 & 0 & 0 \\ 0 & 0 & 0 & 0 & 0 \end{pmatrix} \quad l_3 \leftrightarrow \begin{pmatrix} 0 & 0 & 0 & 0 & 0 \\ 0 & 0 & -1 & 0 & 0 \\ 0 & 1 & 0 & 0 & 0 \\ 0 & 0 & 0 & 0 & 0 \\ 0 & 0 & 0 & 0 & 0 \end{pmatrix}$$

$$k_1 \leftrightarrow \begin{pmatrix} 0 & 1 & 0 & 0 & 0 \\ 1 & 0 & 0 & 0 & 0 \\ 0 & 0 & 0 & 0 & 0 \\ 0 & 0 & 0 & 0 & 0 \\ 0 & 0 & 0 & 0 & 0 \end{pmatrix} \quad k_2 \leftrightarrow \begin{pmatrix} 0 & 0 & 1 & 0 & 0 \\ 0 & 0 & 0 & 0 & 0 \\ 1 & 0 & 0 & 0 & 0 \\ 0 & 0 & 0 & 0 & 0 \\ 0 & 0 & 0 & 0 & 0 \end{pmatrix} \quad k_3 \leftrightarrow \begin{pmatrix} 0 & 0 & 0 & 1 & 0 \\ 0 & 0 & 0 & 0 & 0 \\ 0 & 0 & 0 & 0 & 0 \\ 1 & 0 & 0 & 0 & 0 \\ 0 & 0 & 0 & 0 & 0 \end{pmatrix}$$

$$
t_0 \leftrightarrow \begin{pmatrix} 0 & 0 & 0 & 0 & 1 \\ 0 & 0 & 0 & 0 & 0 \\ 0 & 0 & 0 & 0 & 0 \\ 0 & 0 & 0 & 0 & 0 \\ 0 & 0 & 0 & 0 & 0 \end{pmatrix} \quad
t_1 \leftrightarrow \begin{pmatrix} 0 & 0 & 0 & 0 & 0 \\ 0 & 0 & 0 & 0 & 1 \\ 0 & 0 & 0 & 0 & 0 \\ 0 & 0 & 0 & 0 & 0 \\ 0 & 0 & 0 & 0 & 0 \end{pmatrix}
$$

$$
t_2 \leftrightarrow \begin{pmatrix} 0 & 0 & 0 & 0 & 0 \\ 0 & 0 & 0 & 0 & 0 \\ 0 & 0 & 0 & 0 & 1 \\ 0 & 0 & 0 & 0 & 0 \\ 0 & 0 & 0 & 0 & 0 \end{pmatrix} \quad
t_3 \leftrightarrow \begin{pmatrix} 0 & 0 & 0 & 0 & 0 \\ 0 & 0 & 0 & 0 & 0 \\ 0 & 0 & 0 & 0 & 0 \\ 0 & 0 & 0 & 0 & 1 \\ 0 & 0 & 0 & 0 & 0 \end{pmatrix} \tag{42.1}
$$

We can use this explicit matrix representation to compute the commutators of the infinitesimal translations t_j with the infinitesimal rotations and boosts (l_j, k_j). t_0 commutes with the l_j and t_1, t_2, t_3 transform as a vector under rotations, For rotations one finds

$$
[l_j, t_k] = \epsilon_{jkl} t_l
$$

For boosts one has

$$
[k_j, t_0] = t_j, \quad [k_j, t_j] = t_0, \quad [k_j, t_k] = 0 \text{ if } j \neq k, \; k \neq 0 \tag{42.2}
$$

Note that infinitesimal boosts do not commute with infinitesimal time translation, so after quantization boosts will not commute with the Hamiltonian. Boosts will act on spaces of single-particle wavefunctions in a relativistic theory, and on states of a relativistic quantum field theory, but are not symmetries in the sense of preserving spaces of energy eigenstates.

42.2 Irreducible representations of the Poincaré group

We would like to construct unitary irreducible representations of the Poincaré group. These will be given by unitary operators $u(a, \Lambda)$ on a Hilbert space $\mathcal{H}_1$, which will have an interpretation as a single-particle relativistic quantum state space. In the analogous non-relativistic case, we constructed unitary irreducible representations of $E(3)$ (or its double cover $\widetilde{E(3)}$) as

- The space of wavefunctions of a free particle of mass m, with a fixed energy E (chapter 19). These are solutions to

$$D\psi = E\psi$$

where

$$D = -\frac{1}{2m}\nabla^2$$

and the ψ are single-component wavefunctions. $E(3)$ acts on wavefunctions by

$$\psi \to u(\mathbf{a}, R)\psi(\mathbf{x}) = \psi(R^{-1}(\mathbf{x} - \mathbf{a}))$$

- The space of solutions of the "square root" of the Pauli–Schrödinger equation (see section 34.2). These are solutions to

$$D\psi = \pm\sqrt{2mE}\,\psi$$

where

$$D = -i\boldsymbol{\sigma} \cdot \boldsymbol{\nabla}$$

and the ψ are two-component wavefunctions. $\widetilde{E(3)}$ acts on wavefunctions by

$$\psi \to u(\mathbf{a}, \Omega)\psi(\mathbf{x}) = \Omega\psi(R^{-1}(\mathbf{x} - \mathbf{a}))$$

(where R is the $SO(3)$ element corresponding to $\Omega \in SU(2)$ in the double cover).

In both cases, the group action commutes with the differential operator, or equivalently one has $uDu^{-1} = D$, and this is what ensures that the operators $u(\mathbf{a}, \Omega)$ take solutions to solutions.

To construct representations of $\mathcal{P}$, we would like to generalize this construction from $\mathbf{R}^3$ to Minkowski space M^4. To do this, one begins by defining an action of $\mathcal{P}$ on n-component wavefunctions by

$$\psi \to u(a, \Lambda)\psi(x) = S(\Lambda)\psi(\Lambda^{-1}(x - a))$$

This is the action one gets by identifying n-component wavefunctions with

$$(\text{functions on } M^4) \otimes \mathbf{C}^n$$

and using the induced action on functions for the first factor in the tensor product, on the second factor taking $S(\Lambda)$ to be an n-dimensional representation of the Lorentz group.

One then chooses a differential operator D on n-component wavefunctions, one that commutes with the group action, so

$$u(a,\Lambda)Du(a,\Lambda)^{-1} = D$$

The $u(a,\Lambda)$ then give a representation of $\mathcal{P}$ on the space of solutions to the wave equation

$$D\psi = c\psi$$

for c a constant, and in some cases this will give an irreducible representation. If the space of solutions is not irreducible, an additional set of "subsidiary conditions" can be used to pick out a subspace of solutions on which the representation is irreducible. In later chapters, we will consider several examples of this construction, but now will turn to the general classification of representations of $\mathcal{P}$.

Recall that in the $E(3)$ case, we had two Casimir operators:

$$P^2 = P_1^2 + P_2^2 + P_3^2$$

and

$$\mathbf{J} \cdot \mathbf{P}$$

Here P_j is the representation operator for Lie algebra representation, corresponding to an infinitesimal translation in the j-direction. J_j is the operator for an infinitesimal rotation about the j-axis. The Lie algebra commutation relations of $E(3)$ ensure that these two operators commute with the action of $E(3)$ and thus, by Schur's lemma, act as a scalar on an irreducible representation. Note that the fact that the first Casimir operator is a differential operator in position space and commutes with the $E(3)$ action means that the eigenvalue equation

$$P^2\psi = c\psi$$

has a space of solutions that is a $E(3)$ representation, and potentially irreducible.

In the Poincaré group case, we can easily identify:

Definition (Casimir operator). *The Casimir (or first Casimir) operator for the Poincaré group is the operator*

$$P^2 = -P_0^2 + P_1^2 + P_2^2 + P_3^2$$

A straightforward calculation using the Poincaré Lie algebra commutation relations shows that P^2 is a Casimir operator since one has

$$[P_0, P^2] = [P_j, P^2] = [J_j, P^2] = [K_j, P^2] = 0$$

for $j = 1, 2, 3$. Here J_j is a Lie algebra representation operator corresponding to the l_j, and K_j the operator corresponding to k_j.

The second Casimir operator is more difficult to identify in the Poincaré case than in the $E(3)$ case. To find it, first define:

Definition (Pauli–Lubanski operator). *The Pauli–Lubanski operator is the four-component operator*

$$W_0 = -\mathbf{P} \cdot \mathbf{J}, \quad \mathbf{W} = -P_0 \mathbf{J} + \mathbf{P} \times \mathbf{K}$$

By use of the commutation relations, one can show that the components of W_μ behave like a four-vector, i.e.,

$$[W_\mu, P_\nu] = 0$$

and the commutation relations with the J_j and K_j are the same for W_μ as for P_μ. One can then define:

Definition (Second Casimir operator). *The second Casimir operator for the Poincaré Lie algebra is*

$$W^2 = -W_0^2 + W_1^2 + W_2^2 + W_3^2$$

Use of the commutation relations shows that

$$[P_0, W^2] = [P_j, W^2] = [J_j, W^2] = [K_j, W^2] = 0$$

so W^2 is a Casimir operator.

To classify Poincaré group representations, we have two tools available. We can use the two Casimir operators P^2 and W^2 and characterize irreducible representations by their eigenvalues. In addition, recall from chapter 20 that irreducible representations of semi-direct products $N \rtimes K$ are associated with pairs of a K-orbit $\mathcal{O}_\alpha$ for $\alpha \in \hat{N}$, and an irreducible representation of the corresponding little group K_α.

For the Poincaré group, $\hat{N} = \mathbf{R}^4$ is the space of characters (one-dimensional representations) of the translation group of Minkowski space. Elements α are labeled by

$$p = (p_0, p_1, p_2, p_3)$$

where the p_μ are the eigenvalues of the energy–momentum operators P_μ. For representations on wavefunctions, these eigenvalues will correspond to elements in the representation space with space–time dependence.

$$e^{i(-p_0 x_0 + p_1 x_1 + p_2 x_2 + p_3 x_3)}$$

Given an irreducible representation, the operator P^2 will act by the scalar

$$-p_0^2 + p_1^2 + p_2^2 + p_3^2$$

which can be positive, negative, or zero, so given by $m^2, -m^2, 0$ for various m. The value of the scalar will be the same everywhere on the orbit, so in energy–momentum space, orbits will satisfy one of the three equations

$$-p_0^2 + p_1^2 + p_2^2 + p_3^2 = \begin{cases} -m^2 \\ m^2 \\ 0 \end{cases}$$

The representation can be further characterized in one of two ways:

- By the value of the second Casimir operator W^2.
- By the representation of the stabilizer group K_p on the eigenspace of the momentum operators with eigenvalue p.

At the point p on an orbit, the Pauli–Lubanski operator has components

$$W_0 = -\mathbf{p} \cdot \mathbf{J}, \quad \mathbf{W} = -p_0 + \mathbf{p} \times \mathbf{K}$$

In the next chapter, we will find the possible orbits, then pick a point p on each orbit and see what the stabilizer group K_p and Pauli–Lubanski operator are at that point.

42.3 Classification of representations by orbits

The Lorentz group acts on the energy–momentum space $\mathbf{R}^4$ by

$$p \to \Lambda p$$

and, restricting attention to the $p_0 p_3$ plane, the picture of the orbits looks like this

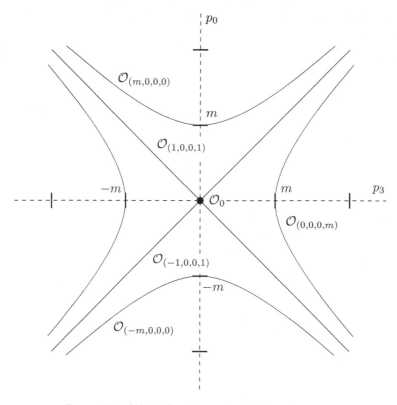

Figure 42.1: Orbits of vectors under the Lorentz group.

Unlike the Euclidean group case, here there are several different kinds of orbits $\mathcal{O}_p$. We will examine them and the corresponding stabilizer groups K_p each in turn and see what can be said about the associated representations.

42.3.1   Positive energy time-like orbits

One way to get negative values $-m^2$ of the Casimir P^2 is to take the vector $p = (m, 0, 0, 0)$, $m > 0$ and generate an orbit $\mathcal{O}_{(m,0,0,0)}$ by acting on it with the Lorentz group. This will be the upper, positive energy, sheet of the hyperboloid of two sheets

$$-p_0^2 + p_1^2 + p_2^2 + p_3^2 = -m^2$$

so

$$p_0 = \sqrt{p_1^2 + p_2^2 + p_3^2 + m^2}$$

The stabilizer group of $K_{(m,0,0,0)}$ is the subgroup of $SO(3,1)$ of elements of the form

$$\begin{pmatrix} 1 & \mathbf{0} \\ \mathbf{0} & R \end{pmatrix}$$

where $R \in SO(3)$, so $K_{(m,0,0,0)} = SO(3)$. Irreducible representations of this group are classified by the spin. For spin 0, points on the hyperboloid can be identified with positive energy solutions to a wave equation called the Klein–Gordon equation, and functions on the hyperboloid both correspond to the space of all solutions of this equation and carry an irreducible representation of the Poincaré group. This case will be studied in detail in chapters 43 and 44. We will study the case of spin $\frac{1}{2}$ in chapter 47, where one must use the double cover $SU(2)$ of $SO(3)$. The Poincaré group representation will be on functions on the orbit that takes values in two copies of the spinor representation of $SU(2)$. These will correspond to solutions of a wave equation called the massive Dirac equation. For choices of higher spin representations of the stabilizer group, one can again find appropriate wave equations and construct Poincaré group representations on their space of solutions (although additional subsidiary conditions are often needed), but we will not enter into this topic.

For $p = (m,0,0,0)$ the Pauli–Lubanski operator will be

$$W_0 = 0, \quad \mathbf{W} = -m\mathbf{J}$$

and the second Casimir operator will be

$$W^2 = m^2 J^2$$

The eigenvalues of W^2 are thus proportional to the eigenvalues of J^2, the Casimir operator for the subgroup of spatial rotations. These are again given by the spin s and will take the values $s(s+1)$. These eigenvalues classify representations consistently with the stabilizer group classification.

42.3.2   Negative energy time-like orbits

Starting instead with the energy–momentum vector $p = (-m,0,0,0)$, $m > 0$, the orbit $\mathcal{O}_{(-m,0,0,0)}$ one gets is the lower, negative energy component of the hyperboloid

$$-p_0^2 + p_1^2 + p_2^2 + p_3^2 = -m^2$$

satisfying

$$p_0 = -\sqrt{p_1^2 + p_2^2 + p_3^2 + m^2}$$

Again, one has the same stabilizer group $K_{(-m,0,0,0)} = SO(3)$ and the same constructions of wave equations of various spins and Poincaré group representations on their solution spaces as in the positive energy case. Since negative energies lead to unstable, unphysical theories, we will see that these representations are treated differently under quantization, corresponding physically not to particles, but to antiparticles.

42.3.3 Space-like orbits

One can get positive values m^2 of the Casimir P^2 by considering the orbit $\mathcal{O}_{(0,0,0,m)}$ of the vector $p = (0,0,0,m)$. This is a hyperboloid of one sheet, satisfying the equation

$$-p_0^2 + p_1^2 + p_2^2 + p_3^2 = m^2$$

It is not too difficult to see that the stabilizer group of the orbit is $K_{(0,0,0,m)} = SO(2,1)$. This is isomorphic to the group $SL(2,\mathbf{R})$, and it has no finite dimensional unitary representations. These orbits correspond physically to "tachyons," particles that move faster than the speed of light, and there is no known way to consistently incorporate them in a conventional theory.

42.3.4 The zero orbit

The simplest case where the Casimir P^2 is zero is the trivial case of a point $p = (0,0,0,0)$. This is invariant under the full Lorentz group, so the orbit $\mathcal{O}_{(0,0,0,0)}$ is just a single point and the stabilizer group $K_{(0,0,0,0)}$ is the entire Lorentz group $SO(3,1)$. For each finite dimensional representation of $SO(3,1)$, one gets a corresponding finite dimensional representation of the Poincaré group, with translations acting trivially. These representations are not unitary, so they are not usable for our purposes. Note that these representations are not distinguished by the value of the second Casimir W^2, which is zero for all of them.

42.3.5 Positive energy null orbits

One has $P^2 = 0$ not only for the zero vector in momentum space, but for a three-dimensional set of energy–momentum vectors, called the null-cone. By the term "cone" one means that if a vector is in the space, so are all products of the vector times a positive number. Vectors $p = (p_0, p_1, p_2, p_3)$ are called "light-like" or "null" when they satisfy

$$p^2 = -p_0^2 + p_1^2 + p_2^2 + p_3^2 = 0$$

One such vector is $p = (|\mathbf{p}|, 0, 0, |\mathbf{p}|)$ and the orbit of the vector under the action of the Lorentz group will be the upper half of the full null-cone, the half with energy $p_0 > 0$, satisfying

$$p_0 = \sqrt{p_1^2 + p_2^2 + p_3^2}$$

It turns out that the stabilizer group $K_{|\mathbf{p}|, 0, 0, |\mathbf{p}|}$ of $p = (|\mathbf{p}|, 0, 0, |\mathbf{p}|)$ is $E(2)$, the Euclidean group of the plane. One way to see this is to use the matrix representation 42.1 which explicitly gives the action of the Poincaré Lie algebra on Minkowski space vectors, and note that

$$l_3, l_1 + k_2, l_2 - k_1$$

each act trivially on $(|\mathbf{p}|, 0, 0, |\mathbf{p}|)$. l_3 is the infinitesimal spatial rotation about the 3-axis. Defining

$$b_1 = \frac{1}{\sqrt{2}}(l_1 + k_2), \quad b_2 = \frac{1}{\sqrt{2}}(l_2 - k_1)$$

and calculating the commutators

$$[b_1, b_2] = 0, \quad [l_3, b_1] = b_2, \quad [l_3, b_2] = -b_1$$

we see that these three elements of the Lie algebra are a basis of a Lie subalgebra isomorphic to the Lie algebra of $E(2)$.

Recall from section 19.1 that there are two kinds of irreducible unitary representations of $E(2)$:

- Representations such that the two translations act trivially. These are irreducible representations of $SO(2)$, so one-dimensional and characterized by an integer n (half-integers when the Poincaré group double cover is used).

- Infinite dimensional irreducible representations on a space of functions on a circle of radius r.

The first of these two cases corresponds to irreducible representations of the Poincaré group labeled by an integer n, which is called the "helicity" of the representation. Given the representation, n will be the eigenvalue of J_3 acting on the energy–momentum eigenspace with energy–momentum $(|\mathbf{p}|, 0, 0, |\mathbf{p}|)$. We will in later chapters consider the cases $n = 0$ (massless scalars, wave equation the Klein–Gordon equation), $n = \pm\frac{1}{2}$ (Weyl spinors, wave equation the Weyl equation), and $n = \pm1$ (photons, wave equation the Maxwell equations). The second sort of representation of $E(2)$ corresponds to representations of the Poincaré group known as "continuous spin" representations, but these seem not to correspond to any known physical phenomena.

Calculating the components of the Pauli–Lubanski operator, one finds

$$W_0 = -|\mathbf{p}|J_3, \quad W_1 = -(J_1 + K_2), \quad W_2 = -(J_2 - K_1), \quad W_3 = -|\mathbf{p}|J_3$$

Defining

$$B_1 = \frac{1}{\sqrt{2}}(J_1 + K_2), \quad B_2 = \frac{1}{\sqrt{2}}(J_2 - K_1)$$

the second Casimir operator is given by

$$W^2 = 2(B_1^2 + B_2^2)$$

which is the Casimir operator for $E(2)$. It takes nonzero values on the continuous spin representations but is zero for the representations where $E(2)$ translations act trivially. It does thus not distinguish between massless Poincaré representations of different helicities.

42.3.6 Negative energy null orbits

Looking instead at the orbit of $p = (-|\mathbf{p}|, 0, 0, |\mathbf{p}|)$, one gets the negative energy part of the null-cone. As with the time-like hyperboloids of nonzero mass m, these will correspond to antiparticles instead of particles, with the same classification as in the positive energy case.

42.4 For further reading

For an extensive discussion of the Poincaré group, its Lie algebra and representations, see [97]. Weinberg [99] (chapter 2) has some discussion of the representations of the Poincaré group on single-particle state spaces that we have classified here. Folland [28] (chapter 4.4) and Berndt [9] (chapter 7.5) discuss the construction of these representations using induced representation methods (as opposed to the construction as solution spaces of wave equations that we will use in following chapters).

Chapter 43
The Klein–Gordon Equation and Scalar Quantum Fields

In the non-relativistic case, we found that it was possible to build a quantum theory describing arbitrary numbers of particles by taking as dual phase space $\mathcal{M}$ the single-particle space $\mathcal{H}_1$ of solutions to the free particle Schrödinger equation. To get the same sort of construction for relativistic systems, one possibility is to take as dual phase space the space of solutions of a relativistic wave equation known as the Klein–Gordon equation.

A major difference with the non-relativistic case is that the equation of motion is second-order in time, so to parametrize solutions in $\mathcal{H}_1$ one needs not just the wavefunction at a fixed time, but also its time derivative. In addition, consistency with conditions of causality and positive energy of states requires making a very different choice of complex structure J, one that is not the complex structure coming from the complex-valued nature of the wavefunction (a choice which may in any case be unavailable, since in the simplest theory Klein–Gordon wavefunctions will be real-valued). In the relativistic case, an appropriate complex structure J_r is defined by complexifying the space of solutions, and then taking J_r to have value $+i$ on positive energy solutions and $-i$ on negative energy solutions. This implies a different physical interpretation than in the non-relativistic case, with a nonnegative energy assignment to states achieved by interpreting negative energy solutions as corresponding to positive energy antiparticle states moving backward in time.

43.1 The Klein–Gordon equation and its solutions

To get a single-particle theory describing an elementary particle with a unitary action of the Poincaré group, one can try to take as single-particle state space any of the irreducible representations classified in chapter 42. There we found that such irreducible representations are characterized in part by the

© Peter Woit 2017
P. Woit, *Quantum Theory, Groups and Representations*,
DOI 10.1007/978-3-319-64612-1_43

scalar value that the Casimir operator P^2 takes on the representation. When this is negative, we have the operator equation

$$P^2 = -P_0^2 + P_1^2 + P_2^2 + P_3^2 = -m^2 \tag{43.1}$$

and we would like to find a state space with momentum operators satisfying this relation. We can use wavefunctions $\phi(x)$ on Minkowski space to get such a state space.

Just as in the non-relativistic case, where we could represent momentum operators as either multiplication operators on functions of the momenta, or differentiation operators on functions of the positions, here we can do the same, with functions now depending on the four space–time coordinates. Taking $P_0 = H = i\frac{\partial}{\partial t}$ as well as the conventional momentum operators $P_j = -i\frac{\partial}{\partial x_j}$, equation 43.1 becomes:

Definition (Klein–Gordon equation). *The Klein–Gordon equation is the second-order partial differential equation*

$$\left(-\frac{\partial^2}{\partial t^2} + \frac{\partial^2}{\partial x_1^2} + \frac{\partial^2}{\partial x_2^2} + \frac{\partial^2}{\partial x_3^2}\right)\phi = m^2\phi$$

or

$$\left(-\frac{\partial^2}{\partial t^2} + \Delta - m^2\right)\phi = 0 \tag{43.2}$$

for functions $\phi(x)$ on Minkowski space (these functions may be real or complex-valued).

This equation is the simplest Lorentz invariant (the Lorentz group acting on functions takes solutions to solutions) wave equation and historically was the one first tried by Schrödinger. He soon realized it could not account for known facts about atomic spectra and instead used the non-relativistic equation that bears his name. In this chapter, we will consider the quantization of the space of real-valued solutions of this equation, with the case of complex-valued solutions appearing later in section 44.1.2.

Taking Fourier transforms by

$$\widetilde{\phi}(p) = \frac{1}{(2\pi)^2}\int_{\mathbf{R}^4} e^{-i(-p_0 x_0 + \mathbf{p}\cdot\mathbf{x})}\phi(x)d^4x$$

the momentum operators become multiplication operators, and the Klein–Gordon equation is now

$$(p_0^2 - p_1^2 - p_2^2 - p_3^2 - m^2)\widetilde{\phi} = (p_0^2 - \omega_{\mathbf{p}}^2)\widetilde{\phi} = 0$$

where

$$\omega_{\mathbf{p}} = \sqrt{p_1^2 + p_2^2 + p_3^2 + m^2}$$

Solutions to this will be distributions that are nonzero only on the hyperboloid

$$p_0^2 - p_1^2 - p_2^2 - p_3^2 - m^2 = 0$$

in energy–momentum space $\mathbf{R}^4$. This hyperboloid has two components, with positive and negative energy

$$p_0 = \pm\omega_{\mathbf{p}}$$

Ignoring one dimension, these look like

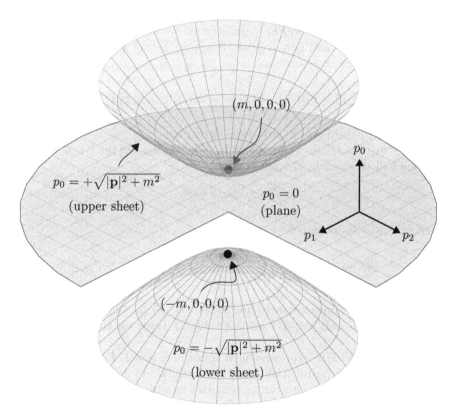

Figure 43.1: Orbits of energy–momentum vectors $(m, 0, 0, 0)$ and $(-m, 0, 0, 0)$ under the Poincaré group action.

and are the orbits of the energy–momentum vectors $(m, 0, 0, 0)$ and $(-m, 0, 0, 0)$ under the Poincaré group action discussed in sections 42.3.1 and 42.3.2.

In the non-relativistic case, a continuous basis of solutions of the free particle Schrödinger equation labeled by $\mathbf{p} \in \mathbf{R}^3$ was given by the functions

$$e^{-i\frac{|\mathbf{p}|^2}{2m}t}e^{i\mathbf{p}\cdot\mathbf{x}}$$

with a general solution a superposition of these given by

$$\psi(\mathbf{x},t) = \frac{1}{(2\pi)^{3/2}} \int_{\mathbf{R}^3} \widetilde{\psi}(\mathbf{p},0)e^{-i\frac{|\mathbf{p}|^2}{2m}t}e^{i\mathbf{p}\cdot\mathbf{x}}d^3\mathbf{p}$$

Besides specifying the function $\psi(\mathbf{x},t)$, elements of the single-particle space $\mathcal{H}_1$ could be uniquely characterized in two other ways by a function on $\mathbf{R}^3$: either the initial value $\psi(\mathbf{x},0)$ or its Fourier transform $\widetilde{\psi}(\mathbf{p},0)$.

In the relativistic case, since the Klein–Gordon equation is second order in time, solutions $\phi(t,\mathbf{x})$ will be parametrized by initial data which, unlike the non-relativistic case, now requires the specification at $t = 0$ of not one, but two functions:

$$\phi(\mathbf{x}) = \phi(0,\mathbf{x}), \quad \dot{\phi}(\mathbf{x}) = \frac{\partial}{\partial t}\phi(t,\mathbf{x})_{|t=0} \equiv \pi(\mathbf{x})$$

the values of the field and its first time derivative.

In the relativistic case, a continuous basis of solutions of the Klein–Gordon equation will be given by the functions

$$e^{\pm i\omega_\mathbf{p}t}e^{i\mathbf{p}\cdot\mathbf{x}}$$

and a general solution can be written

$$\phi(t,\mathbf{x}) = \frac{1}{(2\pi)^{3/2}} \int_{\mathbf{R}^4} \delta(p_0^2 - \omega_\mathbf{p}^2)f(p)e^{i(-p_0t+\mathbf{p}\cdot\mathbf{x})}d^4p \qquad (43.3)$$

for $f(p)$ a complex function satisfying $\overline{f(p)} = f(-p)$ (so that ϕ will be real). The solution will only depend on the values f takes on the hyperboloids $p_0 = \pm\omega_\mathbf{p}$.

The integral 43.3 is expressed in a four-dimensional, Lorentz invariant manner using the delta-function, but this is really an integral over the two-component hyperboloid. This can be rewritten as a three-dimensional integral over $\mathbf{R}^3$ by the following argument. For each $\mathbf{p}$, applying equation 11.9 to the case of the function of p_0 given by

$$g(p_0) = p_0^2 - \omega_\mathbf{p}^2$$

on $\mathbf{R}^4$, and using

$$\frac{d}{dp_0}(p_0^2 - \omega_{\mathbf{p}}^2)|_{p_0=\pm\omega_{\mathbf{p}}} = 2p_0|_{p_0=\pm\omega_{\mathbf{p}}} = \pm 2\omega_{\mathbf{p}}$$

gives

$$\delta(p_0^2 - \omega_{\mathbf{p}}^2) = \frac{1}{2\omega_{\mathbf{p}}}(\delta(p_0 - \omega_{\mathbf{p}}) + \delta(p_0 + \omega_{\mathbf{p}}))$$

We will often in the future use the above to provide a Lorentz invariant measure on the hyperboloids $p_0^2 = \omega_{\mathbf{p}}^2$, which we will write

$$\frac{d^3\mathbf{p}}{2\omega_{\mathbf{p}}}$$

For a function $\varphi(p)$ on these hyperboloids the integral over the hyperboloids can be written in two equivalent ways

$$\int_{\mathbf{R}^4} \delta(p_0^2 - \omega_{\mathbf{p}}^2)\varphi(p)d^4p = \int_{\mathbf{R}^3} (\varphi(\omega_{\mathbf{p}}, \mathbf{p}) + \varphi(-\omega_{\mathbf{p}}, \mathbf{p}))\frac{d^3\mathbf{p}}{2\omega_{\mathbf{p}}} \qquad (43.4)$$

with the left-hand side an explicitly Lorentz invariant measure (since $p_0^2 - \omega_{\mathbf{p}}^2 = -p^2 - m^2$ is Lorentz invariant).

An arbitrary solution of the Klein–Gordon equation (see equation 43.3) can thus be written

$$\phi(t, \mathbf{x}) = \frac{1}{(2\pi)^{3/2}} \int_{\mathbf{R}^4} \frac{1}{2\omega_{\mathbf{p}}}(\delta(p_0 - \omega_{\mathbf{p}}) + \delta(p_0 + \omega_{\mathbf{p}}))f(p)e^{i(-p_0 t + \mathbf{p}\cdot\mathbf{x})}dp_0 d^3\mathbf{p}$$

$$= \frac{1}{(2\pi)^{3/2}} \int_{\mathbf{R}^3} (f(\omega_{\mathbf{p}}, \mathbf{p})e^{-i\omega_{\mathbf{p}}t}e^{i\mathbf{p}\cdot\mathbf{x}} + f(-\omega_{\mathbf{p}}, -\mathbf{p})e^{i\omega_{\mathbf{p}}t}e^{-i\mathbf{p}\cdot\mathbf{x}})\frac{d^3\mathbf{p}}{2\omega_{\mathbf{p}}}$$
$$(43.5)$$

which will be real when

$$f(-\omega_{\mathbf{p}}, -\mathbf{p}) = \overline{f(\omega_{\mathbf{p}}, \mathbf{p})} \qquad (43.6)$$

Instead of the functions f, we will usually instead use

$$\alpha(\mathbf{p}) = \frac{f(\omega_{\mathbf{p}}, \mathbf{p})}{\sqrt{2\omega_{\mathbf{p}}}}, \quad \overline{\alpha}(\mathbf{p}) = \frac{f(-\omega_{\mathbf{p}}, -\mathbf{p})}{\sqrt{2\omega_{\mathbf{p}}}} \qquad (43.7)$$

Other choices of normalization of these complex functions are often used, the motivation for this one is that we will see that it will give simple Poisson bracket relations. With this choice, the Klein–Gordon solutions are

$$\phi(t,\mathbf{x}) = \frac{1}{(2\pi)^{3/2}} \int_{\mathbf{R}^3} (\alpha(\mathbf{p})e^{-i\omega_{\mathbf{p}}t}e^{i\mathbf{p}\cdot\mathbf{x}} + \overline{\alpha}(\mathbf{p})e^{i\omega_{\mathbf{p}}t}e^{-i\mathbf{p}\cdot\mathbf{x}})\frac{d^3\mathbf{p}}{\sqrt{2\omega_{\mathbf{p}}}} \quad (43.8)$$

Such solutions can be specified in terms of their initial data by either the pair of real-valued functions $\phi(\mathbf{x}), \pi(\mathbf{x})$, or their Fourier transforms. We will, however, find it much more convenient to characterize the momentum space initial data by the single complex-valued function $\alpha(\mathbf{p})$. The equations relating these choices of initial data are

$$\phi(\mathbf{x}) = \frac{1}{(2\pi)^{3/2}} \int_{\mathbf{R}^3} (\alpha(\mathbf{p})e^{i\mathbf{p}\cdot\mathbf{x}} + \overline{\alpha}(\mathbf{p})e^{-i\mathbf{p}\cdot\mathbf{x}})\frac{d^3\mathbf{p}}{\sqrt{2\omega_{\mathbf{p}}}} \quad (43.9)$$

$$\pi(\mathbf{x}) = \frac{\partial}{\partial t}\phi(\mathbf{x},t)_{|t=0} = \frac{1}{(2\pi)^{3/2}} \int_{\mathbf{R}^3} (-i\omega_{\mathbf{p}})(\alpha(\mathbf{p})e^{i\mathbf{p}\cdot\mathbf{x}} - \overline{\alpha}(\mathbf{p})e^{-i\mathbf{p}\cdot\mathbf{x}})\frac{d^3\mathbf{p}}{\sqrt{2\omega_{\mathbf{p}}}} \quad (43.10)$$

and

$$\alpha(\mathbf{p}) = \frac{1}{(2\pi)^{3/2}} \int_{\mathbf{R}^3} \frac{1}{\sqrt{2}}\left(\sqrt{\omega_{\mathbf{p}}}\phi(\mathbf{x}) + i\frac{1}{\sqrt{\omega_{\mathbf{p}}}}\pi(\mathbf{x})\right)e^{-i\mathbf{p}\cdot\mathbf{x}}d^3\mathbf{x} \quad (43.11)$$

To construct a relativistic quantum field theory, we would like to proceed as in the non-relativistic case of chapters 36 and 37, but taking the dual phase space $\mathcal{M}$ to be the space of solutions of the Klein–Gordon equation rather than of the Schrödinger equation. As in the non-relativistic case, we have various ways of specifying an element of $\mathcal{M}$:

• $\Phi(\phi(\mathbf{x}), \pi(\mathbf{x}))$: the solution with initial data $\phi(\mathbf{x}), \pi(\mathbf{x})$ at $t = 0$.

• $A(\alpha(\mathbf{p}))$: the solution with initial data at $t = 0$ specified by the complex function $\alpha(\mathbf{p})$ on momentum space, related to $\phi(\mathbf{x}), \pi(\mathbf{x})$ by equation 43.11.

Quantization will take these to operators $\widehat{\Phi}(\phi(\mathbf{x}), \pi(\mathbf{x})), \widehat{A}(\alpha(\mathbf{p}))$.

We can also define versions of the above that are distributional objects corresponding to taking the functions $\phi(\mathbf{x}), \pi(\mathbf{x}), \alpha(\mathbf{p})$ to be delta-functions:

• $\Phi(\mathbf{x})$: the distributional solution with initial data $\phi(\mathbf{x}') = \delta(\mathbf{x}' - \mathbf{x})$, $\pi(\mathbf{x}) = 0$. One then writes

$$\Phi(\phi(\mathbf{x}), 0) = \int_{\mathbf{R}^3} \Phi(\mathbf{x})\phi(\mathbf{x}')d^3\mathbf{x}'$$

• $\Pi(\mathbf{x})$: the distributional solution with initial data $\phi(\mathbf{x}') = 0, \pi(\mathbf{x}') = \delta(\mathbf{x}' - \mathbf{x})$. One then writes

$$\Phi(0, \pi(\mathbf{x})) = \int_{\mathbf{R}^3} \Pi(\mathbf{x})\pi(\mathbf{x}')d^3\mathbf{x}'$$

- $A(\mathbf{p})$: the distributional solution with initial data $\alpha(\mathbf{p}') = \delta(\mathbf{p}' - \mathbf{p})$. One then writes

$$A(\alpha(\mathbf{p})) = \int_{\mathbf{R}^3} A(\mathbf{p})\alpha(\mathbf{p}')d^3\mathbf{p}'$$

43.2 The symplectic and complex structures on $\mathcal{M}$

Taking as dual phase space $\mathcal{M}$ the space of solutions of the Klein–Gordon equation, one way to write elements of this space is as pairs of functions (ϕ, π) on $\mathbf{R}^3$. The symplectic structure is then given by

$$\Omega((\phi_1, \pi_1), (\phi_2, \pi_2)) = \int_{\mathbf{R}^3} (\phi_1(\mathbf{x})\pi_2(\mathbf{x}) - \pi_1(\mathbf{x})\phi_2(\mathbf{x}))d^3\mathbf{x} \qquad (43.12)$$

and the $(\phi(\mathbf{x}), \pi(\mathbf{x}))$ can be thought of as pairs of conjugate coordinates analogous to the pairs q_j, p_j but with a continuous index $\mathbf{x}$ instead of the discrete index j. Also by analogy with the finite dimensional case, the symplectic structure can be written in terms of the distributional fields $\Phi(\mathbf{x}), \Pi(\mathbf{x})$ as

$$\{\Phi(\mathbf{x}), \Pi(\mathbf{x}')\} = \delta^3(\mathbf{x} - \mathbf{x}'), \quad \{\Phi(\mathbf{x}), \Phi(\mathbf{x}')\} = \{\Pi(\mathbf{x}), \Pi(\mathbf{x}')\} = 0 \quad (43.13)$$

We now have a dual phase space $\mathcal{M}$ and a symplectic structure on it, so in principle could quantize using an infinite dimensional version of the Schrödinger representation, treating the values $\phi(\mathbf{x})$ as an infinite number of position-like coordinates, and taking states to be functionals of these coordinates. It is, however, much more convenient, as in the non-relativistic case of chapters 36 and 37, to use the Bargmann–Fock representation, treating the quantum field theory system as an infinite collection of harmonic oscillators. This requires a choice of complex structure, which we will discuss in this section.

By analogy with the non-relativistic case, it is tempting to try and think of solutions of the Klein–Gordon equation as wavefunctions describing a single relativistic particle. A standard physical argument is that a relativistic single-particle theory describing localized particles is not possible, since once the position uncertainty of a particle is small enough, its momentum uncertainty will be large enough to provide the energy needed to create new particles. If you try and put a relativistic particle in a smaller and smaller box, at some point you will no longer have just one particle, and it is this situation that implies that only a many-particle theory will be consistent. This leads one to expect that any attempt to find a consistent relativistic single-particle theory will run into some sort of problem.

One obvious source of trouble are the negative energy solutions. These will cause an instability if the theory is coupled to other physics, by allowing initial positive energy single-particle states to evolve to states with arbitrarily negative energy, transferring positive energy to the other physical system they are coupled to. This can be dealt with by restricting to the space of solutions of positive energy, taking $\mathcal{H}_1$ to be the space of complex solutions with positive energy (i.e., letting ϕ take complex values and setting $f(-\omega_{\mathbf{p}}, -\mathbf{p}) = 0$ in equation 43.5). One then has solutions

$$\phi(t, \mathbf{x}) = \frac{1}{(2\pi)^{3/2}} \int_{\mathbf{R}^3} f(\omega_{\mathbf{p}}, \mathbf{p}) e^{-i\omega_{\mathbf{p}} t} e^{i\mathbf{p} \cdot \mathbf{x}} \frac{d^3 \mathbf{p}}{2\omega_{\mathbf{p}}} \qquad (43.14)$$

parametrized by complex functions f of $\mathbf{p}$.

This choice of $\mathcal{H}_1$ gives a theory much like the non-relativistic free particle Schrödinger case (and which has that theory as a limit if one takes the speed of light to ∞). The factor of $\omega_{\mathbf{p}}$ required by Lorentz invariance, however, leads to the following features (which disappear in the non-relativistic limit):

- There are no states describing localized particles, since one can show that solutions $\phi(t, \mathbf{x})$ of the form 43.14 cannot have compact support in $\mathbf{x}$. The argument is that if they did, the Fourier transforms of both ϕ and its time derivative $\dot{\phi}$ would be analytic functions of $\mathbf{p}$. But the Fourier transforms of solutions would satisfy

$$\frac{\partial}{\partial t} \widetilde{\phi}(t, \mathbf{p}) = -i\omega_{\mathbf{p}} \widetilde{\phi}(t, \mathbf{p})$$

  This leads to a contradiction, since the left-hand side must be analytic, but the right-hand side cannot be (it is a product of an analytic function, and a non-analytic function, $\omega_{\mathbf{p}}$).

- A calculation of the propagator (see section 43.5) shows that it is nonzero for space-like separated points. This implies a potential violation of causality once interactions are allowed, with influence from what happens at one point in space–time traveling to another at faster than the speed of light.

To construct a multiparticle theory with this $\mathcal{H}_1$ along the same lines as the non-relativistic case, we would need to introduce a complex conjugate space $\overline{\mathcal{H}}_1$ and then apply the Bargmann–Fock method. We would then be quantizing a theory with dual phase space a subspace of the solutions of the complex Klein–Gordon equation, those satisfying a condition (positive energy) that is non-local in space–time.

A more straightforward way to construct this theory is to start with dual phase space $\mathcal{M}$ the real-valued solutions of the Klein–Gordon equation. This is a real vector space with no given complex structure, but recall equation 43.11, which describes points in $\mathcal{M}$ by a complex function $\alpha(\mathbf{p})$ rather than a pair of real functions $\phi(\mathbf{x})$ and $\pi(\mathbf{x})$. We can take this as a choice of complex structure, defining:

Definition (Relativistic complex structure). *The relativistic complex structure on the space $\mathcal{M}$ is given by the operator J_r that, extended to $\mathcal{M} \otimes \mathbf{C}$, is $+i$ on the $\alpha(\mathbf{p})$, $-i$ on the $\overline{\alpha}(\mathbf{p})$.*

The $A(\mathbf{p})$ are then a continuous basis of $\mathcal{M}^+_{J_r}$, the $\overline{A}(\mathbf{p})$ a continuous basis of $\mathcal{M}^-_{J_r}$ and

$$\mathcal{M} \otimes \mathbf{C} = \mathcal{M}^+_{J_r} \oplus \mathcal{M}^-_{J_r}$$

A confusing aspect of this setup is that after complexification of $\mathcal{M}$ to get $\mathcal{M} \otimes \mathbf{C}$, $\alpha(\mathbf{p})$ and $\overline{\alpha}(\mathbf{p})$ are not complex conjugates in general, only on the real subspace $\mathcal{M}$. We are starting with a real dual phase space $\mathcal{M}$, which can be parametrized by complex functions

$$f_+(\mathbf{p}) = f(\omega_\mathbf{p}, \mathbf{p}), \quad f_-(\mathbf{p}) = f(-\omega_\mathbf{p}, \mathbf{p})$$

that satisfy the reality condition (see equation 43.6)

$$f_-(\mathbf{p}) = \overline{f_+(-\mathbf{p})}$$

Complexifying, $\mathcal{M} \otimes \mathbf{C}$ will be given by pairs f_+, f_- of complex functions, with no reality condition relating them. The choice of relativistic complex structure J_r is such that $\mathcal{M}^+_{J_r}$ is the space of pairs with $f_- = 0$ (the complex functions on the positive energy hyperboloid), $\mathcal{M}^-_{J_r}$ is the space of pairs with $f_+ = 0$ (complex functions on the negative energy hyperboloid). The conjugation map on $\mathcal{M} \otimes \mathbf{C}$ is not the map conjugating the values of f_- or f_+. It is the map that interchanges

$$(f_+(\mathbf{p}), f_-(\mathbf{p})) \longleftrightarrow (\overline{f_-(-\mathbf{p})}, \overline{f_+(-\mathbf{p})})$$

The $\alpha(\mathbf{p})$ and $\overline{\alpha}(\mathbf{p})$ given by

$$\alpha(\mathbf{p}) = \frac{f_+(\mathbf{p})}{\sqrt{2\omega_\mathbf{p}}}, \quad \overline{\alpha}(\mathbf{p}) = \frac{f_-(-\mathbf{p})}{\sqrt{2\omega_\mathbf{p}}}$$

(see equation 43.7) are related by this nonstandard conjugation (only on real solutions is $\overline{\alpha}(\mathbf{p})$ the complex conjugate of $\alpha(\mathbf{p})$).

Also worth keeping in mind is that, while in terms of the $\alpha(\mathbf{p})$ the complex structure J_r is just multiplication by i, for the basis of field variables $(\phi(\mathbf{x}), \pi(\mathbf{x}))$, J_r is not multiplication by i, but something much more complicated. From equation 43.11, one sees that multiplication by i on the $\alpha(\mathbf{p})$ corresponds to

$$(\phi(\mathbf{x}), \pi(\mathbf{x})) \to \left(\frac{1}{\omega_\mathbf{p}} \pi(\mathbf{x}), -\omega_\mathbf{p} \phi(\mathbf{x}) \right)$$

on the $\phi(\mathbf{x}), \pi(\mathbf{x})$ coordinates. This transformation is compatible with the symplectic structure (preserves the Poisson bracket relations 43.13). As a transformation on position space solutions, the momentum is the differential operator $\mathbf{p} = -i\boldsymbol{\nabla}$, so J_r needs to be thought of as a non-local operation that can be written as

$$(\phi(\mathbf{x}), \pi(\mathbf{x})) \rightarrow \left(\frac{1}{\sqrt{-\boldsymbol{\nabla}^2 + m^2}} \pi(\mathbf{x}), -\sqrt{-\boldsymbol{\nabla}^2 + m^2}\ \phi(\mathbf{x}) \right) \qquad (43.15)$$

Poisson brackets of the continuous basis elements $A(\mathbf{p}), \overline{A}(\mathbf{p})$ are given by

$$\{A(\mathbf{p}), A(\mathbf{p}')\} = \{\overline{A}(\mathbf{p}), \overline{A}(\mathbf{p}')\} = 0, \quad \{A(\mathbf{p}), \overline{A}(\mathbf{p}')\} = i\delta^3(\mathbf{p} - \mathbf{p}') \quad (43.16)$$

Recall from section 26.2 that given a symplectic structure Ω and positive, compatible complex structure J on $\mathcal{M}$, we can define a Hermitian inner product on $\mathcal{M}_J^+$ by

$$\langle u, v \rangle = i\Omega(\overline{u}, v)$$

for $u, v \in \mathcal{M}_J^+$. For the case of $\mathcal{M}$ real solutions to the Klein–Gordon equation, we have on basis elements $A(\mathbf{p})$ of $\mathcal{M}_{J_r}^+$

$$\langle A(\mathbf{p}), A(\mathbf{p}') \rangle = i\Omega(\overline{A}(\mathbf{p}), A(\mathbf{p}')) = i\{\overline{A}(\mathbf{p}), A(\mathbf{p}')\} = \delta^3(\mathbf{p} - \mathbf{p}') \quad (43.17)$$

As a Hermitian inner product on elements $\alpha(\mathbf{p}) \in \mathcal{M}_{J_r}^+$, which is our single-particle state space $\mathcal{H}_1$, equation 43.17 implies that

$$\langle \alpha_1(\mathbf{p}), \alpha_2(\mathbf{p}) \rangle = \int_{\mathbf{R}^3} \overline{\alpha_1(\mathbf{p})} \alpha_2(\mathbf{p}) d^3\mathbf{p} \qquad (43.18)$$

This inner product on $\mathcal{H}_1$ will be positive definite and Lorentz invariant.

Note the difference with the non-relativistic case, where one has the same $E(3)$ invariant inner product on the position space fields or their momentum space Fourier transforms. In the relativistic case, the Hermitian inner product is only the simple one (43.18) on the momentum space initial data for solutions $\overline{\alpha}(\mathbf{p})$ but another quite complicated one on the position space data $\phi(\mathbf{x}), \pi(\mathbf{x})$ (due to the complicated expression 43.15 for J_r there). Unlike the Hermitian inner product and J_r, the symplectic form is simple in both position and momentum space versions, see the Poisson bracket relations 43.13 and 43.16.

Digression. *Quantum field theory textbooks often contain a discussion of a non-positive Hermitian inner product on the space of Klein–Gordon solutions, given by*

$$\langle \phi_1, \phi_2 \rangle = i \int_{\mathbf{R}^3} \left(\overline{\phi}_1(t, \mathbf{x}) \frac{\partial}{\partial t} \phi_2(t, \mathbf{x}) - \left(\frac{\partial}{\partial t} \overline{\phi}_2(t, \mathbf{x}) \right) \phi_1(t, \mathbf{x}) \right) d^3 \mathbf{x}$$

which can be shown to be independent of t. This is defined on $\mathcal{M} \otimes \mathbf{C}$, the complexified Klein–Gordon solutions, and is zero on the real-valued solutions, so does not provide an inner product on those. It does not use the relativistic complex structure. If we start with a theory of complex Klein–Gordon fields, the function $\langle \phi, \phi \rangle$ will be the moment map for the $U(1)$ action by phase transformations on the fields. It will carry an interpretation as charge and give after quantization the charge operator. This will be discussed in section 44.1.2. To compare the formula for the charge to be found there (equation 44.4) with the formula above, use the equation of motion

$$\Pi = \frac{\partial}{\partial t} \overline{\Phi}$$

43.3   Hamiltonian and dynamics of the Klein–Gordon theory

The Klein–Gordon equation for $\phi(t, \mathbf{x})$ in Hamiltonian form is the following pair of first-order equations

$$\frac{\partial}{\partial t} \phi = \pi, \quad \frac{\partial}{\partial t} \pi = (\Delta - m^2) \phi$$

which together imply

$$\frac{\partial^2}{\partial t^2} \phi = (\Delta - m^2) \phi$$

To get these as equations of motion, we need to find a Hamiltonian function h such that

$$\frac{\partial}{\partial t} \phi = \{\phi, h\} = \pi$$

$$\frac{\partial}{\partial t} \pi = \{\pi, h\} = (\Delta - m^2) \phi$$

One can show that two choices of Hamiltonian function with this property are

$$h = \int_{\mathbf{R}^3} \mathcal{H}(\mathbf{x}) d^3 \mathbf{x}$$

where

$$\mathcal{H} = \frac{1}{2}(\pi^2 - \phi\Delta\phi + m^2\phi^2) \quad \text{or} \quad \mathcal{H} = \frac{1}{2}(\pi^2 + (\nabla\phi)^2 + m^2\phi^2)$$

Here the two different integrands $\mathcal{H}(\mathbf{x})$ are related (as in the non-relativistic case) by integration by parts, so these just differ by boundary terms that are assumed to vanish.

In terms of the $A(\mathbf{p}), \overline{A}(\mathbf{p})$, the Hamiltonian will be

$$h = \int_{\mathbf{R}^3} \omega_{\mathbf{p}} A(\mathbf{p})\overline{A}(\mathbf{p})d^3\mathbf{p}$$

The equations of motion are

$$\frac{d}{dt}A = \{A, h\} = \int_{\mathbf{R}^3} \omega_{\mathbf{p}'} A(\mathbf{p}')\{A(\mathbf{p}), \overline{A}(\mathbf{p}')\}d^3\mathbf{p}' = i\omega_{\mathbf{p}} A$$

$$\frac{d}{dt}\overline{A} = \{\overline{A}, h\} = -i\omega_{\mathbf{p}}\overline{A}$$

with solutions

$$A(\mathbf{p}, t) = e^{i\omega_{\mathbf{p}}t}A(\mathbf{p}, 0), \quad \overline{A}(\mathbf{p}, t) = e^{-i\omega_{\mathbf{p}}t}\overline{A}(\mathbf{p}, 0)$$

Digression. *Taking as starting point the Lagrangian formalism, the action for the Klein–Gordon theory is*

$$S = \int_{M^4} \mathcal{L}\, d^4x$$

where

$$\mathcal{L} = \frac{1}{2}\left(\left(\frac{\partial}{\partial t}\phi\right)^2 - (\nabla\phi)^2 - m^2\phi^2\right)$$

This action is a functional of fields on Minkowski space M^4 and is Poincaré invariant. The Euler-Lagrange equations give as equation of motion the Klein–Gordon equation 43.2. One recovers the Hamiltonian formalism by seeing that the canonical momentum for ϕ is

$$\pi = \frac{\partial\mathcal{L}}{\partial\dot{\phi}} = \dot{\phi}$$

and the Hamiltonian density is

$$\mathcal{H} = \pi\dot{\phi} - \mathcal{L} = \frac{1}{2}(\pi^2 + (\nabla\phi)^2 + m^2\phi^2)$$

43.4 Quantization of the Klein–Gordon theory

Given the description we have found in momentum space of real solutions of the Klein–Gordon equation and the choice of complex structure J_r described in the last section, we can proceed to construct a quantum field theory by the Bargmann–Fock method in a manner similar to the non-relativistic quantum field theory case. Quantization takes

$$A(\alpha(\mathbf{p})) \in \mathcal{M}_{J_r}^+ = \mathcal{H}_1 \rightarrow a^\dagger(\alpha(\mathbf{p}))$$

$$\overline{A}(\alpha(\mathbf{p})) \in \mathcal{M}_{J_r}^- = \overline{\mathcal{H}}_1 \rightarrow a(\alpha(\mathbf{p}))$$

Here $A(\alpha(\mathbf{p}))$ is the positive energy solution of the (complexified) Klein–Gordon equation with initial data given by $\alpha(\mathbf{p})$ and $a^\dagger(\alpha(\mathbf{p}))$ is the operator on the Fock space of symmetric tensor products of $\mathcal{H}_1$ given by equation 36.12. $\overline{A}(\alpha(\mathbf{p}))$ is the conjugate negative energy solution and $a(\alpha(\mathbf{p}))$ is the operator given by equation 36.13. All these objects are often written in distributional form, where one has

$$A(\mathbf{p}) \rightarrow a^\dagger(p), \quad \overline{A}(\mathbf{p}) \rightarrow a(\mathbf{p})$$

which one can interpret as corresponding to taking the limit of $\alpha(\mathbf{p}') \rightarrow \delta(\mathbf{p} - \mathbf{p}')$.

The operators $a(\alpha(\mathbf{p}))$ and $a^\dagger(\alpha(\mathbf{p}))$ satisfy the commutation relations

$$[a(\alpha_1(\mathbf{p})), a^\dagger(\alpha_2(\mathbf{p}))] = \langle \alpha_1(\mathbf{p}), \alpha_2(\mathbf{p}) \rangle$$

or in distributional form

$$[a(\mathbf{p}), a^\dagger(\mathbf{p}')] = \delta^3(\mathbf{p} - \mathbf{p}')$$

For the Hamiltonian, we take the normal-ordered form

$$\widehat{H} = \int_{\mathbf{R}^3} \omega_{\mathbf{p}} a^\dagger(\mathbf{p}) a(\mathbf{p}) d^3\mathbf{p}$$

Starting with a vacuum state $|0\rangle$, by applying creation operators one can create arbitrary positive energy multiparticle states of free relativistic particles, with single-particle states having the energy–momentum relation

$$E(\mathbf{p}) = \omega_{\mathbf{p}} = \sqrt{|\mathbf{p}|^2 + m^2}$$

This description of the quantum system is essentially the same as that of the non-relativistic theory, which seems to differ only in the energy–momentum relation, which in that case was $E(\mathbf{p}) = \frac{|\mathbf{p}|^2}{2m}$. The different complex structure used for quantization in the relativistic theory changes the physical meaning of annihilation and creation operators:

- *Non-relativistic theory.* $\mathcal{M} = \mathcal{H}_1$ is the space of complex solutions of the free particle Schrödinger equation, which all have positive energy. $\overline{\mathcal{H}}_1$ is the conjugate space. For each continuous basis element $A(\mathbf{p})$ of $\mathcal{H}_1$ (these are initial data for a positive energy solution with momentum $\mathbf{p}$), quantization takes this to a creation operator $a^\dagger(\mathbf{p})$, which acts with the physical interpretation of addition of a particle with momentum $\mathbf{p}$. Quantization of the complex conjugate $\overline{A}(\mathbf{p})$ in $\overline{\mathcal{H}}_1$ gives an annihilation operator $a(\mathbf{p})$, which removes a particle with momentum $\mathbf{p}$.

- *Relativistic theory.* $\mathcal{M}_{J_r}^+ = \mathcal{H}_1$ is the space of positive energy solutions of the Klein–Gordon equation. It has continuous basis elements $A(\mathbf{p})$ which after quantization become creation operators adding a particle of momentum $\mathbf{p}$ and energy $\omega_{\mathbf{p}}$. $\mathcal{M}_{J_r}^-$ is the space of negative energy solutions of the Klein–Gordon equation. Its continuous basis elements $\overline{A}(\mathbf{p})$ after quantization become annihilation operators for antiparticles of momentum $-\mathbf{p}$ and positive energy $\omega_{\mathbf{p}}$.

To make physical sense of the quanta in the relativistic theory, assigning all non-vacuum states a positive energy, we take such quanta as having two physically equivalent descriptions:

- A positive energy particle moving forward in time with momentum $\mathbf{p}$.
- A positive energy antiparticle moving backward in time with momentum $-\mathbf{p}$.

The operator $a^\dagger(\mathbf{p})$ adds such quanta to a state, and the operator $a(\mathbf{p})$ destroys them. Note that for a theory of quantized real-valued Klein–Gordon fields, the field Φ has components in both $\mathcal{M}_{J_r}^+$ and $\mathcal{M}_{J_r}^-$ so its quantization will both create and destroy quanta.

Just as in the non-relativistic case (see equation 37.1) quantum field operators can be defined using the momentum space decomposition and annihilation and creation operators:

Definition (Real scalar quantum field). *The real scalar quantum field operators are the operator-valued distributions defined by*

$$\widehat{\Phi}(\mathbf{x}) = \frac{1}{(2\pi)^{3/2}} \int_{\mathbf{R}^3} (a(\mathbf{p})e^{i\mathbf{p}\cdot\mathbf{x}} + a^\dagger(\mathbf{p})e^{-i\mathbf{p}\cdot\mathbf{x}}) \frac{d^3\mathbf{p}}{\sqrt{2\omega_{\mathbf{p}}}} \qquad (43.19)$$

$$\widehat{\Pi}(\mathbf{x}) = \frac{1}{(2\pi)^{3/2}} \int_{\mathbf{R}^3} (-i\omega_{\mathbf{p}})(a(\mathbf{p})e^{i\mathbf{p}\cdot\mathbf{x}} - a^\dagger(\mathbf{p})e^{-i\mathbf{p}\cdot\mathbf{x}}) \frac{d^3\mathbf{p}}{\sqrt{2\omega_{\mathbf{p}}}} \qquad (43.20)$$

By essentially the same computation as for Poisson brackets, the commutation relations are

$$[\widehat{\Phi}(\mathbf{x}), \widehat{\Pi}(\mathbf{x}')] = i\delta^3(\mathbf{x} - \mathbf{x}'), \quad [\widehat{\Phi}(\mathbf{x}), \widehat{\Phi}(\mathbf{x}')] = [\widehat{\Pi}(\mathbf{x}), \widehat{\Pi}(\mathbf{x}')] = 0 \qquad (43.21)$$

These can be interpreted as the distributional form of the relations of a unitary representation of a Heisenberg Lie algebra on $\mathcal{M} \oplus \mathbf{R}$, where $\mathcal{M}$ is the space of solutions of the Klein–Gordon equation.

The Hamiltonian operator will be quadratic in the field operators and can be chosen to be

$$\widehat{H} = \int_{\mathbf{R}^3} \frac{1}{2} :(\widehat{\Pi}(\mathbf{x})^2 + (\nabla\widehat{\Phi}(\mathbf{x}))^2 + m^2\widehat{\Phi}(\mathbf{x})^2): d^3\mathbf{x}$$

This operator is normal-ordered, and a computation (see, for instance, chapter 5 of [16]) shows that in terms of momentum space operators this is the expected

$$\widehat{H} = \int_{\mathbf{R}^3} \omega_{\mathbf{p}} a^\dagger(\mathbf{p}) a(\mathbf{p}) d^3\mathbf{p} \qquad (43.22)$$

The dynamical equations of the quantum field theory are now

$$\frac{\partial}{\partial t}\widehat{\Phi} = [\widehat{\Phi}, -i\widehat{H}] = \widehat{\Pi}$$

$$\frac{\partial}{\partial t}\widehat{\Pi} = [\widehat{\Pi}, -i\widehat{H}] = (\Delta - m^2)\widehat{\Phi}$$

which have as solution the following equation for the time-dependent field operator:

$$\widehat{\Phi}(t, \mathbf{x}) = \frac{1}{(2\pi)^{3/2}} \int_{\mathbf{R}^3} (a(\mathbf{p})e^{-i\omega_{\mathbf{p}}t}e^{i\mathbf{p}\cdot\mathbf{x}} + a^\dagger(\mathbf{p})e^{i\omega_{\mathbf{p}}t}e^{-i\mathbf{p}\cdot\mathbf{x}}) \frac{d^3\mathbf{p}}{\sqrt{2\omega_{\mathbf{p}}}} \qquad (43.23)$$

Unlike the non-relativistic case, where fields are non-self-adjoint operators, here the field operator is self-adjoint (and thus an observable) and has both an annihilation operator component and a creation operator component. This gives a theory of positive energy quanta that can be interpreted as either particles moving forward in time or antiparticles of opposite momentum moving backward in time.

43.5 The scalar field propagator

As for any quantum field theory, a fundamental quantity to calculate is the propagator, which for a free quantum field theory actually can be used to calculate all amplitudes between multiparticle states. For the free relativistic scalar field theory, we have

Definition (Propagator, Klein–Gordon theory). *The propagator for the relativistic scalar field theory is the amplitude*

$$U(t_2, \mathbf{x}_2, t_1, \mathbf{x}_1) = \langle 0|\widehat{\Phi}(t_2, \mathbf{x}_2)\widehat{\Phi}(t_1, \mathbf{x}_1)|0\rangle$$

By translation invariance, the propagator will only depend on $t_2 - t_1$ and $\mathbf{x}_2 - \mathbf{x}_1$, so we can just evaluate the case $(t_1, \mathbf{x}_1) = (0, \mathbf{0}), (t_2, \mathbf{x}_2) = (t, \mathbf{x})$, using the formula 43.23 for the time-dependent quantum field to get

$$U(t, \mathbf{x}, 0, \mathbf{0}) = \frac{1}{(2\pi)^3} \int_{\mathbf{R}^3 \times \mathbf{R}^3} \langle 0|(a(\mathbf{p})e^{-i\omega_{\mathbf{p}}t}e^{i\mathbf{p}\cdot\mathbf{x}} + a^\dagger(\mathbf{p})e^{i\omega_{\mathbf{p}}t}e^{-i\mathbf{p}\cdot\mathbf{x}})$$

$$(a(\mathbf{p}') + a^\dagger(\mathbf{p}'))|0\rangle \frac{d^3\mathbf{p}}{\sqrt{2\omega_{\mathbf{p}}}} \frac{d^3\mathbf{p}'}{\sqrt{2\omega_{\mathbf{p}'}}}$$

$$= \frac{1}{(2\pi)^3} \int_{\mathbf{R}^3 \times \mathbf{R}^3} \delta^3(\mathbf{p} - \mathbf{p}')e^{-i\omega_{\mathbf{p}}t}e^{i\mathbf{p}\cdot\mathbf{x}} \frac{d^3\mathbf{p}}{\sqrt{2\omega_{\mathbf{p}}}} \frac{d^3\mathbf{p}'}{\sqrt{2\omega_{\mathbf{p}'}}}$$

$$= \frac{1}{(2\pi)^3} \int_{\mathbf{R}^3} e^{-i\omega_{\mathbf{p}}t}e^{i\mathbf{p}\cdot\mathbf{x}} \frac{d^3\mathbf{p}}{2\omega_{\mathbf{p}}}$$

$$= \frac{1}{(2\pi)^3} \int_{\mathbf{R}^4} \theta(p_0)\delta(p^2 + m^2)e^{-ip_0 t}e^{i\mathbf{p}\cdot\mathbf{x}}d^3\mathbf{p}\,dp_0$$

The last line shows that this distribution is the Minkowski space Fourier transform of the delta-function distribution on the positive energy hyperboloid

$$p^2 = -p_0^2 + |\mathbf{p}|^2 = -m^2$$

As in the non-relativistic case (see section 12.5), this is a distribution that can be defined as a boundary value of an analytic function of complex time. The integral can be evaluated in terms of Bessel functions, and its properties are discussed in all standard quantum field theory textbooks. These include:

- For $\mathbf{x} = 0$, the amplitude is oscillatory in time, a superposition of terms with positive frequency.

- For $t = 0$, the amplitude falls off exponentially as $e^{-m|\mathbf{x}|}$.

The resolution of the potential causality problem caused by the nonzero amplitude at space-like separations between $(t_1, \mathbf{x}_1)$ and $(t_2, \mathbf{x}_2)$ (e.g., for $t_1 = t_2$) is that the condition really needed on observable operators $\mathcal{O}$ localized at points in space–time is that

$$[\mathcal{O}(t_2, \mathbf{x}_2), \mathcal{O}(t_1, \mathbf{x}_1)] = 0$$

for $(t_1, \mathbf{x}_1)$ and $(t_2, \mathbf{x}_2)$ space-like separated (this condition is known as "micro-causality"). This will ensure that measurement of the observable $\mathcal{O}$ at a point will not affect its measurement at a space-like separated point, avoiding potential conflicts with causality. For the field operator $\Phi(t, \mathbf{x})$, one can calculate the commutator by a similar calculation to the one above for $U(t, \mathbf{x}, 0, 0)$, with result

$$[\widehat{\Phi}(t, \mathbf{x}), \widehat{\Phi}(0, \mathbf{0})] = \frac{1}{(2\pi)^3} \int_{\mathbf{R}^4} \delta(p^2 + m^2)(\theta(p_0)e^{ip \cdot x} - \theta(-p_0)e^{-ip \cdot x}) d^3\mathbf{p}\, dp_0$$
$$= U(t, \mathbf{x}, 0, 0) - U(-t, -\mathbf{x}, 0, 0)$$

This will be zero for space-like $(t, \mathbf{x})$, something one can see by noting that the result is Lorentz invariant and is equal to 0 at $t = 0$ by the canonical commutation relation

$$[\widehat{\Phi}(\mathbf{x}_1), \widehat{\Phi}(\mathbf{x}_2)] = 0$$

and any two space-like vectors are related by a Lorentz transformation. Note that the vanishing of this commutator for space-like separations is achieved by cancellation of propagation amplitudes for a particle and antiparticle, showing that the relativistic choice of complex structure for quantization is needed to ensure causality.

One can also study the propagator using Green's function methods as in the single-particle case of section 12.7, now with

$$D = -\frac{\partial^2}{\partial t^2} + \boldsymbol{\nabla}^2 - m^2$$

$$\widehat{D}(p_0, \mathbf{p}) = p_0^2 - (|\mathbf{p}|^2 + m^2)$$

and

$$\widehat{G}(p_0, \mathbf{p}) = \frac{1}{p_0^2 - (|\mathbf{p}|^2 + m^2)} = \frac{1}{(p_0 - \omega_\mathbf{p})(p_0 + \omega_\mathbf{p})}$$

In the non-relativistic case, the Green's function $\widehat{G}$ only had one pole, at $p_0 = |\mathbf{p}|^2/2m$. Two possible choices of how to extend integration over p_0 into the complex plane avoiding the pole gave either a retarded Green's function and propagation from past to future, or an advanced Green's function and propagation from future to past. In the Klein–Gordon case, there are now two poles, at $p_0 = \pm\omega_\mathbf{p}$. Micro-causality requires that the pole at positive energy be treated as a retarded Green's function, with propagation of positive energy particles from past to future, while the one at negative energy must be treated as an advanced Green's function with propagation of negative energy particles from future to past.

43.6 Interacting scalar field theories: some comments

Our discussion so far has dealt purely with a theory of non-interacting quanta, so this theory is called a quantum theory of free fields. The field, however, can be used to introduce interactions between these quanta, interactions which are local in space. The simplest such theory is the one given by adding a quartic term to the Hamiltonian, taking

$$\widehat{H} = \int_{\mathbf{R}^3} :\frac{1}{2}(\widehat{\Pi}(\mathbf{x})^2 + (\nabla\widehat{\Phi}(\mathbf{x}))^2 + m^2\widehat{\Phi}(\mathbf{x})^2) + \lambda\widehat{\Phi}(\mathbf{x})^4: d^3\mathbf{x}$$

This interacting theory is vastly more complicated and much harder to understand than the non-interacting theory. Among the difficult problems that arise are:

- How should one make sense of the expression

$$:\widehat{\Phi}(\mathbf{x})^4:$$

  since it is a product not of operators but of operator-valued distributions?

- How can one construct an appropriate state space on which the interacting Hamiltonian operator will be well defined, with a well-defined ground state?

Quantum field theory textbooks explain how to construct a series expansion in powers of λ about the free field value $\lambda = 0$, by a calculation whose terms are labeled by Feynman diagrams. To get finite results, cutoffs must first be introduced, and then some way found to get a sensible limit as the cutoff is removed (this is the theory of "renormalization"). In this manner finite results can be found for the terms in the series expansion, but the expansion is not convergent, giving only an asymptotic series (for fixed λ, no matter how small, the series will diverge at high enough order).

For known calculational methods not based on the series expansion, again a cutoff must be introduced, making the number of degrees of freedom finite. For a fixed ultraviolet cutoff, corresponding physically to only allowing fields with momentum components smaller than a given value, one can construct a sensible theory with non-trivial interactions (e.g., scattering of one particle by another, which does not happen in the free theory). This gives a sensible theory, for momenta far below the cutoff. However, it appears that, for three or more spatial dimensions, removal of the cutoff will always in the limit give back the non-interacting free field theory. There is thus no known continuum relativistic quantum field theory of scalar fields, other than free field theory, that can be constructed in this way.

43.7 For further reading

Pretty much every quantum field theory textbook has a treatment of the relativistic scalar field with more detail than given here and significantly more physical motivation. A good example is the lectures of Sidney Coleman [15], one that has some detailed versions of the calculations discussed here is chapter 5 of [16]. Chapter 2 of [67] covers the same material, with more discussion of the propagator and the causality question.

For a more detailed rigorous construction of the Klein–Gordon theory in terms of Fock space, closely related to the outline given in this chapter, three sources are

- Chapter X.7 of [73].
- Chapter 5.2 of [28].
- Chapter 8.2.2 of [17].

For a general axiomatic mathematical treatment of relativistic quantum fields as distribution-valued operators, some standard references are [87] and [12]. The second of these includes a rigorous construction of the Klein–Gordon theory.

For a treatment of relativistic quantum field theory relatively close to ours, not only for the Klein–Gordon theory of this chapter, but also for the spin-$\frac{1}{2}$ theories of later chapters, see [33].

Chapter 44
Symmetries and Relativistic Scalar Quantum Fields

Just as for non-relativistic quantum fields, the theory of free relativistic scalar quantum fields starts by taking as phase space an infinite dimensional space of solutions of an equation of motion. Quantization of this phase space proceeds by constructing field operators which provide a representation of the corresponding Heisenberg Lie algebra, using an infinite dimensional version of the Bargmann–Fock construction. In both cases, the equation of motion has a representation-theoretical significance: It is an eigenvalue equation for the Casimir operator of a group of space–time symmetries, picking out an irreducible representation of that group. In the non-relativistic case, the Laplacian Δ was the Casimir operator, the symmetry group was the Euclidean group $E(3)$, and one got an irreducible representation for fixed energy. In the relativistic case, the Casimir operator is the Minkowski space version of the Laplacian

$$-\frac{\partial^2}{\partial t^2} + \Delta$$

the space–time symmetry group is the Poincaré group, and the eigenvalue of the Casimir is m^2.

The Poincaré group acts on the phase space of solutions to the Klein–Gordon equation, preserving the Poisson bracket. The same general methods as in the finite dimensional and non-relativistic quantum field theory cases can be used to get a representation of the Poincaré group by intertwining operators for the Heisenberg Lie algebra representation (the representation given by the field operators). These methods give a representation of the Lie algebra of the Poincaré group in terms of quadratic combinations of the field operators.

We will begin though with the case of an even simpler group action on the phase space, that coming from an "internal symmetry" of multicomponent scalar fields with an orthogonal group or unitary group acting on the real or complex vector space in which the classical fields take their values. For the

© Peter Woit 2017
P. Woit, *Quantum Theory, Groups and Representations*,
DOI 10.1007/978-3-319-64612-1_44

simplest case of fields taking values in $\mathbf{R}^2$ or $\mathbf{C}$, one gets a theory of charged relativistic particles, with antiparticles now distinguishable from particles (they have opposite charge).

44.1 Internal symmetries

The relativistic real scalar field theory of chapter 43 lacks one important feature of the non-relativistic theory, which is an action of the group $U(1)$ by phase changes on complex fields. This is needed to provide a notion of "charge" and allow the introduction of electromagnetic forces into the theory (see chapter 45). In the real scalar field theory, there is no distinction between states describing particles and states describing antiparticles. To get a theory with such a distinction, we need to introduce fields with more components. Two possibilities are to consider real fields with m components, in which case we will have a theory with $SO(m)$ symmetry, or to consider complex fields with n components, in which case we have a theory with $U(n)$ symmetry. Identifying $\mathbf{C}$ with $\mathbf{R}^2$ using the standard complex structure, we find $SO(2) = U(1)$, and two equivalent ways of getting a theory with $U(1)$ symmetry, using two real or one complex scalar field.

44.1.1 SO(m) symmetry and real scalar fields

Starting with the case $m = 2$, and taking as dual phase space $\mathcal{M}$ the space of pairs ϕ_1, ϕ_2 of real solutions to the two-component Klein–Gordon equation, elements $g(\theta)$ of the group $SO(2)$ will act on the fields by

$$\begin{pmatrix} \Phi_1(\mathbf{x}) \\ \Phi_2(\mathbf{x}) \end{pmatrix} \rightarrow g(\theta) \cdot \begin{pmatrix} \Phi_1(\mathbf{x}) \\ \Phi_2(\mathbf{x}) \end{pmatrix} = \begin{pmatrix} \cos\theta & \sin\theta \\ -\sin\theta & \cos\theta \end{pmatrix} \begin{pmatrix} \Phi_1(\mathbf{x}) \\ \Phi_2(\mathbf{x}) \end{pmatrix}$$

$$\begin{pmatrix} \Pi_1(\mathbf{x}) \\ \Pi_2(\mathbf{x}) \end{pmatrix} \rightarrow g(\theta) \cdot \begin{pmatrix} \Pi_1(\mathbf{x}) \\ \Pi_2(\mathbf{x}) \end{pmatrix} = \begin{pmatrix} \cos\theta & \sin\theta \\ -\sin\theta & \cos\theta \end{pmatrix} \begin{pmatrix} \Pi_1(\mathbf{x}) \\ \Pi_2(\mathbf{x}) \end{pmatrix}$$

Here, $\Phi_1(\mathbf{x}), \Phi_2(\mathbf{x}), \Pi_1(\mathbf{x}), \Pi_2(\mathbf{x})$ are the continuous basis elements for the space of two-component Klein–Gordon solutions, determined by their initial values at $t = 0$.

This group action on $\mathcal{M}$ breaks up into a direct sum of an infinite number (one for each value of $\mathbf{x}$) of identical copies of the case of rotations in a configuration space plane, as discussed in section 20.3.1. We will use the calculation there, where we found that for a basis element L of the Lie algebra of $SO(2)$, the corresponding quadratic function on the phase space with coordinates q_1, q_2, p_1, p_2 was

$$\mu_L = q_1 p_2 - q_2 p_1$$

For the case here, we take

$$q_1, q_2, p_1, p_2 \to \Phi_1(\mathbf{x}), \Phi_2(\mathbf{x}), \Pi_1(\mathbf{x}), \Pi_2(\mathbf{x})$$

and integrate the analog of μ_L over $\mathbf{R}^3$ to get an appropriate moment map for the field theory case. This gives a quadratic functional on the fields that will have the desired Poisson bracket with the fields for each value of $\mathbf{x}$. We will denote the result by Q, since it is an observable that will have a physical interpretation as electric charge when this theory is coupled to the electromagnetic field (see chapter 45):

$$Q = \int_{\mathbf{R}^3} (\Pi_2(\mathbf{x})\Phi_1(\mathbf{x}) - \Pi_1(\mathbf{x})\Phi_2(\mathbf{x}))d^3\mathbf{x}$$

One can use the field Poisson bracket relations

$$\{\Phi_j(\mathbf{x}), \Pi_k(\mathbf{x}')\} = \delta_{jk}\delta(\mathbf{x} - \mathbf{x}')$$

to check that

$$\left\{Q, \begin{pmatrix} \Phi_1(\mathbf{x}) \\ \Phi_2(\mathbf{x}) \end{pmatrix}\right\} = \begin{pmatrix} -\Phi_2(\mathbf{x}) \\ \Phi_1(\mathbf{x}) \end{pmatrix}, \quad \left\{Q, \begin{pmatrix} \Pi_1(\mathbf{x}) \\ \Pi_2(\mathbf{x}) \end{pmatrix}\right\} = \begin{pmatrix} -\Pi_2(\mathbf{x}) \\ \Pi_1(\mathbf{x}) \end{pmatrix}$$

Quantization of the classical field theory gives a unitary representation U of $SO(2)$ on the multiparticle state space, with

$$U'(L) = -i\widehat{Q} = -i\int_{\mathbf{R}^3} (\widehat{\Pi}_2(\mathbf{x})\widehat{\Phi}_1(\mathbf{x}) - \widehat{\Pi}_1(\mathbf{x})\widehat{\Phi}_2(\mathbf{x}))d^3\mathbf{x}$$

The operator

$$U(\theta) = e^{-i\theta\widehat{Q}}$$

will act by conjugation on the fields:

$$U(\theta)\begin{pmatrix} \widehat{\Phi}_1(\mathbf{x}) \\ \widehat{\Phi}_2(\mathbf{x}) \end{pmatrix} U(\theta)^{-1} = \begin{pmatrix} \cos\theta & \sin\theta \\ -\sin\theta & \cos\theta \end{pmatrix} \begin{pmatrix} \widehat{\Phi}_1(\mathbf{x}) \\ \widehat{\Phi}_2(\mathbf{x}) \end{pmatrix}$$

$$U(\theta)\begin{pmatrix} \widehat{\Pi}_1(\mathbf{x}) \\ \widehat{\Pi}_2(\mathbf{x}) \end{pmatrix} U(\theta)^{-1} = \begin{pmatrix} \cos\theta & \sin\theta \\ -\sin\theta & \cos\theta \end{pmatrix} \begin{pmatrix} \widehat{\Pi}_1(\mathbf{x}) \\ \widehat{\Pi}_2(\mathbf{x}) \end{pmatrix}$$

It will also give a representation of $SO(2)$ on states, with the state space decomposing into sectors each labeled by the integer eigenvalue of the operator $\widehat{Q}$ (which will be called the "charge" of the state).

Using the definitions of $\widehat{\Phi}$ and $\widehat{\Pi}$ (43.19 and 43.20), $\widehat{Q}$ can be computed in terms of annihilation and creation operators, with the result

$$\widehat{Q} = i \int_{\mathbf{R}^3} (a_2^\dagger(\mathbf{p})a_1(\mathbf{p}) - a_1^\dagger(\mathbf{p})a_2(\mathbf{p}))d^3\mathbf{p} \qquad (44.1)$$

One expects that since the time evolution action on the classical field space commutes with the $SO(2)$ action, the operator $\widehat{Q}$ should commute with the Hamiltonian operator $\widehat{H}$. This can readily be checked by computing $[\widehat{H}, \widehat{Q}]$ using

$$\widehat{H} = \int_{\mathbf{R}^3} \omega_{\mathbf{p}}(a_1^\dagger(\mathbf{p})a_1(\mathbf{p}) + a_2^\dagger(\mathbf{p})a_2(\mathbf{p}))d^3\mathbf{p}$$

Note that the vacuum state $|0\rangle$ is an eigenvector for $\widehat{Q}$ and $\widehat{H}$ with both eigenvalues 0: It has zero energy and zero charge. States $a_1^\dagger(\mathbf{p})|0\rangle$ and $a_2^\dagger(\mathbf{p})|0\rangle$ are eigenvectors of $\widehat{H}$ with eigenvalue and thus energy $\omega_{\mathbf{p}}$, but these are not eigenvectors of $\widehat{Q}$, so do not have a well-defined charge.

All of this can be generalized to the case of $m > 2$ real scalar fields, with a larger group $SO(m)$ now acting instead of the group $SO(2)$. The Lie algebra is now multidimensional, with a basis the elementary antisymmetric matrices ϵ_{jk}, with $j, k = 1, 2, \cdots, m$ and $j < k$, which correspond to infinitesimal rotations in the jk planes. Group elements can be constructed by multiplying rotations $e^{\theta \epsilon_{jk}}$ in different planes. Instead of a single operator $\widehat{Q}$, we get multiple operators

$$-i\widehat{Q}_{jk} = -i \int_{\mathbf{R}^3} (\widehat{\Pi}_k(\mathbf{x})\widehat{\Phi}_j(\mathbf{x}) - \widehat{\Pi}_j(\mathbf{x})\widehat{\Phi}_k(\mathbf{x}))d^3\mathbf{x}$$

and conjugation by

$$U_{jk}(\theta) = e^{-i\theta\widehat{Q}_{jk}}$$

rotates the field operators in the jk plane. These also provide unitary operators on the state space, and, taking appropriate products of them, a unitary representation of the full group $SO(m)$ on the state space. The $\widehat{Q}_{jk}$ commute with the Hamiltonian (generalized to the m-component case), so the energy eigenstates of the theory break up into irreducible representations of $SO(m)$ (a subject we have not discussed for $m > 3$).

44.1.2 U(1) symmetry and complex scalar fields

Instead of describing a scalar field system with $SO(2)$ symmetry using a pair Φ_1, Φ_2 of real fields, it is sometimes more convenient to work with complex scalar fields and a $U(1)$ symmetry. This will also allow the use of field operators and annihilation and creation operators for states with a definite value of the charge observable. Taking as $\mathcal{M}$ the complex vector space of complex solutions to the Klein–Gordon equation, however, is confusing, since the Bargmann–Fock quantization method requires that we complexify $\mathcal{M}$, and the complexification of a complex vector space is a notion that requires some care. More simply, here one can think of the space $\mathcal{M}$ of solutions to the Klein–Gordon equation for a pair of real fields as having two different complex structures:

- The relativistic complex structure J_r, which is $+i$ on positive energy solutions in $\mathcal{M} \otimes \mathbf{C}$ and $-i$ on negative energy solutions in $\mathcal{M} \otimes \mathbf{C}$.

- The "charge" complex structure J_C, which is $+i$ on positive charge solutions and $-i$ on negative charge solutions.

The operators J_r and J_C will commute, so we can simultaneously diagonalize them on $\mathcal{M} \otimes \mathbf{C}$ and decompose the positive energy solution space into $\pm i$ eigenspaces of J_C, so

$$\mathcal{H}_1 = \mathcal{M}_{J_r}^+ = \mathcal{H}_1^+ \oplus \mathcal{H}_1^-$$

where $\mathcal{H}_1^+$ will be positive energy solutions with $J_C = +i$ and $\mathcal{H}_1^-$ will be positive energy solutions with $J_C = -i$. Taking as before $\alpha_1(\mathbf{p}), \alpha_2(\mathbf{p})$ for the momentum space initial data for elements of $\mathcal{H}_1$, we will define

$$\alpha(\mathbf{p}) = \frac{1}{\sqrt{2}}(\alpha_1(\mathbf{p}) - i\alpha_2(\mathbf{p})) \in \mathcal{H}_1^+, \quad \beta(\mathbf{p}) = \frac{1}{\sqrt{2}}(\alpha_1(\mathbf{p}) + i\alpha_2(\mathbf{p})) \in \mathcal{H}_1^-$$

The negative energy solution space can be decomposed as

$$\mathcal{M}_{J_r}^- = \overline{\mathcal{H}}_1 = \overline{\mathcal{H}}_1^+ \oplus \overline{\mathcal{H}}_1^-$$

and

$$\overline{\alpha}(\mathbf{p}) = \frac{1}{\sqrt{2}}(\overline{\alpha}_1(\mathbf{p}) + i\overline{\alpha}_2(\mathbf{p})) \in \mathcal{H}_1^+, \quad \overline{\beta}(\mathbf{p}) = \frac{1}{\sqrt{2}}(\overline{\alpha}_1(\mathbf{p}) - i\overline{\alpha}_2(\mathbf{p})) \in \mathcal{H}_1^-$$

We will write $A(\mathbf{p}), B(\mathbf{p})$ for the solutions α, β with initial data delta-functions at $\mathbf{p}$, $\overline{A}(\mathbf{p}), \overline{B}(\mathbf{p})$ for their conjugates, and quantization will take

$$A(\mathbf{p}) \to a^\dagger(\mathbf{p}) = \frac{1}{\sqrt{2}}(a_1^\dagger(\mathbf{p}) - ia_2^\dagger(\mathbf{p}))$$

$$B(\mathbf{p}) \to b^\dagger(\mathbf{p}) = \frac{1}{\sqrt{2}}(a_1^\dagger(\mathbf{p}) + ia_2^\dagger(\mathbf{p}))$$

$$\overline{A}(\mathbf{p}) \to a(\mathbf{p}) = \frac{1}{\sqrt{2}}(a_1(\mathbf{p}) + ia_2(\mathbf{p}))$$

$$\overline{B}(\mathbf{p}) \to b(\mathbf{p}) = \frac{1}{\sqrt{2}}(a_1(\mathbf{p}) - ia_2(\mathbf{p}))$$

with the nonzero commutation relations between these operators given by

$$[a(\mathbf{p}), a^\dagger(\mathbf{p}')] = \delta(\mathbf{p} - \mathbf{p}'), \quad [b(\mathbf{p}), b^\dagger(\mathbf{p}')] = \delta(\mathbf{p} - \mathbf{p}')$$

The state space of this theory is a tensor product of two copies of the state space of a real scalar field. The operators $a^\dagger(\mathbf{p}), a(\mathbf{p})$ act on the state space by creating or annihilating a positively charged particle of momentum $\mathbf{p}$, whereas the $b^\dagger(\mathbf{p}), b(\mathbf{p})$ create or annihilate antiparticles of negative charge. The vacuum state will satisfy

$$a(\mathbf{p})|0\rangle = b(\mathbf{p})|0\rangle = 0$$

The Hamiltonian operator for this theory will be

$$\widehat{H} = \int_{\mathbf{R}^3} \omega_{\mathbf{p}}(a^\dagger(\mathbf{p})a(\mathbf{p}) + b^\dagger(\mathbf{p})b(\mathbf{p}))d^3\mathbf{p}$$

and the charge operator is

$$\widehat{Q} = \int_{\mathbf{R}^3} (a^\dagger(\mathbf{p})a(\mathbf{p}) - b^\dagger(\mathbf{p})b(\mathbf{p}))d^3\mathbf{p}$$

Using these creation and annihilation operators, we can define position space field operators analogous to the ones given by equations 43.19 and 43.20 in the real scalar field case. Now, $\widehat{\Phi}(\mathbf{x})$ will not be self-adjoint, but its adjoint will be a field $\widehat{\Phi}^\dagger(\mathbf{x})$ which will act on states by increasing the charge by 1, with one term that creates particles and another that annihilates antiparticles. We define

Definition (Complex scalar quantum field). *The complex scalar quantum field operators are the operator-valued distributions defined by*

$$\widehat{\Phi}(\mathbf{x}) = \frac{1}{(2\pi)^{3/2}} \int_{\mathbf{R}^3} (a(\mathbf{p})e^{i\mathbf{p}\cdot\mathbf{x}} + b^\dagger(\mathbf{p})e^{-i\mathbf{p}\cdot\mathbf{x}}) \frac{d^3\mathbf{p}}{\sqrt{2\omega_{\mathbf{p}}}}$$

$$\widehat{\Phi}^\dagger(\mathbf{x}) = \frac{1}{(2\pi)^{3/2}} \int_{\mathbf{R}^3} (b(\mathbf{p})e^{i\mathbf{p}\cdot\mathbf{x}} + a^\dagger(\mathbf{p})e^{-i\mathbf{p}\cdot\mathbf{x}}) \frac{d^3\mathbf{p}}{\sqrt{2\omega_{\mathbf{p}}}}$$

$$\widehat{\Pi}(\mathbf{x}) = \frac{1}{(2\pi)^{3/2}} \int_{\mathbf{R}^3} (-i\omega_{\mathbf{p}})(a(\mathbf{p})e^{i\mathbf{p}\cdot\mathbf{x}} - b^\dagger(\mathbf{p})e^{-i\mathbf{p}\cdot\mathbf{x}}) \frac{d^3\mathbf{p}}{\sqrt{2\omega_{\mathbf{p}}}}$$

$$\widehat{\Pi}^\dagger(\mathbf{x}) = \frac{1}{(2\pi)^{3/2}} \int_{\mathbf{R}^3} (-i\omega_{\mathbf{p}})(b(\mathbf{p})e^{i\mathbf{p}\cdot\mathbf{x}} - a^\dagger(\mathbf{p})e^{-i\mathbf{p}\cdot\mathbf{x}}) \frac{d^3\mathbf{p}}{\sqrt{2\omega_{\mathbf{p}}}}$$

These satisfy the commutation relations

$$[\widehat{\Phi}(\mathbf{x}), \widehat{\Pi}^\dagger(\mathbf{x}')] = [\widehat{\Pi}(\mathbf{x}), \widehat{\Pi}^\dagger(\mathbf{x}')] = [\widehat{\Phi}(\mathbf{x}), \widehat{\Pi}^\dagger(\mathbf{x}')] = [\widehat{\Phi}^\dagger(\mathbf{x}), \widehat{\Pi}(\mathbf{x}')] = 0$$

$$[\widehat{\Phi}(\mathbf{x}), \widehat{\Pi}(\mathbf{x}')] = [\widehat{\Phi}^\dagger(\mathbf{x}), \widehat{\Pi}^\dagger(\mathbf{x}')] = i\delta^3(\mathbf{x} - \mathbf{x}') \qquad (44.2)$$

In terms of these field operators, the Hamiltonian operator will be

$$\widehat{H} = \int_{\mathbf{R}^3} :(\widehat{\Pi}^\dagger(\mathbf{x})\widehat{\Pi}(\mathbf{x}) + (\nabla\widehat{\Phi}^\dagger(\mathbf{x}))(\nabla\widehat{\Phi}(\mathbf{x})) + m^2\widehat{\Phi}^\dagger(\mathbf{x})\widehat{\Phi}(\mathbf{x})): d^3\mathbf{x}$$

and the charge operator will be

$$\widehat{Q} = -i \int_{\mathbf{R}^3} :(\widehat{\Pi}(\mathbf{x})\widehat{\Phi}(\mathbf{x}) - \widehat{\Pi}^\dagger(\mathbf{x})\widehat{\Phi}^\dagger(\mathbf{x})): d^3\mathbf{x}$$

Taking $L = i$ as a basis element for $\mathfrak{u}(1)$, one gets a unitary representation U of $U(1)$ using

$$U'(L) = -i\widehat{Q}$$

and

$$U(\theta) = e^{-i\theta\widehat{Q}}$$

U acts by conjugation on the fields:

$$U(\theta)\widehat{\Phi}U(\theta)^{-1} = e^{-i\theta}\widehat{\Phi}, \quad U(\theta)\widehat{\Phi}^\dagger U(\theta)^{-1} = e^{i\theta}\widehat{\Phi}^\dagger$$

$$U(\theta)\widehat{\Pi}U(\theta)^{-1} = e^{i\theta}\widehat{\Pi}, \quad U(\theta)\widehat{\Pi}^\dagger U(\theta)^{-1} = e^{-i\theta}\widehat{\Pi}^\dagger$$

It will also give a representation of $U(1)$ on states, with the state space decomposing into sectors each labeled by the integer eigenvalue of the operator $\widehat{Q}$.

Instead of starting in momentum space with solutions given by $A(\mathbf{p}), B(\mathbf{p})$, we could instead have considered position space initial data and distributional fields

$$\Phi(\mathbf{x}) = \frac{1}{\sqrt{2}}(\Phi_1(\mathbf{x}) + i\Phi_2(x)), \quad \Pi(\mathbf{x}) = \frac{1}{\sqrt{2}}(\Pi_1(\mathbf{x}) - i\Pi_2(\mathbf{x})) \qquad (44.3)$$

and their complex conjugates $\overline{\Phi}(\mathbf{x}), \overline{\Pi}(\mathbf{x})$. The Poisson bracket relations on such complex fields will be

$$\{\Phi(\mathbf{x}), \overline{\Phi}(\mathbf{x}')\} = \{\Pi(\mathbf{x}), \overline{\Pi}(\mathbf{x}')\} = \{\Phi(\mathbf{x}), \overline{\Pi}(\mathbf{x}')\} = \{\overline{\Phi}(\mathbf{x}), \Pi(\mathbf{x}')\} = 0$$

$$\{\Phi(\mathbf{x}), \Pi(\mathbf{x}')\} = \{\overline{\Phi}(\mathbf{x}), \overline{\Pi}(\mathbf{x}')\} = \delta(\mathbf{x} - \mathbf{x}')$$

and the classical Hamiltonian is

$$h = \int_{\mathbf{R}^3} (|\Pi|^2 + |\nabla \Phi|^2 + m^2 |\Phi|^2) d^3\mathbf{x}$$

The charge function Q would be given by

$$Q = -i \int_{\mathbf{R}^3} (\Pi(\mathbf{x})\Phi(\mathbf{x}) - \overline{\Pi}(\mathbf{x})\overline{\Phi}(\mathbf{x})) d^3\mathbf{x} \qquad (44.4)$$

satisfying

$$\{Q, \Phi(\mathbf{x})\} = i\Phi(\mathbf{x}), \quad \{Q, \overline{\Phi}(\mathbf{x})\} = -i\overline{\Phi}(\mathbf{x})$$

44.2 Poincaré symmetry and scalar fields

Returning to the case of a single real relativistic field, the dual phase space $\mathcal{M}$ carries an action of the Poincaré group $\mathcal{P}$, and the quantum field theory will come with a unitary representation of this group, in much the same way that the non-relativistic case came with a representation of the Euclidean group $E(3)$ (see section 38.3). The Poincaré group acts on the space of solutions to the Klein–Gordon equation since its action on functions on space–time commutes with the Casimir operator

$$P^2 = \frac{\partial^2}{\partial t^2} - \frac{\partial^2}{\partial x_1^2} - \frac{\partial^2}{\partial x_2^2} - \frac{\partial^2}{\partial x_3^2}$$

This Poincaré group action on Klein–Gordon solutions is by the usual action on functions

$$\phi \rightarrow u(a, \Lambda)\phi = \phi(\Lambda^{-1}(x - a)) \qquad (44.5)$$

induced from the group action on Minkowski space. On fields $\Phi(\mathbf{x})$ the action is

$$\Phi \rightarrow u(a, \Lambda)\Phi = \Phi(\Lambda x + a)$$

Quantization should give unitary operators $U(a, \Lambda)$, which act on field operators by

$$\widehat{\Phi} \to U(a, \Lambda)\widehat{\Phi}(x)U(a, \Lambda)^{-1} = \widehat{\Phi}(\Lambda x + a) \qquad (44.6)$$

The $U(a, \Lambda)$ will provide a unitary representation of the Poincaré group on the quantum field theory state space, acting by intertwining operators of the sort discussed in the finite dimensional context in chapter 20. We would like to construct these operators by the usual method: using the moment map to get a quadratic polynomial on phase space, quantizing to get Lie algebra representation operators, and then exponentiating to get the $U(a, \Lambda)$.

This will require that the symplectic structure on the phase space $\mathcal{H}_1$ be Poincaré invariant. The Poisson bracket relations on the position space fields

$$\{\Phi(\mathbf{x}), \Pi(\mathbf{x}')\} = \delta^3(\mathbf{x} - \mathbf{x}'), \quad \{\Phi(\mathbf{x}), \Phi(\mathbf{x}')\} = \{\Pi(\mathbf{x}), \Pi(\mathbf{x}')\} = 0$$

are easily seen to be invariant under the action of the Euclidean group of spatial translations and rotations by

$$\Phi(\mathbf{x}) \to \Phi(R\mathbf{x} + \mathbf{a}), \quad \Pi(\mathbf{x}) \to \Pi(R\mathbf{x} + \mathbf{a})$$

(since the delta-function is). Things are not so simple for the rest of the Poincaré group, since the definition of the $\Phi(\mathbf{x}), \Pi(\mathbf{x})$ is based on a choice of the distinguished $t = 0$ hyperplane. In addition, the complicated form of the relativistic complex structure J_r in these coordinates (see equation 43.15) makes it difficult to see if this is invariant under Poincaré transformations.

Taking Fourier transforms, recall that solutions to the Klein–Gordon equation can be written as (see 43.3)

$$\phi(t, \mathbf{x}) = \frac{1}{(2\pi)^{3/2}} \int_{\mathbf{R}^4} \delta(p_0^2 - \omega_{\mathbf{p}}^2) f(p) e^{i(-p_0 t + \mathbf{p} \cdot \mathbf{x})} d^4 p$$

so these are given by functions (actually distributions) $f(p)$ on the positive and negative energy hyperboloids. The complex structure J_r is $+i$ on functions on the negative energy hyperboloid, $-i$ on functions on the positive energy hyperboloid. The action of the Poincaré group preserves J_r, since it acts separately on the negative and positive energy hyperboloids. It also preserves the Hermitian inner product (see equations 43.17 and 43.18) and thus gives a unitary action on $\mathcal{M}_{J_r}^+ = \mathcal{H}_1$.

Just as for the finite dimensional case in chapter 25 and the non-relativistic quantum field theory case in section 38.3, we can find for each element L of the Lie algebra of the group acting (here the Poincaré group $\mathcal{P}$) a quadratic expression in the $A(\mathbf{p}), \overline{A}(\mathbf{p})$ (this is the moment map μ_L). Quantization then gives a corresponding normal-ordered quadratic operator in terms of the operators $a^\dagger(\mathbf{p}), a(\mathbf{p})$.

44.2.1 Translations

For time translations, we have already found the Hamiltonian operator $\widehat{H}$, which gives the infinitesimal translation action on fields by

$$\frac{\partial \widehat{\Phi}}{\partial t} = [\widehat{\Phi}, -i\widehat{H}]$$

The behavior of the field operator under time translation is given by the standard Heisenberg picture relation for operators

$$\widehat{\Phi}(t + a_0) = e^{ia_0\widehat{H}} \widehat{\Phi}(t) e^{-ia_0\widehat{H}}$$

For the infinitesimal action of spatial translations on $\mathcal{H}_1$, the momentum operator is the usual

$$\mathbf{P} = -i\boldsymbol{\nabla}$$

(the convention for the Hamiltonian is the opposite sign $H = i\frac{\partial}{\partial t}$). On fields, the infinitesimal action will be given by an operator $\widehat{\mathbf{P}}$ satisfying the commutation relations

$$[-i\widehat{\mathbf{P}}, \widehat{\Phi}] = \boldsymbol{\nabla}\widehat{\Phi}$$

(see the discussion for the non-relativistic case in section 38.3 and equation 38.10). Finite spatial translations by $\mathbf{a}$ will act by

$$\widehat{\Phi}(\mathbf{x}) \rightarrow e^{-i\mathbf{a}\cdot\widehat{\mathbf{P}}} \widehat{\Phi}(\mathbf{x}) e^{i\mathbf{a}\cdot\widehat{\mathbf{P}}} = \widehat{\Phi}(\mathbf{x} + \mathbf{a})$$

The operator needed is the quadratic operator

$$\widehat{\mathbf{P}} = \int_{\mathbf{R}^3} \mathbf{p} a^\dagger(\mathbf{p}) a(\mathbf{p}) d^3\mathbf{p} \tag{44.7}$$

which in terms of fields is given by

$$\widehat{\mathbf{P}} = -\int_{\mathbf{R}^3} :\widehat{\Pi}(\mathbf{x})\boldsymbol{\nabla}\widehat{\Phi}(\mathbf{x}): d^3\mathbf{x}$$

One can see that this is the correct operator by showing that it satisfies the commutation relation 38.10 with $\widehat{\Phi}$, using the canonical commutation relations for $\widehat{\Phi}$ and $\widehat{\pi}$.

Note that here again moment map methods could have been used to find the expression for the momentum operator. This is a similar calculation to that of section 38.3 although one needs to keep track of a factor of $-i$ caused by the fact that the basic Poisson bracket relations are

$$\{\Phi(\mathbf{x}), \Pi(\mathbf{x})\} = \delta^3(\mathbf{x} - \mathbf{x}') \quad \text{versus} \quad \{\Psi(\mathbf{x}), \overline{\Psi}(\mathbf{x})\} = i\delta^3(\mathbf{x} - \mathbf{x}')$$

44.2.2 Rotations

We can use the same method as for translations to find the quadratic combinations of coordinates on $\mathcal{M}$ corresponding to the Lie algebra of the rotation group, which after quantization will provide the angular momentum operators. The action on Klein–Gordon solutions will be given by the operators

$$\begin{aligned} \mathbf{L} = \mathbf{X} \times \mathbf{P} &= \mathbf{x} \times -i\boldsymbol{\nabla} && \text{position space} \\ &= -i\boldsymbol{\nabla}_p \times \mathbf{p} && \text{momentum space} \end{aligned}$$

The corresponding quadratic operators will be

$$\begin{aligned} \widehat{\mathbf{L}} &= -\int_{\mathbf{R}^3} :\widehat{\pi}(\mathbf{x})(\mathbf{x} \times \boldsymbol{\nabla})\widehat{\phi}(\mathbf{x}): d^3\mathbf{x} \\ &= \int_{\mathbf{R}^3} a^\dagger(\mathbf{p})(\mathbf{p} \times i\boldsymbol{\nabla}_{\mathbf{p}})a(\mathbf{p})d^3\mathbf{p} \end{aligned} \tag{44.8}$$

which, again, could be found using the moment map method, although we will not work that out here. One can check using the canonical commutation relations that the components of this operator satisfy the $\mathfrak{so}(3)$ commutation relations

$$[-i\widehat{L}_j, -i\widehat{L}_k] = \epsilon_{jkl}(-i\widehat{L}_l) \tag{44.9}$$

and that, together with the momentum operators $\widehat{P}$, they give a Lie algebra representation of the Euclidean group $E(3)$ on the multiparticle state space.

Note that the operators $\widehat{\mathbf{L}}$ commute with the Hamiltonian $\widehat{H}$, and so will act on the energy eigenstates of the state space, providing unitary representations of the group $SO(3)$ on these energy eigenspaces. Energy eigenstates will be characterized by the irreducible representation of $SO(3)$ they are in, so as spin $s = 0, s = 1, \ldots$ states.

44.2.3 Boosts

From the point of view that a symmetry of a physical theory corresponds to a group action on the theory that commutes with time translation, Lorentz boosts are not symmetries because they do not commute with time translations (see the commutators in equation 42.2). From the Lagrangian point of view though, boosts are symmetries because the Lagrangian is invariant under them. From our Hamiltonian point of view, they act on phase space, preserving the symplectic structure. They thus have a moment map, and quantization will give a quadratic expression in the field operators which, when exponentiated, will give a unitary action on the multiparticle state space.

Note that boosts preserve not only the symplectic structure, but also the relativistic complex structure J_r, since they preserve the decomposition of momentum space coordinates into separate coordinates on the positive and negative energy hyperboloids. As a result, when expressed in terms of creation and annihilation operators, the quadratic boost operators $\widehat{\mathbf{K}}$ will have the same form as the operators $\widehat{\mathbf{P}}$ and $\widehat{\mathbf{L}}$, an integral of a product involving one creation and one annihilation operator. The boost operators will be given in momentum space by

$$\widehat{\mathbf{K}} = i \int_{\mathbf{R}^3} \omega_{\mathbf{p}} a^\dagger(\mathbf{p}) \nabla_{\mathbf{p}} a(\mathbf{p}) d^3\mathbf{p} \qquad (44.10)$$

One can check that this gives a Poincaré Lie algebra representation on the multiparticle state space, by evaluating first the commutators for the Lorentz group Lie algebra, which, together with 44.9, are (recall the Lie bracket relations 40.1 and 40.2)

$$[-i\widehat{L}_j, -i\widehat{K}_k] = \epsilon_{jkl}(-i\widehat{K}_l), \quad [-i\widehat{K}_j, -i\widehat{K}_k] = -\epsilon_{jkl}(-i\widehat{K}_l)$$

The commutators with the momentum and Hamiltonian operators

$$[-i\widehat{K}_j, -i\widehat{P}_j] = -i\widehat{H}, \quad [-i\widehat{K}_j, -i\widehat{H}] = -i\widehat{P}_j$$

show that the rest of the nonzero Poincaré Lie algebra bracket relations (equation 42.2) are satisfied. All of these calculations are easily performed using the expressions 43.22, 44.7, 44.8, and 44.10, for $\widehat{H}, \widehat{\mathbf{P}}, \widehat{\mathbf{L}}, \widehat{\mathbf{K}}$ and theorem 25.2 (generalized from a sum to an integral), which reduces the calculation to that of the commutators of

$$\omega_{\mathbf{p}}, \ \mathbf{p}, \ \mathbf{p} \times i\nabla_{\mathbf{p}}, \ i\omega_{\mathbf{p}}\nabla_{\mathbf{p}}$$

44.3 For further reading

The operators corresponding to various symmetries of scalar quantum fields described in this chapter are discussed in many quantum field theory books, with a typical example chapter 4 of [35]. In these books, the form of the operators is typically derived from an invariance of the Lagrangian via Noether's theorem rather than by the Hamiltonian moment map methods used here.

Chapter 45
$U(1)$ Gauge Symmetry
and Electromagnetic Fields

We have now constructed both relativistic and non-relativistic quantum field theories for free scalar particles. In the non-relativistic case, we had to use complex-valued fields and found that the theory came with an action of a $U(1)$ group, the group of phase transformations on the fields. In the relativistic case, real-valued fields could be used, but if we took complex-valued ones (or used pairs of real-valued fields), again there was an action of a $U(1)$ group of phase transformations. This is the simplest example of a so-called "internal symmetry," and it is reflected in the existence of an operator $\widehat{Q}$ called the "charge."

In this chapter, we will see how to go beyond the theory of free quantized charged particles, by introducing background electromagnetic fields that the charged particles will interact with. It turns out that this can be done using the $U(1)$ group action, but now acting independently at each point in space-time, giving a large, infinite dimensional group called the "gauge group." This requires introducing a new sort of space–time-dependent field, called a "vector potential" by physicists, a "connection" by mathematicians. Use of this field allows the construction of a Hamiltonian dynamics invariant under the gauge group. This fixes the way charged particles interact with electromagnetic fields, which are described by the vector potential.

Most of our discussion will be for the case of the $U(1)$ group, but we will also indicate how this generalizes to the case of non-Abelian groups such as $SU(2)$.

45.1 U(1) gauge symmetry

In sections 38.2.1 and 44.1, we saw that the existence of a $U(1)$ group action by overall phase transformations on the complex field values led to the existence of an operator $\widehat{Q}$, which commuted with the Hamiltonian and acted

© Peter Woit 2017
P. Woit, *Quantum Theory, Groups and Representations*,
DOI 10.1007/978-3-319-64612-1_45

with integral eigenvalues on the space of states. Instead of acting on fields by multiplication by a constant phase $e^{i\varphi}$, one can imagine multiplying by a phase that varies with the coordinates x. Such phase transformations are called "gauge transformations" and form an infinite dimensional group under point-wise multiplication:

Definition (Gauge group). *The group $\mathcal{G}$ of functions on M^4 with values in the unit circle $U(1)$, with group law given by point-wise multiplication*

$$e^{ie\varphi_1(x)} \cdot e^{ie\varphi_2(x)} = e^{ie(\varphi_1(x)+\varphi_2(x))}$$

is called the $U(1)$ gauge group, or group of $U(1)$ gauge transformations.

Here $x = (x_0, \mathbf{x}) \in M^4$, e is a constant, and φ is a real-valued function

$$\varphi(x) : M^4 \to \mathbf{R}$$

The vector space of such functions is the Lie algebra *Lie* $\mathcal{G}$, with a trivial Lie bracket. The constant e determines the normalization of the Lie algebra-valued $\varphi(x)$, with its appearance here a standard convention (such that it does not appear in the Hamiltonian, Poisson brackets, or equations of motion).

The group $\mathcal{G}$ acts on complex functions ψ of space–time as

$$\psi(x) \to e^{ie\varphi(x)}\psi(x), \quad \overline{\psi}(x) \to e^{-ie\varphi(x)}\overline{\psi}(x)$$

Note that in quantum mechanics, this is a group action on the wavefunctions, and it does not correspond to any group action on the finite dimensional phase space of coordinates and momenta, so has no classical interpretation. In quantum field theory though, where these wavefunctions make up the phase space to be quantized, this is a group action on the phase space, preserving the symplectic structure.

Terms in the Hamiltonian that just involve $|\psi(x)|^2 = \overline{\psi}(x)\psi(x)$ will be invariant under the group $\mathcal{G}$, but terms with derivatives such as

$$|\nabla\psi|^2$$

will not, since when

$$\psi \to e^{ie\varphi(x)}\psi(x)$$

one has the inhomogeneous behavior

$$\frac{\partial\psi(x)}{\partial x_\mu} \to \frac{\partial}{\partial x_\mu}(e^{ie\varphi(x)}\psi(x)) = e^{ie\varphi(x)}\left(ie\frac{\partial\varphi(x)}{\partial x_\mu} + \frac{\partial}{\partial x_\mu}\right)\psi(x)$$

To deal with this problem, one introduces a new kind of field:

Definition (Connection or vector potential). *A U(1) connection (mathematician's terminology) or vector potential (physicist's terminology) is a function A on space–time M^4 taking values in $\mathbf{R}^4$, with its components denoted*

$$A_\mu(x) = (A_0(t, \mathbf{x}), \mathbf{A}(t, \mathbf{x}))$$

The gauge group $\mathcal{G}$ acts on the space of $U(1)$ connections by

$$A_\mu(t, \mathbf{x}) \to A_\mu^\varphi(t, \mathbf{x}) \equiv A_\mu(t, \mathbf{x}) + \frac{\partial \varphi(t, \mathbf{x})}{\partial x_\mu} \tag{45.1}$$

The vector potential allows one to define a new kind of derivative, such that the derivative of the field ψ has the same homogeneous transformation properties under $\mathcal{G}$ as ψ itself:

Definition (Covariant derivative). *Given a connection A, the associated covariant derivative in the μ direction is the operator*

$$D_\mu^A = \frac{\partial}{\partial x_\mu} - ieA_\mu(x)$$

With this definition, the effect of a gauge transformation is

$$D_\mu^A \psi \to \left(D_\mu^A - ie\frac{\partial \varphi}{\partial x_\mu} \right) e^{ie\varphi(x)} \psi = e^{ie\varphi(x)} \left(D_\mu^A + ie\frac{\partial}{\partial x_\mu} - ie\frac{\partial}{\partial x_\mu} \right) \psi$$

$$= e^{ie\varphi(x)} D_\mu^A \psi$$

If one replaces derivatives by covariant derivatives, terms in a Hamiltonian such as

$$|\nabla \psi|^2 = \sum_{j=1}^{3} \overline{\frac{\partial \psi}{\partial x_j}} \frac{\partial \psi}{\partial x_j}$$

will become

$$\sum_{j=1}^{3} \overline{(D_j^A \psi)}(D_j^A \psi)$$

which will be invariant under the infinite dimensional group $\mathcal{G}$. The procedure of starting with a theory of complex fields and then introducing a connection while changing derivatives to covariant derivatives in the equations of motion is called the "minimal coupling prescription." It determines how a theory of complex free fields describing charged particles can be turned into a theory of fields coupled to a background electromagnetic field, in the simplest or "minimal" way.

45.2 Curvature, electric and magnetic fields

While the connection or vector potential A is the fundamental geometrical quantity needed to construct theories with gauge symmetry, one often wants to work instead with certain quantities derived from A that are invariant under gauge transformations. To a mathematician this is the curvature of a connection, and to a physicist these are the electric and magnetic field strengths derived from a vector potential. The definition is:

Definition (Curvature of a connection, electromagnetic field strengths). *The curvature of a connection A_μ is given by*

$$F_{\mu\nu} = \frac{i}{e}[D_\mu^A, D_\nu^A]$$

which can more explicitly be written

$$F_{\mu\nu} = \frac{\partial A_\nu}{\partial x_\mu} - \frac{\partial A_\mu}{\partial x_\nu}$$

Note that while D_μ^A is a differential operator, $[D_\mu^A, D_\nu^A]$ and thus the curvature is just a multiplication operator.

The electromagnetic field strengths break up into those components with a time index and those without:

Definition (Electric and magnetic fields). *The electric and magnetic fields are two functions from $\mathbf{R}^4$ to $\mathbf{R}^3$, with components given by*

$$E_j = F_{j0} = -\frac{\partial A_j}{\partial t} + \frac{\partial A_0}{\partial x_j}$$

$$B_j = \frac{1}{2}\epsilon_{jkl}F_{kl} = \epsilon_{jkl}\frac{\partial A_l}{\partial x_k}$$

or, in vector notation

$$\mathbf{E} = -\frac{\partial \mathbf{A}}{\partial t} + \boldsymbol{\nabla} A_0, \quad \mathbf{B} = \boldsymbol{\nabla} \times \mathbf{A}$$

$\mathbf{E}$ *is called the electric field, and* $\mathbf{B}$ *the magnetic field.*

These are invariant under gauge transformations since

$$\mathbf{E} \to \mathbf{E} - \frac{\partial}{\partial t}\boldsymbol{\nabla}\varphi + \boldsymbol{\nabla}\frac{\partial\varphi}{\partial t} = \mathbf{E}$$

$$\mathbf{B} \rightarrow \mathbf{B} + \boldsymbol{\nabla} \times \boldsymbol{\nabla}\varphi = \mathbf{B}$$

Here, we use the fact that

$$\boldsymbol{\nabla} \times \boldsymbol{\nabla}f = 0 \tag{45.2}$$

for any function f.

45.3 Field equations with background electromagnetic fields

The minimal coupling method described above can be used to write down field equations for our free particle theories, now coupled to electromagnetic fields. They are:

- The Schrödinger equation for a non-relativistic particle coupled to a background electromagnetic field is

$$i\left(\frac{\partial}{\partial t} - ieA_0\right)\psi = -\frac{1}{2m}\sum_{j=1}^{3}\left(\frac{\partial}{\partial x_j} - ieA_j\right)^2 \psi$$

A special case of this is the Coulomb potential problem discussed in Chapter 21, which corresponds to the choice of background field

$$A_0 = \frac{1}{r}, \quad \mathbf{A} = 0$$

Another exactly solvable special case is that of a constant magnetic field $\mathbf{B} = (0, 0, B)$, for which one possible choice of vector potential is

$$A_0 = 0, \quad \mathbf{A} = (-By, 0, 0)$$

- The Pauli–Schrödinger equation (34.3) describes a free spin$\frac{1}{2}$ non-relativistic quantum particle. Replacing derivatives by covariant derivatives, one gets

$$i\left(\frac{\partial}{\partial t} - ieA_0\right)\begin{pmatrix}\psi_1(\mathbf{x}) \\ \psi_2(\mathbf{x})\end{pmatrix} = -\frac{1}{2m}(\boldsymbol{\sigma} \cdot (\boldsymbol{\nabla} - ie\mathbf{A}))^2 \begin{pmatrix}\psi_1(\mathbf{x}) \\ \psi_2(\mathbf{x})\end{pmatrix}$$

Using the anticommutation

$$\sigma_j\sigma_k + \sigma_k\sigma_j = 2\delta_{jk}$$

and commutation

$$\sigma_j \sigma_k - \sigma_k \sigma_j = i\epsilon_{jkl}\sigma_l$$

relations, one finds

$$\sigma_j \sigma_k = \delta_{jk} + \frac{1}{2}i\epsilon_{jkl}\sigma_l$$

This implies that

$$(\boldsymbol{\sigma} \cdot (\boldsymbol{\nabla} - ie\mathbf{A}))^2 = \sum_{j=1}^{3} \left(\frac{\partial}{\partial x_j} - ieA_j\right)^2 +$$

$$\sum_{j,k=1}^{3} \left(\frac{\partial}{\partial x_j} - ieA_j\right)\left(\frac{\partial}{\partial x_k} - ieA_k\right)\frac{i}{2}\epsilon_{jkl}\sigma_l$$

$$= \sum_{j=1}^{3} \left(\frac{\partial}{\partial x_j} - ieA_j\right)^2 + e\boldsymbol{\sigma} \cdot \mathbf{B}$$

and the Pauli–Schrödinger equation can be written

$$i(\frac{\partial}{\partial t} - ieA_0)\begin{pmatrix}\psi_1(\mathbf{x})\\\psi_2(\mathbf{x})\end{pmatrix} = -\frac{1}{2m}(\sum_{j=1}^{3}(\frac{\partial}{\partial x_j} - ieA_j)^2 + e\boldsymbol{\sigma} \cdot \mathbf{B})\begin{pmatrix}\psi_1(\mathbf{x})\\\psi_2(\mathbf{x})\end{pmatrix}$$

This two-component equation is just two copies of the standard Schrödinger equation and an added term coupling the spin and magnetic field which is exactly the one studied in chapter 7. Comparing to the discussion there, we see that the minimal coupling prescription here is equivalent to a choice of gyromagnetic ratio $g = 1$.

• With minimal coupling to the electromagnetic field, the Klein–Gordon equation becomes

$$\left(-\left(\frac{\partial}{\partial t} - ieA_0\right)^2 + \sum_{j=1}^{3}\left(\frac{\partial}{\partial x_j} - ieA_j\right)^2 - m^2\right)\phi = 0$$

The first two equations are for non-relativistic theories, and one can interpret these equations as describing a single quantum particle (with spin $\frac{1}{2}$ in the second case) moving in a background electromagnetic field. In the relativistic Klein–Gordon case, here we are in the case of a complex Klein–Gordon field, as discussed in section 44.1.2. In all three cases, in principle a quantum field theory can be defined by taking the space of solutions of the equation as phase space, and applying the Bargmann–Fock quantization method (in practice this is difficult, since in general there is no translation invariance and no plane-wave basis of solutions).

45.4 The geometric significance of the connection

The information contained in a connection $A_\mu(x)$ can be put in a different form, using it to define a phase for any curve γ between two points in M^4:

Definition (Path-dependent phase factor). *Given a connection $A_\mu(x)$, one can define for any curve γ parametrized by $\tau \in [0,1]$, with position at time τ given by $x(\tau)$, the path-dependent phase factor*

$$\int_\gamma A \equiv \int_0^1 \sum_{\mu=1}^4 A_\mu(x) \frac{dx_\mu}{d\tau} d\tau$$

The effect of a gauge transformation φ is

$$\int_\gamma A \to \int_0^1 \sum_{\mu=1}^4 \left(A_\mu(x) + \frac{\partial \varphi}{\partial x_\mu} \right) \frac{dx_\mu}{d\tau} d\tau = \int_\gamma A + \varphi(\gamma(1)) - \varphi(\gamma(0))$$

Note that if γ is a closed curve, with $\gamma(1) = \gamma(0)$, then the path-dependent phase factor $\int_\gamma A$ is gauge invariant.

Digression. *For readers familiar with differential forms, A can be thought of as an element of $\Omega^1(M^4)$, the space of 1-forms on space–time M^4. The path-dependent phase $\int_\gamma A$ is then the standard integral of a 1-form along a curve γ. The curvature of A is simply the 2-form $F = dA$, where d is the de Rham differential. The gauge group acts on connections by*

$$A \to A + d\varphi$$

and the curvature is gauge invariant since

$$F \to F + d(d\varphi)$$

and d satisfies $d^2 = 0$.

    Stokes theorem for differential forms implies that if γ is a closed curve, and γ is the boundary of a surface S ($\gamma = \partial S$), then

$$\int_\gamma A = \int_S F$$

Note that if $F = 0$, then $\int_\gamma A = 0$ for any closed curve γ, and this can be used to show that path-dependent phase factors do not depend on the path. To see this, consider any two paths γ_1 and γ_2 from $\gamma(0)$ to $\gamma(1)$, and $\gamma = \gamma_1 - \gamma_2$

the closed curve that goes from $\gamma(0)$ *to* $\gamma(1)$ *along* γ_1, *and then back to* $\gamma(0)$
along γ_2. *Then,*

$$\int_S F = 0 \implies \int_\gamma A = \int_{\gamma_1} A - \int_{\gamma_2} A = 0$$

so

$$\int_{\gamma_1} A = \int_{\gamma_2} A$$

The path-dependent phase factors $\int_\gamma A$ allow comparison of the values
of the complex field ψ at different points in a gauge invariant manner. To
compare the value of a field ψ at $\gamma(0)$ to that of the field at $\gamma(1)$ in a gauge
invariant manner, we just need to consider the path-dependent quantity

$$e^{ie \int_\gamma A} \psi(\gamma(0))$$

where γ is a curve from $\gamma(0)$ to $\gamma(1)$. Under a gauge transformation, this will
change as

$$e^{ie \int_\gamma A} \psi(\gamma(0)) \to e^{ie(\int_\gamma A + \varphi(\gamma(1)) - \varphi(\gamma(0)))} e^{ie\varphi(\gamma(0))} \psi(\gamma(0)) = e^{ie\varphi(\gamma(1))} e^{ie \int_\gamma A} \psi(\gamma(0))$$

which is the same transformation property as that of $\psi(\gamma(1))$.

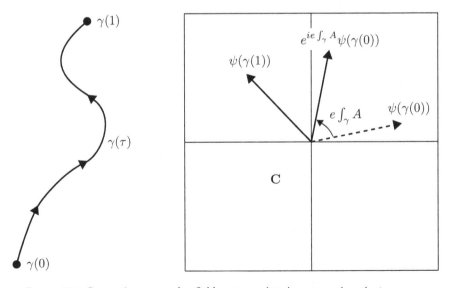

Figure 45.1: Comparing a complex field at two points in a gauge invariant manner.

In the path integral formalism (see section 35.3), the minimal coupling of a single particle to a background electromagnetic field described by a vector potential A_μ can be introduced by weighting the integral over paths by the path-dependent phase factor. This changes the formal path integral by

$$\int D\gamma \; e^{\frac{i}{\hbar}S[\gamma]} \rightarrow \int D\gamma \; e^{\frac{i}{\hbar}S[\gamma]} e^{\frac{ie}{\hbar}\int_\gamma A}$$

and a path integral with such a weighting of paths then must be sensibly defined. This method only works for the single-particle theory, with minimal coupling for a quantum theory of fields given by the replacement of derivatives by covariant derivatives described earlier.

45.5 The non-Abelian case

We saw in section 38.2.2 that quantum field theories with a $U(m)$ group acting on the fields can be constructed by taking m-component complex fields and a Hamiltonian that is the sum of the single complex field Hamiltonians for each component. The constructions of this chapter generalize from the $U(1)$ to $U(m)$ case, getting a gauge group $\mathcal{G}$ of maps from $\mathbf{R}^4$ to $U(m)$, as well as a generalized notion of connection and curvature. In this section, we will outline how this works, without going into full detail. The non-Abelian $U(m)$ case is a relatively straightforward generalization of the $U(1)$ case, except for the definition of the curvature, where new terms with different behavior arise, due to the non-commutative nature of the group.

The diagonal $U(1) \subset U(m)$ subgroup is treated using exactly the same formalism for the vector potential, covariant derivative, electric and magnetic fields as above. It is only the $SU(m) \subset U(m)$ subgroup which requires a separate treatment. We will do this just for the case $m = 2$, which is known as the Yang-Mills case, since it was first investigated by the physicists Yang and Mills in 1954. We can think of the $i\varphi(x)$ in the $U(1)$ case as a function valued in the Lie algebra $\mathfrak{u}(1)$ and replace it by a matrix-valued function, taking values in the Lie algebra $\mathfrak{su}(2)$ for each x. This can be written in terms of three functions φ_a as

$$i\varphi(x) = i \sum_{a=1}^{3} \varphi_a(x)\frac{\sigma_a}{2}$$

using the Pauli matrices. The gauge group becomes the group $\mathcal{G}_{YM}$ of maps from space–time to $SU(2)$, with Lie algebra the maps $i\varphi(x)$ from space–time to $\mathfrak{su}(2)$. Unlike the $U(1)$ case, this Lie algebra has a non-trivial Lie bracket, given by the point-wise $\mathfrak{su}(2)$ Lie bracket (the commutator of matrices).

In the $SU(2)$ case, the analog of the real-valued function A_μ will now be matrix-valued and one can write

$$A_\mu(x) = \sum_{a=1}^{3} A_\mu^a(x) \frac{\sigma_a}{2}$$

Instead of one vector potential function for each space–time direction μ, we now have three (the $A_\mu^a(x)$), and we will refer to these functions as the connection or "gauge field." The complex fields ψ are now two-component fields, and the covariant derivative is

$$D_\mu^A \begin{pmatrix} \psi_1 \\ \psi_2 \end{pmatrix} = \left(\frac{\partial}{\partial x_\mu} - ie \sum_{a=1}^{3} A_\mu^a(x) \frac{\sigma_a}{2} \right) \begin{pmatrix} \psi_1 \\ \psi_2 \end{pmatrix}$$

For the theories of complex fields with $U(m)$ symmetry discussed in chapters 38 and 44, this is the $m = 2$ case. Replacing derivatives by covariant derivatives yields non-relativistic and relativistic theories of matter particles coupled to background gauge fields.

In the Yang-Mills case, the curvature or field strengths can still be defined as a commutator of covariant derivatives, but now this is a commutator of matrix-valued differential operators. The result will as in the $U(1)$ case be a multiplication operator, but it will be matrix-valued. The curvature can be defined as

$$\sum_{a=1}^{3} F_{\mu\nu}^a \frac{\sigma_a}{2} = \frac{i}{e} [D_\mu^A, D_\nu^A]$$

which can be calculated much as in the Abelian case, except now the term involving the commutator of A_μ and A_ν no longer cancels. Distinguishing electric and magnetic field strength components as in the $U(1)$ case, the equations for matrix-valued electric and magnetic fields are:

$$E_j(x) = \sum_{a=1}^{3} E_j^a(x) \frac{\sigma_a}{2} = -\frac{\partial A_j}{\partial t} + \frac{\partial A_0}{\partial x_j} - ie[A_j, A_0] \qquad (45.3)$$

and

$$B_j(x) = \sum_{a=1}^{3} B_j^a(x) \frac{\sigma_a}{2} = \epsilon_{jkl} \left(\frac{\partial A_l}{\partial x_k} - \frac{\partial A_k}{\partial x_l} - ie[A_k, A_l] \right) \qquad (45.4)$$

The Yang-Mills theory thus comes with electric and magnetic fields that now are valued in $\mathfrak{su}(2)$ and can be written as 2 by 2 matrices, or in terms of the Pauli matrix basis, as fields $\mathbf{E}^a(x)$ and $\mathbf{B}^a(x)$ indexed by $a = 1, 2, 3$. These fields are no longer linear in the A_μ fields, but have extra quadratic terms.

These non-quadratic terms will introduce nonlinearities into the equations of motion for Yang-Mills theory, making its study much more difficult than the $U(1)$ case.

45.6   For further reading

Most electromagnetism textbooks in physics will have some discussion of the vector potential, electric and magnetic fields, and gauge transformations. A textbook covering the geometry of connections and curvature as it occurs in physics is [29]. [30] is a recent textbook aimed at mathematicians that covers the subject of electromagnetism in detail.

Chapter 46
Quantization of the Electromagnetic Field: the Photon

Understanding the classical field theory of coupled dynamical scalar fields and vector potentials is rather difficult, with the quantized theory even more so, due to the fact that the Hamiltonian is no longer quadratic in the field variables and the field equations are nonlinear. Simplifying the problem by ignoring the scalar fields and only considering the vector potentials gives a theory with quadratic Hamiltonian that can be readily understood and quantized. The classical equations of motion are the linear Maxwell equations in a vacuum, with solutions electromagnetic waves. The corresponding quantum field theory will be a relativistic theory of free, massless particles of helicity ± 1, the photons.

To get a physically sensible theory of photons, the infinite dimensional group $\mathcal{G}$ of gauge transformations that acts on the classical phase space of solutions to the Maxwell equations must be taken into account. We will describe several methods for doing this, carrying out the quantization in detail using one of them. All of these methods have various drawbacks, with unitarity and explicit Lorentz invariance seemingly impossible to achieve simultaneously. The reader should be warned that due to the much greater complexities involved, this chapter and succeeding ones will be significantly sketchier than most earlier ones.

46.1 Maxwell's equations

We saw in chapter 45 that our quantum theories of free particles could be coupled to a background electromagnetic field by introducing vector potential fields $A_\mu = (A_0, \mathbf{A})$. Electric and magnetic fields are defined in terms of A_μ by the equations

$$\mathbf{E} = -\frac{\partial \mathbf{A}}{\partial t} + \boldsymbol{\nabla} A_0, \quad \mathbf{B} = \boldsymbol{\nabla} \times \mathbf{A}$$

© Peter Woit 2017
P. Woit, *Quantum Theory, Groups and Representations*,
DOI 10.1007/978-3-319-64612-1_46

In this chapter, we will see how to make the A_μ fields dynamical variables, although restricting to the special case of free electromagnetic fields, without interaction with matter fields. The equations of motion will be:

Definition (Maxwell's equations in vacuo). *The Maxwell equations for electromagnetic fields in the vacuum are*

$$\nabla \cdot \mathbf{B} = 0 \qquad\qquad (46.1)$$

$$\nabla \times \mathbf{E} = -\frac{\partial \mathbf{B}}{\partial t} \qquad\qquad (46.2)$$

$$\nabla \times \mathbf{B} = \frac{\partial \mathbf{E}}{\partial t} \qquad\qquad (46.3)$$

and Gauss's law:

$$\nabla \cdot \mathbf{E} = 0 \qquad\qquad (46.4)$$

Digression. *In terms of differential forms, these equations can be written very simply as*

$$dF = 0, \quad d * F = 0$$

where the first equation is equivalent to 46.1 and 46.2, and the second (which uses the Hodge star operator for the Minkowski metric) is equivalent to 46.3 and 46.4. Note that the first equation is automatically satisfied, since by definition $F = dA$, and the d operator satisfies $d^2 = 0$.

Writing out equation 46.1 in terms of the vector potential gives

$$\nabla \cdot \nabla \times \mathbf{A} = 0$$

which is automatically satisfied for any vector field $\mathbf{A}$. Similarly, in terms of the vector potential, equation 46.2 is

$$\nabla \times \left(-\frac{\partial \mathbf{A}}{\partial t} + \nabla A_0 \right) = -\frac{\partial}{\partial t}(\nabla \times \mathbf{A})$$

which is automatically satisfied since

$$\nabla \times \nabla f = 0$$

for any function f.

Note that since Maxwell's equations only depend on A_μ through the gauge invariant fields $\mathbf{B}$ and $\mathbf{E}$, if $A_\mu = (A_0, \mathbf{A})$ is a solution, so is the gauge transform

$$A_\mu^\varphi = \left(A_0 + \frac{\partial \varphi}{\partial t}, \mathbf{A} + \nabla \varphi \right)$$

46.2   The Hamiltonian formalism for electromagnetic fields

In order to quantize the electromagnetic field, we first need to express Maxwell's equations in Hamiltonian form. These equations are second-order differential equations in t, so we expect to parametrize solutions in terms of initial data

$$(A_0(0, \mathbf{x}), \mathbf{A}(0, \mathbf{x})) \quad \text{and} \quad \left(\frac{\partial A_0}{\partial t}(0, \mathbf{x}), \frac{\partial \mathbf{A}}{\partial t}(0, \mathbf{x}) \right) \tag{46.5}$$

The problem with this is that gauge invariance implies that if A_μ is a solution with this initial data, so is its gauge transform A_μ^φ (see equation 45.1) for any function $\varphi(t, \mathbf{x})$ such that

$$\varphi(0, \mathbf{x}) = 0, \quad \frac{\partial \varphi}{\partial t}(t, \mathbf{x}) = 0$$

This implies that solutions are not uniquely determined by the initial data of the vector potential and its time derivative at $t = 0$, and thus, this initial data will not provide coordinates on the space of solutions.

One way to deal with this problem is to try and find conditions on the vector potential which will remove this freedom to perform such gauge transformations and then take as phase space the subspace of initial data satisfying the conditions. This is called making a "choice of gauge." We will begin with:

Definition (Temporal gauge). *A vector potential A_μ is said to be in temporal gauge if $A_0 = 0$.*

Note that given any vector potential A_μ, we can find a gauge transformation φ such that the gauge transformed vector potential will have $A_0 = 0$ by solving the equation

$$\frac{\partial \varphi}{\partial t}(t, \mathbf{x}) = A_0(t, \mathbf{x})$$

which has solution

$$\varphi(t, \mathbf{x}) = \int_0^t A_0(\tau, \mathbf{x}) d\tau + \varphi_0(\mathbf{x}) \tag{46.6}$$

where $\varphi_0(\mathbf{x}) = \varphi(0, \mathbf{x})$ is any function of the spatial variables $\mathbf{x}$.

In temporal gauge, initial data for a solution to Maxwell's equations is given by a pair of functions

$$\left(\mathbf{A}(\mathbf{x}), \frac{\partial \mathbf{A}}{\partial t}(\mathbf{x}) \right)$$

and we can take these as our coordinates on the phase space of solutions. The electric field is now

$$\mathbf{E} = -\frac{\partial \mathbf{A}}{\partial t}$$

so we can also write our coordinates on phase space as

$$(\mathbf{A}(\mathbf{x}), -\mathbf{E}(\mathbf{x}))$$

Requiring that these coordinates behave just like position and momentum coordinates in the finite dimensional case, we can specify the Poisson bracket and thus the symplectic form by

$$\{A_j(\mathbf{x}), A_k(\mathbf{x}')\} = \{E_j(\mathbf{x}), E_k(\mathbf{x}')\} = 0$$

$$\{A_j(\mathbf{x}), E_k(\mathbf{x}')\} = -\delta_{jk}\delta^3(\mathbf{x} - \mathbf{x}') \tag{46.7}$$

If we then take as Hamiltonian function

$$h = \frac{1}{2}\int_{\mathbf{R}^3}(|\mathbf{E}|^2 + |\mathbf{B}|^2)d^3\mathbf{x} \tag{46.8}$$

Hamilton's equations become

$$\frac{\partial \mathbf{A}}{\partial t}(t, \mathbf{x}) = \{\mathbf{A}(t, \mathbf{x}), h\} = -\mathbf{E}(\mathbf{x}) \tag{46.9}$$

and

$$\frac{\partial \mathbf{E}}{\partial t}(t, \mathbf{x}) = \{\mathbf{E}(t, \mathbf{x}), h\}$$

which one can show is just the Maxwell equation 46.3

$$\frac{\partial \mathbf{E}}{\partial t} = \mathbf{\nabla} \times \mathbf{B}$$

The final Maxwell's equation, Gauss's law (46.4), does not appear in the Hamiltonian formalism as an equation of motion. In later sections, we will see several different ways of dealing with this problem.

For the Yang–Mills case, in temporal gauge we can again take as initial data $(\mathbf{A}(\mathbf{x}), -\mathbf{E}(\mathbf{x}))$, where these are now matrix-valued. For the Hamiltonian, we can use the trace function on matrices and take

$$h = \frac{1}{2}\int_{\mathbf{R}^3} tr(|\mathbf{E}|^2 + |\mathbf{B}|^2)d^3\mathbf{x} \tag{46.10}$$

since

$$\langle X_1, X_2 \rangle = tr(X_1 X_2)$$

is a non-degenerate, positive, $SU(2)$ invariant inner product on $\mathfrak{su}(2)$. One of Hamilton's equations is then equation 46.9, which is also just the definition of the Yang–Mills electric field when $A_0 = 0$ (see equation 45.3).

The other Hamilton's equation can be shown to be

$$\frac{\partial E_j}{\partial t}(t, \mathbf{x}) = \{E_j(t, \mathbf{x}), h\}$$
$$= (\mathbf{\nabla} \times \mathbf{B})_j - ie\epsilon_{jkl}[A_k, B_l] \qquad (46.11)$$

where $\mathbf{B}$ is the Yang–Mills magnetic field (45.4). If a covariant derivative acting on fields valued in $\mathfrak{su}(2)$ is defined by

$$\mathbf{\nabla}^A(\cdot) = \mathbf{\nabla}(\cdot) - ie[\mathbf{A}, \cdot]$$

then equation 46.11 can be written

$$\frac{\partial \mathbf{E}}{\partial t}(t, \mathbf{x}) = \mathbf{\nabla}^A \times \mathbf{B}$$

The problem with these equations is that they are nonlinear equations in $\mathbf{A}$, so the phase space of solutions is no longer a linear space, and different methods are needed for quantization of the theory.

46.3 Gauss's law and time-independent gauge transformations

Returning to the $U(1)$ case, a problem with the temporal gauge is that Gauss's law (equation 46.4) is not necessarily satisfied. At the same time, the group $\mathcal{G}_0 \subset \mathcal{G}$ of time-independent gauge transformation will act non-trivially on the phase space of initial data of Maxwell's equations, preserving the temporal gauge condition $A_0 = 0$ (see equation 46.6). We will see that the condition of invariance under this group action can be used to impose Gauss's law.

It is a standard fact from the theory of electromagnetism that

$$\mathbf{\nabla} \cdot \mathbf{E} = \rho(\mathbf{x})$$

is the generalization of Gauss's law to the case of a background electric charge density $\rho(\mathbf{x})$. A failure of Gauss's law can thus be interpreted physically as due to the inclusion of states with background electric charge, rather than just electromagnetic fields in the vacuum.

There are two different ways to deal with this kind of problem:

- Before quantization, impose Gauss's law as a condition on the phase space.

- After quantization, impose Gauss's law as a condition on the states, defining the physical state space $\mathcal{H}_{phys} \subset \mathcal{H}$ as the subspace of states satisfying

$$\boldsymbol{\nabla} \cdot \widehat{\mathbf{E}} |\psi\rangle = 0$$

where $\widehat{\mathbf{E}}$ is the quantized electric field.

To understand what happens if one tries to implement one of these choices, consider first a much simpler example, that of a non-relativistic particle in 3 dimensions, with a potential that does not depend on one configuration variable, say q_3. The system has a symmetry under translations in the 3 direction, and the condition

$$P_3|\psi\rangle = 0 \tag{46.12}$$

on states will commute with time evolution since $[P_3, H] = 0$. In the Schrödinger representation, since

$$P_3 = -i\frac{\partial}{\partial q_3}$$

if we define $\mathcal{H}_{phys}$ as the subset of states satisfying 46.12, this can be identified with the space of wavefunctions of two position variables q_1, q_2. One technical problem that appears at this point is that the original inner product includes an integral over the q_3 coordinate, which will diverge since the wavefunction will be independent of this coordinate.

If the condition $p_3 = 0$ is instead imposed before quantization (i.e., on the phase space coordinates), the phase space will now be five dimensional, with coordinates q_1, q_2, q_3, p_1, p_2, and it will no longer have a non-degenerate symplectic form. It is clear that what we need to do to get a phase space whose quantization will have state space $\mathcal{H}_{phys}$ is to remove the dependence on the coordinate q_3.

In general, if we have a group G acting on a phase space M, we can define:

Definition (Symplectic reduction). *Given a group G acting on phase space M, preserving the Poisson bracket, with moment map*

$$M \to \mathfrak{g}^*$$

the symplectic reduction $M//G$ is the quotient space $\mu^{-1}(0)/G$.

We will not show this here, but under appropriate conditions the space $M//G$ will have a non-degenerate symplectic form. It can be thought of as the phase space describing the G-invariant degrees of freedom of the phase space M. What one would like to be true is that "quantization commutes with reduction": Quantization of $M//G$ gives a quantum system with state space $\mathcal{H}_{phys}$ identical to the G-invariant subspace of the state space $\mathcal{H}$ of the quantization of M. Rarely are both M and $M//G$ the sort of linear phase spaces that we know how to quantize, so this should be thought of as a desirable property for schemes that allow quantization of more general symplectic manifolds.

For the case of a system invariant under translations in the 3-direction, $M = \mathbf{R}^6$, $G = \mathbf{R}$, $\mathfrak{g} = \mathbf{R}$) and the moment map takes as value (see equation 15.12) the element of $\mathfrak{g}^*$ given by $\mu(\mathbf{q}, \mathbf{p})$ where

$$\mu(\mathbf{q}, \mathbf{p})(a) = a p_3$$

$\mu^{-1}(0)$ will be the subspace of phase space with $p_3 = 0$. On this space, the translation group acts by translating the coordinate q_3, so we can identify

$$M//\mathbf{R} = \mu^{-1}(0)/\mathbf{R} = \mathbf{R}^4$$

with the phase space with coordinates q_1, q_2, p_1, p_2. In this case, quantization will commute with reduction since imposing $P_3|\psi\rangle = 0$ or quantizing $M//\mathbf{R}$ give the same space of states (in the Schrödinger representation, the wavefunctions of position variables q_1, q_2).

This same principle can be applied in the infinite dimensional example of the temporal gauge phase space with coordinates

$$(\mathbf{A}(\mathbf{x}), -\mathbf{E}(\mathbf{x}))$$

and an action of the group $\mathcal{G}_0$ of time-independent gauge transformations, with Lie algebra the functions $\varphi(\mathbf{x})$. In this case, the condition $\mu = 0$ will just be Gauss's law. To see this, note that the moment map $\mu \in (Lie\ \mathcal{G}_0)^*$ will be given by

$$\mu(\mathbf{A}(\mathbf{x}), -\mathbf{E}(\mathbf{x}))(\varphi(\mathbf{x})) = \int_{\mathbf{R}^3} \varphi(\mathbf{x}')\nabla \cdot \mathbf{E}\ d^3 \mathbf{x}'$$

since

$$\{\mu, \mathbf{A}(\mathbf{x})\} = \{\int_{\mathbf{R}^3} \varphi(\mathbf{x}')\nabla \cdot \mathbf{E}\ d^3 \mathbf{x}', \mathbf{A}(\mathbf{x})\}$$
$$= \{\int_{\mathbf{R}^3} (-\nabla\varphi(\mathbf{x}'))\mathbf{E}(\mathbf{x}')\ d^3 \mathbf{x}', \mathbf{A}(\mathbf{x})\}$$
$$= -\nabla\varphi(\mathbf{x})$$

(using integration by parts in the first step, the Poisson bracket relations in the second). This agrees with the definition in section 15.3 of the moment map, since $\boldsymbol{\nabla}\varphi(\mathbf{x})$ is the infinitesimal change in $\mathbf{A}(\mathbf{x})$ for an infinitesimal gauge transformation $\varphi(\mathbf{x})$. One can similarly show that, as required since $\mathbf{E}$ is gauge invariant, μ satisfies

$$\{\mu, \mathbf{E}(\mathbf{x})\} = 0$$

46.4 Quantization in Coulomb gauge

A different method for dealing with the time-independent gauge transformations is to impose an additional gauge condition. For any vector potential satisfying the temporal gauge condition $A_0 = 0$, a gauge transformation can be found such that the transformed vector potential satisfies:

Definition (Coulomb gauge). *A vector potential A_μ is said to be in Coulomb gauge if* $\boldsymbol{\nabla} \cdot \mathbf{A} = 0$.

To see that this is possible, note that under a gauge transformation one has

$$\boldsymbol{\nabla} \cdot \mathbf{A} \to \boldsymbol{\nabla} \cdot \mathbf{A} + \boldsymbol{\nabla}^2 \varphi$$

so such a gauge transformation will put a vector potential in Coulomb gauge if we can find a solution to

$$\boldsymbol{\nabla}^2 \varphi = -\boldsymbol{\nabla} \cdot \mathbf{A} \tag{46.13}$$

Using Green's function methods like those of section 12.7, this equation for φ can be solved, with the result

$$\varphi(\mathbf{x}) = \frac{1}{4\pi} \int_{\mathbf{R}^3} \frac{1}{|\mathbf{x} - \mathbf{x}'|} (\boldsymbol{\nabla} \cdot \mathbf{A}(\mathbf{x}')) d^3\mathbf{x}'$$

Since in temporal gauge

$$\boldsymbol{\nabla} \cdot \mathbf{E} = -\frac{\partial}{\partial t} \boldsymbol{\nabla} \cdot \mathbf{A}$$

the Coulomb gauge condition automatically implies that Gauss's law (46.4) will hold. We thus can take as phase space the solutions to the Maxwell equations satisfying the two conditions $A_0 = 0, \boldsymbol{\nabla} \cdot \mathbf{A} = 0$, with phase space coordinates the pairs $(\mathbf{A}(\mathbf{x}), -\mathbf{E}(\mathbf{x}))$ satisfying the constraints $\boldsymbol{\nabla} \cdot \mathbf{A} = 0$, $\boldsymbol{\nabla} \cdot \mathbf{E} = 0$.

In Coulomb gauge, the one Maxwell equation (46.3) that is not automatically satisfied is, in terms of the vector potential

$$-\frac{\partial^2 \mathbf{A}}{\partial t^2} = \mathbf{\nabla} \times (\mathbf{\nabla} \times \mathbf{A})$$

Using the vector calculus identity

$$\mathbf{\nabla} \times (\mathbf{\nabla} \times \mathbf{A}) = \mathbf{\nabla}(\mathbf{\nabla} \cdot \mathbf{A}) - \mathbf{\nabla}^2 \mathbf{A}$$

and the Coulomb gauge condition, this becomes the wave equation

$$\left(\frac{\partial^2}{\partial t^2} - \mathbf{\nabla}^2\right) \mathbf{A} = 0 \qquad (46.14)$$

This is just three copies of the real Klein–Gordon equation for mass $m = 0$, although it needs to be supplemented by the Coulomb gauge condition.

One can proceed exactly as for the Klein–Gordon case, using the Fourier transform to identify solutions with functions on momentum space and quantizing with annihilation and creation operators. The momentum space solutions are given by the Fourier transforms $\widetilde{A}_j(\mathbf{p})$, and a classical solution can be written in terms of them by a simple generalization of equation 43.8 for the scalar field case

$$A_j(t, \mathbf{x}) = \frac{1}{(2\pi)^{3/2}} \int_{\mathbf{R}^3} (\alpha_j(\mathbf{p}) e^{-i\omega_{\mathbf{p}} t} e^{i\mathbf{p}\cdot\mathbf{x}} + \overline{\alpha}_j(\mathbf{p}) e^{i\omega_{\mathbf{p}} t} e^{-i\mathbf{p}\cdot\mathbf{x}}) \frac{d^3\mathbf{p}}{\sqrt{2\omega_{\mathbf{p}}}} \qquad (46.15)$$

where

$$\alpha_j(\mathbf{p}) = \frac{\widetilde{A}_{j,+}(\mathbf{p})}{\sqrt{2\omega_{\mathbf{p}}}}, \quad \overline{\alpha}_j(\mathbf{p}) = \frac{\widetilde{A}_{j,-}(-\mathbf{p})}{\sqrt{2\omega_{\mathbf{p}}}}$$

Here, $\omega_{\mathbf{p}} = |\mathbf{p}|$, and $\widetilde{A}_{j,+}, \widetilde{A}_{j,-}$ are the Fourier transforms of positive and negative energy solutions of 46.14.

The three components $\alpha_j(\mathbf{p})$ make up a vector-valued function $\boldsymbol{\alpha}(\mathbf{p})$. Solutions must satisfy the Coulomb gauge condition $\mathbf{\nabla} \cdot \mathbf{A} = 0$, which in momentum space is

$$\mathbf{p} \cdot \boldsymbol{\alpha}(\mathbf{p}) = 0 \qquad (46.16)$$

The space of solutions of this will be two dimensional for each value of $\mathbf{p}$, and we can choose some orthonormal basis

$$\epsilon_1(\mathbf{p}), \epsilon_2(\mathbf{p})$$

of such solutions (there is a topological obstruction to doing this continuously, but a continuous choice is not necessary). Here, the $\epsilon_\sigma(\mathbf{p}) \in \mathbf{R}^3$ for $\sigma = 1, 2$ are called "polarization vectors" and satisfy

$$\mathbf{p} \cdot \epsilon_1(\mathbf{p}) = \mathbf{p} \cdot \epsilon_2(\mathbf{p}) = 0, \quad \epsilon_1(\mathbf{p}) \cdot \epsilon_2(\mathbf{p}) = 0, \quad |\epsilon_1(\mathbf{p})|^2 = |\epsilon_2(\mathbf{p})|^2 = 1$$

They provide an orthonormal basis of the tangent space at $\mathbf{p}$ to the sphere of radius $|\mathbf{p}|$.

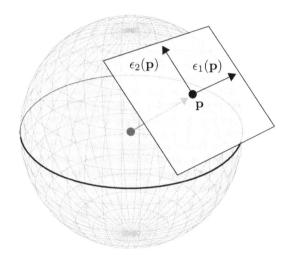

Figure 46.1: Polarization vectors at a point $\mathbf{p}$ in momentum space

The space of solutions is thus two copies of the space of solutions of the massless Klein–Gordon case. The quantum field for the theory of photons is then

$$\widehat{\mathbf{A}}(t, \mathbf{x}) = \frac{1}{(2\pi)^{3/2}} \int_{\mathbf{R}^3} \sum_{\sigma=1,2} (\epsilon_\sigma(\mathbf{p}) a_\sigma(\mathbf{p}) e^{-i\omega_\mathbf{p} t} e^{i\mathbf{p} \cdot \mathbf{x}}$$

$$+ \epsilon_\sigma(\mathbf{p}) a_\sigma^\dagger(\mathbf{p}) e^{i\omega_\mathbf{p} t} e^{-i\mathbf{p} \cdot \mathbf{x}}) \frac{d^3\mathbf{p}}{\sqrt{2\omega_\mathbf{p}}} \quad (46.17)$$

where $a_\sigma, a_\sigma^\dagger$ are annihilation and creation operators satisfying

$$[a_\sigma(\mathbf{p}), a_{\sigma'}^\dagger(\mathbf{p}')] = \delta_{\sigma\sigma'} \delta^3(\mathbf{p} - \mathbf{p}')$$

The state space of the theory will describe an arbitrary number of particles for each value of the momentum $\mathbf{p}$ (called photons), obeying the energy–momentum relation $\omega_\mathbf{p} = |\mathbf{p}|$, with a two-dimensional degree of freedom describing their polarization.

Note the appearance here of the following problem: Unlike the scalar field case (equation 43.8) where the Fourier coefficients were unconstrained functions, here they satisfy a condition (equation 46.16), and the $\alpha_j(\mathbf{p})$ cannot simply be quantized as independent annihilation operators for each j. Solving equation 46.16 and reducing the number of degrees of freedom by introducing the polarization vectors $\boldsymbol{\epsilon}_\sigma$ involves an arbitrary choice and makes the properties of the theory under the action of the Lorentz group much harder to understand. A similar problem for solutions to the Dirac equation will appear in chapter 47.

46.5   Space–time symmetries

The choice of Coulomb gauge nicely isolates the two physical degrees of freedom that describe photons and allows a straightforward quantization in terms of two copies of the previously studied relativistic scalar field. It does, however, do this in a way which makes some aspects of the Poincaré group action on the theory hard to understand, in particular the action of boost transformations. Our choice of continuous basis elements for the space of solutions of the Maxwell equations was not invariant under boost transformations (since it uses initial data at a fixed time, see equation 46.5), but making the gauge choice $A_0 = 0$, $\boldsymbol{\nabla} \cdot \mathbf{A} = 0$ creates another fundamental problem. Acting by a boost on a solution in this gauge will typically take it to a solution no longer satisfying the gauge condition.

For some indication of the difficulties introduced by the non-Lorentz invariant Coulomb gauge choice, the field commutators can be computed, with the result

$$[\widehat{A}_j(\mathbf{x}), \widehat{A}_k(\mathbf{x}')] = [\widehat{E}_j(\mathbf{x}), \widehat{E}_k(\mathbf{x}')] = 0$$

$$[\widehat{A}_j(\mathbf{x}), \widehat{E}_k(\mathbf{x}')] = -\frac{i}{(2\pi)^3} \int_{\mathbf{R}^3} \left(\delta_{jk} - \frac{p_j p_k}{|\mathbf{p}|^2} \right) e^{i\mathbf{p}\cdot(\mathbf{x}-\mathbf{x}')} d^3\mathbf{p} \qquad (46.18)$$

The right-hand side in the last case is not just the expected delta-function, but includes a term that is non-local in position space.

In section 46.6, we will discuss what happens with a Lorentz invariant gauge choice, but for now will just consider the Poincaré subgroup of space-time translations and spatial rotations, which do preserve the Coulomb gauge choice. Such group elements can be labeled by (a, R), where $a = (a_0, \mathbf{a})$ is a translation in space–time, and $R \in SO(3)$ is a spatial rotation. Generalizing the scalar field case (equation 44.6), we want to construct a unitary representation of the group of such elements by operators $U(a, R)$ on the state space, with the $U(a, R)$ also acting as intertwining operators on the field operators, by:

$$\widehat{\mathbf{A}}(t, \mathbf{x}) \rightarrow U(a, R)\widehat{\mathbf{A}}(t, \mathbf{x})U^{-1}(a, R) = R^{-1}\widehat{\mathbf{A}}(t + a_0, R\mathbf{x} + \mathbf{a}) \qquad (46.19)$$

To construct $U(a, R)$, we will proceed as for the scalar field case (see section 44.2) to identify the Lie algebra representation operators that satisfy the needed commutation relations, skipping some details (these can be found in most quantum field theory textbooks).

46.5.1   Time translations

For time translations, as usual one just needs to find the Hamiltonian operator $\widehat{H}$, and then

$$U(a_0, \mathbf{1}) = e^{ia_0\widehat{H}}$$

Taking

$$\widehat{H} = \frac{1}{2}\int_{\mathbf{R}^3} :(|\widehat{\mathbf{E}}|^2 + |\widehat{\mathbf{B}}|^2): d^3\mathbf{x} = \frac{1}{2}\int_{\mathbf{R}^3} : \left(|\frac{\partial\widehat{\mathbf{A}}}{\partial t}|^2 + |\nabla \times \widehat{\mathbf{A}}|^2\right): d^3\mathbf{x}$$

$$= \int_{\mathbf{R}^3} \omega_{\mathbf{p}}(a_1^\dagger(\mathbf{p})a_1(\mathbf{p}) + a_2^\dagger(\mathbf{p})a_2(\mathbf{p}))d^3\mathbf{p}$$

one can show, using equations 46.17, 46.18 and properties of the polarization vectors $\epsilon_j(\mathbf{p})$, that one has as required

$$\frac{\partial\widehat{\mathbf{A}}}{\partial t} = [\widehat{\mathbf{A}}, -i\widehat{H}]$$

$$\frac{\partial}{\partial t}(a_\sigma(\mathbf{p})e^{-i\omega_{\mathbf{p}}t})_{|t=0} = -i\omega_{\mathbf{p}}a_\sigma(\mathbf{p}) = [a_\sigma(\mathbf{p}), -i\widehat{H}]$$

The first of these uses the position space expression in terms of fields and the second the momentum space expression in terms of annihilation and creation operators.

46.5.2   Spatial translations

For spatial translations, we have

$$U(\mathbf{a}, \mathbf{1}) = e^{-i\mathbf{a}\cdot\widehat{\mathbf{P}}}$$

where $\widehat{\mathbf{P}}$ is the momentum operator. It has the momentum space expression

$$\widehat{\mathbf{P}} = \int_{\mathbf{R}^3} \mathbf{p}(a_1^\dagger(\mathbf{p})a_1(\mathbf{p}) + a_2^\dagger(\mathbf{p})a_2(\mathbf{p}))d^3\mathbf{p}$$

which satisfies

$$\boldsymbol{\nabla}(a_\sigma(\mathbf{p})e^{i\mathbf{P}\cdot\mathbf{x}}) = i\mathbf{p}a_\sigma(\mathbf{p})e^{i\mathbf{P}\cdot\mathbf{x}} = [-i\widehat{\mathbf{P}}, a_\sigma(\mathbf{p})e^{i\mathbf{P}\cdot\mathbf{x}}]$$

In terms of position space fields, one has

$$\widehat{\mathbf{P}} = \int_{\mathbf{R}^3} :\widehat{\mathbf{E}} \times \widehat{\mathbf{B}}: d^3\mathbf{x}$$

satisfying

$$\boldsymbol{\nabla}\widehat{A}_j = [-i\widehat{\mathbf{P}}, \widehat{A}_j]$$

One way to derive this is to use the fact that, for the classical theory,

$$\mathbf{P}_{EM} = \int_{\mathbf{R}^3} \mathbf{E} \times \mathbf{B}\ d^3\mathbf{x}$$

is the momentum of the electromagnetic field, since one can use the Poisson bracket relations 46.7 to show that

$$\{\mathbf{P}_{EM}, A_j(\mathbf{x})\} = \nabla A_j(\mathbf{x})$$

46.5.3 Rotations

We will not go through the exercise of constructing the angular momentum operator $\widehat{\mathbf{J}}$ for the electromagnetic field that gives the action of rotations on the theory. Details of how to do this can for instance be found in chapters 6 and 7 of [35]. The situation is similar to that of the spin $\frac{1}{2}$ Pauli equation case in section 34.2. There we found that $\mathbf{J} = \mathbf{L} + \mathbf{S}$ where the first term is the "orbital" angular momentum, due to the action of rotations on space, while $\mathbf{S}$ was the infinitesimal counterpart of the $SU(2)$ action on two-component spinors.

Much the same thing happens in this case, with $\mathbf{L}$ due to the action on spatial coordinates $\mathbf{x}$ and $\mathbf{S}$ due to the action of $SO(3)$ rotations on the 3 components of the vector $\mathbf{A}$ (see equation 46.19). We have seen that in the Coulomb gauge, the field $\widehat{\mathbf{A}}(\mathbf{x})$ decomposes into two copies of fields behaving much like the scalar Klein–Gordon theory, corresponding to the two basis vectors $\epsilon_1(\mathbf{p}), \epsilon_2(\mathbf{p})$ of the plane perpendicular to the vector $\mathbf{p}$. The subgroup

$SO(2) \subset SO(3)$ of rotations about the axis $\mathbf{p}$ acts on this plane, and its basis vectors in exactly the same way as the internal $SO(2)$ symmetry acted on pairs of real Klein–Gordon fields (see section 44.1.1). In that case, the same $SO(2)$ acts in the same way at each point in space–time, whereas here this $SO(2) \subset SO(3)$ varies depending on the momentum vector.

In the internal symmetry case, we found an operator $\widehat{Q}$ with integer eigenvalues (the charge). The analogous operator in this case is the helicity operator. The massless Poincaré group representations described in section 42.3.5 are the ones that occur here, for the case of helicity ± 1. Just as in the internal symmetry case, where complexification allowed diagonalization of $\widehat{Q}$ on the single-particle space, getting charges ± 1, here complexification of the $\epsilon_1(\mathbf{p}), \epsilon_2(\mathbf{p})$ diagonalizes the helicity, getting so-called "left circularly polarized" and "right circularly polarized" photon states.

46.6 Covariant gauge quantization

The methods used so far to handle gauge invariance suffer from various problems that can make their use awkward, most obviously the problem that they break Lorentz invariance by imposing non-Lorentz invariant conditions $(A_0 = 0, \boldsymbol{\nabla} \cdot \mathbf{A} = 0)$. Lorentz invariance can be maintained by use of a Lorentz invariant gauge condition, for example (note that the name is not a typo):

Definition (Lorenz gauge). *A vector potential A_μ is said to be in Lorenz gauge if*

$$\chi(A) \equiv -\frac{\partial A_0}{\partial t} + \boldsymbol{\nabla} \cdot \mathbf{A} = 0$$

Besides Lorentz invariance (Lorentz transforms of vector potentials in Lorenz gauge remain in Lorenz gauge), this gauge has the attractive feature that Maxwell's equations become just the standard massless wave equation. Since

$$\boldsymbol{\nabla} \times \mathbf{B} - \frac{\partial \mathbf{E}}{\partial t} = \boldsymbol{\nabla} \times \boldsymbol{\nabla} \times \mathbf{A} + \frac{\partial^2 \mathbf{A}}{\partial t^2} - \frac{\partial}{\partial t} \boldsymbol{\nabla} A_0$$

$$= \boldsymbol{\nabla}(\boldsymbol{\nabla} \cdot \mathbf{A}) - \nabla^2 \mathbf{A} + \frac{\partial^2 \mathbf{A}}{\partial t^2} - \frac{\partial}{\partial t} \boldsymbol{\nabla} A_0$$

$$= \boldsymbol{\nabla}\left(\boldsymbol{\nabla} \cdot \mathbf{A} - \frac{\partial}{\partial t} A_0\right) - \nabla^2 \mathbf{A} + \frac{\partial^2 \mathbf{A}}{\partial t^2}$$

$$= -\nabla^2 \mathbf{A} + \frac{\partial^2 \mathbf{A}}{\partial t^2}$$

the Maxwell equation 46.3 is the massless Klein–Gordon equation for the spatial components of $\mathbf{A}$.

Similarly,

$$\boldsymbol{\nabla} \cdot \mathbf{E} = \boldsymbol{\nabla} \left(-\frac{\partial \mathbf{A}}{\partial t} + \boldsymbol{\nabla} A_0 \right)$$

$$= -\frac{\partial}{\partial t} \boldsymbol{\nabla} \cdot \mathbf{A} + \nabla^2 A_0^2$$

$$= -\frac{\partial^2 A_0}{\partial t^2} + \nabla^2 A_0$$

so Gauss's law becomes the massless Klein–Gordon equation for A_0.

Like the temporal gauge, the Lorenz gauge does not completely remove the gauge freedom. Under a gauge transformation

$$-\frac{\partial A_0}{\partial t} + \boldsymbol{\nabla} \cdot \mathbf{A} \to -\frac{\partial A_0}{\partial t} + \boldsymbol{\nabla} \cdot \mathbf{A} - \frac{\partial^2 \varphi}{\partial t^2} + \nabla^2 \varphi$$

so φ that satisfy the wave equation

$$\frac{\partial^2 \varphi}{\partial t^2} = \nabla^2 \varphi \tag{46.20}$$

will give gauge transformations that preserve the Lorenz gauge condition $\chi(A) = 0$.

The four components of A_μ can be treated as four separate solutions of the massless Klein–Gordon equation and the theory then quantized in a Lorentz covariant manner. The field operators will be

$$\widehat{A}_\mu(t, \mathbf{x}) = \frac{1}{(2\pi)^{3/2}} \int_{\mathbf{R}^3} (a_\mu(\mathbf{p}) e^{-i\omega_{\mathbf{p}} t} e^{i\mathbf{p} \cdot \mathbf{x}} + a_\mu^\dagger(\mathbf{p}) e^{i\omega_{\mathbf{p}} t} e^{-i\mathbf{p} \cdot \mathbf{x}}) \frac{d^3 \mathbf{p}}{\sqrt{2\omega_{\mathbf{p}}}}$$

using annihilation and creation operators that satisfy

$$[a_\mu(\mathbf{p}), a_\nu^\dagger(\mathbf{p}')] = \pm \delta_{\mu\nu} \delta(\mathbf{p} - \mathbf{p}') \tag{46.21}$$

where $\pm$ is $+1$ for spatial coordinates, -1 for the time coordinate.

Two sorts of problems, however, arise:

- The Lorenz gauge condition is needed to get Maxwell's equations, but it cannot be imposed as an operator condition

$$-\frac{\partial \widehat{A}_0}{\partial t} + \boldsymbol{\nabla} \cdot \widehat{\mathbf{A}} = 0$$

since this is inconsistent with the canonical commutation relation, because

$$
\left[\widehat{A}_0(\mathbf{x}), -\frac{\partial \widehat{A}_0(\mathbf{x}')}{\partial t} + \boldsymbol{\nabla} \cdot \widehat{\mathbf{A}}(\mathbf{x}')\right] = \left[\widehat{A}_0(\mathbf{x}), -\frac{\partial \widehat{A}_0(\mathbf{x}')}{\partial t}\right] = i\delta(\mathbf{x} - \mathbf{x}') \neq 0
$$

- The commutation relations 46.21 for the operators $a_0(\mathbf{p}), a_0^\dagger(\mathbf{p})$ have the wrong sign. Recall from the discussion in section 26.3 that the positive sign is required in order for Bargmann–Fock quantization to give a unitary representation on a harmonic oscillator state space, with a an annihilation operator and $a^\dagger$ a creation operator.

We saw in section 46.3 that in $A_0 = 0$ gauge, there was an analog of the first of these problems, with $\boldsymbol{\nabla} \cdot \mathbf{E} = 0$ an inconsistent operator equation. There $\boldsymbol{\nabla} \cdot \mathbf{E}$ played the role of a moment map for the group of time-independent gauge transformations. One can show that similarly, $\chi(A)$ plays the role of a moment map for the group of gauge transformations φ satisfying the wave equation 46.20. In the $A_0 = 0$ gauge case, we saw that $\boldsymbol{\nabla} \cdot \widehat{\mathbf{E}} = 0$ could be treated not as an operator equation, but as a condition on states, determining the physical state space $\mathcal{H}_{phys} \subset \mathcal{H}$.

This will not work for the Lorenz gauge condition, since it can be shown that there will be no states such that

$$
\left(-\frac{\partial \widehat{A}_0}{\partial t} + \boldsymbol{\nabla} \cdot \widehat{\mathbf{A}}\right) |\psi\rangle = 0
$$

The problem is that, unlike in the Gauss's law case, the complex structure J_r used for quantization ($+i$ on the positive energy single-particle states, $-i$ on the negative energy ones) does not commute with the Lorenz gauge condition. The gauge condition needs to be implemented not on the dual phase space $\mathcal{M}$ (here the space of A_μ satisfying the massless wave equation), but on $\mathcal{H}_1 = \mathcal{M}_{J_r}^+$, where

$$
\mathcal{H}_1 \otimes \mathbf{C} = \mathcal{M}_{J_r}^+ \oplus \mathcal{M}_{J_r}^-
$$

is the decomposition of the complexification of $\mathcal{H}_1$ into negative and positive energy subspaces. The condition we want is thus

$$
\chi(A)^+ = 0
$$

where $\chi(A)^+$ is the positive energy part of the decomposition of $\chi(A)$ into positive and negative energy components.

This sort of gauge condition can be implemented either before or after quantization, as follows:

- One can take elements of $\mathcal{H}_1$ to be $\mathbf{C}^4$-valued functions $\alpha_\mu(\mathbf{p})$ on $\mathbf{R}^3$ with Lorentz invariant indefinite inner product

$$\langle \alpha_\mu(\mathbf{p}), \alpha_\nu(\mathbf{p}') \rangle = \int_{\mathbf{R}^3} (-|\alpha_0(\mathbf{p})|^2 + |\boldsymbol{\alpha}(\mathbf{p})|^2) d^3\mathbf{p} \qquad (46.22)$$

Here, each $\alpha_\mu(\mathbf{p})$ is defined as in equation 43.7 for the single-component field case. The subspace satisfying $\chi(A)^+ = 0$ will be the subspace $\mathcal{H}_1' \subset \mathcal{H}_1$ of α_μ satisfying

$$-p_0 \alpha_0(\mathbf{p}) + \sum_{j=1}^{3} p_j \alpha_j(\mathbf{p}) = 0$$

This subspace will in turn have a subspace $\mathcal{H}_1^{+\prime\prime} \subset \mathcal{H}_1^{+\prime}$ corresponding to A_μ that are gauge transforms of 0, i.e., with Fourier coefficients satisfying

$$\alpha_0(\mathbf{p}) = p_0 f(\mathbf{p}), \quad \alpha_j(\mathbf{p}) = p_j f(\mathbf{p})$$

for some function $f(\mathbf{p})$. Both of these subspaces carry an action of the Lorentz group and so does the quotient space

$$\mathcal{H}_1^{+\prime}/\mathcal{H}_1^{+\prime\prime}$$

One can show that the indefinite inner product 46.22 is nonnegative on $\mathcal{H}_1^{+\prime}$ and null on $\mathcal{H}_1^{+\prime\prime}$, so positive definite on the quotient space. Note that this is an example of a symplectic reduction, although in the context of an action of an infinite dimensional complex group (the positive energy gauge transformations satisfying the massless wave equation). One can construct the quantum theory by applying the Bargmann–Fock method to this quotient space.

- One can instead implement the gauge condition after quantization, first quantizing the four components of A_μ as massless fields, getting a state space $\mathcal{H}$ (which will not have a positive-definite inner product), and then defining $\mathcal{H}' \subset \mathcal{H}$ to be the subspace of states satisfying

$$\left(-\frac{\partial \widehat{A}_0}{\partial t} + \boldsymbol{\nabla} \cdot \widehat{\mathbf{A}} \right)^+ |\psi\rangle = 0$$

where the positive energy part of the operator is taken. The state space $\mathcal{H}'$ in turn has a subspace $\mathcal{H}''$ of states of zero norm, and one can define

$$\mathcal{H}_{phys} = \mathcal{H}'/\mathcal{H}''$$

$\mathcal{H}_{phys}$ will have a positive-definite Hermitian inner product and carry a unitary action of the Poincaré group. It can be shown to be isomorphic to the physical state space of transverse photons constructed using the Coulomb gauge.

This sort of covariant quantization method is often referred to as the "Gupta–Bleuler" method and is described in more detail in many quantum field theory textbooks.

In the Yang–Mills case, each of the methods that we have discussed for dealing with the gauge symmetry runs into problems:

- In the $A_0 = 0$ gauge, there again is a symmetry under the group of time-independent gauge transformations, and a moment map μ, with $\mu = 0$ the Yang–Mills version of Gauss's law. The symplectic reduction, however, is now a nonlinear space, so the quantization method we have developed does not apply. Gauss's law can instead be imposed on the states, but it is difficult to explicitly characterize the physical state space that this gives.

- A Yang–Mills analog of the Coulomb gauge condition can be defined, but then the analog of equation 46.13 will be a nonlinear equation without a unique solution (this problem is known as the "Gribov ambiguity").

- The combination of fields $\chi(A)$ now no longer satisfies a linear wave equation, and one cannot consistently restrict to a positive energy subspace and use the Gupta–Bleuler covariant quantization method.

Digression. *There is a much more sophisticated Lorentz covariant quantization method (called the "BRST method") for dealing with gauge symmetry that uses quite different techniques. The theory is first extended by the addition of non-physical ("ghost") fermionic fields, giving a theory of coupled bosonic and fermionic oscillators of the sort we studied in section 33.1. This includes an operator analogous to Q_1, with the property that $Q_1^2 = 0$. One can arrange things in the electromagnetic field case such that*

$$\mathcal{H}_{phys} = \frac{\{|\psi\rangle : Q_1|\psi\rangle = 0\}}{\{|\psi\rangle : |\psi\rangle = Q_1|\psi'\rangle\}}$$

In the BRST method, one works with unconstrained Lorentz covariant fields, but non-unitary state spaces, with unitarity only achieved on a quotient such as $\mathcal{H}_{phys}$. This construction is related to the Gupta–Bleuler method in the case of electromagnetic fields, but unlike that method generalizes to the Yang–Mills case. It can also be motivated by considerations of what happens when one imposes a gauge condition in a path integral (this is called the "Faddeev–Popov method").

46.7 For further reading

The topic of this chapter is treated in some version in every quantum field textbook. Some examples for Coulomb gauge quantization are chapter 14 of [10] or chapter 9 of [16] and for covariant Lorenz gauge quantization see chapter 7 of [35] or chapter 9 of [78]. [17] has a mathematically careful discussion of both the Coulomb gauge quantization and covariant quantization in Lorenz gauge. For a general discussion of constrained Hamiltonian systems and their quantization, with details for the cases of the electromagnetic field and Yang–Mills theory, see [89]. The homological BRST method for dealing with gauge symmetries is treated in detail in the Hamiltonian formalism in [46].

Chapter 47
The Dirac Equation and Spin $\frac{1}{2}$ Fields

The space of solutions to the Klein–Gordon equation gives an irreducible representation of the Poincaré group corresponding to a relativistic particle of mass m and spin zero. Elementary matter particles (quarks and leptons) are spin $\frac{1}{2}$ particles, and we would like to have a relativistic wave equation that describes them, suitable for building a quantum field theory.

This is provided by a remarkable construction that uses the Clifford algebra and its action on spinors to find a square root of the Klein–Gordon equation, the Dirac equation. We will begin with the case of real-valued spinor fields, for which the quantum field theory describes spin $\frac{1}{2}$ neutral massive relativistic fermions, known as Majorana fermions. In the massless case, it turns out that the Dirac equation decouples into two separate equations for two-component complex fields, the Weyl equations, and quantization leads to a relativistic theory of massless helicity $\pm\frac{1}{2}$ particles which can carry a charge, the Weyl fermions. Pairs of Weyl fermions of opposite helicity can be coupled together to form a theory of charged, massive, spin $\frac{1}{2}$ particles, the Dirac fermions. In the low energy, non-relativistic limit, the Dirac fermion theory becomes the Pauli–Schrödinger theory discussed in chapter 34.

47.1 The Dirac equation in Minkowski space

Recall from section 34.4 that for any real vector space $\mathbf{R}^{r+s}$ with an inner product of signature (r, s) we can use the Clifford algebra $\mathrm{Cliff}(r, s)$ to define a first-order differential operator, the Dirac operator $\partial\!\!\!/$. For the Minkowski space case of signature $(3, 1)$, the Clifford algebra $\mathrm{Cliff}(3, 1)$ is generated by elements $\gamma_0, \gamma_1, \gamma_2, \gamma_3$ satisfying

$$\gamma_0^2 = -1, \quad \gamma_1^2 = \gamma_2^2 = \gamma_3^2 = +1, \quad \gamma_j\gamma_k + \gamma_k\gamma_j = 0 \text{ for } j \neq k$$

© Peter Woit 2017
P. Woit, *Quantum Theory, Groups and Representations*,
DOI 10.1007/978-3-319-64612-1_47

Cliff$(3, 1)$ is isomorphic to the algebra $M(4, \mathbf{R})$ of 4 by 4 real matrices. Several conventional identifications of the generators γ_j with 4 by 4 complex matrices satisfying the relations of the algebra were described in chapter 41. Each of these gives an identification of Cliff$(3, 1)$ with a specific subset of the complex matrices $M(4, \mathbf{C})$ and of the complexified Clifford algebra Cliff$(3, 1) \otimes \mathbf{C}$ with $M(4, \mathbf{C})$ itself. The Dirac operator in Minkowski space is thus

$$\partial\!\!\!/ = \gamma_0 \frac{\partial}{\partial x_0} + \gamma_1 \frac{\partial}{\partial x_1} + \gamma_2 \frac{\partial}{\partial x_2} + \gamma_3 \frac{\partial}{\partial x_3} = \gamma_0 \frac{\partial}{\partial x_0} + \boldsymbol{\gamma} \cdot \boldsymbol{\nabla}$$

and it will act on four-component functions $\psi(t, \mathbf{x}) = \psi(x)$ on Minkowski space. These functions take values in the four-dimensional vector space that the Clifford algebra elements act on (which can be $\mathbf{R}^4$ if using real matrices, $\mathbf{C}^4$ in the complex case).

We have seen in chapter 42 that $-P_0^2 + P_1^2 + P_2^2 + P_3^2$ is a Casimir operator for the Poincaré group. Acting on four-component wavefunctions $\psi(x)$, the Dirac operator provides a square root of (minus) this Casimir operator since

$$\partial\!\!\!/^2 = -\frac{\partial^2}{\partial x_0^2} + \frac{\partial^2}{\partial x_1^2} + \frac{\partial^2}{\partial x_2^2} + \frac{\partial^2}{\partial x_3^2} = -(-P_0^2 + P_1^2 + P_2^2 + P_3^2)$$

For irreducible representations of the Poincaré group, the Casimir operator acts as a scalar (0 for massless particles, $-m^2$ for particles of mass m). Using the Dirac operator, we can rewrite this condition as follows:

$$(-\frac{\partial^2}{\partial x_0^2} + \Delta - m^2)\psi = (\partial\!\!\!/ + m)(\partial\!\!\!/ - m)\psi = 0 \tag{47.1}$$

This motivates the following definition of a new wave equation:

Definition 1 (Dirac equation). *The Dirac equation is the differential equation*

$$(\partial\!\!\!/ - m)\psi(x) = 0 \tag{47.2}$$

for four-component functions on Minkowski space.

Using equation 40.3, for the Minkowski space Fourier transform, the Dirac equation in energy–momentum space is

$$(i\not{p} - m)\widetilde{\psi}(p) = (i(-\gamma_0 p_0 + \gamma_1 p_1 + \gamma_2 p_2 + \gamma_3 p_3) - m)\widetilde{\psi}(p) = 0 \tag{47.3}$$

Note that solutions to this Dirac equation are also solutions to equation 47.1, but in a sense only half of them. The Dirac equation is first order in time, so solutions are determined by the initial value data

$$\psi(\mathbf{x}) = \psi(0, \mathbf{x})$$

of ψ at a fixed time, while equation 47.1 is second-order, with solutions determined by specifying both ψ and its time derivative.

The Dirac equation

$$(\gamma_0 \frac{\partial}{\partial x_0} + \boldsymbol{\gamma} \cdot \boldsymbol{\nabla} - m)\psi(x) = 0$$

can be written in the form of a Schrödinger equation as follows:

$$i\frac{\partial}{\partial t}\psi(t, \mathbf{x}) = H_D \psi(t, \mathbf{x}) \tag{47.4}$$

with Hamiltonian

$$H_D = i\gamma_0(\boldsymbol{\gamma} \cdot \boldsymbol{\nabla} - m) \tag{47.5}$$

Fourier transforming, in momentum space the energy eigenvalue equation is

$$-\gamma_0(\boldsymbol{\gamma} \cdot \mathbf{p} + im)\widetilde{\psi}(\mathbf{p}) = E\widetilde{\psi}(\mathbf{p})$$

The square of the left-hand side of this equation is

$$
\begin{aligned}
(-\gamma_0(\boldsymbol{\gamma} \cdot \mathbf{p} + im))^2 &= \gamma_0(\boldsymbol{\gamma} \cdot \mathbf{p} + im)\gamma_0(\boldsymbol{\gamma} \cdot \mathbf{p} + im) \\
&= (\boldsymbol{\gamma} \cdot \mathbf{p} - im)(\boldsymbol{\gamma} \cdot \mathbf{p} + im) \\
&= (\boldsymbol{\gamma} \cdot \mathbf{p})^2 + m^2 = |\mathbf{p}|^2 + m^2
\end{aligned}
$$

This shows that solutions to the Dirac equation have the expected relativistic energy–momentum relation

$$E = \pm\omega_{\mathbf{p}} = \pm\sqrt{|\mathbf{p}|^2 + m^2}$$

For each $\mathbf{p}$, there will be a two-dimensional space of solutions $\widetilde{\psi}_+(\mathbf{p})$ to

$$-\gamma_0(\boldsymbol{\gamma} \cdot \mathbf{p} + im)\widetilde{\psi}_+(\mathbf{p}) = \omega_{\mathbf{p}}\widetilde{\psi}_+(\mathbf{p}) \tag{47.6}$$

(the positive energy solutions) and a two-dimensional space of solutions $\widetilde{\psi}_-(\mathbf{p})$ to

$$-\gamma_0(\boldsymbol{\gamma} \cdot \mathbf{p} + im)\widetilde{\psi}_-(\mathbf{p}) = -\omega_{\mathbf{p}}\widetilde{\psi}_-(\mathbf{p}) \tag{47.7}$$

(the negative energy solutions). Solutions to the Dirac equation can be identified with either

- Four-component functions $\psi(\mathbf{x})$, initial value data at a time $t = 0$.

- Four-component functions $\widetilde{\psi}(\mathbf{p})$, Fourier transforms of the initial value data. These can be decomposed as follows:

$$\widetilde{\psi}(\mathbf{p}) = \widetilde{\psi}_+(\mathbf{p}) + \widetilde{\psi}_-(\mathbf{p}) \tag{47.8}$$

into positive (solutions of 47.6) and negative (solutions of 47.7) energy components.

The four-dimensional Fourier transform of a solution is of the form

$$\widetilde{\psi}(p) = \frac{1}{(2\pi)^2} \int_{\mathbf{R}^4} e^{-i(-p_0 x_0 + \mathbf{p} \cdot \mathbf{x})} \psi(x) d^4 x$$
$$= \theta(p_0)\delta(-p_0^2 + |\mathbf{p}|^2 + m^2)\widetilde{\psi}_+(\mathbf{p}) + \theta(-p_0)\delta(-p_0^2 + |\mathbf{p}|^2 + m^2)\widetilde{\psi}_-(\mathbf{p})$$

The Poincaré group acts on solutions to the Dirac equation by

$$\psi(x) \to u(a, \Lambda)\psi(x) = S(\Lambda)\psi(\Lambda^{-1} \cdot (x - a)) \tag{47.9}$$

or, in terms of Fourier transforms, by

$$\widetilde{\psi}(p) \to \widetilde{u}(a, \Lambda)\widetilde{\psi}(p) = e^{-i(-p_0 a_0 + \mathbf{p} \cdot \mathbf{a})} S(\Lambda)\widetilde{\psi}(\Lambda^{-1} \cdot p) \tag{47.10}$$

Here Λ is in $Spin(3,1)$, the double cover of the Lorentz group and $\Lambda \cdot x$ means the action of $Spin(3,1)$ on Minkowski space vectors. $S(\Lambda)$ is the spin representation, realized explicitly as 4 by 4 matrices by exponentiating quadratic combinations of the Clifford algebra generators (using a chosen identification of the γ_j with 4 by 4 matrices). Spinor fields ψ can be interpreted as elements of the tensor product of the spinor representation space ($\mathbf{R}^4$ or $\mathbf{C}^4$) and functions on Minkowski space. Then equation 47.9 means that $S(\Lambda)$ acts on the spinor factor, and the action on functions is the one induced from the Poincaré action on Minkowski space.

Recall that (equation 29.4) conjugation by $S(\Lambda)$ takes vectors v to their Lorentz transform $v' = \Lambda \cdot v$, in the sense that

$$S(\Lambda)^{-1}\slashed{v}S(\Lambda) = \slashed{v}'$$

so

$$\slashed{p}S(\Lambda)\widetilde{\psi}(\Lambda^{-1} \cdot p) = S(\Lambda)\slashed{p}'S^{-1}(\Lambda)S(\Lambda)\widetilde{\psi}(p') = S(\Lambda)\slashed{p}'\widetilde{\psi}(p')$$

where $p' = \Lambda^{-1} \cdot p$. As a result, the action 47.10 takes solutions of the Dirac equation 47.3 to solutions, since

$$(i\slashed{p} - m)\widetilde{u}(a, \Lambda)\widetilde{\psi}(p) = (i\slashed{p} - m)e^{-i(-p_0 a_0 + \mathbf{p} \cdot \mathbf{a})} S(\Lambda)\widetilde{\psi}(\Lambda^{-1} \cdot p)$$
$$= e^{-i(-p_0 a_0 + \mathbf{p} \cdot \mathbf{a})} S(\Lambda)(i\slashed{p}' - m)\widetilde{\psi}(p') = 0$$

47.2 Majorana spinors and the Majorana field

The analog for spin $\frac{1}{2}$ of the real scalar field is known as the Majorana spinor field and can be constructed using a choice of real-valued matrices for the generators γ_0, γ_j, acting on a four-component real-valued field ψ. Such a choice was given explicitly in section 41.2 and can be rewritten in terms of 2 by 2 block matrices, using the real matrices

$$\sigma_1 = \begin{pmatrix} 0 & 1 \\ 1 & 0 \end{pmatrix}, \quad i\sigma_2 = \begin{pmatrix} 0 & 1 \\ -1 & 0 \end{pmatrix}, \quad \sigma_3 = \begin{pmatrix} 1 & 0 \\ 0 & -1 \end{pmatrix}$$

as follows:

$$\gamma_0^M = \begin{pmatrix} 0 & -i\sigma_2 \\ -i\sigma_2 & 0 \end{pmatrix}, \quad \gamma_1^M = \begin{pmatrix} \sigma_3 & 0 \\ 0 & \sigma_3 \end{pmatrix}$$

$$\gamma_2^M = \begin{pmatrix} 0 & i\sigma_2 \\ -i\sigma_2 & 0 \end{pmatrix}, \quad \gamma_3^M = \begin{pmatrix} -\sigma_1 & 0 \\ 0 & -\sigma_1 \end{pmatrix}$$

Quadratic combinations of Clifford generators have a basis

$$\gamma_0^M \gamma_1^M = \begin{pmatrix} 0 & \sigma_1 \\ \sigma_1 & 0 \end{pmatrix}, \quad \gamma_0^M \gamma_2^M = \begin{pmatrix} -1 & 0 \\ 0 & 1 \end{pmatrix}, \quad \gamma_0^M \gamma_3^M = \begin{pmatrix} 0 & \sigma_3 \\ \sigma_3 & 0 \end{pmatrix}$$

$$\gamma_1^M \gamma_2^M = \begin{pmatrix} 0 & \sigma_1 \\ -\sigma_1 & 0 \end{pmatrix}, \quad \gamma_2^M \gamma_3^M = \begin{pmatrix} 0 & -\sigma_3 \\ \sigma_3 & 0 \end{pmatrix}, \quad \gamma_1^M \gamma_3^M = \begin{pmatrix} -i\sigma_2 & 0 \\ 0 & -i\sigma_2 \end{pmatrix}$$

and one has

$$\gamma_5^M = i\gamma_0^M \gamma_1^M \gamma_2^M \gamma_3^M = \begin{pmatrix} \sigma_2 & 0 \\ 0 & -\sigma_2 \end{pmatrix}$$

The quantized Majorana field can be understood as an example of a quantization of a pseudo-classical fermionic oscillator system (as described in section 30.3.2), by the fermionic analog of the Bargmann–Fock quantization method (as described in section 31.3). We take as dual pseudo-classical phase space $\mathcal{V}$ the real-valued solutions of the Dirac equation in the Majorana representation. Using values of the solutions at $t = 0$, continuous basis elements of $\mathcal{V}$ are given by the four-component distributional field $\Psi(\mathbf{x})$, with components $\Psi_a(\mathbf{x})$ for $a = 1, 2, 3, 4$.

This space $\mathcal{V}$ comes with an inner product

$$(\psi, \phi) = \int_{\mathbf{R}^3} \psi^T(\mathbf{x})\phi(\mathbf{x}) d^3\mathbf{x} \tag{47.11}$$

In this form, the invariance under translations and under spatial rotations is manifest, with the $S(\Lambda)$ acting by orthogonal transformations on the Majorana spinors when Λ is a rotation. One way to see this is to note that the $S(\Lambda)$ in this case are exponentials of linear combinations of the antisymmetric matrices $\gamma_1^M \gamma_2^M, \gamma_2^M \gamma_3^M, \gamma_1^M \gamma_3^M$ and thus are orthogonal matrices.

As we have seen in chapter 30, the fermionic Poisson bracket of a pseudo-classical system is determined by an inner product, with the one above giving in this case

$$\{\Psi_a(\mathbf{x}), \Psi_b(\mathbf{x}')\}_+ = \delta^3(\mathbf{x} - \mathbf{x}')\delta_{ab}$$

The Hamiltonian that will give a pseudo-classical system evolving according to the Dirac equation is

$$h = \frac{1}{2} \int_{\mathbf{R}^3} \Psi^T(\mathbf{x})\gamma_0(\boldsymbol{\gamma} \cdot \boldsymbol{\nabla} - m)\Psi(\mathbf{x})d^3\mathbf{x}$$

One can see this by noting that the operator $\gamma_0(\boldsymbol{\gamma} \cdot \boldsymbol{\nabla} - m)$ is minus its adjoint with respect to the inner product 47.11, since γ_0 is an antisymmetric matrix, the $\gamma_0\boldsymbol{\gamma}$ are symmetric, and the derivative is antisymmetric. Applying the finite dimensional theorem 30.1 in this infinite dimensional context, one finds that the pseudo-classical equation of motion is

$$\frac{\partial}{\partial t}\Psi(\mathbf{x}) = \{\Psi(\mathbf{x}), h\}_+ = \gamma_0(\boldsymbol{\gamma} \cdot \boldsymbol{\nabla} - m)\Psi(\mathbf{x})$$

which is the Dirac equation in Hamiltonian form (see equations 47.4 and 47.5). The antisymmetry of the operator $\gamma_0(\boldsymbol{\gamma} \cdot \boldsymbol{\nabla} - m)$ that generates time evolution corresponds to the fact that time evolution gives for each t an (infinite dimensional) orthogonal group action on the space of solutions $\mathcal{V}$, preserving the inner product 47.11.

Corresponding to the Poincaré group action 47.9 on solutions to the Dirac equation, at least for translations by $\mathbf{a}$ and rotations Λ, one has a corresponding action on the fields, written

$$\Psi(\mathbf{x}) \to u(\mathbf{a}, \Lambda)\Psi(\mathbf{x}) = S(\Lambda)^{-1}\Psi(\Lambda \cdot \mathbf{x} + \mathbf{a})$$

The quadratic pseudo-classical moment map that generates the action of spatial translations on solutions is the momentum

$$\mathbf{P} = -\frac{1}{2} \int_{\mathbf{R}^3} \Psi^T(\mathbf{x})\boldsymbol{\nabla}\Psi(\mathbf{x})d^3\mathbf{x}$$

since it satisfies (generalizing equation 30.1)

$$\{\mathbf{P}, \Psi(\mathbf{x})\}_+ = \boldsymbol{\nabla}\Psi(\mathbf{x})$$

For rotations, the moment map is the angular momentum

$$\mathbf{J} = -\frac{1}{2}\int_{\mathbf{R}^3}\Psi^T(\mathbf{x})(\mathbf{x}\times\boldsymbol{\nabla} - \mathbf{s})\Psi(\mathbf{x})d^3\mathbf{x}$$

which satisfies

$$\{\mathbf{J}, \Psi(\mathbf{x})\}_+ = (\mathbf{x}\times\boldsymbol{\nabla} - \mathbf{s})\Psi(\mathbf{x})$$

Here, the components s_j of $\mathbf{s}$ are the matrices

$$s_j = \frac{1}{2}\epsilon_{jkl}\gamma_k\gamma_l$$

Our use here of the fixed-time fields $\Psi(\mathbf{x})$ as continuous basis elements on the phase space $\mathcal{V}$ comes with two problematic features:

- One cannot easily implement Lorentz transformations that are boosts, since these change the fixed-time hypersurface used to define the $\Psi(\mathbf{x})$.

- The relativistic complex structure on $\mathcal{V}$ needed for a consistent quantization is defined by a splitting of $\mathcal{V}\otimes\mathbf{C}$ into positive and negative energy solutions, but this decomposition is only easily made in momentum space, not position space.

47.2.1   Majorana spinor fields in momentum space

Recall that in the case of the real relativistic scalar field studied in chapter 43, we had the following expression (equation 43.8) for a solution to the Klein–Gordon equation

$$\phi(t,\mathbf{x}) = \frac{1}{(2\pi)^{3/2}}\int_{\mathbf{R}^3}(\alpha(\mathbf{p})e^{-i\omega_{\mathbf{p}}t}e^{i\mathbf{p}\cdot\mathbf{x}} + \overline{\alpha}(\mathbf{p})e^{i\omega_{\mathbf{p}}t}e^{-i\mathbf{p}\cdot\mathbf{x}})\frac{d^3\mathbf{p}}{\sqrt{2\omega_{\mathbf{p}}}}$$

with $\alpha(\mathbf{p})$ and $\overline{\alpha}(\mathbf{p})$ parametrizing positive and negative energy subspaces of the complexified phase space $\mathcal{M}\otimes\mathbf{C}$. This was quantized by an infinite dimensional version of the Bargmann–Fock quantization described in chapter 26, with dual phase space $\mathcal{M}$ the space of real-valued solutions of the Klein–Gordon equation, and the complex structure J_r the relativistic one discussed in section 43.2.

For the Majorana theory, one can write four-component Majorana spinor solutions to the Dirac equation as follows:

$$\psi(t,\mathbf{x}) = \frac{1}{(2\pi)^{3/2}}\int_{\mathbf{R}^3}(\alpha(\mathbf{p})e^{-i\omega_{\mathbf{p}}t}e^{i\mathbf{p}\cdot\mathbf{x}} + \overline{\alpha}(\mathbf{p})e^{i\omega_{\mathbf{p}}t}e^{-i\mathbf{p}\cdot\mathbf{x}})\frac{d^3\mathbf{p}}{\sqrt{2\omega_{\mathbf{p}}}} \quad (47.12)$$

where $\alpha(\mathbf{p})$ is a four-component complex vector, satisfying

$$-\gamma_0^M(\gamma^M \cdot \mathbf{p} + im)\alpha(\mathbf{p}) = \omega_{\mathbf{p}}\alpha(\mathbf{p}) \tag{47.13}$$

These $\alpha(\mathbf{p})$ are the positive energy solutions $\widetilde{\psi}_+(\mathbf{p})$ of 47.6, with the conjugate equation for $\overline{\alpha}(\mathbf{p})$ giving the negative energy solutions of 47.7 (with the sign of $\mathbf{p}$ interchanged, $\overline{\alpha}(\mathbf{p}) = \widetilde{\psi}_-(-\mathbf{p})$).

For each $\mathbf{p}$, there is a two-dimensional space of solutions to equation 47.13. One way to choose a basis $u_+(\mathbf{p})$, $u_-(\mathbf{p})$ of this space is by first considering the case $\mathbf{p} = 0$. The equation 47.13 becomes

$$\gamma_0^M \alpha(\mathbf{0}) = i\alpha(\mathbf{0})$$

which will have a basis of solutions

$$u_+(\mathbf{0}) = \frac{1}{2}\begin{pmatrix} 1 \\ 0 \\ 0 \\ i \end{pmatrix}, \quad u_-(\mathbf{0}) = \frac{1}{2}\begin{pmatrix} 0 \\ 1 \\ -i \\ 0 \end{pmatrix}$$

These will satisfy

$$-\frac{i}{2}\gamma_1^M\gamma_2^M u_+(\mathbf{0}) = \frac{1}{2}u_+(\mathbf{0}), \quad -\frac{i}{2}\gamma_1^M\gamma_2^M u_-(\mathbf{0}) = -\frac{1}{2}u_-(\mathbf{0})$$

The two solutions

$$u_+(\mathbf{0})e^{-imt} + \overline{u_+(\mathbf{0})}e^{imt} = \begin{pmatrix} \cos(mt) \\ 0 \\ 0 \\ \sin(mt) \end{pmatrix}$$

$$u_-(\mathbf{0})e^{-imt} + \overline{u_-(\mathbf{0})}e^{imt} = \begin{pmatrix} 0 \\ \cos(mt) \\ -\sin(mt) \\ 0 \end{pmatrix}$$

correspond physically to a relativistic spin $\frac{1}{2}$ particle of mass m at rest, with the first having spin "up" in the 3-direction, the second spin "down."

The Majorana spinor field theory comes with a significant complication with respect to the case of scalar fields. The complex four-component $\alpha(\mathbf{p})$ provide twice as many basis elements as one needs to describe the solutions of the Dirac equation (put differently, they are not independent, but satisfy the relation 47.13). Quantizing using four sets of annihilation and creation operators (one for each component of α) would produce a quantum field theory with too many degrees of freedom by a factor of two. The standard solution to this problem is to make a choice of basis elements of the space of solutions for each value of $\mathbf{p}$ by defining polarization vectors

$$u_+(\mathbf{p}) = L(\mathbf{p})u_+(\mathbf{0}), \quad u_-(\mathbf{p}) = L(\mathbf{p})u_-(\mathbf{0})$$

Here, $L(\mathbf{p})$ is an element of $SL(2, \mathbf{C})$ chosen so that, acting by a Lorentz transformation on energy–momentum vectors it takes $(m, \mathbf{0})$ to $(\omega_{\mathbf{p}}, \mathbf{p})$. More explicitly, using equation 40.4, one has

$$L(\mathbf{p}) \begin{pmatrix} m & 0 \\ 0 & m \end{pmatrix} L^\dagger(\mathbf{p}) = \begin{pmatrix} \omega_{\mathbf{p}} + p_3 & p_1 - ip_2 \\ p_1 + ip_2 & \omega_{\mathbf{p}} - p_3 \end{pmatrix}$$

Such a choice is not unique and is a matter of convention. Explicit choices are discussed in most quantum field theory textbooks (although in a different representation of the γ-matrices), see, for instance, chapter 3.3 of [67]. Note that these polarization vectors are not the same as the Bloch sphere polarization vectors used in earlier chapters. They are defined on the positive mass hyperboloid, not on the sphere, and for these, there is no topological obstruction to a continuous definition.

Solutions are then written as follows:

$$\psi(t, \mathbf{x}) = \frac{1}{(2\pi)^{3/2}} \int_{\mathbf{R}^3} \sum_{s=\pm} (\alpha_s(\mathbf{p})u_s(\mathbf{p})e^{-i\omega_{\mathbf{p}}t}e^{i\mathbf{p}\cdot\mathbf{x}} + \overline{\alpha}_s(\mathbf{p})\overline{u_s(\mathbf{p})}e^{i\omega_{\mathbf{p}}t}e^{-i\mathbf{p}\cdot\mathbf{x}}) \frac{d^3\mathbf{p}}{\sqrt{2\omega_{\mathbf{p}}}}$$

$$(47.14)$$

One now has the correct number of functions to parametrize pseudo-classical complexified dual phase space $\mathcal{V} \otimes \mathbf{C}$. These are the single-component complex functions $\alpha_+(\mathbf{p}), \alpha_-(\mathbf{p})$, providing elements of $\mathcal{V}^+_{J_r} = \mathcal{H}_1$ and their conjugates $\overline{\alpha}_+(\mathbf{p}), \overline{\alpha}_-(\mathbf{p})$, which provide elements of $\mathcal{V}^-_{J_r}$.

To quantize the Majorana field in a way that allows a simple understanding of the action of the full Poincaré group, we need a positive-definite Poincaré invariant inner product on the space of solutions of the Dirac equation. We have already seen what the right inner product is (see equation 47.11), but unfortunately this is not written in a way that makes Lorentz invariance manifest. Unlike the case of the Klein–Gordon equation, working in momentum space does not completely resolve the problem. Using the $\alpha_+(\mathbf{p}), \alpha_-(\mathbf{p})$ allows for an explicitly positive-definite inner product, which is just two copies of the Klein–Gordon one for scalars, see equation 43.18. In the next section, we will quantize the theory using these. This inner product is not, however, manifestly Lorentz invariant (due to the dependence on the choice of polarization vectors $u_\pm(\mathbf{p})$).

47.2.2 Quantization of the Majorana field

Quantization of the dual pseudo-classical phase space $\mathcal{M}$, using the fermionic Bargmann-Fock method, the relativistic complex structure and the functions $\alpha_{\pm}(\mathbf{p})$ from equation 47.14 is given by annihilation and creation operators $a_{\pm}(\mathbf{p}), a_{\pm}^{\dagger}(\mathbf{p})$ that anticommute, except for the relations

$$[a_{+}(\mathbf{p}), a_{+}^{\dagger}(\mathbf{p}')]_{+} = \delta^3(\mathbf{p} - \mathbf{p}'), \quad [a_{-}(\mathbf{p}), a_{-}^{\dagger}(\mathbf{p}')]_{+} = \delta^3(\mathbf{p} - \mathbf{p}')$$

The field operator is then constructed using these, giving

Definition 2 (Majorana field operator). *The Majorana field operator is given by*

$$\widehat{\Psi}(\mathbf{x}, t) = \frac{1}{(2\pi)^{3/2}} \int_{\mathbf{R}^3} \sum_{s=\pm} (a_s(\mathbf{p}) u_s(\mathbf{p}) e^{-i\omega_{\mathbf{p}}t} e^{i\mathbf{p}\cdot\mathbf{x}} + a_s^{\dagger}(\mathbf{p}) \overline{u_s(\mathbf{p})} e^{i\omega_{\mathbf{p}}t} e^{-i\mathbf{p}\cdot\mathbf{x}}) \frac{d^3\mathbf{p}}{\sqrt{2\omega_{\mathbf{p}}}}$$

If one uses commutation instead of anticommutation relations, the Hamiltonian operator will have eigenstates with arbitrarily negative energy, and there will be problems with causality due to observable operators at space-like separated points not commuting. These two problems are resolved by the use of anticommutation instead of commutation relations. The multiparticle state space for the theory has occupation numbers 0 or 1 for each value of $\mathbf{p}$ and for each value of $s = \pm$. Like the case of the real scalar field, the particles described by these states are their own antiparticles. Unlike the case of the real scalar field, each particle state has a $\mathbf{C}^2$ degree of freedom corresponding to its spin $\frac{1}{2}$ nature.

One can show that the Hamiltonian and momentum operators are given by

$$\widehat{H} = \int_{\mathbf{R}^3} \omega_{\mathbf{p}}(a_{+}^{\dagger}(\mathbf{p})a_{+}(\mathbf{p}) + a_{-}^{\dagger}(\mathbf{p})a_{-}(\mathbf{p}))d^3\mathbf{p}$$

and

$$\widehat{\mathbf{P}} = \int_{\mathbf{R}^3} \mathbf{p}(a_{+}^{\dagger}(\mathbf{p})a_{+}(\mathbf{p}) + a_{-}^{\dagger}(\mathbf{p})a_{-}(\mathbf{p}))d^3\mathbf{p}$$

The angular momentum and boost operators are much more complicated to describe, again due to the dependence of the $\alpha_{\pm}(\mathbf{p})$ on a choice of polarization vectors $u_{\pm}(\mathbf{p})$.

Note that, as in the case of the real scalar field, the theory of a single Majorana field has no internal symmetry group acting, so no way to introduce a charge operator and couple the theory to electromagnetic fields.

47.3 Weyl spinors

For the case $m = 0$ of the Dirac equation, it turns out that there is an interesting operator acting on the space of solutions:

Definition 3 (Chirality). *The operator*

$$\gamma_5 = i\gamma_0\gamma_1\gamma_2\gamma_3$$

is called the chirality operator. It has eigenvalues ± 1, and its eigenstates are said to have chirality ± 1. States with chirality $+1$ are called "right-handed," those with chirality -1 are called "left-handed."

Note that the operator $J_W = -i\gamma_5 = \gamma_0\gamma_1\gamma_2\gamma_3$ satisfies $J_W^2 = -1$ and provides a choice of complex structure on the space $\mathcal{V}$ of real-valued solutions of the Dirac equation. We can complexify such solutions and write

$$\mathcal{V} \otimes \mathbf{C} = \mathcal{V}_L \oplus \mathcal{V}_R \tag{47.15}$$

where $\mathcal{V}_L$ is the $+i$ eigenspace of J_W (the negative or left-handed chirality solutions), and $\mathcal{V}_R$ is the $-i$ eigenspace of J_W (the positive or right-handed chirality solutions).

To work with J_W eigenvectors, it is convenient to adopt a choice of γ-matrices in which γ_5 is diagonal. This cannot be done with real matrices but requires complexification. One such choice was already described in 41.2, the chiral or Weyl representation. In this choice, the γ-matrices can be written in 2 by 2 block form as

$$\gamma_0 = -i\begin{pmatrix} 0 & 1 \\ 1 & 0 \end{pmatrix}, \gamma_1 = -i\begin{pmatrix} 0 & \sigma_1 \\ -\sigma_1 & 0 \end{pmatrix}, \gamma_2 = -i\begin{pmatrix} 0 & \sigma_2 \\ -\sigma_2 & 0 \end{pmatrix}, \gamma_3 = -i\begin{pmatrix} 0 & \sigma_3 \\ -\sigma_3 & 0 \end{pmatrix}$$

and the chirality operator is diagonal

$$\gamma_5 = \begin{pmatrix} -1 & 0 \\ 0 & 1 \end{pmatrix}$$

We can thus write (complexified) solutions in terms of chiral eigenstates as

$$\psi = \begin{pmatrix} \psi_L \\ \psi_R \end{pmatrix}$$

where ψ_L and ψ_R are two-component wavefunctions, of left and right chirality, respectively.

The Dirac equation 47.2 is then

$$-i \begin{pmatrix} -im & \frac{\partial}{\partial t} + \boldsymbol{\sigma} \cdot \boldsymbol{\nabla} \\ \frac{\partial}{\partial t} - \boldsymbol{\sigma} \cdot \boldsymbol{\nabla} & -im \end{pmatrix} \begin{pmatrix} \psi_L \\ \psi_R \end{pmatrix} = 0$$

or, in terms of two-component functions

$$\left(\frac{\partial}{\partial t} + \boldsymbol{\sigma} \cdot \boldsymbol{\nabla} \right) \psi_R = im\psi_L$$

$$\left(\frac{\partial}{\partial t} - \boldsymbol{\sigma} \cdot \boldsymbol{\nabla} \right) \psi_L = im\psi_R$$

When $m = 0$ the equations decouple and one can consistently restrict attention to just right-handed or left-handed solutions, giving:

Definition 4 (Weyl equations). *The Weyl wave equations for two-component spinors are*

$$\left(\frac{\partial}{\partial t} + \boldsymbol{\sigma} \cdot \boldsymbol{\nabla} \right) \psi_R = 0 \tag{47.16}$$

$$\left(\frac{\partial}{\partial t} - \boldsymbol{\sigma} \cdot \boldsymbol{\nabla} \right) \psi_L = 0 \tag{47.17}$$

Also in the massless case, the chirality operator satisfies

$$[\gamma_5, H_D] = 0$$

(H_D is the Dirac Hamiltonian 47.5) since for each j

$$[\gamma_5, \gamma_0 \gamma_j] = 0$$

This follows from the fact that commuting γ_0 through γ_5 gives three minus signs, commuting γ_j through γ_5 gives another three. In this case, chirality is a conserved quantity, and the complex structure $J_W = -i\gamma_5$ commutes with H_D. J_W then takes positive energy solutions to positive energy solutions, negative energy to negative energy solutions, and thus commutes with the relativistic complex structure J_r.

We now have two commuting complex structures J_W and J_r on $\mathcal{V}$, and they can be simultaneously diagonalized (much like the situation in section 44.1.2). We get a decomposition

$$\mathcal{V}_{J_r}^+ = \mathcal{H}_1 = \mathcal{H}_{1,L} \oplus \mathcal{H}_{1,R}$$

of the positive energy solutions into $+i$ ($\mathcal{H}_{1,L}$) and $-i$ ($\mathcal{H}_{1,R}$) eigenspaces of J_W. Restricting to the solutions of equation 47.17 we get a decomposition

$$\mathcal{V}_L = \mathcal{H}_{1,L} \oplus \overline{\mathcal{H}}_{1,L}$$

into positive and negative energy left-handed solutions. We can then take Weyl spinor fields to be two-component objects

$$\Psi_L(\mathbf{x}), \overline{\Psi}_L(\mathbf{x})$$

that are continuous basis elements of $\mathcal{H}_{1,L}$ and $\overline{\mathcal{H}}_{1,L}$, respectively. The action of the (double cover of the) Poincaré group on space–time-dependent Weyl fields will be given by

$$\Psi_L(x) \to (a, \Lambda)\Psi_L(x) = S(\Lambda)^{-1}\Psi_L(\Lambda \cdot x + a)$$

where Λ is an element of $Spin(3,1)$, and $S(\Lambda) \in SL(2, \mathbf{C})$ is the $(\frac{1}{2}, 0)$ representation (see chapter 41, where $S(\Lambda) = \Omega$). $\Lambda \cdot x$ is the $Spin(3,1)$ action on Minkowski space vectors.

Just as in the Majorana case, parametrizing the space of solutions using fixed-time fields does not allow one to see the action of boosts on the fields. In addition, we know that relativistic field quantization requires use of the relativistic complex structure J_r, which is not simply expressed in terms of the fixed-time fields. To solve both problems, we need to study the solutions in momentum space. To find solutions in momentum space, we Fourier transform using

$$\psi(t, \mathbf{x}) = \frac{1}{(2\pi)^2} \int d^4p \; e^{i(-p_0 t + \mathbf{p} \cdot \mathbf{x})} \widetilde{\psi}(p_0, \mathbf{p})$$

and see that the Weyl equations are

$$(p_0 - \boldsymbol{\sigma} \cdot \mathbf{p})\widetilde{\psi}_R = 0$$

$$(p_0 + \boldsymbol{\sigma} \cdot \mathbf{p})\widetilde{\psi}_L = 0$$

Since

$$(p_0 + \boldsymbol{\sigma} \cdot \mathbf{p})(p_0 - \boldsymbol{\sigma} \cdot \mathbf{p}) = p_0^2 - (\boldsymbol{\sigma} \cdot \mathbf{p})^2 = p_0^2 - |\mathbf{p}|^2$$

both $\widetilde{\psi}_R$ and $\widetilde{\psi}_L$ satisfy

$$(p_0^2 - |\mathbf{p}|^2)\widetilde{\psi} = 0$$

so are functions with support on the positive ($p_0 = |\mathbf{p}|$) and negative ($p_0 = -|\mathbf{p}|$) energy null-cone. These are Fourier transforms of solutions to the massless Klein–Gordon equation

$$\left(-\frac{\partial^2}{\partial x_0^2} + \frac{\partial^2}{\partial x_1^2} + \frac{\partial^2}{\partial x_2^2} + \frac{\partial^2}{\partial x_3^2} \right) \psi = 0$$

In the two-component formalism, one can define:

Definition 5 (Helicity). *The operator*

$$\frac{1}{2} \frac{\boldsymbol{\sigma} \cdot \mathbf{p}}{|\mathbf{p}|} \tag{47.18}$$

on the space of solutions to the Weyl equations is called the helicity operator. It has eigenvalues $\pm\frac{1}{2}$, and its eigenstates are said to have helicity $\pm\frac{1}{2}$.

The helicity operator is the component of the spin operator $\mathbf{S} = \frac{1}{2}\boldsymbol{\sigma}$ along the direction of the momentum of a particle. Single-particle helicity eigenstates of eigenvalue $+\frac{1}{2}$ are said to have "right-handed helicity," and described as having spin in the same direction as their momentum, those with helicity eigenvalue $-\frac{1}{2}$ are said to have "left-handed helicity" and spin in the opposite direction to their momentum.

A continuous basis of solutions to the Weyl equation for ψ_L is given by the wavefunctions

$$u_L(\mathbf{p})e^{i(-p_0 x_0 + \mathbf{p}\cdot\mathbf{x})}$$

where the polarization vector $u_L(\mathbf{p}) \in \mathbf{C}^2$ satisfies

$$\boldsymbol{\sigma} \cdot \mathbf{p}\, u_L(\mathbf{p}) = -p_0 u_L(\mathbf{p})$$

Note that the $u_L(\mathbf{p})$ are the same basis elements of a specific $\mathbf{p}$-dependent $\mathbf{C} \subset \mathbf{C}^2$ subspace first seen in the case of the Bloch sphere in section 7.5, and later in the case of solutions to the Pauli equation (the $u_-(\mathbf{p})$ of equation 34.8 is $u_L(\mathbf{p})$ for positive energy, the $u_+(\mathbf{p})$ of equation 34.7 is $u_L(\mathbf{p})$ for negative energy). The positive energy ($p_0 = |\mathbf{p}|$) solutions have negative helicity, while the negative energy ($p_0 = -|\mathbf{p}|$) solutions have positive helicity. After quantization, this wave equation leads to a quantum field theory describing massless left-handed helicity particles and right-handed helicity antiparticles. Unlike the case of the Majorana field, the theory of the Weyl field comes with a non-trivial internal symmetry, due to the action of the group $U(1)$ on solutions by multiplication by a phase, and this allows the introduction of a charge operator.

Recall that our general analysis of irreducible representations of the Poincaré group in chapter 42 showed that we expected to find such representations by looking at functions on the positive and negative energy null-cones, with values in representations of $SO(2)$, the group of rotations preserving the vector $\mathbf{p}$. Acting on solutions to the Weyl equations, the generator of this group is given by the helicity operator (equation 47.18). The solution space to the Weyl equations provides the expected irreducible representations of helicity $\pm\frac{1}{2}$ and of either positive or negative energy.

47.4 Dirac spinors

In section 47.2, we saw that four-component real Majorana spinors could be used to describe neutral massive spin $\frac{1}{2}$ relativistic particles, while in section 47.3, we saw that with two-component complex Weyl spinors one could describe charged massless spin $\frac{1}{2}$ particles. To get a theory of charged massive spin $\frac{1}{2}$ particles, one needs to double the number of degrees of freedom, which one can do in two different ways, with equivalent results:

- In section 44.1.2, we saw that one could get a theory of charged scalar relativistic particles, by taking scalar fields valued in $\mathbf{R}^2$. One can do much the same thing for Majorana fields, having them take values in $\mathbf{R}^4 \otimes \mathbf{R}^2$ rather than $\mathbf{R}^4$. The theory will then, as in the scalar case, have an internal $SO(2) = U(1)$ symmetry by rotations of the $\mathbf{R}^2$ factor, a charge operator, and potential coupling to an electromagnetic field (using the covariant derivative). It will describe charged massive spin $\frac{1}{2}$ particles, with antiparticle states that are now distinguishable from particle states.

- Instead of a pair of Majorana spinor fields, one can take a pair ψ_L, ψ_R of Weyl fields, with opposite signs of eigenvalue for J_W. This will describe massive spin $\frac{1}{2}$ particles and antiparticles. The $U(1)$ symmetry acts in the same way on the ψ_L and the ψ_R, so they have the same charge. One can also consider the $U(1)$ action that is the inverse on ψ_L of that on ψ_R, but it is only in the case $m = 0$ that the corresponding charge commutes with the Hamiltonian (in which case, the theory is said to have an "axial symmetry").

The conventional starting point in physics textbooks is that of $\mathbf{C}^4$-valued spinor fields, a point of view we have avoided in order to keep straight the various ways in which complex numbers enter the theory. One should consult any quantum field theory textbook for an extensive discussion of this case, something we will not try and reproduce here. A standard topic in such textbooks is to show how the Pauli–Schrödinger theory of section 34 is recovered in the non-relativistic limit, where $|\mathbf{p}|/m$ goes to zero.

47.5 For further reading

The material of this chapter is discussed in detail in every textbook on relativistic quantum field theory. These discussions usually start with the massive Dirac spinor case, only later restricting to the massless case and Weyl spinors. They may or may not contain some discussion of the Majorana spinor case, usually described by imposing a condition on the Dirac case removing half the degrees of freedom (see, for instance, chapter 48 of [53]).

Chapter 48
An Introduction to the Standard Model

The theory of fundamental particles and their non-gravitational interactions is encapsulated by an extremely successful quantum field theory known as the Standard Model. This quantum field theory is determined by a particular set of quantum fields and a particular Hamiltonian, which we will outline in this chapter. It is an interacting quantum field theory, not solvable by the methods we have seen so far. In the non-interacting approximation, it includes just the sorts of free quantum field theories that we have studied in earlier chapters. This chapter gives only a very brief sketch of the definition of the theory; for details, one needs to consult a conventional particle physics textbook.

After outlining the basic structures of the Standard Model, we will indicate the major issues that it does not address, issues that one might hope will someday find a resolution through a better understanding of the mathematical structures that underlie this particular example of a quantum field theory.

48.1 Non-Abelian gauge fields

The Standard Model includes gauge fields for a $U(1) \times SU(2) \times SU(3)$ gauge group, with Hamiltonian given by the sum 46.10

$$h_{YM} = \sum_{j=1}^{3} \frac{1}{2} \int_{\mathbf{R}^3} tr_j(|\mathbf{E}_j|^2 + |\mathbf{B}_j|^2) d^3\mathbf{x} \tag{48.1}$$

The $\mathbf{E}_j$ and $\mathbf{B}_j$ take values in the Lie algebras $\mathfrak{u}(1), \mathfrak{su}(2), \mathfrak{su}(3)$ for $j = 1, 2, 3$, respectively. tr_j indicates a choice of an adjoint-invariant inner product for each of these Lie algebras, which could be defined in terms of the trace in some representation. Such an invariant inner product is unique up to a choice

© Peter Woit 2017

P. Woit, *Quantum Theory, Groups and Representations*,
DOI 10.1007/978-3-319-64612-1_48

of normalization, and this introduces three parameters into the theory, which we will call g_1, g_2, g_3.

Some method must be found to deal appropriately with the gauge-invariance problems associated with quantization of gauge fields discussed in section 46.6. Interacting non-Abelian gauge field theory remains incompletely understood outside of perturbation theory. The theory of interacting quantum fields shows that, to get a well-defined theory, one should think of the parameters g_j as being dependent on the distance scale at which the physics is being probed, and one can calculate the form of this scale-dependence. The $SU(2)$ and $SU(3)$ gauge field dynamics is "asymptotically free," meaning that g_2, g_3 can be defined so as to go to zero at short-distance scales, with behavior of the theory approaching that of a free field theory. This indicates that one should be able to consistently remove short-distance cutoffs necessary to define the theory, at least for those two terms in the Hamiltonian.

48.2 Fundamental fermions

The Standard Model includes both left-handed and right-handed Weyl spinor fields, each coming in multiple copies and transforming under the $U(1) \times SU(2) \times SU(3)$ group according to a very specific choice of representations. They are described by the following terms in the Hamiltonian

$$h = \int_{\mathbf{R}^3} \left(\sum_a (\Psi_{L,a}^\dagger \boldsymbol{\sigma} \cdot (\nabla - i\mathbf{A}_L)\Psi_{L,a}) - \sum_b (\Psi_{R,b}^\dagger \boldsymbol{\sigma} \cdot (\nabla - i\mathbf{A}_R)\Psi_{R,b}) \right) d^3\mathbf{x}$$

Here the left-handed fermions take values in three copies (called "generations") of the representation

$$(-1 \otimes \mathbf{2} \otimes \mathbf{1}) \oplus (\frac{1}{3} \otimes \mathbf{2} \otimes \mathbf{3})$$

while the right-handed fermions use three copies of

$$(-2 \otimes \mathbf{1} \otimes \mathbf{1}) \oplus (-\frac{2}{3} \otimes \mathbf{1} \otimes \mathbf{3}) \oplus (\frac{4}{3} \otimes \mathbf{1} \otimes \mathbf{3})$$

with the first term in each tensor product giving the representation of $U(1)$ (the fractions indicate a cover is needed), the second the representation of $SU(2)$, the third the representation of $SU(3)$. $\mathbf{1}$ is the trivial representation, $\mathbf{2}$ the defining representation for $SU(2)$, $\mathbf{3}$ that for $SU(3)$.

The $\mathbf{A}_R$ and $\mathbf{A}_R$ are vector potential fields acting by the representations given above, scaled by a constant g_j corresponding to the appropriate term in the gauge group.

48.3 Spontaneous symmetry breaking

The Higgs field is an $\mathbf{R}^4$-valued scalar field, with a complex structure chosen so that it can be taken to be $\mathbf{C}^2$-valued $\Phi(\mathbf{x})$, with $U(1)$ and $SU(2)$ acting by the defining representations. The Hamiltonian is

$$ h_{Higgs} = \int_{\mathbf{R}^3} (|\Pi|^2 + |(\nabla - i\mathbf{A})\Phi|^2 - m^2|\Phi|^2 + \lambda|\Phi|^4)d^3\mathbf{x} $$

where $\mathbf{A}$ are vector potential fields for $U(1) \otimes SU(2)$ acting in the defining representation, scaled by the coefficients g_1, g_2.

Note the unusual sign of the mass term and the existence of a quartic term. This means that the dynamics of such a theory cannot be analyzed by the methods we have used so far. Thinking of the mass term and quartic term as a potential energy for the field, for constant fields this will have a minimum at some nonzero values of the field. To analyze the physics, one shifts the field Φ by such a value, and approximates the theory by a quadratic expansion of the potential energy about that point. Such a theory will have states corresponding to a new scalar particle (the Higgs particle) but will also require a new analysis of the gauge symmetry, since gauge transformations act non-trivially on the space of minima of the potential energy. For how this "Anderson–Higgs mechanism" affects the physics, one should consult a standard textbook.

Finally, the Higgs field and the spinor fields are coupled by cubic terms called Yukawa terms, of a general form such as

$$ h_{Yukawa} = \int_{\mathbf{R}^3} M\Phi\Psi^\dagger\Psi d^3x $$

where $\Psi, \Psi^\dagger$ are the spinor fields, Φ the Higgs field, and M a complicated matrix. When one expresses this in terms of the shifted Higgs field, the constant term in the shift gives terms quadratic in the fermion fields. These determine the masses of the spin $\frac{1}{2}$ particles as well as the so-called "mixing angles" that appear in the coupling of these particles to the gauge fields.

48.4 Unanswered questions and speculative extensions

While the Standard Model has been hugely successful, with no conflicting experimental evidence yet found, it is not a fully satisfactory theory, leaving unanswered a short list of questions that one would expect a fundamental theory to address. These are:

48.4.1   Why these gauge groups and couplings?

We have seen that the theory has a $U(1) \times SU(2) \times SU(3)$ gauge group acting on it, and this motivates the introduction of gauge fields that take values
in the Lie algebra of this group. An obvious question is that of why this precise pattern of groups appears. When one introduces the Hamiltonian 48.1
for these gauge fields, one gets three different coupling constants g_1, g_2, g_3.
Why do these have their measured values? Such coupling constants should be
thought of as energy-scale dependent, and one of them (g_1) is not asymptotically free, raising the question of whether there is a short-distance problem
with the definition of this part of the theory.

One attempt to answer these questions is the idea of a "Grand Unified
Theory (GUT)," based on a large Lie group that includes $U(1) \times SU(2) \times$
$SU(3)$ as subgroups (typical examples are $SU(5)$ and $SO(10)$). The question
of "why this group?" remains, but in principle, one now only has one coupling
constant instead of three. A major problem with this idea is that it requires
introduction of some new fields (a new set of Higgs fields), with dynamics
designed to leave a low-energy $U(1) \times SU(2) \times SU(3)$ gauge symmetry. This
introduces a new set of problems in place of the original one of the three
coupling constants.

48.4.2   Why these representations?

We saw in section 48.2 that the fundamental left- and right-handed spin
$\frac{1}{2}$ fermionic fields carry specific representations under the $U(1) \times SU(2) \times$
$SU(3)$ gauge group. We would like some sort of explanation for this particular pattern. An additional question is whether there is a fundamental
right-handed neutrino, with the gauge groups acting trivially on it. Such a
field would have quanta that do not directly interact with the known non-
gravitational forces.

The strongest argument for the $SO(10)$ GUT scenario is that a distinguished representation of this group, the 16-dimensional spinor representation, restricts on the $U(1) \times SU(2) \times SU(3) \subset SO(10)$ subgroup to precisely
the representation corresponding to a single generation of fundamental fermions (including the right-handed neutrino as a trivial representation).

48.4.3   Why three generations?

The pattern of fundamental fermions occurs with a three-fold multiplicity,
the "generations." Why three? In principle, there could be other generations,

but these would have to have all their particles at masses too high to have been observed, including their neutrinos. These would be quite different from the known three generations, where the neutrino masses are light.

48.4.4   Why the Higgs field?

As described in section 48.3, the Higgs field is an elementary scalar field, transforming as the standard $\mathbf{C}^2$ representation of the $U(1) \times SU(2)$ part of the gauge group. As a scalar field, it has quite different properties and presumably a different origin than that of fundamental fermion and gauge fields, but what this might be remains a mystery. Besides the coupling to gauge fields, its dynamics is determined by its potential function, which depends on two parameters. Why do these have their measured values?

48.4.5   Why the Yukawas?

The fundamental fermion masses and mixing angles in the Standard Model are determined by Yukawa terms in the Hamiltonian coupling the Higgs field to the fermions. These terms involve matrices with a significant number of parameters and the origin of these parameters is unknown. This is related to the mystery of the Higgs field itself, with our understanding of the nature of the Higgs field not able to constrain these parameters.

48.4.6   What is the dynamics of the gravitational field?

Our understanding of gravitational forces in classical physics is based on Einstein's theory of general relativity, which has fundamental degrees of freedom that describe the geometry of space–time (which is no longer just Minkowski space–time). These degrees of freedom can be chosen so as to include a connection (called the "spin connection") and its curvature, much like the connection variables of gauge theory. The fields of the Standard Model can be consistently coupled to the space–time geometry by a minimal coupling prescription using the spin connection. The Hamiltonian with Einstein's equations as equations of motion is, however, not of the Yang–Mills form. Applying standard perturbation theory and renormalization methods to this Hamiltonian leads to problems with defining the theory (it is not asymptotically free). There are a number of proposals for how to deal with this problem and consistently handle quantization of the space–time degrees of freedom, but none so far have any compelling evidence in their favor.

48.5 For further reading

The details of the Standard Model are described in just about every text-book on high energy physics, and most modern textbooks (such as [67]) of relativistic quantum field theory.

Chapter 49
Further Topics

There is a long list of other topics that belong in a more complete discussion of the general subject of this volume. Some of these are standard topics which are well covered in many physics or mathematics textbooks. In this chapter, We will just give a short list of some of the most important things that have been left out.

Several of these have to do with quantum field theory:

- **Lower-dimensional quantum field theories**. The simplest examples of quantum field theories are those with just one dimension of space (and one dimension of time, so often described as "1 + 1"-dimensional). While it would have been pedagogically a good idea to first examine in detail this case, keeping the length of this volume under control led to the decision to not take the time to do this, but to directly go to the physical case of "3 +1"-dimensional theories. The case of two spatial dimensions is another lower-dimensional case with simpler behavior than the physical case.

- **Topological quantum field theories**. One can also formulate quantum field theories on arbitrary manifolds. An important class of such quantum field theories has Hamiltonian $H = 0$ and observables that only depend on the topology of the manifold. The observables of such "topological quantum field theories" provide new sorts of topological invariants of manifolds, and such theories are actively studied by mathematicians and physicists. We have already mentioned a simple supersymmetrical quantum mechanical model of this kind in section 33.3.

© Peter Woit 2017   627
P. Woit, *Quantum Theory, Groups and Representations*,
DOI 10.1007/978-3-319-64612-1_49

49.1 Connecting quantum theories to experimental results

Our emphasis has been on the fundamental mathematical structures that occur in quantum theory, but using these to derive results that can be compared to real-world experiments requires additional techniques. Such techniques are the main topics of typical standard physics textbooks dealing with quantum mechanics and quantum field theory. Among the most important are:

- **Scattering theory.** In the usual single-particle quantum mechanics, one can study solutions to the Schrödinger equation that in the far past and future correspond to free particle solutions, while interacting with a potential at some intermediate finite times. This corresponds to the situation analyzed experimentally through the study of scattering processes.

  In quantum field theory, one generalizes this to the case of "inelastic scattering," where particles are being produced as well as scattered. Such calculations are of central importance in high energy physics, where most experimental results come from colliding accelerated particles and studying these sorts of scattering and particle production processes.

- **Perturbation methods.** Rarely can one find exact solutions to quantum mechanical problems, so one needs to have at hand an array of approximation techniques. The most important is perturbation theory, the study of how to construct series expansions about exact solutions. This technique can be applied to a wide variety of situations, as long as the system in question is not too dramatically of a different nature than one for which an exact solution exists. In practice, this means that one studies in this manner Hamiltonians that consist of a quadratic term (and thus exactly solvable by the methods we have discussed) plus a higher-order term multiplied by a small parameter λ. Various methods are available to compute the terms in a power series solution of the theory about $\lambda = 0$, and such calculational methods are an important topic of most quantum mechanics and quantum field theory textbooks.

49.2 Other important mathematical physics topics

There are quite a few important mathematical topics which go beyond those discussed here, but which have significant connections to fundamental physical theories. These include:

- **Higher rank simple Lie groups**. The representation theory of groups like $SU(3)$ has many applications in physics, and is also a standard topic in the graduate-level mathematics curriculum, part of the general theory of finite dimensional representations of semi-simple Lie groups and Lie algebras. This theory uses various techniques to reduce the problem to the cases of $SU(2)$ and $U(1)$ that we have studied. Historically, the recognition of the approximate $SU(3)$ symmetry of the strong interactions (because of the relatively light masses of the up, down, and strange quarks) led to the first widespread use of more sophisticated representation theory techniques in the physics community.

- **Euclidean methods**. Quantum field theories, especially in the path integral formalism, are analytically best behaved in Euclidean rather than Minkowski space–time signature. Analytic continuation methods can then be used to extract the Minkowski space behavior from the Euclidean space formulation of the theory. Such analytic continuation methods, using a complexification of the Lorentz group, can be used to understand some very general properties of relativistic quantum field theories, including the spin-statistics and CPT theorems.

- **Conformal geometry and the conformal group**. For theories of massless particles, it is useful to study the group $SU(2,2)$ that acts on Minkowski space by conformal transformations, with the Poincaré group as a subgroup. The complexification of this is the group $SL(4, \mathbf{C})$. The complexification of (conformally compactified) Minkowski space turns out to be a well-known mathematical object, the Grassmannian manifold of complex two-dimensional subspaces of $\mathbf{C}^4$. The theory of twistors exploits this sort of geometry of $\mathbf{C}^4$, with spinor fields appearing in a "tautological" manner: a point of space–time is a $\mathbf{C}^2 \subset \mathbf{C}^4$, and the spinor field takes values in that $\mathbf{C}^2$.

- **Infinite dimensional groups**. We have seen that infinite dimensional gauge groups play an important role in physics, but unfortunately the representation theory of such groups is poorly understood. Much is known if one takes space to be one-dimensional. For periodic boundary conditions, such one-dimensional gauge groups are loop groups, groups of maps from the circle to a finite dimensional Lie group G. The Lie algebras of such groups are called affine Lie algebras, and their representation theory can be studied by a combination of relatively conventional mathematical methods and quantum field theory methods, with the anomaly phenomenon playing a crucial role. The infinite dimensional group of diffeomorphisms of the circle and its Lie algebra (the Virasoro algebra) also play a role in this context. From the two-dimensional space–time point of view, many such theories have an infinite dimensional group action corresponding to conformal transformations of the space–time. The study of such conformal field theories is an important topic in mathematical physics, with representation theory methods a central part of that subject.

Appendix A
Conventions

I have attempted to stay close to the conventions used in the physics litera-ture, leading to the choices listed here. Most of the time, units are chosen so that $\hbar = c = 1$.

A.1   Bilinear forms

Parentheses

$$(\cdot, \cdot)$$

will be used for a non-degenerate symmetric bilinear form (inner product) on a vector space $\mathbf{R}^n$, with the same symbol also used for the complex linear extension to a bilinear form on $\mathbf{C}^n$. The group of linear transformations preserving the inner product will be $O(n)$ or $O(n, \mathbf{C})$, respectively.

Angle brackets

$$\langle \cdot, \cdot \rangle$$

will be used for non-degenerate symmetric sesquilinear forms (Hermitian inner products) on vector spaces $\mathbf{C}^n$. These are antilinear in the first entry and linear in the second. When the inner product is positive, it will be pre-served by the group $U(n)$, but the indefinite case with $n = 2d$ and group $U(d, d)$ will also occur. The quantum mechanical state space comes with such a positive Hermitian inner product, but may be infinite dimensional and a Hilbert space.

Non-degenerate antisymmetric forms (symplectic forms) will be denoted by

$$\omega(\cdot, \cdot) \quad \text{or} \quad \Omega(\cdot, \cdot)$$

with the first used for the symplectic form on a real phase space M of dimen-sion $2d$ and the second for the corresponding form on the dual phase space

© Peter Woit 2017
P. Woit, *Quantum Theory, Groups and Representations*,
DOI 10.1007/978-3-319-64612-1

$\mathcal{M}$. The group preserving these bilinear forms is $Sp(2d, \mathbf{R})$. The same symbol will also be used for the complex linear extension of these forms to $\mathbf{C}^{2d}$, where the bilinear form is preserved by the group $Sp(2d, \mathbf{C})$.

A.2   Fourier transforms

The Fourier transform is defined by

$$\widetilde{f}(k) = \frac{1}{\sqrt{2\pi}} \int_{-\infty}^{+\infty} f(q) e^{-ikq} dk$$

except for the case of functions of a time variable t, for which we use the opposite sign in the exponent (and a $\widehat{}$ instead of $\widetilde{}$), i.e.,

$$\widehat{f}(\omega) = \frac{1}{\sqrt{2\pi}} \int_{-\infty}^{+\infty} f(t) e^{i\omega t} d\omega$$

A.3   Symplectic geometry and quantization

The Lie bracket on the space of functions on phase space M is given by the Poisson bracket, determined by

$$\{q, p\} = 1$$

Quantization takes $1, q, p$ to self-adjoint operators $\mathbf{1}, Q, P$. To make this a unitary representation of the Heisenberg Lie algebra $\mathfrak{h}_3$, multiply the self-adjoint operators by $-i$, so they satisfy

$$[-iQ, -iP] = -i\mathbf{1}, \quad \text{or} \quad [Q, P] = i\mathbf{1}$$

In other words, our quantization map is the unitary representation of $\mathfrak{h}_3$ that satisfies

$$\Gamma'(q) = -iQ, \quad \Gamma'(p) = -iP, \quad \Gamma'(1) = -i\mathbf{1}$$

Dynamics is determined classically by the Hamiltonian function h as follows

$$\frac{d}{dt} f = \{f, h\}$$

After quantization, this becomes the equation

$$\frac{d}{dt}\mathcal{O}(t) = [\mathcal{O}, -iH]$$

for the dynamics of Heisenberg picture operators, which implies

$$\mathcal{O}(t) = e^{itH}\mathcal{O}e^{-itH}$$

where $\mathcal{O}$ is the Schrödinger picture operator. In the Schrödinger picture, states evolve according to the Schrödinger equation

$$-iH|\psi\rangle = \frac{d}{dt}|\psi\rangle$$

If a group G acts on a space M, the representation one gets on functions on M is given by

$$\pi(g)(f(x)) = f(g^{-1} \cdot x)$$

Examples include

- Space translation $(q \to q + a)$. On states, one has

$$|\psi\rangle \to e^{-iaP}|\psi\rangle$$

which in the Schrödinger representation is

$$e^{-ia(-i\frac{d}{dq})}\psi(q) = e^{-a\frac{d}{dq}}\psi(q) = \psi(q - a)$$

So, the Lie algebra action is given by the operator $-iP = -\frac{d}{dq}$. Note that this has opposite sign to the time translation. On operators, one has

$$\mathcal{O}(a) = e^{iaP}\mathcal{O}e^{-iaP}$$

or infinitesimally

$$\frac{d}{da}\mathcal{O}(a) = [\mathcal{O}, -iP]$$

- The classical expressions for angular momentum quadratic in q_j, p_j, for example

$$l_1 = q_2 p_3 - q_3 p_2$$

under quantization go to the self-adjoint operator

$$L_1 = Q_2 P_3 - Q_3 P_2$$

and $-iL_1$ will be the skew-adjoint operator giving a unitary representation of the Lie algebra $\mathfrak{so}(3)$. The three such operators will satisfy the Lie bracket relations of $\mathfrak{so}(3)$, for instance

$$[-iL_1, -iL_2] = -iL_3$$

A.4 Complex structures and Bargmann–Fock quantization

We define complex coordinates on phase space by

$$z_j = \frac{1}{\sqrt{2}}(q_j - ip_j), \quad \overline{z}_j = \frac{1}{\sqrt{2}}(q_j + ip_j)$$

The standard choice of complex structure on phase space M is given by

$$J_0 \frac{\partial}{\partial q_j} = -\frac{\partial}{\partial p_j}, \quad J_0 \frac{\partial}{\partial p_j} = \frac{\partial}{\partial q_j}$$

and on coordinate basis vectors q_j, p_j of the dual space $\mathcal{M}$ by

$$J_0 q_j = p_j, \quad J_0 p_j = -q_j$$

The complex coordinates satisfy

$$J_0 z_j = iz_j, \quad J_0 \overline{z}_j = -i\overline{z}_j$$

so the z_j are a basis of $\mathcal{M}_{J_0}^+$, the $\overline{z}_j$ of $\mathcal{M}_{J_0}^-$. They have Poisson brackets

$$\{z_j, \overline{z}_k\} = i\delta_{jk}$$

In the Bargmann–Fock quantization, the state space is taken to be polynomials in the z_j, with annihilation and creation operators

$$a_j = \frac{\partial}{\partial z_j}, \quad a_j^\dagger = z_j$$

A.5 Special relativity

The "mostly plus" convention for the Minkowski inner product is used, so four-vectors $x = (x_0, x_1, x_2, x_3)$ satisfy

$$(x, x) = ||x||^2 = -x_0^2 + x_1^2 + x_2^2 + x_3^2$$

The relativistic energy–momentum relation is then

$$p^2 = -E^2 + ||\mathbf{p}||^2 = -m^2$$

A.6 Clifford algebras and spinors

The Clifford algebra associated with an inner product $(\cdot, \cdot)$ satisfies the relation

$$uv + vu = 2(u, v)$$

With the choice of signature for the Minkowski inner product above, the Clifford algebra is isomorphic to a real matrix algebra

$$\text{Cliff}(3, 1) = M(4, \mathbf{R})$$

Under this isomorphism, basis elements of Minkowski space correspond to γ matrices, which satisfy

$$\gamma_0^2 = -1, \quad \gamma_1^2 = \gamma_2^2 = \gamma_3^2 = 1$$

Explicit choices of these matrices are described in sections 41.2, 47.2, and 47.3.

The Dirac equation is taken to be

$$(\slashed{\partial} - m)\psi(x) = 0$$

see section 47.1.

Appendix B
Exercises

B.1 Chapters 1 and 2

Problem 1:

Consider the group S_3 of permutations of 3 objects. This group acts on the set of 3 elements. Consider the representation $(\pi, \mathbf{C}^3)$ this gives on the vector space $\mathbf{C}^3$ of complex-valued functions on the set of 3 elements (as defined in section 1.3.2). Choose a basis of this set of functions, and find the matrices $\pi(g)$ for each element $g \in S_3$.

Is this representation irreducible? If not, can you give its decomposition into irreducibles and find a basis in which the representation matrices are block diagonal?

Problem 2:

Use a similar argument to that of theorem 2.3 for $G = U(1)$ to classify the irreducible differentiable representations of the group $\mathbf{R}$ under the group law of addition. Which of these are unitary?

Problem 3:

Consider the group $SO(2)$ of 2 by 2 real orthogonal matrices of determinant one. What are the complex irreducible representations of this group? (A hint: how are $SO(2)$ and $U(1)$ related?)

There is an obvious representation of $SO(2)$ on $\mathbf{R}^2$ given by matrix multiplication on real 2-vectors. If you replace the real 2-vectors by complex 2-vectors, but use the same representation matrices, you get a two-complex dimensional representation (this is called "complexification"). How does this decompose as a direct sum of irreducibles?

Problem 4:

Consider a quantum mechanical system with state space $\mathcal{H} = \mathbf{C}^3$ and Hamiltonian operator

© Peter Woit 2017
P. Woit, *Quantum Theory, Groups and Representations*,
DOI 10.1007/978-3-319-64612-1

$$H = \begin{pmatrix} 0 & 1 & 0 \\ 1 & 0 & 0 \\ 0 & 0 & 2 \end{pmatrix}$$

Solve the Schrödinger equation for this system to find its state vector $|\Psi(t)\rangle$ at any time $t > 0$, given that the state vector at $t = 0$ was

$$\begin{pmatrix} \psi_1 \\ \psi_2 \\ \psi_3 \end{pmatrix}$$

with $\psi_i \in \mathbf{C}$.

B.2   Chapters 3 and 4

Problem 1:
   Calculate the exponential e^{tM} for

$$\begin{pmatrix} 0 & \pi & 0 \\ -\pi & 0 & 0 \\ 0 & 0 & 0 \end{pmatrix}$$

by two different methods:

- Diagonalize the matrix M (i.e., write as PDP^{-1}, for D diagonal) and then show that
$$e^{tPDP^{-1}} = Pe^{tD}P^{-1}$$
  and use this to compute e^{tM}.

- Calculate e^{tM} using the Taylor series expansion for the exponential, as well as the series expansions for the sine and cosine.

Problem 2:
   Consider a two-state quantum system, with Hamiltonian

$$H = -B_x \sigma_1$$

(this is the Hamiltonian for a spin $\frac{1}{2}$ system subjected to a magnetic field in the x-direction).

- Find the eigenvectors and eigenvalues of H. What are the possible energies that can occur in this quantum system?

- If the system starts out at time $t = 0$ in the state

$$|\psi(0)\rangle = \begin{pmatrix} 1 \\ 0 \end{pmatrix}$$

(i.e., spin "up"), find the state at later times.

Problem 3:

By using the fact that any unitary matrix can be diagonalized by conjugation by a unitary matrix, show that all unitary matrices can be written as e^X, for X a skew-adjoint matrix in $\mathfrak{u}(n)$.

By contrast, show that

$$A = \begin{pmatrix} -1 & 1 \\ 0 & -1 \end{pmatrix}$$

is in the group $SL(2, \mathbf{C})$, but is not of the form e^X for any $X \in \mathfrak{sl}(2, \mathbf{C})$ (this Lie algebra is all 2 by 2 matrices with trace zero).

Hint: For 2 by 2 matrices X, one can show (this is the Cayley–Hamilton theorem: Matrices X satisfy their own characteristic equation $\det(\lambda \mathbf{1} - X) = 0$, and for 2 by 2 matrices, this equation is $\lambda^2 - tr(X)\lambda + \det(X) = 0$)

$$X^2 - tr(X)X + \det(X)\mathbf{1} = 0$$

For $X \in \mathfrak{sl}(2, \mathbf{C})$, $tr(X) = 0$, so here $X^2 = -\det(X)\mathbf{1}$. Use this to show that

$$e^X = \cos(\sqrt{\det(X)})\mathbf{1} + \frac{\sin(\sqrt{\det(X)})}{\sqrt{\det(X)}}X$$

Try to use this for $e^X = A$ and derive a contradiction (taking the trace of the equation, what is $\cos(\sqrt{\det(X)})$?)

Problem 4:

- Show that M is an orthogonal matrix iff its rows are orthonormal vectors for the standard inner product (this is also true for the columns).

- Show that M is a unitary matrix iff its columns are orthonormal vectors for the standard Hermitian inner product (this is also true for the rows).

B.3   Chapters 5 to 7

Problem 1:

On the Lie algebras $\mathfrak{g} = \mathfrak{su}(2)$ and $\mathfrak{g} = \mathfrak{so}(3)$, one can define the Killing form $K(\cdot, \cdot)$ by

$$(X, Y) \in \mathfrak{g} \times \mathfrak{g} \to K(X, Y) = tr(XY)$$

1. For both Lie algebras, show that this gives a bilinear, symmetric form, negative definite, with the basis vectors X_j in one case and l_j in the other providing an orthogonal basis if one uses $K(\cdot,\cdot)$ as an inner product.

2. Another possible way to define the Killing form is as

$$K'(X,Y) = tr(ad(X) \circ ad(Y))$$

Here, the Lie algebra adjoint representation $(ad, \mathfrak{g})$ gives for each $X \in \mathfrak{g}$ a linear map

$$ad(X) : \mathbf{R}^3 \to \mathbf{R}^3$$

and thus a 3 by 3 real matrix. This K' is determined by taking the trace of the product of two such matrices. How are K and K' related?

Problem 2:
Under the homomorphism

$$\Phi : Sp(1) \to SO(3)$$

of section 6.2.3, what elements of $SO(3)$ do the quaternions $\mathbf{i}, \mathbf{j}, \mathbf{k}$ (unit length, so elements of $Sp(1)$) correspond to? Note that this is not the same question as that of evaluating Φ' on $\mathbf{i}, \mathbf{j}, \mathbf{k}$.

Problem 3:
In special relativity, we consider space and time together as $\mathbf{R}^4$, with an inner product such that $(v, v) = -v_0^2 + v_1^2 + v_2^2 + v_3^2$, where $v = (v_0, v_1, v_2, v_3) \in \mathbf{R}^4$. The group of linear transformations of determinant one preserving this inner product is written $SO(3,1)$ and known as the Lorentz group. Show that, just as $SO(4)$ has a double cover $Spin(4) = Sp(1) \times Sp(1)$, the Lorentz group has a double cover $SL(2, \mathbf{C})$, with action on vectors given by identifying $\mathbf{R}^4$ with 2 by 2 Hermitian matrices according to

$$(v_0, v_1, v_2, v_3) \leftrightarrow \begin{pmatrix} v_0 + v_3 & v_1 - iv_2 \\ v_1 + iv_2 & v_0 - v_3 \end{pmatrix}$$

and using the conjugation action of $SL(2, \mathbf{C})$ on these matrices (*hint*: use determinants).

Note that the Lorentz group has a spinor representation, but it is not unitary.

Problem 4:
Consider a spin $\frac{1}{2}$ particle, with a state $|\psi(t)\rangle$ evolving in time under the influence of a magnetic field of strength $B = |\mathbf{B}|$ in the 3-direction. If the state is an eigenvector for S_1 at $t = 0$, what are the expectation values

$$\langle\psi(t)|S_j|\psi(t)\rangle$$

at later times for the observables S_j (recall that $S_j = \frac{\sigma_j}{2}$)?

B.4   Chapter 8

Problem 1:
Using the definition

$$\langle f,g\rangle = \frac{1}{\pi^2}\int_{\mathbf{C}^2}\overline{f(z_1,z_2)}g(z_1,z_2)e^{-(|z_1|^2+|z_2|^2)}dx_1dy_1dx_2dy_2$$

for an inner product on polynomials on homogeneous polynomials on $\mathbf{C}^2$

- Show that the representation π on such polynomials given in section 8.2 (induced from the $SU(2)$ representation on $\mathbf{C}^2$) is a unitary representation with respect to this inner product.

- Show that the monomials

$$\frac{z_1^j z_2^k}{\sqrt{j!k!}}$$

  are orthonormal with respect to this inner product (*hint*: break up the integrals into integrals over the two complex planes and use polar coordinates).

- Show that the differential operator $\pi'(S_3)$ is self-adjoint. Show that $\pi'(S_-)$ and $\pi'(S_+)$ are adjoints of each other.

Problem 2:
Using the formulas for the $Y_1^m(\theta,\phi)$ and the inner product of equation 8.3, show that

- The Y_1^1, Y_1^0, Y_1^{-1} are orthonormal.
- Y_1^1 is a highest weight vector.
- Y_1^0 and Y_1^{-1} can be found by repeatedly applying L_- to a highest weight vector.

Problem 3:
Recall that the Casimir operator L^2 of $\mathfrak{so}(3)$ is the operator that in any representation ρ is given by

$$L^2 = L_1^2 + L_2^2 + L_3^2$$

Show that this operator commutes with the $\rho'(X)$ for all $X \in \mathfrak{so}(3)$. Use this to show that L^2 has the same eigenvalue on all vectors in an irreducible representation of $\mathfrak{so}(3)$.

Problem 4:

For the case of the $SU(2)$ representation π on polynomials on $\mathbf{C}^2$ given in the notes, find the Casimir operator

$$L^2 = \pi'(S_1)\pi'(S_1) + \pi'(S_2)\pi'(S_2) + \pi'(S_3)\pi'(S_3)$$

as an explicit differential operator. Show that homogeneous polynomials are eigenfunctions, and calculate the eigenvalues.

B.5   Chapter 9

Problem 1:

Consider the action of $SU(2)$ on the tensor product $V^1 \otimes V^1$ of two spin representations. According to the Clebsch–Gordan decomposition, this breaks up into irreducibles as $V^0 \oplus V^2$.

1. Show that

$$\frac{1}{\sqrt{2}}\left(\begin{pmatrix} 1 \\ 0 \end{pmatrix} \otimes \begin{pmatrix} 0 \\ 1 \end{pmatrix} - \begin{pmatrix} 0 \\ 1 \end{pmatrix} \otimes \begin{pmatrix} 1 \\ 0 \end{pmatrix} \right)$$

   is a basis of the V^0 component of the tensor product, by computing first the action of $SU(2)$ on this vector and then the action of $\mathfrak{su}(2)$ on the vector (i.e., compute the action of $\pi'(X)$ on this vector, for π the tensor product representation and X basis elements of $\mathfrak{su}(2)$).

2. Show that

$$\begin{pmatrix} 1 \\ 0 \end{pmatrix} \otimes \begin{pmatrix} 1 \\ 0 \end{pmatrix}, \frac{1}{\sqrt{2}}\left(\begin{pmatrix} 1 \\ 0 \end{pmatrix} \otimes \begin{pmatrix} 0 \\ 1 \end{pmatrix} + \begin{pmatrix} 0 \\ 1 \end{pmatrix} \otimes \begin{pmatrix} 1 \\ 0 \end{pmatrix} \right), \begin{pmatrix} 0 \\ 1 \end{pmatrix} \otimes \begin{pmatrix} 0 \\ 1 \end{pmatrix}$$

   give a basis for the irreducible representation V^2, by showing that they are eigenvectors of $\pi'(S_3)$ with the right eigenvalues (weights) and computing the action of the raising and lowering operators for $\mathfrak{su}(2)$ on these vectors.

Problem 2:

Prove that the algebra $S^*(V^*)$ is isomorphic to the algebra of polynomial functions on the vector space V.

B.6 Chapters 10 to 12

Problem 1:
Consider a quantum system describing a free particle in one spatial dimension, of size L (the wavefunction satisfies $\psi(q,t) = \psi(q+L,t)$). If the wavefunction at time $t=0$ is given by

$$\psi(q,0) = C\left(\sin\left(\frac{6\pi}{L}q\right) + \cos\left(\frac{4\pi}{L}q + \phi_0\right)\right)$$

where C is a constant and ϕ_0 is an angle, find the wavefunction for all t. For what values of C is this a normalized wavefunction ($\int |\psi(q,t)|^2 dq = 1$)?

Problem 2:
Consider a state at $t=0$ of the one-dimensional free particle quantum system given by a Gaussian peaked at $q=0$

$$\psi(q,0) = \sqrt{\frac{C}{\pi}}e^{-Cq^2}$$

where C is a real positive constant.

Show that the wavefunction $\psi(q,t)$ for $t>0$ remains a Gaussian, but one with an increasing width.

Now consider the case of an initial state $\psi(q,0)$ with Fourier transform peaked at $k=k_0$

$$\tilde{\psi}(k,0) = \sqrt{\frac{C}{\pi}}e^{-C(k-k_0)^2}$$

What is the initial wavefunction $\psi(q,0)$?

Show that at later times $|\psi(q,t)|^2$ is peaked about a point that moves with velocity $\frac{\hbar k}{m}$.

Problem 3:
Show that the limit as $T \to 0$ of the propagator

$$U(T, q_T - q_0) = \sqrt{\frac{m}{i2\pi T}}e^{-\frac{m}{i2T}(q_T-q_0)^2}$$

is a δ-function distribution.

Problem 4:
Use the Cauchy integral formula method of section 12.6 to derive equation 12.9 for the propagator from equation 12.13.

Problem 5:
In chapter 10, we described the quantum system of a free non-relativistic particle of mass m in $\mathbf{R}^3$. Using tensor products, how would you describe a

system of two identical such particles? Find the Hamiltonian and momentum operators. Find a basis for the energy and momentum eigenstates for such a system, first under the assumption that the particles are bosons and then under the assumption that the particles are fermions.

B.7   Chapters 14 to 16

Problem 1:

Consider a particle moving in two dimensions, with the Hamiltonian function

$$h = \frac{1}{2m}((p_1 - Bq_2)^2 + p_2^2)$$

- Find the vector field X_h associated with this function.
- Show that the quantities

$$p_1 \quad \text{and} \quad p_2 - Bq_1$$

are conserved.

- Write down Hamilton's equations for this system, and find the general solutions for the trajectories $(q(t), p(t))$.

This system describes a particle moving in a plane, experiencing a magnetic field orthogonal to the plane. You should find that the trajectories are circles in the plane, with a frequency called the Larmor frequency.

Problem 2:

Consider the action of the group $SO(3)$ on phase space $\mathbf{R}^6$ by simultaneously rotating position and momentum vectors.

- For the three basis elements l_j of $\mathfrak{so}(3)$, show that the momentum map gives functions μ_{l_j} that are just the components of the angular momentum.
- Show that the maps

$$l_j \in \mathfrak{so}(3) \to \mu_{l_j}$$

give a Lie algebra homomorphism from $\mathfrak{so}(3)$ to the Lie algebra of functions on phase space (with Lie bracket on such functions the Poisson bracket).

Problem 3:

In the same context as problem 2, compute the Poisson brackets

$$\{\mu_{l_j}, q_k\}$$

between the angular momentum functions μ_{l_j} and the configuration space coordinates q_j. Compare this calculation to the calculation of

$$\pi'(l_j)\mathbf{e}_k$$

for π the spin-1 representation of $SO(3)$ on $\mathbf{R}^3$ (the vector representation).

Problem 4:

Consider the symplectic group $Sp(2d, \mathbf{R})$ of linear transformations of phase space $\mathbf{R}^{2d}$ that preserve Ω.

- Consider the group of linear transformations of phase space $\mathbf{R}^{2d}$ that act in the same way on positions and momenta, preserving the standard inner products on position and momentum space. Show that this group is a subgroup of $Sp(2d, \mathbf{R})$, isomorphic to $O(d)$.

- Using the identification between $Sp(2d, \mathbf{R})$ and matrices satisfying equation 16.10, which matrices give the subgroup above?

- Again in terms of matrices, what is the Lie algebra of this subgroup?

- Identifying the Lie algebra of $Sp(2d, \mathbf{R})$ with quadratic functions of the coordinates and momenta, which such quadratic functions are in the Lie algebra of the $SO(d)$ subgroup?

- Consider the function

$$\frac{1}{2}\sum_{j=1}^{d}(q_j^2 + p_j^2)$$

What matrix does this correspond to as an element of the Lie algebra of $Sp(2d, \mathbf{R})$? Show that one gets an $SO(2)$ subgroup of $Sp(2d, \mathbf{R})$ by taking exponentials of this matrix. Is this $SO(2)$ a subgroup of the $SO(d)$ above?

B.8   Chapter 17

Problem 1:

This is part of a proof of the Groenewold–van Hove theorem.

- Show that one can write $q^2 p^2$ in two ways as a Poisson bracket

$$q^2 p^2 = \frac{1}{3}\{q^2 p, p^2 q\} = \frac{1}{9}\{q^3, p^3\}$$

- Assume that we can quantize any polynomial in q, p by a Lie algebra homomorphism π' that takes polynomials in q, p with the Poisson bracket to polynomials in Q, P with the commutator, in a way that extends the standard Schrödinger representation

$$\pi'(q) = -iQ, \quad \pi'(p) = -iP, \quad \pi'(1) = -i\mathbf{1}$$

Further assume that the following relations are satisfied for low-degree polynomials (it actually is possible to prove that these are necessary):

$$\pi'(qp) = -i\frac{1}{2}(QP + PQ), \quad \pi'(q^2) = -iQ^2, \quad \pi'(p^2) = -iP^2$$

$$\pi'(q^3) = -iQ^3, \quad \pi'(p^3) = -iP^3$$

Then, show that

$$\pi'(q^2p) = -i\frac{1}{2}(Q^2P + PQ^2)$$

(Hint: use $\{q^3, p^2\} = 6q^2p$)

- Also show that
$$\pi'(qp^2) = -i\frac{1}{2}(QP^2 + P^2Q)$$

- Finally, show that

$$\pi'\left(\frac{1}{9}\{q^3, p^3\}\right) = -i\left(-\frac{2}{3}\mathbf{1} - 2iQP + Q^2P^2\right)$$

and

$$\pi'\left(\frac{1}{3}\{q^2p, p^2q\}\right) = -i\left(-\frac{1}{3}\mathbf{1} - 2iQP + Q^2P^2\right)$$

which demonstrates a contradiction.

B.9 Chapters 18 and 19

Problem 1:
Starting with the Lie algebra $\mathfrak{so}(3)$, with basis l_1, l_2, l_3, consider new basis elements given by leaving l_3 alone and rescaling

$$l_1 \to l_1/R, \quad l_2 \to -l_2/R$$

where R is a real parameter. Show that in the limit $R \to \infty$ these new basis elements satisfy the Lie bracket relations for the Lie algebra of $E(2)$. Because of this, the group $E(2)$ is sometimes said to be a "contraction" of $SO(3)$.

Problem 2:
For the case of the group $E(2)$, show that in any representation π' of its Lie algebra, there is a Casimir operator

$$|\mathbf{P}|^2 = \pi'(p_1)\pi'(p_1) + \pi'(p_2)\pi'(p_2)$$

that commutes with all the Lie algebra representation operators (i.e., with $\pi'(p_1)$, $\pi'(p_2)$, $\pi'(l)$).

For the case of the group $E(3)$, similarly show that there are two Casimir operators.

$$|\mathbf{P}|^2 = \pi'(p_1)\pi'(p_1) + \pi'(p_2)\pi'(p_2) + \pi'(p_3)\pi'(p_3)$$

and

$$\mathbf{L} \cdot \mathbf{P} = \pi'(l_1)\pi'(p_1) + \pi'(l_2)\pi'(p_2) + \pi'(l_3)\pi'(p_3)$$

that commute with all the Lie algebra representation operators.

Problem 3:
Show that the $E(3)$ Casimir operator $\mathbf{L} \cdot \mathbf{P}$ acts trivially on the $E(3)$ representation on free particle wave functions of energy $E > 0$.

B.10   Chapters 21 and 22

Problem 1:
Consider the classical Hamiltonian function for a particle moving in a central potential

$$h = \frac{1}{2m}(p_1^2 + p_2^2 + p_3^2) + V(r)$$

where

$$r^2 = q_1^2 + q_2^2 + q_3^2$$

- Show that the angular momentum functions l_j satisfy

$$\{l_j, h\} = 0$$

and note that this implies that the l_j are conserved functions along classical trajectories.

- Show that in the quantized theory, the angular momentum operators and the $SO(3)$ Casimir operator satisfy

$$[L_j, H] = 0, \quad [L^2, H] = 0$$

- Show that for a fixed energy E, the subspace $\mathcal{H}_E \subset \mathcal{H}$ of states of energy E will be a Lie algebra representation of $SO(3)$. Decomposing into irreducibles, this can be characterized by the various spin values l that occur, together with their multiplicity.

- Show that if a state of energy E lies in a spin-l irreducible representation of $SO(3)$ at time $t = 0$, it will remain in a spin-l irreducible representation at later times.

Problem 2:

   If

$$\mathbf{w} = \frac{1}{m}(\mathbf{l} \times \mathbf{p}) + e^2 \frac{\mathbf{q}}{|\mathbf{q}|}$$

is the Lenz vector, show that its components satisfy

$$\{w_j, h\} = 0$$

for the Hydrogen atom Hamiltonian h.

Problem 3:

   For the one-dimensional quantum harmonic oscillator:

- Compute the expectation values in the energy eigenstate $|n\rangle$ of the following operators

$$Q, \ P, \ Q^2, \ P^2$$

   and

$$Q^4$$

- Use these to find the standard deviations in the statistical distributions of observed values of q and p in these states. These are

$$\Delta Q = \sqrt{\langle n|Q^2|n\rangle - \langle n|Q|n\rangle^2}, \quad \Delta P = \sqrt{\langle n|P^2|n\rangle - \langle n|P|n\rangle^2}$$

- For two energy eigenstates $|n\rangle$ and $|n'\rangle$, find

$$\langle n'|Q|n\rangle \text{ and } \langle n'|P|n\rangle$$

Problem 4:

   Show that the functions $1, z, \bar{z}, z\bar{z}$ of section 22.4 give a basis of a Lie algebra (with Lie bracket the Poisson bracket of that section). Show that this is a semi-direct product Lie algebra and that the harmonic oscillator state space gives a unitary representation of this Lie algebra.

B.11   Chapter 23

Problem 1:
  For the coherent state $|\alpha\rangle$, compute

$$\langle \alpha | Q | \alpha \rangle$$

and

$$\langle \alpha | P | \alpha \rangle$$

Show that coherent states are not eigenstates of the number operator $N = a^\dagger a$ and compute

$$\langle \alpha | N | \alpha \rangle$$

Problem 2:
  Show that the propagator 23.12 for the harmonic oscillator satisfies the Schrödinger equation for the harmonic oscillator Hamiltonian.

B.12   Chapters 24 to 26

Problem 1:
  Consider the harmonic oscillator in two dimensions, with the Hamiltonian

$$H = \frac{1}{2m}(P_1^2 + P_2^2) + \frac{1}{2}m\omega^2(Q_1^2 + Q_2^2)$$

There are two different $U(1) = SO(2)$ groups acting on the phase space of this system as symmetries, with corresponding operators:

- The rotation action on position space, with a simultaneous rotation action on momentum space. The operator here will be the $d = 2$ angular momentum operator $Q_1 P_2 - Q_2 P_1$.

- Simultaneous rotations in the q_1, p_1 and q_2, p_2 planes. The operator here will be the Hamiltonian.

For each case, the state space $\mathcal{H} = \mathcal{F}_2$ will be a representation of the group $U(1) = SO(2)$. For each energy eigenspace, which irreducible representations (weights) occur? What are the corresponding joint eigenfunctions of the two operators?

Problem 2:
Consider the harmonic oscillator in three dimensions, with the Hamiltonian

$$H = \frac{1}{2m}(P_1^2 + P_2^2 + P_3^2) + \frac{1}{2}m\omega^2(Q_1^2 + Q_2^2 + Q_3^2)$$

- The group $SO(3)$ acts on the system by rotations of the position space $\mathbf{R}^3$, and the corresponding Lie algebra action on the state space $\mathcal{F}_3$ is given in section 25.4.2 as the operators

$$U'_{l_1}, U'_{l_2}, U'_{l_3}$$

  Exponentiating to get an $SO(3)$ representation by operators $U(g)$, show that acting by such operators on the a_j by conjugation

$$a_j \rightarrow U(g)a_j U(g)^{-1}$$

  one gets the same action as the standard action of a rotation on coordinates on $\mathbf{R}^3$.

- The energy eigenspaces are the subspaces $\mathcal{H}^n \subset \mathcal{H}$ with total number of eigenvalue n. These are irreducible representations of $SU(3)$. They are also representations of the $SO(3)$ rotation action. Derive the rule for which irreducibles of $SO(3)$ will occur in $\mathcal{H}^n$.

Problem 3:
Prove the relation of equation 26.16.

Problem 4:
Compute

$$_\tau\langle 0|N|0\rangle_\tau$$

as a function of τ, for $|0\rangle_\tau$ the squeezed state of equation 26.19 and N the usual number operator.

B.13   Chapters 27 and 28

Problem 1:
Consider the fermionic oscillator, for $d = 3$ degrees of freedom, with Hamiltonian

$$H = \frac{1}{2}\sum_{j=1}^{3}(a_{Fj}^\dagger a_{Fj} - a_{Fj}a_{Fj}^\dagger)$$

- Use fermionic annihilation and creation operators to construct a representation of the Lie algebra $\mathfrak{u}(3) = \mathfrak{u}(1) + \mathfrak{su}(3)$ on the fermionic state space $\mathcal{H}_F$. Which irreducible representations of $\mathfrak{su}(3)$ occur in this state space? Picking a basis X_j of $\mathfrak{u}(3)$ and bases for each irreducible representation you find, what are the representation matrices (for each X_j) for each such irreducible representation?

- Consider the subgroup $SO(3) \subset U(3)$ of real orthogonal matrices and the Lie algebra representation of $\mathfrak{so}(3)$ on $\mathcal{H}_F$ one gets by restriction of the above representation. Which irreducible representations of $SO(3)$ occur in the state space?

Problem 2:
Prove that, as algebras over $\mathbf{C}$,

- $\mathrm{Cliff}(2d, \mathbf{C})$ is isomorphic to $M(2^d, \mathbf{C})$
- $\mathrm{Cliff}(2d + 1, \mathbf{C})$ is isomorphic to $M(2^d, \mathbf{C}) \oplus M(2^d, \mathbf{C})$

B.14   Chapters 29 to 31

Problem 1:
Show that for vectors $\mathbf{v} \in \mathbf{R}^n$ and ϵ_{jk} the basis element of $\mathfrak{so}(n)$ corresponding to an infinitesimal rotation in the jk plane, one has

$$e^{-\frac{\theta}{2}\gamma_j\gamma_k}\gamma(\mathbf{v})e^{\frac{\theta}{2}\gamma_j\gamma_k} = \gamma(e^{\theta\epsilon_{jk}}\mathbf{v})$$

and

$$\left[-\frac{1}{2}\gamma_j\gamma_k, \gamma(\mathbf{v})\right] = \gamma(\epsilon_{jk}\mathbf{v})$$

Problem 2:
Prove the following change of variables formula for the fermionic integral

$$\int F(\boldsymbol{\xi})d\xi_1 d\xi_2 \cdots d\xi_n = \frac{1}{\det A}\int F(A\boldsymbol{\xi}')d\xi_1' d\xi_2' \cdots d\xi_n'$$

where $\boldsymbol{\xi} = A\boldsymbol{\xi}'$, i.e.,

$$\xi_j = \sum_{k=1}^{n} A_{jk}\xi_k'$$

for any invertible matrix A with entries A_{jk}.

For a skew-symmetric matrix A, and $n = 2d$ even, show that one can evaluate the fermionic version of the Gaussian integral as

$$\int e^{\frac{1}{2}\sum_{j,k=1}^{n} A_{jk}\xi_j\xi_k} d\xi_1 d\xi_2 \cdots d\xi_n = Pf(A)$$

where

$$Pf(A) = \frac{1}{d!2^d}\sum_{\sigma}(-1)^{|\sigma|}A_{\sigma(1)\sigma(2)}A_{\sigma(3)\sigma(4)}\cdots A_{\sigma(n-1)\sigma(n)}$$

Here, the sum is over all permutations σ of the n indices. $Pf(A)$ is called the Pfaffian of the matrix A.

Problem 3:

For the fermionic oscillator construction of the spinor representation in dimension $n = 2d$, with number operator $N_F = \sum_{j=1}^{d} a_{Fj}^\dagger a_{Fj}$, define

$$\Gamma = e^{i\pi N_F}$$

Show that

- $$\Gamma = \prod_{j=1}^{d}(1 - 2a_{Fj}^\dagger a_{Fj})$$

- $$\Gamma = c\gamma_1\gamma_2\cdots\gamma_{2d}$$

  for some constant c. Compute c.

- $$\gamma_j\Gamma + \Gamma\gamma_j = 0$$

  for all j.

- $$\Gamma^2 = 1$$

- $$P_\pm = \frac{1}{2}(1\pm\Gamma)$$

  are projection operators onto subspaces $\mathcal{H}^+$ and $\mathcal{H}^-$ of $\mathcal{H}_F$.

- Show that $\mathcal{H}^+$ and $\mathcal{H}^-$ are each separately representations of $\mathfrak{spin}(n)$ (i.e., the representation operators commute with $P_\pm$).

Problem 4:

Using the fermionic analog of Bargmann–Fock to construct spinors, and the inner product 31.3, show that the operators a_{Fj} and $a_{Fj}^\dagger$ are adjoints with respect to this inner product.

B.15   Chapters 33 and 34

Problem 1:

Consider a two-dimensional version of the Pauli equation that includes a coupling to an electromagnetic field, with Hamiltonian

$$H = \frac{1}{2m}\left((P_1 - eA_1)^2 + (P_2 - eA_2)^2\right) - \frac{e}{2m}B\sigma_3$$

where A_1 and A_2 are functions of q_1, q_2 and

$$B = \frac{\partial A_2}{\partial q_1} - \frac{\partial A_1}{\partial q_2}$$

Show that this is a supersymmetric quantum mechanics system, by finding operators Q_1, Q_2 that satisfy the relations 33.1.

Problem 2:

For the three choices of inner product given in section 34.3, show that the inner product is invariant under the action of the group $\widetilde{E(3)}$ on the space of solutions.

B.16   Chapter 36

Problem 1:

When the single-particle state space $\mathcal{H}_1$ is a complex vector space with Hermitian inner product, one has an infinite dimensional case of the situation of section 26.4. In this case, one can write annihilation and creation operators acting on the multiparticle state spaces $S^*(\mathcal{H}_1)$ or $\Lambda^*(\mathcal{H}_1)$ in a basis-independent manner as follows:

$$a^\dagger(f)P^\pm(g_1 \otimes g_2 \otimes \cdots \otimes g_n) = \sqrt{n+1}P^\pm(f \otimes g_1 \otimes g_2 \otimes \cdots \otimes g_n)$$

$$a(f)P^\pm(g_1 \otimes g_2 \otimes \cdots \otimes g_n) = \frac{1}{\sqrt{n}}\sum_{j=1}^{n}(\pm1)^{j+1}\langle f, g_j\rangle P^\pm(g_1 \otimes g_2 \cdots \otimes \widehat{g_j} \otimes \cdots \otimes g_n)$$

Here, $f, g_j \in \mathcal{H}_1$, $P^\pm$ is the operation of summing over permutations used in section 9.6 that produces symmetric or antisymmetric tensor products, and $\widehat{g_j}$ means omit the g_j term in the tensor product.

Show that, for f orthonormal basis elements of $\mathcal{H}_1$, these annihilation and creation operators satisfy the CCR ($+$ case) or CAR ($-$ case).

Problem 2:

In the fermionic case of problem 1, show that the inner product on $\Lambda^*(\mathcal{H}_1)$ that, for an orthonormal basis $e_1, \cdots, e_n$ of $\mathcal{H}_1$, makes the $e_{i_1} \wedge e_{i_2} \wedge \cdots \wedge e_{i_k}$ orthonormal for $i_1 < i_2 < \cdots < i_k$ can be written in an basis independent way as

$$\langle f_1 \wedge f_2 \wedge \cdots \wedge f_k, g_1 \wedge g_2 \wedge \cdots \wedge g_k \rangle = \det M$$

where M is the k by k matrix with lm entry $\langle f_l, g_m \rangle$.

When $\mathcal{H}_1$ is a space of wavefunctions (in position or momentum space), then taking f_j to be some single-particle wavefunctions, and the g_j to be delta-functions in position or momentum space, this construction is known as the "Slater determinant" construction giving antisymmetric wavefunctions. For the bosonic case, a similar construction exists for symmetric tensor products, using instead of the determinant of the matrix, something called the "permanent" of the matrix.

B.17   Chapters 37 and 38

Problem 1:

Show that if one takes the quantum field theory Hamiltonian operator to be

$$\widehat{H} = \int_{-\infty}^{\infty} \widehat{\Psi}^\dagger(x) \left(-\frac{1}{2m} \frac{d^2}{dx^2} + V(x) \right) \widehat{\Psi}(x) dx$$

the field operators will satisfy the conventional Schrödinger equation for the case of a potential $V(x)$.

Problem 2:

A quantum system corresponding to indistinguishable particles interacting with each other with an interaction energy $v(x - y)$ (where x, y are the positions of the particles) is given by adding a term

$$\frac{1}{2} \int_{-\infty}^{\infty} \int_{-\infty}^{\infty} \widehat{\Psi}^\dagger(x) \widehat{\Psi}^\dagger(y) v(y - x) \widehat{\Psi}(y) \widehat{\Psi}(x) dx dy$$

to the free particle Hamiltonian. Just as the free particle Hamiltonian has an expression as a momentum space integral involving products of annihilation and creation operators, can you write this interaction term as a momentum space integral involving products of annihilation and creation operators (in terms of the Fourier transform of $v(x - y)$)?

Problem 3:

For non-relativistic quantum field theory of a free particle in three dimensions, show that the momentum operators $\widehat{\mathbf{P}}$ (equation 38.11) and angular momentum operators $\widehat{\mathbf{L}}$ (equation 38.12) satisfy the commutation relations for the Lie algebra of $E(3)$ (equations 38.13).

Problem 4:
Show that the total angular momentum operators for the non-relativistic theory of spin $\frac{1}{2}$ fermions discussed in section 38.3.3 satisfy the commutation relations for the Lie algebra of $SU(2)$.

B.18 Chapters 40 to 42

Problem 1:
If P_0, P_j, L_j, K_j are the operators in any Lie algebra representation of the Poincaré group corresponding to the basis elements t_0, t_j, l_j, k_j of the Lie algebra of the group, show that the operator

$$-P_0^2 + P_1^2 + P_2^2 + P_3^2$$

commutes with P_0, P_j, L_j, K_j and thus is a Casimir operator for the Poincaré Lie algebra.

Problem 2:
Show that the Lie algebra $\mathfrak{so}(4, \mathbf{C})$ is $\mathfrak{sl}(2, \mathbf{C}) \oplus \mathfrak{sl}(2, \mathbf{C})$. Within this Lie algebra, identify the sub-Lie algebras of the groups $Spin(4)$, $Spin(3, 1)$, and $Spin(2, 2)$.

Problem 3:
Find an explicit realization of the Clifford algebra Cliff$(4, 0)$ in terms of 4 by 4 matrices (γ matrices for this case), and use this to realize the group $Spin(4)$ as a group of 4 by 4 matrices (*hint*: recall that the Lie algebra of the spin group is given by products of two generators). Use these matrices to explicitly construct the representations of $Spin(4)$ on two kinds of half-spinors, on complexified vectors ($\mathbf{C}^4$), and the adjoint representation on the Lie algebra.

Problem 4:
The Pauli–Lubanski operator is the four-component operator

$$W_0 = -\mathbf{P} \cdot \mathbf{L}, \quad \mathbf{W} = -P_0 \mathbf{L} + \mathbf{P} \times \mathbf{K}$$

(same notation as in problem 1) Show that

$$W^2 = -W_0 W_0 + W_1 W_1 + W_2 W_2 + W_3 W_3$$

commutes with the energy–momentum operator $(P_0, \mathbf{P})$
Show that W^2 is a Casimir operator for the Poincaré Lie algebra.

B.19   Chapters 43 and 44

Problem 1:
Show that, assuming the standard Poisson brackets

$$\{\phi(\mathbf{x}), \pi(\mathbf{x}')\} = \delta(\mathbf{x} - \mathbf{x}'), \quad \{\phi(\mathbf{x}), \phi(\mathbf{x}')\} = \{\pi(\mathbf{x}), \pi(\mathbf{x}')\} = 0$$

Hamilton's equations for the Hamiltonian

$$h = \int_{\mathbf{R}^3} \frac{1}{2}(\pi^2 + (\nabla\phi)^2 + m^2\phi^2)d^3\mathbf{x}$$

are equivalent to the Klein–Gordon equation for a classical field ϕ.

Problem 2:
In section 44.1.2, we studied the theory of a relativistic complex scalar field, with a $U(1)$ symmetry, and found the charge operator $\widehat{Q}$ that gives the action of the Lie algebra of $U(1)$ on the state space of this theory.

- Show that the charge operator $\widehat{Q}$ has the following commutators with the fields

$$[\widehat{Q}, \widehat{\phi}] = -\widehat{\phi}, \quad [\widehat{Q}, \widehat{\phi}^\dagger] = \widehat{\phi}^\dagger$$

  and thus that $\widehat{\phi}$ on charge eigenstates reduces the charge eigenvalue by 1, whereas $\widehat{\phi}^\dagger$ increases the charge eigenvalue by 1.

- Consider the theory of two identical complex free scalar fields, and show that this theory has a $U(2)$ symmetry. Find the four operators that give the Lie algebra action for this symmetry on the state space, in terms of a basis for the Lie algebra of $U(2)$.
  Note that this is the field content and symmetry of the Higgs sector of the Standard Model (where the difference is that the theory is not free, but interacting, and has a lowest energy state not invariant under the symmetry).

B.20   Chapters 45 and 46

Problem 1:
Show that the Yang–Mills equations (46.9 and 46.11) are Hamilton's equations for the Yang–Mills Hamiltonian 46.10.

Problem 2:
Show that the matrix $P_\perp$ with entries

$$(P_\perp)_{jk} = \delta_{jk} - \frac{p_j p_k}{|\mathbf{p}|^2}$$

acts on momentum vectors in $\mathbf{R}^3$ by orthogonal projection on the plane perpendicular to $\mathbf{p}$. Use this to explain why one expects to get the commutation relations of equation 46.18 in Coulomb gauge (the condition $\nabla \cdot \mathbf{A} = 0$ in momentum space says that the vector potential is perpendicular to the momentum).

Problem 3:
Show that the group $SO(2)$ acts on the space of solutions of Maxwell's equations by

$$\mathbf{E}(\mathbf{x}) \to \cos\theta\ \mathbf{E}(\mathbf{x}) + \sin\theta\ \mathbf{B}(\mathbf{x})$$

$$\mathbf{B}(\mathbf{x}) \to -\sin\theta\ \mathbf{E}(\mathbf{x}) + \cos\theta\ \mathbf{B}(\mathbf{x})$$

For $\theta = \frac{\pi}{2}$, this symmetry interchanges $\mathbf{E}$ and $\mathbf{B}$ fields and is known as electric-magnetic duality. A much harder problem is to see what the corresponding operator acting on states is (it turns out to be the helicity operator).

B.21   Chapter 47

Problem 1:
Show that complex-valued solutions of the Dirac equation 47.2 correspond in the non-relativistic limit (energies small compared to the mass) to solutions of the Pauli–Schrödinger equation 34.3. *Hint*: write solutions in the form

$$\psi(t, \mathbf{x}) = e^{-imt}\phi(t, \mathbf{x})$$

Problem 2:
For two copies of the Majorana fermion theory, with the same mass m and field operators $\widehat{\Psi}_1, \widehat{\Psi}_2$, show that the theory has an $SO(2)$ symmetry, and find the Lie algebra representation operator $\widehat{Q}$ for this symmetry. Compute the commutators

$$[\widehat{Q}, \widehat{\Psi}_1], \quad [\widehat{Q}, \widehat{\Psi}_2]$$

Problem 3:

Show that, for $m = 0$, the Majorana fermion theory has an $SO(2)$ symmetry given by the action

$$\Psi(\mathbf{x}) \to \cos\theta\ \Psi(\mathbf{x}) + \sin\theta\ \gamma_0\gamma_1\gamma_2\gamma_3\Psi(\mathbf{x})$$

What is the corresponding Lie algebra representation operator (this is called the "axial charge")?

Bibliography

1. O. Alvarez, Lectures on quantum mechanics and the index theorem, *Geometry and Quantum Field Theory (Park City, UT, 1991)*, vol. 1, IAS/Park City Mathematics Series (American Mathematical Society, 1995), pp. 271–322
2. V.I. Arnold, *Mathematical Methods of Classical Mechanics*, vol. 60, 2nd edn., Graduate Texts in Mathematics (Springer, Berlin, 1989)
3. M. Artin, *Algebra* (Prentice Hall Inc, New Delhi, 1991)
4. J.C. Baez, I.E. Segal, Z.-F. Zhou, *Introduction to Algebraic and Constructive Quantum Field Theory, Princeton Series in Physics* (Princeton University Press, Princeton, 1992)
5. G. Baym, *Lectures on Quantum Mechanics (Lecture Notes and Supplements in Physics)* (The Benjamin/Cummings Publishing Company, 1969)
6. F.A. Berezin, *The Method of Second Quantization*, vol. 24 (Pure and Applied Physics (Academic Press, San Diego, 1966)
7. F.A. Berezin, M.S. Marinov, Particle spin dynamics as the Grassmann variant of classical mechanics. Ann. Phys. **104**(2), 336–362 (1977)
8. R. Berndt, *An Introduction to Symplectic Geometry*, vol. 26 (Graduate Studies in Mathematics (American Mathematical Society, Providence, 2001)
9. R. Berndt, *Representations of Linear Groups* (Vieweg, New York, 2007)
10. J.D. Bjorken, S.D. Drell, *Relativistic Quantum Fields* (McGraw-Hill Book Co., New York, 1965)
11. M. Blasone, G. Vitiello, P. Jizba, *Quantum Field Theory and its Macroscopic Manifestations* (Imperial College Press, London, 2011)
12. N. Bogolubov, A. Logunov, I. Todorov, *Introduction to Axiomatic Quantum Field Theory* (W. A. Benjamin, Reading, 1975)
13. A. Cannas da Silva, *Lectures on Symplectic Geometry*, vol. 1764 (Lecture Notes in Mathematics (Springer, Berlin, 2001)
14. R. Carter, G. Segal, I. Macdonald, *Lectures on Lie Groups and Lie Algebras*, vol. 32 (London Mathematical Society Student Texts (Cambridge University Press, Cambridge, 1995)
15. S. Coleman, *Lectures on Quantum Field Theory* (World Scientific, New Jersey, 2017)
16. A. Das, *Lectures on Quantum Field Theory* (World Scientific, Singapore, 2008)
17. J. Dimock, *Quantum Mechanics and Quantum Field Theory* (Cambridge University Press, Cambridge, 2011)

© Peter Woit 2017

P. Woit, *Quantum Theory, Groups and Representations*,

DOI 10.1007/978-3-319-64612-1

18. I. Dolgachev, *A Brief Introduction to Physics for Mathematicians* (1995-6), http://www.math.lsa.umich.edu/~idolga/physicsbook.pdf

19. J. Earman, D. Fraser, Haag's theorem and its implications for the foundations of quantum field theory. Erkenntnis **64**(3), 305–344 (2006)

20. L.D. Faddeev, O.A. Yakubovskiĭ, *Lectures on Quantum Mechanics for Mathematics Students*, vol. 47 (Student Mathematical Library (American Mathematical Society, Providence, 2009)

21. R.P. Feynman, Space-time approach to non-relativistic quantum mechanics. Rev. Mod. Phys. **20**, 367–387 (1948)

22. R.P. Feynman, *The Character of Physical Law* (M.I.T. Press, Cambridge, 1967), p. 129

23. R.P. Feynman, *Statistical Mechanics*, Advanced Book Classics (Perseus Books, Advanced Book Program, Reading, 1998). A set of lectures, Reprint of the 1972 original

24. R.P. Feynman, A.R. Hibbs, *Quantum Mechanics and Path Integrals*, emended edn. (Dover Publications Inc, New York, 2010)

25. R.P. Feynman, R.B. Leighton, M. Sands, *The Feynman Lectures on Physics. Vol. 3: Quantum Mechanics* (Addison-Wesley Publishing Co. Inc., 1965)

26. G.B. Folland, *Harmonic Analysis in Phase Space*, vol. 122 (Princeton, Annals of Mathematics Studies (Princeton University Press, 1989)

27. G.B. Folland, *Fourier Analysis and its Applications* (Wadsworth & Brooks/Cole Advanced Books & Software, Pacific Grove, 1992)

28. G.B. Folland, *Quantum Field Theory*, vol. 149 (Mathematical Surveys and Monographs (American Mathematical Society, Providence, 2008)

29. T. Frankel, *The Geometry of Physics*, 3rd edn. (Cambridge University Press, Cambridge, 2012)

30. T.A. Garrity, *Electricity and Magnetism for Mathematicians* (Cambridge University Press, Cambridge, 2015)

31. L.E. Gendenshteĭn, I.V. Krive, Supersymmetry in quantum mechanics. Uspekhi Fiz. Nauk **146**(4), 553–590 (1985)

32. H. Georgi, *Lie Algebras in Particle Physics*, vol. 54, Frontiers in Physics (Benjamin/Cummings Publishing Co. Inc., Advanced Book Program, Reading, 1982)

33. R. Geroch, *Quantum Field Theory: 1971 Lecture Notes* (Minkowski Institute Press, Montreal, 2013)

34. F. Gieres, Mathematical surprises and Dirac's formalism in quantum mechanics. Rep. Prog. Phys. **63**(12), 1893 (2000)

35. W. Greiner, J. Reinhardt, *Field Quantization* (Springer, Berlin, 1996)

36. W. Greub, *Multilinear Algebra*, 2nd edn. (Springer, Berlin, 1978)

37. V. Guillemin, S. Sternberg, *Symplectic Techniques in Physics*, 2nd edn. (Cambridge University Press, Cambridge, 1990)

38. V. Guillemin, S. Sternberg, *Variations on a Theme by Kepler*, vol. 42 (American Mathematical Society Colloquium Publications (American Mathematical Society, Providence, 1990)

39. D. Gurarie, *Symmetries and Laplacians*, vol. 174, North-Holland Mathematics Studies (North-Holland Publishing Co., Amsterdam, 1992)

40. B. Hall, An elementary introduction to groups and representations (2000), arXiv:math-ph/0005032

41. B. Hall, *Quantum Theory for Mathematicians*, vol. 267 (Graduate Texts in Mathematics (Springer, Berlin, 2013)

42. B. Hall, *Lie Groups, Lie Algebras, and Representations*, vol. 222, 2nd edn., Graduate Texts in Mathematics (Springer, Berlin, 2015)

43. K. Hannabuss, *An Introduction to Quantum Theory*, vol. 1 (Oxford Graduate Texts in Mathematics (Oxford University Press, Oxford, 1997)

44. S. Haroche, J.-M. Raimond, *Exploring the Quantum, Oxford Graduate Texts* (Oxford University Press, Oxford, 2006)
45. B. Hatfield, *Quantum Field Theory of Point Particles and Strings*, vol. 75 (Frontiers in Physics (Addison-Wesley Publishing Company, Advanced Book Program, Redwood City, 1992)
46. M. Henneaux, C. Teitelboim, *Quantization of Gauge Systems* (Princeton University Press, Princeton, 1992)
47. R. Hermann, *Lie Groups for Physicists* (W. A. Benjamin Inc., New York, 1966)
48. M.W. Hirsch, S. Smale, *Differential Equations, Dynamical Systems, and Linear Algebra* (Academic Press, New York, 1974)
49. R. Howe, On the role of the Heisenberg group in harmonic analysis. Bull. Amer. Math. Soc. (N.S.) **3**(2), 821–843 (1980)
50. I. Khavkine, Covariant phase space, constraints, gauge and the Peierls formula. Int. J. Modern Phys. A **29**(5), 1430009 (2014)
51. A.A. Kirillov, *Lectures on the Orbit Method*, vol. 64 (Graduate Studies in Mathematics (American Mathematical Society, Providence, 2004)
52. B. Kostant, Quantization and unitary representations. I. Prequantization, *Lectures in Modern Analysis and Applications, III*, vol. 170, Lecture Notes in Mathematics (Springer, Berlin, 1970), pp. 87–208
53. T. Lancaster, S.J. Blundell, *Quantum Field Theory for the Gifted Amateur* (Oxford University Press, Oxford, 2014)
54. N.P. Landsman, *Between classical and quantum, Philosophy of Physics: Part A, Handbook of the Philosophy of Science* (Elsevier, Amsterdam, 2007)
55. H.B. Lawson Jr., M.-L. Michelsohn, *Spin Geometry*, vol. 38 (Princeton, Princeton Mathematical Series (Princeton University Press, 1989)
56. G. Lion, M. Vergne, *The Weil Representation, Maslov Index and Theta Series*, vol. 6 (Progress in Mathematics (Birkhäuser, Boston, 1980)
57. G.W. Mackey, *The Mathematical Foundations of Quantum Mechanics: A Lecturenote Volume* (W. A. Benjamin Inc., New York, 1963)
58. G.W. Mackey, *Unitary Group Representations in Physics, Probability, and Number Theory*, 2nd edn. (Advanced Book Classics (Addison-Wesley Publishing Company, Advanced Book Program, Reading, 1989)
59. E. Meinrenken, *Clifford Algebras and Lie Theory*, vol. 58, Ergebnisse der Mathematik und ihrer Grenzgebiete. 3. Folge. A Series of Modern Surveys in Mathematics (Springer, Berlin, 2013)
60. A. Messiah, *Quantum Mechanics* (Dover, New York, 1999)
61. YuA Neretin, *Categories of Symmetries and Infinite-dimensional Groups*, vol. 16 (London Mathematical Society Monographs, New Series (Oxford University Press, Oxford, 1996)
62. Y.A. Neretin, *Lectures on Gaussian Integral Operators and Classical Groups* (London Mathematical Society Monographs, New Series (European Mathematical Society, Zurich, 2011)
63. D.E. Neuenschwander, *Emmy Noether's Wonderful Theorem* (Johns Hopkins University Press, Baltimore, 2011)
64. J.T. Ottesen, *Infinite Dimensional Groups and Algebras in Quantum Physics*, vol. 27 (Lecture Notes in Physics Monographs (Springer, Berlin, 1995)
65. R. Penrose, *The Road to Reality* (Alfred A. Knopf Inc., New York, 2005)
66. A. Perelomov, *Generalized Coherent States and Their Applications, Texts and Monographs in Physics* (Springer, Berlin, 1986)
67. M.E. Peskin, D.V. Schroeder, *An Introduction to Quantum Field Theory* (Addison-Wesley Publishing Company, Advanced Book Program, Redwood City, 1995)
68. I.R. Porteous, *Clifford Algebras and the Classical Groups*, vol. 50 (Cambridge Studies in Advanced Mathematics (Cambridge University Press, Cambridge, 1995)

69. J. Preskill, Quantum computation course notes (1997–2016), http://www.theory. caltech.edu/people/preskill/ph219/

70. A. Pressley, G. Segal, *Loop Groups, Oxford Mathematical Monographs* (Oxford University Press, Oxford, 1986)

71. P. Ramond, *Group Theory, a Physicist's Survey* (Cambridge University Press, Cambridge, 2010)

72. J. Rauch, *Partial Differential Equations*, vol. 128 (Graduate Texts in Mathematics (Springer, Berlin, 1991)

73. M. Reed, B. Simon, *Methods of Modern Mathematical Physics* (Self-adjointness (Academic Press, New York, II. Fourier Analysis, 1975)

74. J. Rosenberg, A selective history of the Stone-von Neumann theorem, *Operator Algebras, Quantization, and Noncommutative Geometry*, vol. 365, Contemporary Mathematics (American Mathematical Society, Providence, 2004), pp. 331–353

75. M. Schlosshauer, *Decoherence and the Quantum-to-Classical Transition* (Springer, Berlin, 2007)

76. M. Schlosshauer, *Elegance and Enigma: The Quantum Interviews* (Springer, Berlin, 2011)

77. L.S. Schulman, *Techniques and Applications of Path Integration* (Wiley, New York, 1981)

78. S.S. Schweber, *An Introduction to Relativistic Quantum Field Theory* (Row, Peterson and Company, Evanston, 1961)

79. D. Shale, Linear symmetries of free boson fields. Trans. Amer. Math. Soc. **103**, 149–167 (1962)

80. D. Shale, W. Forrest, Stinespring, States of the Clifford algebra. Ann. Math. **80**(2), 365–381 (1964)

81. R. Shankar, *Principles of Quantum Mechanics* (Springer, Berlin, 2008). Corrected reprint of the second (1994) edition

82. S.F. Singer, *Linearity, Symmetry, and Prediction in the Hydrogen Atom, Undergraduate Texts in Mathematics* (Springer, Berlin, 2005)

83. E.M. Stein, R. Shakarchi, *Fourier Analysis*, vol. 1 (Princeton, Princeton Lectures in Analysis (Princeton University Press, 2003)

84. S. Sternberg, *Group Theory and Physics* (Cambridge University Press, Cambridge, 1994)

85. J. Stillwell, *Naive Lie theory, Undergraduate Texts in Mathematics* (Springer, Berlin, 2008)

86. M. Stone, *The Physics of Quantum Fields* (Springer, Berlin, 2000)

87. R.F. Streater, A.S. Wightman, *PCT, spin and Statistics, and All That* (Princeton University Press, Princeton, 2000). Corrected third printing of the 1978 edition

88. R.S. Strichartz, *A Guide to Distribution Theory and Fourier Transforms* (World Scientific, River Edge, 2003)

89. K. Sundermeyer, *Constrained Dynamics*, vol. 169 (Lecture Notes in Physics (Springer, Berlin, 1982)

90. L.A. Takhtajan, *Quantum Mechanics for Mathematicians*, vol. 95 (Graduate Studies in Mathematics (American Mathematical Society, Providence, 2008)

91. M. Talagrand, *What is a Quantum Field Theory?* (Cambridge University Press, Cambridge, to appear)

92. J.D. Talman, *Special Functions: A Group Theoretic Approach* (W. A. Benjamin Inc., New York, 1968)

93. K. Tapp, *Matrix Groups for Undergraduates*, 2nd edn. (American Mathematical Society, Providence, 2016)

94. M.E. Taylor, *Noncommutative Harmonic Analysis*, vol. 22 (Mathematical Surveys and Monographs (American Mathematical Society, Providence, 1986)

95. C. Teleman, Representation theory course notes (2005), http://math.berkeley.edu/ ~teleman/math/RepThry.pdf

96. J. Townsend, *A Modern Approach to Quantum Mechanics* (University Science Books, Sausalito, 2000)

97. W.-K. Tung, *Group Theory in Physics* (World Scientific, Singapore, 1985)

98. F.W. Warner, *Foundations of Differentiable Manifolds and Lie Groups*, vol. 94 (Graduate Texts in Mathematics (Springer, Berlin, 1983)

99. S. Weinberg, *The Quantum Theory of Fields*, vol. I (Cambridge University Press, Cambridge, 2005)

100. H. Weyl, *The Theory of Groups and Quantum Mechanics* (Dover Publications, Inc., New York, 1950). Reprint of the 1931 English translation

101. E.P. Wigner, The unreasonable effectiveness of mathematics in the natural sciences. Comm. Pure Appl. Math. **13**, 1–14 (1960)

102. E. Witten, Supersymmetry and Morse theory. J. Differ. Geom. **17**(4), 661–692 (1982)

103. N.M.J. Woodhouse, *Geometric Quantization*, 2nd edn. (Oxford Mathematical Monographs (Oxford University Press, Oxford, 1992)

104. N.M.J. Woodhouse, *Special Relativity, Springer Undergraduate Mathematics Series* (Springer, Berlin, 2003)

105. A. Zee, *Quantum Field Theory in a Nutshell*, 2nd edn. (Princeton University Press, Princeton, 2010)

106. A. Zee, *Group Theory in a Nutshell for Physicists* (Princeton University Press, Princeton, 2016)

107. E. Zeidler, *Quantum Field Theory ii* (Springer, Berlin, 2008)

108. V. Zelevinsky, *Quantum Physics: Volume 1 - From Basics to Symmetries and Perturbations* (Wiley-VCH, 2010)

109. J. Zinn-Justin, *Path Integrals in Quantum Mechanics, Oxford Graduate Texts* (Oxford University Press, Oxford, 2010)

110. W. Zurek, Decoherence and the transition from quantum to classical revisited. Los Alamos Sci. **27**, 86–109 (2002)

111. W. Zurek, Quantum Darwinism. Nat. Phys. **5**, 181–188 (2009)

Index

L

Lagrangian, 433
Laplace operator in spherical coordinates, 124, 276
Leibniz property, 192
Lenz vector, 280
Lie algebra, 23, 56, 67
  automorphism, 222
  derivation, 223
  orthogonal group, 60
  structure constants, 59
  unitary group, 61
Lie bracket, 58
Lie superalgebra, 388
Little group, 271
Lorentz group, 506
  representations, 517
Lowering operator, 109

M

Magnetic field, 37, 92, 576
Majorana spinors, 609
Maxwell's equations, 586
Metaplectic group, 231, 261, 316
Metaplectic representation, 231, 315
Minimal coupling, 575
Minkowski space, 504
Möbius transformations, 98
Moment map, 208
  for rotations, 211
  for spatial translations, 210
Momentum eigenstates, 167
Momentum operator, 143
  and spatial translations, 143
  Klein-Gordon theory, 570
  Majorana fermion theory, 614
  Maxwell theory, 597
  non-relativistic quantum field theory, 486
Momentum space representation, 167
Multiplicity, 108

N

NMR spectroscopy, 93
Noether's theorem, 438
Normal-ordered product, 321
Number operator, 292

O

Observable, 4
O(n), 49
Orthogonal group, 49, 375

P

Path integrals, 441
Pauli equation, 423
Pauli–Lubanski operator, 533
Pauli matrices, 28
  anticommutation relations, 31
  commutation relations, 34
Pauli principle, 129
Perturbation methods, 628
Phase space, 190
Plancherel theorem, 155
Planck's constant, 4, 90
Poincaré group, 528
  representations, 533
Poisson bracket, 190
  fermionic, 387
Poisson structure, 212
Position eigenstates, 167
Position operator, 165
Propagator
  free particle, 172
  harmonic oscillator, 308
  non-relativistic QFT, 471
  relativistic QFT, 556
Pseudo-classical mechanics, 390

Q

Quantum field operators
  complex relativistic scalars, 566
  for photons, 594
  Majorana case, 614
  non-relativistic case, 466
  non-relativistic spin 1/2, 489
  real relativistic scalars, 555
Quaternions, 77
Qubit, 27, 93

R

Raising operator, 109
Real projective space, 95
Renormalization, 500
Representation, 9, 16
  irreducible, 16
  Lie algebra, 63
  unitary, 10
Resolution of the identity, 52
Riemann sphere, 96
R (translation group), 140
  representations, 141

Printed in the United States
By Bookmasters